Gmelin Handbook of Inorganic Chemistry

8th Edition

Gmelin Handbook of Inorganic Chemistry

8th Edition

Gmelin Handbuch der Anorganischen Chemie

Achte, völlig neu bearbeitete Auflage

Prepared and issued by: Gmelin-Institut für Anorganische Chemie der Max-Planck-Gesellschaft zur Förderung der Wissenschaften

Director: Ekkehard Fluck

Founded by: Leopold Gmelin

8th Edition: 8th Edition begun under the auspices of the Deutsche Chemische Gesellschaft by R. J. Meyer

Continued by: E. H. E. Pietsch and A. Kotowski, and by Margot Becke-Goehring

Springer-Verlag
Berlin · Heidelberg · New York · London · Paris · Tokyo 1990

Volumes published on "Sulfur" (Syst. No. 9)

Sulfur A 1
History (in German) – 1942

Sulfur A 2
Occurrence. Technology of Sulfur and Its Compounds. Colloidal Sulfur. Toxicity (in German) – 1953

Sulfur A 3
The Element. Preparation in Pure State. Properties (in German) – 1953

Sulfur B 1
Hydrides and Oxides of Sulfur (in German) – 1953

Sulfur B 2
Sulfur-Oxygen Acids (in German) – 1960

Sulfur B 3
Compounds (concluded) (in German) – 1963

Sulfur-Nitrogen Compounds 1
Compounds with Sulfur of Oxidation Number VI (in German) – 1977

Sulfur Suppl. Vol. 1
Thionyl Halides (in German) – 1978

Sulfur Suppl. Vol. 2
Sulfur Halides (in German) – 1978

Sulfur Suppl. Vol. 3
Sulfur Oxides (in German) – 1980

Sulfur Suppl. Vol. 4 a/b
Sulfanes – 1983

Sulfur-Nitrogen Compounds 2
Compounds with Sulfur of Oxidation Number IV – 1985

Sulfur-Nitrogen Compounds 3
Compounds with Sulfur of Oxidation Number IV – 1987

Sulfur-Nitrogen Compounds 4
Compounds with Sulfur of Oxidation Number IV – 1987

Sulfur-Nitrogen Compounds 5
Compounds with Sulfur of Oxidation Number IV – 1990 (present volume)

Gmelin Handbook of Inorganic Chemistry

8th Edition

S

Sulfur-Nitrogen Compounds

Part 5

Compounds with Sulfur of Oxidation Number IV

With 69 illustrations

AUTHORS Hans-Jürgen Fachmann, Reimund Jotter, Alfons Kubny, Joachim Wagner

EDITORS Norbert Baumann, Gerhard Czack, Brigitte Heibel, Peter Merlet, Joachim Wagner

CHIEF EDITOR Alfons Kubny

System Number 9

Springer-Verlag
Berlin · Heidelberg · New York · London · Paris · Tokyo 1990

LITERATURE CLOSING DATE: 1988
IN MANY CASES MORE RECENT DATA HAVE BEEN CONSIDERED

Library of Congress Catalog Card Number: Agr 25-1383

ISBN 3-540-93599-1 Springer-Verlag, Berlin · Heidelberg · New York · London · Paris · Tokyo
ISBN 0-387-93599-1 Springer-Verlag, New York · Heidelberg · Berlin · London · Paris · Tokyo

Typesetting, printing, and bookbinding: LN-Druck Lübeck

Preface

The present volume is the first of a series describing acyclic sulfur-nitrogen compounds with sulfur of oxidation number IV. The acyclic S^{IV}–N compounds are arranged according to the coordination number of the sulfur. Neutral compounds are described before ions and complex compounds. The preceding series "Sulfur-Nitrogen Compounds" Parts 2, 3, and 4 covers the cyclic S^{IV}–N compounds.

In this volume, the first section deals with sulfur-nitrogen compounds with 1-coordinate sulfur and begins with the sulfur nitride (thiazyl) radical, SN. This transient molecule was observed in its electronic ground state and several electronically excited states. The descriptions of the sulfur nitride (thiazyl) ions SN^+ and SN^- follow. The SN^+ ion was studied in the gas phase as well as in the solid state where it forms salts. Thionitrosyl complexes containing the SN ligand as a terminal linear unit are described at the end of the first section.

The second section concerns sulfur-nitrogen compounds with 2-coordinate sulfur and starts with the description of poly(sulfur nitride), $(SN)_x$. The preparation, crystal structure, and metallic and superconducting properties of $(SN)_x$, which were extensively studied, fill a large part of the volume. Halogen-modified poly(sulfur nitride) such as the widely studied $(SNBr_{0.4})_x$ and Na-modified poly(sulfur nitride) are dealt with in the following chapters.

Sulfur nitride (thiazyl) compounds such as the well-characterized molecules FSN, ClSN, BrSN, *cis*-HOSN, $(CF_3)_2NOSN$, and F_5TeOSN are described subsequently. Metal complexes of sulfur nitride (thiazyl) compounds are considered in the next two chapters. In the thiazyl halide complexes, the ligands are bonded in a relatively loose manner through the nitrogen atom to the metal. In the halothionitrene complexes the ClSN or BrSN ligands are strongly coordinated corresponding to a metal-nitrogen double bond.

The nomenclature of acyclic S^{IV}–N compounds is not uniform in the original literature. Special rules for naming inorganic sulfur-nitrogen compounds are not recommended by IUPAC. Therefore the name for each compound has been selected from those given in the literature and synonyms are given.

The literature up to 1988 has been fully evaluated. Frequently even more recent publications are included.

The series will be continued by describing other acyclic S^{IV}–N compounds with 2-coordinate sulfur, such as sulfinyl imides, thiosulfinyl imides, and sulfur diimides. In the last volume of this series a formula index will be included.

Frankfurt am Main
November 1989

Alfons Kubny

Remarks on Abbreviations and Standards

Physical data are cited in short form using abbreviations:

D_m	measured density
D_x	density calculated from X-ray data
m.p.	melting point; dec.: melting with decomposition
b.p.	boiling point, often measured at low pressure (given in Torr behind the b.p.)
IR	(infrared) spectrum. The medium of measurement (e.g., KBr or a solvent) is given in parentheses. For the absorption maxima (vibration frequencies ν in cm^{-1}), intensity and shape are abbreviated as usual: s (strong), vs (very strong), w (weak), vw (very weak), vvw (very very weak), m (medium), br (broad), sh (shoulder). The assigned bands are usually labeled with the symbols ν for stretching vibration and δ for deformation vibration.
FT-IR	Fourier transform infrared
Ra	(Raman) spectrum. For abbreviations, see IR.
UV	(ultraviolet) spectrum includes the visible spectrum. The medium of measurement is set in parentheses. In addition to λ_{max} (wavelength of absorption maximum in nm) in many cases ε (extinction coefficient in $L \cdot mol^{-1} \cdot cm^{-1}$) is given.
XPS	X-ray photoelectron spectroscopy
UPS	ultraviolet photoelectron spectroscopy
NMR	(nuclear magnetic resonance) spectra. Solvent and standard are given in parentheses. For δ (chemical shift in ppm) downfield shift is indicated by a positive sign as recommended by IUPAC. Standard substances (if not otherwise cited) are TMS (= $Si(CH_3)_4$) as internal standard for ^1H and ^{13}C NMR, neat CH_3NO_2 as external standard for ^{14}N NMR, $CFCl_3$ as internal standard for ^{19}F NMR, and 85% H_3PO_4 as external standard for ^{31}P NMR. The multiplicity of the signal is given in parentheses: s (singlet), d (doublet), t (triplet), q (quartet), quint (quintet), sept (septet), m (multiplet). The assignment is given behind the multiplicity; if necessary for clarity the assigned atoms are underlined. J (coupling constant in Hz) is fixed by the number of bonds involved and the coupling nuclei (e.g., $^1J(^{13}C,H)$), or by listing the coupling groups (e.g., $^3J(CH_2,CH_3)$)
ESR	electron spin resonance
MS	(mass spectrum). m/e = mass/charge; M^+ = molecular ion (1+); relative intensity in parentheses
EIMS	electron impact mass spectrum
SCE	standard calomel electrode
NHE	normal hydrogen electrode

Table of Contents

Page

1 Sulfur-Nitrogen Compounds with 1-Coordinate Sulfur

1.1 Sulfur Nitride Radical, Thiazyl Radical, Thionitrosyl Radical, Nitrogen Sulfide Radical, SN

1.1.1 Occurrence. Existence

SN was detected as an interstellar molecule in Sagittarius B2 [1, 2] and in the core of the Orion A molecule cloud [3] by its rotational emission-line spectrum.

The SN radical polymerizes so readily that it is not possible to isolate it as a monomer solid or liquid. Even in the gaseous phase it has only a transient existence. It was studied in its electronic ground state and several electronically excited states.

The SN radical exists in its electronic ground state ($X\,^2\Pi$) for several ms when it is formed by flash photolysis of an OCS–N_2O–He gas mixture [4]. It disappeared by 250 μs after formation by flash photolysis of an OCS–NF_3–Ar mixture [5].

References:

[1] Gottlieb, C. A.; Ball, J. A.; Gottlieb, E. W.; Lada, C. J.; Penfield, H. (Astrophys. J. **200** [1975] L147/L149; C. A. **83** [1975] No. 186093).

[2] Gottlieb, C. A.; Gottlieb, E. W.; Litvak, M. M.; Ball, J. A.; Penfield, H. (Astrophys. J. **219** [1978] 77/94).

[3] Blake, G. A.; Sutton, E. C.; Masson, C. R.; Phillips, T. G, (Astrophys. J. Suppl. Ser. **60** [1986] 357/74; C. A. **104** [1986] No. 11922).

[4] Donovan, R. J.; Breckenridge, W. H. (Chem. Phys. Letters **11** [1971] 520/2).

[5] McGrath, W. D.; Morrow, T. (Nature **212** [1966] 746).

1.1.2 Formation

1.1.2.1 Formation from the Elements

Formation from Active Nitrogen and Sulfur Vapor

SN radicals form by the reaction of active nitrogen generated by subjecting N_2 at a flow rate of 70 to 150 μmol/s to an electrodeless discharge from a 2450 MHz microwave generator and sulfur vapor consisting of S_2 and S_8 molecules in N_2 at about 400°C. The flow rate of sulfur vapor was about 10^{-2} μmol/s. The kinetic measurements of the SN($B\,^2\Pi \rightarrow X\,^2\Pi$) emission at 250 to 500 nm showed that the emission was proportional to N and S atom concentrations. The $B\,^2\Pi$ state of SN is selectively formed by the three-body recombination reaction of N and S atoms, $N(^4S^\circ) + S(^3P) + M \rightarrow SN(B\,^2\Pi) + M$. The highest vibrational level of the $B\,^2\Pi$ state observed is v'=12. Weak emissions of the SN($A\,^2\Delta \rightarrow X\,^2\Pi$) and SN($C\,^2\Sigma^+ \rightarrow X\,^2\Pi$) systems were also observed [1].

SN is produced from active nitrogen and sulfur vapor in a flow system for studying the SN($B\,^2\Pi \rightarrow X\,^2\Pi$) emission [2] and the photoelectron spectrum of SN($X\,^2\Pi$) [3]. The reactions $N + S_2 \rightarrow SN + S$ and $N + NS_2 \rightarrow 2\,SN$ were proposed for the formation of SN from active nitrogen and sulfur vapor [4, 5].

Formation from N_2–Sulfur Mixtures by Discharge

The SN radical in various electronically excited states is produced by microwave discharge at 2450 MHz of N_2–sulfur vapor mixtures in flow systems or sealed tubes at low pressure [6 to 12].

SN forms by radiofrequency discharge, near 50 MHz, of an N_2–sulfur vapor mixture at a pressure of 20 mTorr. The millimeter wave spectra of SN ($X^2\Pi$) were measured in vibrational levels $v \leq 5$ [13].

The formation of SN by electric discharges through mixtures of N_2 and sulfur at low pressure was detected from the emission spectra of the electronic transitions $A^2\Delta \rightarrow X^2\Pi$ and $C^2\Sigma^+ \rightarrow X^2\Pi$ [14 to 18] and the photoelectron spectrum of SN ($X^2\Pi$) [3].

1.1.2.2 Formation from Sulfur Atoms and Nitrogen Compounds

Formation from Sulfur Atoms and N_2O

The production of SN radicals following the flash photolysis in the vis-UV region of OCS in the presence of excess N_2O and He as diluent gas (p(OCS) = 0.27 kPa, p(N_2O) = 1.6 kPa, p(He) = 9.1 kPa) was observed with kinetic absorption spectroscopy. The SN radical showed the absorption spectrum of the $A \leftarrow X$ system at 250 nm, the more intense $C \leftarrow X$ system at 230 nm, and the (0,1) band of the $C \leftarrow X$ system. In the presence of SF_6 as diluent gas, vibrationally excited SN was not observed which must be attributed to rapid vibration-vibration energy transfer. When Ar was employed as the diluent gas, the spectrum of SN was completely suppressed as a consequence of quenching of $S(^1D_2)$ atoms by Ar. From these observations it was deduced that SN is formed in the reaction $S(^1D_2) + N_2O \rightarrow SN\ (X^2\Pi) + NO\ (X^2\Pi)$. The formation of the SN radical was found to follow the integrated form of the photolysis flash and thus an absolute rate constant for this reaction could not be directly obtained. The decay of SN is relatively slow and the radical is seen to persist for several ms. This decay is second order in SN concentration and shows a first-order dependence on the total pressure of the system [19].

Formation from Sulfur Atoms and NF_3

The presence of an appreciable concentration of SN was observed (by the absorption spectra in the range 220 to 280 nm) when a mixture of OCS, NF_3, and Ar in the ratio 1:2:25 was flash photolyzed using 2 Torr partial pressure of OCS. The SN reached maximum intensity at about 50 μs and had disappeared by 250 μs. The initial reaction of $S(^1D)$ produced by flash photolysis of OCS with NF_3 possibly gives a hot SNF_3 molecule which then decomposes [20].

1.1.2.3 Formation from Nitrogen and Sulfur Compounds

Formation from N and HS

The production of SN by reaction of HS with N in the gas phase is discussed in a model of molecular processes in dense interstellar clouds [21].

Formation from Active Nitrogen and H_2S

SN is generated in active nitrogen–H_2S flames in a continuous flow system at low pressures. Active nitrogen was produced with an electrodeless discharge from a 2450 MHz microwave generator [22 to 26] and a high-voltage pulsed electric discharge using tungsten electrodes

[22] or aluminium electrodes [27, 28, 29]. The emission from the $B^2\Pi \rightarrow X^2\Pi$ transition was observed [22, 23]. The EPR spectrum was attributed to $SN(X^2\Pi_{3/2}, J = 3/2)$ [24, 25, 26].

The intensity of the SN ($B^2\Pi \rightarrow X^2\Pi$) (3,0) emission band at 310.0 nm was found to be dependent on the flow rate of H_2S when introduced into the active nitrogen flow. The mechanism that results in the production of SN and its subsequent emission can be summarized as $H_2S + N \rightarrow NH + HS$, $HS + N + M \rightarrow NH + S + M$, $N + S + M \rightarrow SN(B^2\Pi) + M$; $SN(B^2\Pi) \rightarrow SN(X^2\Pi) + h\nu$ [23]. The reaction of active nitrogen with H_2S yields large amounts of the polymer $(SN)_x$ [27, 30]. Intermediate SN formation was proposed: $N + H_2S \rightarrow N \cdot H_2S \rightarrow SN + H_2$ [27].

Formation from Active Nitrogen and SO_2

SN is produced by reaction of active nitrogen generated by a continuous microwave discharge in low-pressure N_2 and SO_2 which is introduced to the active nitrogen flow. The intensity of the $SN(B^2\Pi \rightarrow X^2\Pi)$ (3,0) emission band at 310 nm was found to be dependent on the flow rate of SO_2 [23].

Formation from Active Nitrogen and SF_6

SN forms by introducing SF_6 into a stream of active nitrogen generated by microwave discharge in N_2 in a flow system at low pressure. It was found that the intensity of the $SN(B^2\Pi \rightarrow X^2\Pi)$ (3,0) emission band at 310 nm is dependent on the flow rate of SF_6 when introduced into the active nitrogen flow. The detection limit for SF_6 using the SN emission band is 175 ng of sulfur/s [23].

Formation from N_2–SF_6 Mixtures by Discharge

SN was detected by its emission spectrum originating from the $B^2\Pi \rightarrow X^2\Pi$ electronic transition in N_2–SF_6 mixtures discharged at 2450 MHz in a flow system [31, 32]. A flow of N_2 at 1.1 mmol/s and 0.73 Torr and of SF_6 at 14 to 15 µmol/s was used [31].

Formation from Active Nitrogen and SCl Radicals

SN formation by the reaction of SCl radicals with discharged N_2 in a flow system was detected by mass spectrometry. The rate constant $k = (3.9 \pm 1.1) \times 10^{-11}$ $cm^3 \cdot molecule^{-1} \cdot s^{-1}$ for the reaction $SCl + N(^4S) \rightarrow SN + Cl$ at 293 ± 2 K was determined under pseudo-first-order conditions with $[N] > [SCl]_0$. Thus $\ln([SCl]_0/[SCl]_t) = k[N]t$. The reaction enthalpy $\Delta H^\circ_{298} = -244$ kJ/mol was calculated [41].

Formation from Active Nitrogen and SCl_2 or S_2Cl_2

The formation of SN by the reaction of active nitrogen with SCl_2 [33 to 37, 53, 54] and S_2Cl_2 [1, 22, 38 to 41, 53] in a flow system at low pressure was observed in the emission spectrum of the electronic transition $B^2\Pi \rightarrow X^2\Pi$ [1, 22, 33 to 37, 40], in the electron paramagnetic resonance spectrum of the $X^2\Pi_{3/2}$ state [38, 39, 54], and in the microwave spectrum [53]. Active nitrogen was generated by a 2450 MHz microwave discharge or a high-tension pulsed electric discharge in N_2; see e.g. [22]. SN formation was formerly supposed in the reaction of active nitrogen with SCl_2 which gave polymer $(SN)_x$ [30].

The concentration of N and S atoms and the intensity of the $B^2\Pi \rightarrow X^2\Pi$ emission were measured as a function of the partial pressure of S_2Cl_2. Under the conditions of excess N atoms ($[S_2Cl_2]_0/[N]_0 \leq 3 \times 10^{-3}$) where the selective excitation of the $B^2\Pi$ state was observed, the variation of the emission intensity of the $B^2\Pi$ (v' = 6) state with $[S_2Cl_2]_0$ follows the product [N]·[S]. These results suggest that the $B^2\Pi$ state is produced by a three-body recombination reaction: $N + S + M \rightarrow SN(B^2\Pi) + M$. On the other hand, in the region of high-partial pressure of S_2Cl_2 ($[S_2Cl_2]_0 > 2.2 \times 10^{-5}$ Torr), the emission intensity of the $B^2\Pi$ (v' = 6) state decreases slowly with increasing partial pressure of S_2Cl_2 in spite of a rapid decrease of the product [N]·[S], and the emission from the $A^2\Delta$ and $C^2\Sigma^+$ states appears. These facts indicate that the different excitation mechanism from the three-body recombination reaction would be operative in this region [1].

Formation from N Atoms or Active Nitrogen and OCS

$SN(B^2\Pi)$ was produced by the reaction $N(^2D) + OCS(X^1\Sigma^+) \rightarrow SN(B^2\Pi) + CO(X^1\Sigma^+)$ in a crossed molecular beam. The emission spectrum of the $SN(B^2\Pi \rightarrow X^2\Pi)$ band was observed [55, 56].

SN forms by reaction of active nitrogen generated by microwave discharge in N_2 and OCS in a continuous flow system at low pressure. The gas-phase electron paramagnetic resonance spectrum of the $X^2\Pi_{3/2}$ state in the J = 3/2 level was measured [42].

Formation from N Atoms or Active Nitrogen and CS_2

$SN(B^2\Pi)$ was produced by the reaction $N(^2P) + CS_2(X^1\Sigma^+) \rightarrow SN(B^2\Pi) + CS(X^1\Sigma^+)$ in a crossed molecular beam. The emission spectrum of the $SN(B^2\Pi \rightarrow X^2\Pi)$ band was observed [55].

The formation of SN by reaction of active nitrogen with CS_2 vapor in a continuous flow system at low pressure was supposed. Large amounts of the polymer $(SN)_x$ form on the walls. Active nitrogen was generated by pulsed electric discharge in N_2 at high tension [27,30]. The reaction was studied in the range of 174 to 394°C. The reaction mechanism $N + CS_2 \rightarrow N \cdot CS_2 \rightarrow SN + CS$ was proposed [27].

1.1.2.4 Formation from Sulfur–Nitrogen Compounds

Formation from SN^- by Electron Detachment

Intersection of an SN^- ion beam accelerated to 680 eV by the 488 nm intracavity radiation of an Ar II laser causes electron detachment from several vibrational levels to give $SN(X^2\Pi)$ according to $SN^-(^3\Sigma, v'' = 0) \rightarrow SN(X^2\Pi, v' = 0, 1, 2, 3)$ and $SN^-(^3\Sigma, v'' = 1, 2, 3) \rightarrow SN(X^2\Pi, v' = 0)$ [43].

Formation by Pyrolysis of S_4N_4

SN is produced by passing S_4N_4 vapors over quartz wool [44, 45] and silver wool [44, 46] at 80 to 400°C [44, 45] and 900 and 1100 K [46] at low pressure. An examination of the mass spectra and the results of phase-angle mass spectrometry showed that the major neutral species resulting from the pyrolysis of S_4N_4 over quartz wool and silver wool at 80 to 400°C are S_4N_2, S_3N_3, S_2N_2, S_2, SN, and N_2. At > 300°C SN, S_2, and N_2 become the major pyrolysis products [44, 45]. Photoelectron spectroscopy shows that, at 900 and 1100 K, all S_4N_4 is first converted predominantly into S_2N_2, which decomposes preferentially into two SN molecules at 900 K, whereas S_2 and N_2 are favored at 1100 K. A crude MNDO hypersurface study rationalizes the two reaction channels [46].

Formation by Pyrolysis of $(SN)_x$

SN forms by evaporating $(SN)_x$ at 140°C in the high vacuum in a mole fraction of 3% of the neutral vapor species. This results from field ionization and phase-angle mass spectra [47]. The formation of SN by heating $(SN)_x$ to about 160°C at 10^{-4} Torr was suggested from the IR spectrum of the condensed vapor species trapped in an Ar matrix at an M/A ratio of about 1000 at 10 K [48].

Formation by Pyrolysis of S_7NH

SN forms by heating S_7NH to approximately 500 K into flowing He. The photoelectron spectrum of SN($X^2\Pi$) could be observed [3].

Formation from ClS≡N – He Mixtures by Discharge

SN is produced by a microwave discharge of ClS≡N – He mixtures in a flow system at low pressure. The photoelectron spectrum of SN($X^2\Pi$) could be observed [3].

1.1.2.5 Formation in Flames

Laser-induced fluorescence (LIF) was used [49] to detect the SN radical in sulfur and nitrogen doped CH_4 flames. The SN($C^2\Sigma^+$, v' = 0—$X^2\Pi$, v'' = 0) system was excited and the strong fluorescence of the ($C^2\Sigma^+$, v' = 0—$X^2\Pi$, v'' = 1) band near 237.1 nm was observed. SN forms in rich CH_4–O_2 flames with the equivalence ratio $\Phi = 1.2$ doped with 3.6 mol% of NH_3 and 0.22 mol% of H_2S in the concentration of 3 ppm or 0.001 of the added H_2S concentration. With more than ~1 mol% of added NH_3, the SN concentration is proportional to the amount of added H_2S.

In a rich CH_4–air flame ($\Phi = 2.1$) with >2 mol% of added NH_3 and 0.3 and 0.5 mol% of added H_2S, the SN concentration was measured to be 0.0002 of the added H_2S. In rich CH_4–N_2O flames ($\Phi = 2.0$) at atmospheric pressure, the SN concentration was proportional to the added H_2S for 0.1 and 0.4 mol% H_2S. In both flames the SN concentration was equal to 0.001 of the added H_2S concentration. When NH_3 was added to this flame the SN concentration dropped with increasing NH_3 concentration by ≤50%. NH_3 addition of above ~4 mol% had no further effect on the SN concentration.

A rich CH_4–N_2O flame ($\Phi = 1.75$) doped with SF_6 between 0.01 and 0.02 mol% produced SN in concentrations equal to 0.001 of the added SF_6 concentration. In a rich commercial natural gas flame with N_2O as oxidizer ($\Phi = 2.3$) with added SF_6 at 0.04 mol% the SN concentration was 0.0006 that of the added SF_6. However, with no added SF_6 there is still sufficient sulfur to produce observable amounts of SN from CH_3SH in the natural gas. In all these flames SN was present only in the flame front.

The SN production rate is given by $k_b \cdot [P_S] \cdot [P_N]$, where k_b is the bimolecular reaction rate constant, P_S is the sulfur precursor molecule, and P_N the nitrogen radical. The value for a fast chemical reaction is $k_b = 10^{-11}$ $cm^3 \cdot s^{-1}$. Thus, with $[P_S] \geq 0.001$ mole fraction, the nitrogen radical species needs only to have a fractional concentration of 2 ppm.

In addition to the exothermic reaction N + SH → SN + H with $\Delta H = -33$ kcal/mol there are exothermic reactions of N, NH, and NH_2 with the reduced sulfur compounds CS_2, S_2, H_2S, and SH which produce either SN or HNS. SN production from either S + NO or N + SO is substantially endothermic, and these are likely to be quite slow reactions. Thus for the rapid production observed, the nitrogen radical reactions cited above are the likely production mechanism for SN [49].

In a kinetic model for the formation of NO and N_2 in the postflame gases of rich ($\Phi = 2.12$) premixed laminar moist CO–Ar–O_2 flames, SN formation is involved: $N + SO \rightarrow NO + S \rightarrow SN + O \rightarrow SO + N$. Fuel nitrogen as C_2N_2 and fuel sulfur as SO_2 were added to these flames [50].

Calculations on the equilibrium product distribution in flames of hydrocarbon–air mixtures containing hetero atoms such as Cl, N, and S for the temperature range of 100 to 2100°C predict the formation of SN at elevated temperatures in oxygen-lean hydrogen-lean and in oxygen-lean hydrogen-rich systems [51].

For the high temperatures of combustion reactions, estimates of the rate constants for the SN formation reactions $S + NX \rightarrow (S \cdots N \cdots X)^{\neq} \rightarrow SN + X$ and $N + SX \rightarrow (N \cdots S \cdots X)^{\neq} \rightarrow NS + X$, with X = H, O, N, S, and C, have been made. From transition state theory and estimates of the heat capacity of activation for atom transfer reactions between atoms and diatomic molecules, it could be shown that the form of the rate constant is $k = AT^{0.5} \exp(-C/RT)$ where $A = 10^{16.6} \exp(\Delta S^{\circ\neq}_{300}/R)$ in $cm^3 \cdot mol^{-1} \cdot s^{-1}$ ($\Delta S^{\circ\neq}_{300}$ = the entropy of activation at 300 K); $C = -(\Delta H^{\circ\neq}_{300} + 1)$ in kcal/mol ($\Delta H^{\circ\neq}_{300}$ = the heat of activation at 300 K). The triatomic intermediate, the reaction equation, and the values of log A, C, and log k at 1500 K for these reactions are presented below [52]:

intermediate	reaction	log (A/cm³ · mol⁻¹ · s⁻¹)	C in kcal/mol	log (k/cm³ · mol⁻¹ · s⁻¹) at 1500 K
NS_2	$N + S_2 \rightarrow NS + S$	11.8	8.0	12.2
N_2S	$S + N_2 \rightarrow SN + N$	12.6	109.7	−1.8
HNS	$S + NH \rightarrow SN + H$	11.8	0	13.4
HNS	$N + SH \rightarrow NS + H$	11.8	0.8	13.4
NOS	$S + NO \rightarrow SN + O$	12.0	34.7	8.6
NOS	$N + SO \rightarrow NS + O$	12.2	16.4	11.4
CNS	$S + NC \rightarrow SN + C$	12.3	63.6	4.6
CNS	$N + SC \rightarrow NS + C$	12.6	73.9	3.5

References:

[1] Murai, H.; Yagi, T.; Obi, K.; Tanaka, I. (Chem. Phys. Letters **61** [1979] 513/7).
[2] Petropulos, B.; Dessaux, O.; Goudmand, P. (Compt. Rend. C **270** [1970] 1223/6).
[3] Dyke, J. M.; Morris, A.; Trickle, I. R. (J. Chem. Soc. Faraday Trans. II **73** [1977] 147/51).
[4] Bett, J.A.S.; Winkler, C.A. (J. Phys. Chem. **68** [1964] 2501/8).
[5] Bett, J.A.S.; Winkler, C.A. (J. Phys. Chem. **68** [1964] 2735/6).
[6] Vervloet, M.; Jenouvrier, A. (Can. J. Phys. **54** [1976] 1909/23).
[7] Jenouvrier, A.; Pascat, B. (Can. J. Phys. **51** [1973] 2143/61).
[8] Jenouvrier, A.; Pascat, B. (Can. J. Phys. **58** [1980] 1275/90).
[9] Narasimham, N. A.; Balasubramaniam, T. K. (J. Mol. Spectrosc. **40** [1971] 511/8).
[10] Narasimham, N. A.; Subramanian, B. T. K. (J. Mol. Spectrosc. **29** [1969] 294/304).

[11] Narasimham, N. A.; Srikameswaran, K. (Proc. Indian Acad. Sci. A **59** [1964] 227/40).
[12] Narasimham, N. A.; Srikameswaran, K. (Nature **197** [1963] 370).
[13] Anacona, J. R.; Bogey, M.; Davies, P. B.; Demuynck, C.; Destombes, J. L. (Mol. Phys. **59** [1986] 81/8).
[14] Zeeman, P. B. (Can. J. Phys. **29** [1951] 174/85).
[15] Dressler, K. (Helv. Phys. Acta **28** [1955] 563/90).

[16] Fowler, A.; Bakker, C. J. (Proc. Roy. Soc. [London] A **136** [1932] 28/6).
[17] Joshi, K. C. (Z. Physik **191** [1966] 126/36).
[18] Patel, M. M. (Z. Physik **173** [1963] 347/51).
[19] Donovan, R. J.; Breckenridge, W. H. (Chem. Phys. Letters **11** [1971] 520/2).
[20] McGrath, W. D.; Morrow, T. (Nature **212** [1966] 746).

[21] Prasad, S. S.; Huntress, W. T., Jr. (Astrophys. J. **260** [1982] 590/8).
[22] Smith, J. J.; Meyer, B. (J. Mol. Spectrosc. **14** [1964] 160/72).
[23] Hood, W. H.; Niemczyk, T. M. (Anal. Chem. **58** [1986] 210/5).
[24] Carrington, A. (Proc. Roy. Soc. [London] A **302** [1968] 291/304, 300).
[25] Carrington, A.; Levy, D. H. (J. Chem. Phys. **44** [1966] 1298/9).
[26] Carrington, A.; Levy, D. H. (J. Phys. Chem. **71** [1967] 2/11, 11/2).
[27] Westbury, R. A.; Winkler, C. A. (Can. J. Chem. **38** [1960] 334/42).
[28] Pannetier, G.; Goudmand, P.; Dessaux, O.; Tavernier, N. (J. Chim. Phys. **61** [1964] 395/406).
[29] Pannetier, G.; Goudmand, P.; Dessaux, O.; Tavernier, N. (Compt. Rend. **255** [1962] 91/3).
[30] Strutt, R. J. (Proc. Roy. Soc. [London] A **88** [1913] 539/49).

[31] Sutton, D. G.; Melzer, J. E.; Heidner, R. F., III (J. Appl. Phys. **49** [1978] 844/7).
[32] Peyron, M.; My, Lam Thanh (J. Chim. Phys. **64** [1967] 129/34).
[33] Goudmand, P.; Dessaux, O. (J. Chim. Phys. **64** [1967] 135/40).
[34] Pannetier, G.; Dessaux, O.; Arditi, I.; Goudmand, P. (Bull. Soc. Chim. France **1966** 313/8).
[35] Pannetier, G.; Dessaux, O.; Arditi, I.; Goudmand, P. (Compt. Rend. **259** [1964] 2198/9).
[36] Pannetier, G.; Goudmand, P.; Dessaux, O.; Arditi, I. (Compt. Rend. Congr. Natl. Soc. Savantes Sect. Sci. **90** [1965/66] 117/27; C. A. **67** [1967] No. 77697).
[37] Petropoulos, B.; Dessaux, O.; Chaffiol, D.; Goudmand, P. (Compt. Rend. B **265** [1967] 355/8).
[38] Byfleet, C. R.; Carrington, A.; Russel, D. K. (Mol. Phys. **20** [1971] 271/7).
[39] Carrington, A.; Howard, B. J.; Levy, D. H.; Robertson, J. C. (Mol. Phys. **15** [1968] 187/200).
[40] Matsumi, Y.; Munakata, T.; Kasuya, T. (J. Phys. Chem. **88** [1984] 264/9).

[41] Murrells, T. P. (J. Chem. Soc. Faraday Trans. II **84** [1988] 85/94).
[42] Carrington, A. (Proc. Colloq. AMPERE **15** [1968/69] 23/41; C. A. **71** [1969] No. 130594).
[43] Burnett, S. M.; Feigerle, C. S.; Stevens, A. E.; Lineberger, W. C. (J. Phys. Chem. **86** [1982] 4486/8).
[44] Smith, R. D. (J. Chem. Soc. Dalton Trans. **1979** 478/81).
[45] Smith, R. D. (Chem. Phys. Letters **55** [1978] 590/2).
[46] Bock, H.; Solouki, B.; Roesky, H. W. (Inorg. Chem. **24** [1985] 4425/7).
[47] Smith, R. D.; Wyatt, J. R.; DeCorpo, J. J.; Saalfeld, F. E.; Moran, M. J.; MacDiarmid, A. G. (J. Am. Chem. Soc. **99** [1977] 1726/30).
[48] Teichmann, R. A., III; Nixon, E. R. (Inorg. Chem. **15** [1976] 1993/4).
[49] Jeffries, J. B.; Crosley, D. R. (Combust. Flame **64** [1986] 55/64).
[50] Wendt, J. O. L.; Wootan, E. C.; Corley, T. L. (Combust. Flame **49** [1983] 261/74, 273).

[51] Yang, M.; Karra, S. B.; Senkan, S. M. (Hazard. Waste Hazard. Mater. **4** [1987] 55/68, 63/6).
[52] Benson, S. W.; Golden, D. M.; Lawrence, R. W.; Shaw, R.; Woolfolk, R. W. (EPA-600-2-75-019 [1975] 6/38).
[53] Amano, T.; Saito, S.; Hirota, E.; Morino, Y. (J. Mol. Spectrosc. **32** [1969] 97/107).
[54] Uehara, H.; Morino, Y. (Mol. Phys. **17** [1969] 239/48).
[55] Tabayashi, K.; Shobatake, K. (J. Chem. Phys. **87** [1987] 7344/5).
[56] Obase, H.; Nagano, I.; Tsuji, M.; Nishimura, Y. (J. Chem. Phys. **89** [1988] 257/61).

1.1.3 Molecular Properties. Spectra

Numerical values generally refer to the by far most abundant isotopic species $^{32}S^{14}N$, if not explicitly stated otherwise. Where the last digit is written as a subscript, the uncertainty is larger than ±10 units of that last decimal place.

1.1.3.1 Electronic States

A general survey, comprising a table (Table 1, p. 9) and several graphs of potential curves (**Fig. 1**, p. 10), is given first. Details, including the fine-structure constants A (spin-orbit splitting) and γ (spin-rotation splitting), are reported in two further sections, devoted to the ground state $X^2\Pi$ (p. 11) and to electronically excited states (p. 13). Hyperfine-structure, Λ-doubling, rotational, and vibrational constants are dealt with later in separate sections (pp. 18, 21, 23, 27). Lifetimes and quenching rates have been incorporated into the chapter on the electronic emission spectra (p. 32).

1.1.3.1.1 Survey

Table 1, p. 9, lists altogether 14 states. Besides the regular ground state $X^2\Pi_r$, ten electronically excited doublet states (including the inverted $H^2\Pi_i$) were observed from the respective electronic emission or absorption spectra, the ground state being the lower state in each case (see pp. 32/7). Two low-lying quartet states were not directly observed but were inferred from perturbations executed by them on the regular doublet $B^2\Pi_r$ (see [1]). They were calculated as being stable by three ab initio configuration-interaction (CI) studies, i.e., an older full-valence CI calculation [2] and two more recent multi-reference double-excitation (MRD)–CI calculations [3, 4]. Finally, a doublet $^2\Phi$ was assumed to be responsible for perturbations of the observed $A^2\Delta$ state [5] and was shown theoretically to be stable [2, 3, 4].

Four observed states are of Rydberg character, which is indicated by "R" in the second column of the table. Their electronic configurations were taken from [6]. One state, $B'^2\Sigma^+$, is a valence state ("V") with some Rydberg character [1]. The other states are valence states, with configurations resulting from that of the ground state by excitations $3\pi \leftarrow 2\pi$ (excited states $^{2,4}\Pi$ and $^2\Phi$ [5, 7, 8]) and $3\pi \leftarrow 7\sigma$ (states $^{2,4}\Sigma$ and $^2\Delta$ [5, 7, 8, 9]). The leading configurations at the equilibrium internuclear distance, obtained by the CI calculations [2, 3, 4], are identical to these valence state configurations.

Term energies T_e in cm^{-1} of the observed doublet states up to $I^2\Sigma^+$ are taken from a compilation [10], and of the highest three Rydberg states from [6]. Energies of doublet components are followed by the quantum number of the total electronic angular momentum about the molecular axis (Ω). Energies of the two quartets are from [1]. Values in brackets are not of a true T_e character; for their definition, see the remarks to the table. Values in parentheses are uncertain. Rounded values for T_e in eV (see [4]) have been added for convenience.

The electronic states of the dissociation products (S and N) were taken from correlation tables [4] and calculated potential curves for the valence states $X^2\Pi$, $a^4\Pi$, $b^4\Sigma^-$, $B^2\Pi$, $^2\Phi$, $A^2\Delta$, $G^2\Sigma^-$, $H^2\Pi$, and $I^2\Sigma^+$ [2, 3, 4] and for $B'^2\Sigma^+$ and $C^2\Sigma^+$ [4]. Deviating correlations for three states of $^2\Sigma^+$symmetry (C, I, and J) in [3] are apparently due to the circumstance that the state $B'^2\Sigma^+$ was not found in these calculations [3]. Theoretical potential curves for the $^2\Sigma^+$ states B′, C, and I and for $A^2\Delta$, see also [11] (legends to figures for SN, CCl, and SiF are interchanged).

Potential energy curves based on spectroscopically observed constants are shown in Fig. 1, p. 10. Curves for $X^2\Pi$ and $B^2\Pi$ in Fig. 1a), from [12], were derived by the Rydberg-Klein-Rees (RKR) method [13], those for $A^2\Delta$, $C^2\Sigma^+$, and $G^2\Sigma^-$ by the Morse method [14]. Fig. 1b), from [1], and Fig. 1c), from [6], show Morse curves. RKR curves for $X^2\Pi$, $B^2\Pi$, $A^2\Delta$, $C^2\Sigma^+$, $G^2\Sigma^-$, and $H^2\Pi$ are also in [7], for $X^2\Pi$ and $B^2\Pi$ also in [15]. Further Morse curves for $X^2\Pi$ are in [16, 17, 18], for $B^2\Pi$, $A^2\Delta$, and $C^2\Sigma^+$ in [5, 17, 18], and for $G^2\Sigma^-$, $H^2\Pi$, and $I^2\Sigma^+$ in [5].

Table 1

SN Electronic States with Their Term Energies T_e.

state	character	configuration	T_e in cm^{-1}	T_e in eV	S + N	remark
$F^2\Delta$	R	$\ldots7\sigma^22\pi^4(S3d)\delta$	[56677.58]	[7.03]		a)
$J^2\Sigma^+$	R	$\ldots7\sigma^22\pi^4(S4p)\sigma$	[~56280]	[~6.98]		a)
$E^2\Pi_r$	R	$\ldots7\sigma^22\pi^4(S4p)\pi$	[51900] (3/2)			a), b)
			[51805] (1/2)	[6.43]		a), b)
$I^2\Sigma^+$	V	$\ldots7\sigma2\pi^43\pi^2$	(44400)	(5.50)	$^3P + ^2D^o$	
$H^2\Pi_i$	V	$\ldots7\sigma^22\pi^33\pi^2$	(44049) (1/2)		$^3P + ^2D^o$	
			43876 (3/2)	5.44	$^3P + ^2D^o$	
$G^2\Sigma^-$	V	$\ldots7\sigma2\pi^43\pi^2$	43346	5.37	$^3P + ^2D^o$	
$C^2\Sigma^+$	R	$\ldots7\sigma^22\pi^4(S4s)\sigma$	43290	5.36	$^3P + ^2D^o$	
$A^2\Delta_r$	V	$\ldots7\sigma2\pi^43\pi^2$	(40046) (5/2)		$^3P + ^2D^o$	c)
			40005 (3/2)	4.96	$^3P + ^2D^o$	c)
$^2\Phi$	V	$\ldots7\sigma^22\pi^33\pi^2$	see remark		$^3P + ^2D^o$	d)
$B'^2\Sigma^+$	V(+R)	$\ldots7\sigma^22\pi^48\sigma$	(36255)	(4.49)	$^3P + ^4S^o$	
$B^2\Pi_r$	V	$\ldots7\sigma^22\pi^33\pi^2$	30384.1 (3/2)		$^3P + ^2D^o$	e)
			30294.9 (1/2)	3.76	$^3P + ^2D^o$	e)
$b^4\Sigma^-$	V	$\ldots7\sigma2\pi^43\pi^2$	[29311]	[3.64]	$^1D + ^4S^o$	f)
$a^4\Pi_i$	V	$\ldots7\sigma^22\pi^33\pi^2$	[~27005]	[~3.35]	$^3P + ^4S^o$	f)
$X^2\Pi_r$	V	$\ldots7\sigma^22\pi^43\pi$	221.5 (3/2)		$^3P + ^4S^o$	
			0 (1/2)	0	$^3P + ^4S^o$	

a) The term energy belongs to the lowest vibrational level (v = 0) and is referred to the average of the minima of $X^2\Pi_{1/2}$ and $X^2\Pi_{3/2}$. – b) The levels v = 0 are strongly perturbed (probably by higher levels of $H^2\Pi_i$). Deperturbed energies are 51375.9 ± 5.0 (3/2) and 51330.4 ± 2.8 cm^{-1} (1/2) [6]. – c) Strongly perturbed (by $^2\Phi$, see above), especially the $\Delta_{5/2}$ component [10]. – d) MRD-CI calculations yielded T_e = 4.77 [4] and 4.60 eV (37117 cm^{-1}) [3]. – e) For more recent term energies, see p. 13. – f) Term energies referred to the lowest level (v = 0, J = 1/2) of $X^2\Pi_{1/2}$.

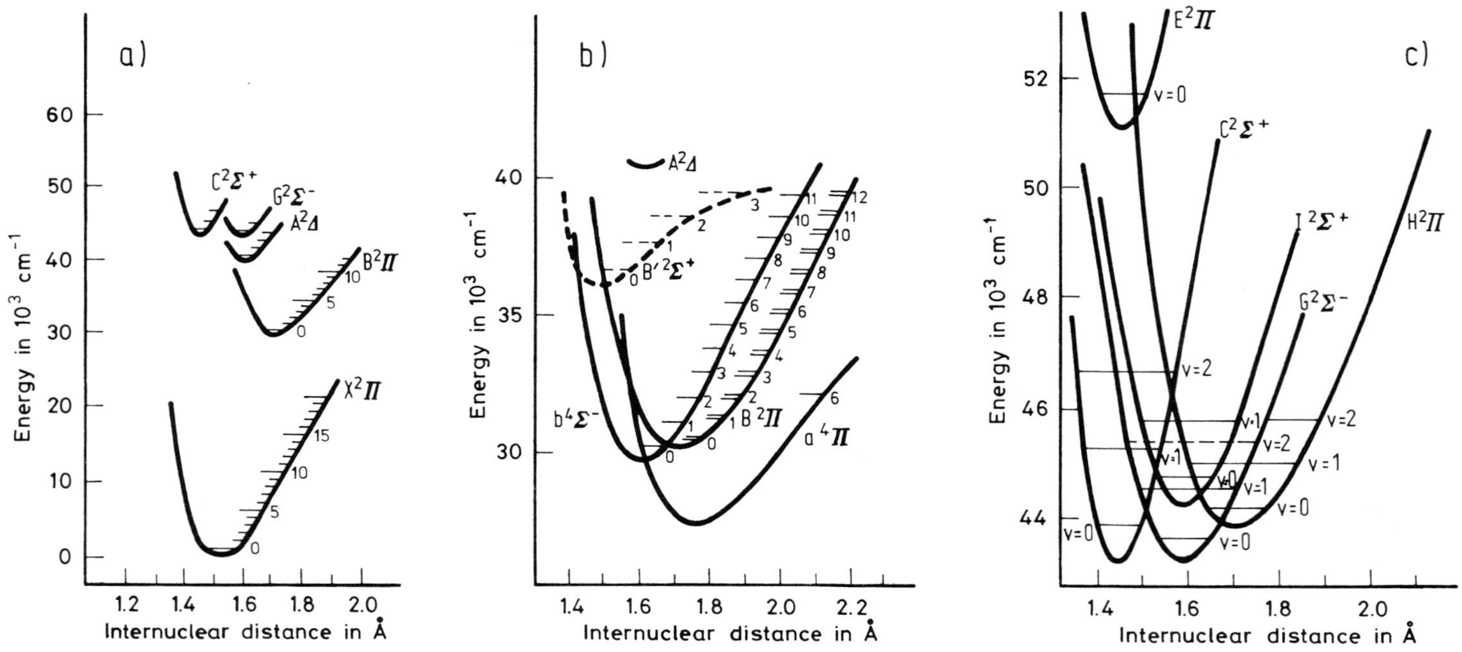

Fig. 1. Potential energy curves for the ground state and electronically excited states of SN based on spectroscopically observed constants (a) from [12]; b) from [1]; c) from [6]).

References:

[1] Jenouvrier, A.; Pascat, B. (Can. J. Phys. **58** [1980] 1275/90, 1289).
[2] Bialski, M.; Grein, F. (J. Mol. Spectrosc. **61** [1976] 321/31, 324).
[3] Lie, G. C.; Peyerimhoff, S. D.; Buenker, R. J. (J. Chem. Phys. **82** [1985] 2672/8).
[4] Karna, S. P.; Grein, F. (J. Mol. Spectrosc. **120** [1986] 284/91).
[5] Jenouvrier, A.; Pascat, B. (Can. J. Phys. **51** [1973] 2143/61, 2160/1).
[6] Vervloet, M.; Jenouvrier, A. (Can. J. Phys. **54** [1976] 1909/23, 1916, 1920/3).
[7] Chiu, C. L.; Silvers, S. J. (J. Chem. Phys. **63** [1975] 1095/8).
[8] Smith, J. J.; Meyer, B. (J. Mol. Spectrosc. **14** [1964] 160/72, 170).
[9] Barrow, R. F.; Drummond, G.; Zeeman, P. B. (Proc. Phys. Soc. [London] A **67** [1954] 365/77, 375/6).
[10] Huber, K. P.; Herzberg, G. (Molecular Spectra and Molecular Structure, Vol. 4, Constants of Diatomic Molecules, Van Nostrand Reinhold, New York 1979, pp. 484/7).

[11] Karna, S. P.; Grein, F. (Intern. J. Quantum Chem. **29** [1986] 755/66, 761).
[12] Jeffries, J. B.; Crosley, D. R.; Smith, G. P. (J. Phys. Chem. **93** [1989] 1082/90).
[13] Rees, A. L. G. (Proc. Phys. Soc. [London] **59** [1947] 998/1008).
[14] Morse, P. M. (Phys. Rev. [2] **34** [1929] 57/64).
[15] Peyron, M.; My, Lam Thanh (J. Chim. Phys. Physicochim. Biol. **64** [1967] 129/34).
[16] Dyke, J. M.; Morris, A.; Trickle, I. R. (J. Chem. Soc. Faraday Trans. II **73** [1977] 147/51).
[17] Goudmand, P.; Dessaux, O. (J. Chim. Phys. Physicochim. Biol. **64** [1967] 135/40).
[18] Pannetier, G.; Dessaux, O.; Arditi, I.; Goudmand, P. (Bull. Soc. Chim. France **1966** 313/8).

1.1.3.1.2 Ground State $X^2\Pi_r$

It was first observed as the common lower state of two electronic emission systems called "β" (from $A^2\Delta$; in original paper $^2\Pi$) and "γ" (from $C^2\Sigma^{(+)}$) in analogy to NO bands [1]. An absorption study of the γ system [2, 3] then showed that the lower state of the β and γ emissions was indeed the ground state, and the types of all states involved were confirmed by rotational analyses of the β [4] and the γ system [5]. The ground state has since been observed as the lower state of all electronic emission systems (see p. 32).

The ground-state **configuration** was given as KKL $z\sigma^2\, y\sigma^2\, x\sigma^2\, w\pi^4 v\pi$ in analogy to the similar NO radical [4]. This configuration or $1\sigma^2 \ldots 4\sigma^2 1\pi^4 5\sigma^2 6\sigma^2 7\sigma^2 2\pi^4 3\pi$, see Table 1, p. 9, was confirmed by ab initio SCF calculations (restricted HF [6, 7, 8], for unrestricted HF, see [9]), by SCF–Xα–SW (scattered wave) calculations [10, 11] (spin unrestricted, see [12]), and by EH (extended Hückel) calculations [10].

The following **bonding** properties of the valence orbitals were derived from spin-unrestricted HF calculations (yielding orbital energies ε_i as functions of the internuclear distance r) [9]: 5σ and 6σ are roughly the bonding and antibonding combinations, respectively, of the S 3s and N 2s atomic orbitals. 7σ is the (weakly) bonding combination of S $3p_z$ and N $2p_z$ (z = molecular axis). 2π and 3π are the bonding and antibonding combinations of S $3p_{x,y}$ and N $2p_{x,y}$ (finer details see original paper) [9]. These properties may also be derived from earlier restricted HF calculations [6] and are also in agreement with an early statement, based on the analogous NO radical, that there were one σ bond and one and one-half π bonds [4].

A transfer of negative **charge** from S to N was found in most theoretical work (see also "Dipole Moment μ", p. 20). It amounted to 0.16 e from a Mulliken population analysis in an unrestricted HF calculation [7] and ~0.5 e from Xα–SW and EH calculations [10, 12]. An ionic character of the bond of 24% was derived [6] from an experimental value of the dipole moment.

Total energies E_t (in E_h) from ab initio MRD–CI calculations were about −452.21 [13] and −452.20 [14] (both read from graphs of potential energy curves). Another CI value was calculated for a minimal basis [15]. A few (rounded) values from SCF calculations using a restricted HF procedure are $E_t = -451.93$ (large basis set; experimental equilibrium internuclear distance r_e) [6], −451.91 (for the largest basis used; calculated r_e) [16], −451.58 (basis of better than double-zeta quality; experimental r_e) [8]. A minimal basis was used in [7]. The unrestricted HF method yielded $E_t = -451.93$ [8], −451.92 [17], and −451.91 [9].

The **spin-orbit splitting** constant A (in cm^{-1}, if no other units are given) was allowed to depend on the vibrational quantum number, $A_v = A_e - \alpha_e^A(v+1/2)$ (see, e.g., [18, 19]). $\alpha_e^A (= A_0 - A_1) = 0.1863(22)\ cm^{-1}$ was obtained from a study of the $v=1 \leftarrow 0$ vibration-rotation bands in $X^2\Pi_{1/2}$ and $X^2\Pi_{3/2}$ by a tunable diode laser [20]. $A_e = 223.37\ cm^{-1}$ (6696434 MHz in original paper) was based [21] upon this α_e^A (given as 5585 MHz) and on $A_1 = 223.09 \pm 0.12$ from [22]. This latter value and $A_0 = 222.94 \pm 0.17$ (not fitting, however, the above v dependence) had both been derived [22] from the rotational structures of the $C^2\Sigma^+(v=0) \rightarrow X^2\Pi(v=1, 0)$ emission bands measured by Zeeman [5]. His own value, $A_1 = 223.03\ cm^{-1}$, was based upon an analysis of the $v=0 \rightarrow 1$ band [5]. $A_0 = 223.15$ from the $A^2\Delta_{3/2}$ (v = 2) → $X^2\Pi_{3/2,1/2}$ (v = 0) emission band [23] was later slightly changed into $A_0 = 222.98$ [24]; $A_2 = 222.62$ was derived from the emission band $I^2\Sigma^+$ (v = 0) → $X^2\Pi_{3/2,1/2}$ (v = 2) [23]. A rough value, $A_0 = 210 \pm 20$, was extracted from the photoelectron spectrum [25].

The spin-orbit splitting constant was also corrected by a centrifugal distortion term A_J (see [26, 27, 28]). The vibrational dependence of this constant was expressed by a Dunham-type series, $A_J(v) = \Sigma A_J^{(n)}(v+1/2)^n$. Microwave spectra measured in $X^2\Pi_{1/2}$ (v = 0 to 5) and $X^2\Pi_{3/2}$ (v = 0 to 4) then yielded $A_J^{(0)} = 2.5654(69)$ MHz and $A_J^{(1)} = -0.14669(261)$ MHz (and $A_J^{(n)} = 0$ for n = 2, 3). An independent fitting of the data for the two levels v = 0 and 1 yielded $A_J(0) = 2.4944(84)$ and $A_J(1) = 2.3395(93)$ MHz [21]. To arrive at the above A_J values, one of the Λ-doubling constants (p_v^*, see p. 21) and the spin-rotation splitting constant (γ_v, see below) had to be fixed to zero, as was explained in [29]. An explicit dependence of A on the internuclear distance r was derived from a microwave spectrum for v = 0, $A' = (r \cdot dA/dr)_{r_e} = 130 \pm 56\ cm^{-1}$ [22].

The **spin-rotation splitting** constant γ_v may be estimated by $\gamma_v \approx m \cdot A_v/\mu$, with m = electron mass and μ = reduced mass of the molecule (see [18, 26, 30, 31]). Thus, $\gamma_0 < 400$ MHz was used in [22].

References:

[1] Fowler, A.; Bakker, C. J. (Proc. Roy. Soc. [London] A **136** [1932] 28/36).
[2] Laird, R. K. (from Barrow, R. F., Discussions Faraday Soc. No. 9 [1950] 81).
[3] Barrow, R. F.; Downie, A. R.; Laird, R. K. (Proc. Phys. Soc. [London] A **65** [1952] 70/1).
[4] Barrow, R. F.; Drummond, G.; Zeeman, P. B. (Proc. Phys. Soc. [London] A **67** [1954] 365/77).
[5] Zeeman, P. B. (Can. J. Phys. **29** [1951] 174/85).
[6] O'Hare, P. A. G. (J. Chem. Phys. **52** [1970] 2992/6).
[7] Deutsch, P. W.; Curtiss, L. A. (Chem. Phys. Letters **51** [1977] 125/31).
[8] Findlay, R. H.; Palmer, M. H.; Downs, A. J.; Egdell, R. G.; Evans, R. (Inorg. Chem. **19** [1980] 1307/14).
[9] Török, F.; Pulay, P.; Szondy, T.; Nagy, P. (Acta Chim. [Budapest] **80** [1974] 139/45).
[10] Salahub, D. R.; Messmer, R. P. (Phys. Rev. [3] B **14** [1976] 2592/602, 2599).

[11] Ringers, D. A. (AD-A037802 [1977] 1/18, 11; C.A. **87** [1977] No. 90996).
[12] Salahub, D. R.; Messmer, R. P. (J. Chem. Phys. **64** [1976] 2039/47).
[13] Karna, S. P.; Grein, F. (J. Mol. Spectrosc. **120** [1986] 284/91).

[14] Lie, G. C.; Peyerimhoff, S. D.; Buenker, R. J. (J. Chem. Phys. **82** [1985] 2672/8).
[15] Bialski, M.; Grein, F. (J. Mol. Spectrosc. **61** [1976] 321/31, 324).
[16] Karpfen, A.; Schuster, P.; Petkov, J.; Lischka, H. (J. Chem. Phys. **68** [1978] 3884/90).
[17] Jafri, J. A.; Newton, M. D.; Pakkanen, T. A.; Whitten, J. L. (J. Chem. Phys. **66** [1977] 5167/72).
[18] Amano, T.; Saito, S.; Hirota, E.; Morino, Y.; Johnson, D. R.; Powell, F. X. (J. Mol. Spectrosc. **30** [1969] 275/89, 277/8, 285).
[19] Brown, J. M.; Watson, J. K. G. (J. Mol. Spectrosc. **65** [1977] 65/74, 68).
[20] Matsumura, K.; Kawaguchi, K.; Nagai, K.; Yamada, C.; Hirota, E. (J. Mol. Spectrosc. **84** [1980] 68/73).

[21] Anacona, J. R.; Bogey, M.; Davies, P. B.; Demuynck, C.; Destombes, J. L. (Mol. Phys. **59** [1986] 81/8).
[22] Amano, T.; Saito, S.; Hirota, E.; Morino, Y. (J. Mol. Spectrosc. **32** [1969] 97/107, 103/4).
[23] Jenouvrier, A.; Pascat, B. (Can. J. Phys. **51** [1973] 2143/61, 2144, 2154, 2160).
[24] Jenouvrier, A.; Pascat, B. (Can. J. Phys. **58** [1980] 1275/90, 1289).
[25] Dyke, J. M.; Morris, A.; Trickle, I. R. (J. Chem. Soc. Faraday Trans. II **73** [1977] 147/51).
[26] James, T. C. (J. Chem. Phys. **41** [1964] 631/41, 635).
[27] Veseth, L. (J. Phys. B **3** [1970] 1677/91, 1682).
[28] Veseth, L. (J. Mol. Spectrosc. **38** [1971] 228/42, 231).
[29] Endo, Y.; Saito, S.; Hirota, E. (J. Mol. Spectrosc. **92** [1982] 443/50).
[30] Coxon, J. A. (J. Mol. Spectrosc. **58** [1975] 1/28, 22).

[31] Van Vleck, J. H. (Phys. Rev. [2] **33** [1929] 467/506, 499).

1.1.3.1.3 Electronically Excited States

Inverted Quartet $a^4\Pi_i$

All spectroscopic data given for this state (see below, Table 1, p. 9, and pp. 24, 27) were inferred [1] from perturbations, observed in the $B^2\Pi \rightarrow X^2\Pi$ emission system and partly caused by interactions between vibrational levels of $B^2\Pi$ and $a^4\Pi$. $T_e \approx 30000$ cm^{-1} was an older estimation [2]. $T_e \approx 18000$ cm^{-1} had been predicted [5] from a comparison with NO.

A spin-orbit splitting constant of $A \approx -40$ cm^{-1} was given [1].

Quartet $b^4\Sigma^-$

As for $a^4\Pi_i$ above, the spectroscopic constants were based on a study of the $B^2\Pi \rightarrow X^2\Pi$ emission [1]. $T_e \approx 32000$ cm^{-1} had been predicted from a comparison with NO [5]. An ab initio SCF calculation was performed for this state [4] to obtain the molecular orbitals needed for CI calculations of potential curves for several valence and Rydberg states (see p. 8).

Regular Doublet $B^2\Pi_r$

The state was first correctly identified in [6, 7, 8] and in [5], and later also in [9, 10, 11]. A more recent study of perturbations in the $B^2\Pi \rightarrow X^2\Pi$ emission (caused by $a^4\Pi$, $b^4\Sigma^-$, and $B'\,^2\Sigma^+$) yielded the energies 29687.19 and 29778.29 cm^{-1} for the minima of the spin-orbit split components $B^2\Pi_{1/2}$ and $B^2\Pi_{3/2}$, respectively, both referred to the lowest level ($v=0$, $J=1/2$) of $X^2\Pi_{1/2}$ [1]. Previously, $T_e = 30296.4$ and 30386.8 cm^{-1} had been measured [12]. Partly differing were $T_e = 30297.9$ and 30363.3 cm^{-1} [13].

The electron configuration ...$7\sigma^2\ 2\pi^3 3\pi^2$, corresponding to the promotion of a bonding 2π to an antibonding 3π electron, appeared to be consistent with an enlarged internuclear distance r (vs. that of the ground state, see pp. 25/6) [7]. Identical configurations of $B^2\Pi_r$ and $H^2\Pi_i$ (compare Table 1, p. 9) are suggested by the similar constants r_e (see p. 26) and ω_e (see p. 27) [12].

A spin-orbit splitting constant of $A = 90.4\ cm^{-1}$ was given [12].

Predissociation seems to occur above an energy of 39935 cm^{-1}, corresponding to the level $J = 51/2$, $v = 12$ in the component $B^2\Pi_{1/2}$ (for $^{32}S^{14}N$; $J = 65/2$ for $^{32}S^{15}N$) [1].

Semidiffuse Doublet $B'^2\Sigma^+$

This state was detected by Narasimham et al. [14]. Their T_0 value was corrected in two later studies, i.e., one of the $B'^2\Sigma^+ \rightarrow X^2\Pi$ emission for both isotopic species $^{32}S^{14}N$ and $^{32}S^{15}N$ [15] and another of perturbations excecuted by $B'^2\Sigma^+$ on the $B^2\Pi_{1/2} \rightarrow X^2\Pi_{1/2}$ emission [16]. A more extensive study of such perturbations yielded a "deperturbed" $T_0 = 36163.83(8)\ cm^{-1}$ and an energetic distance of 35624.25 cm^{-1} between the potential minimum and the lowest level ($v = 0$, $J = 1/2$) of $X^2\Pi_{1/2}$ [1]. Previously, $T_0 = 36162.5\ cm^{-1}$ [16] and 36175 cm^{-1} (presumably not deperturbed) [15] had been given.

The state is considered to be a valence state with some Rydberg character (in view of a slightly smaller equilibrium distance than that of the ground state, see p. 26), having an electron configuration with an antibonding 3π orbital being replaced by an antibonding 8σ orbital [1]; for the configuration see also [14]. A "semidiffuse" nature of the state was inferred from the diffuse character of the 8σ orbital (for a detailed discussion, see original paper) [17], see also [18]. The state had not been found in similar MRD-CI calculations [4] and had been calculated to be repulsive by an earlier full-valence CI method [3].

A spin-rotation splitting constant of $\gamma = -0.016(6)\ cm^{-1}$ was given [1].

Unobserved Doublet $^2\Phi$

An early estimation for T_e was $\sim 24000\ cm^{-1}$ from a comparison with NO [5]. The state was suggested by Lie et al. [4] to be identical with $B'^2\Sigma^+$, which was not found in their MRD-CI calculations. For critical remarks see, however, [18].

Regular Doublet $A^2\Delta_r$

The state was first shown by Barrow et al. [19, 20] to be the upper state of the "β" emission system (see p. 35), previously thought [21, 22] to arise from a $^2\Pi$ upper state. Both spin-orbit split components are strongly perturbed, the $A^2\Delta_{3/2}$ substate showing only rotational, the $A^2\Delta_{5/2}$ substate also vibrational perturbations [12], see also [20, 23, 24]. Two rotational analyses of the β bands led for $A^2\Delta_{3/2}$ to $T_0 = 39912.7$ [12] and 39875.70 cm^{-1} and for $A^2\Delta_{5/2}$ to $T_0 = 39911.08\ cm^{-1}$ [20]. Vibrational analyses of β band heads also yielded T_0 values for isotopic species. Thus, for $A^2\Delta_{3/2}$: 39882.2 [25] and 39881.8 cm^{-1} [13] (both for $^{32}S^{14}N$), 39884.5 ($^{32}S^{15}N$) [25], and 39883.8 cm^{-1} ($^{34}S^{14}N$) [13]; the $^{32}S^{14}N$ data agreeing with band head positions reported earlier [20, 21, 22].

The electronic configuration ...$7\sigma 2\pi^4 3\pi^2$, first discussed in [20], should be the same for the three states $A^2\Delta$, $G^2\Sigma^-$, and $I^2\Sigma^+$ (see Table 1, p. 9), in view of similar molecular constants r_e (see p. 26) and ω_e (see p. 27) [12].

A spin-orbit splitting constant of $A \approx 30\ cm^{-1}$ was given [12]. Strongly v-dependent values A_v were given earlier, $A_0 = 35.3_8$, $A_1 = 25.9_4$, and $A_2 = 18._1\ cm^{-1}$ [20].

Rydberg Doublet $C^2\Sigma^+$

It was first identified as the upper state of the "γ" emission bands (see p. 36) [21]. The state was also seen in absorption [19, 26]. A rotational analysis of emission bands revealed numerous perturbations (for the vibrational levels v=1 and 2). Term energies of the lowest vibrational level (v=0), referred to the average of the ground-state minima, were 43883.79 ($^{32}S^{14}N$) and 43887.80 cm^{-1} ($^{32}S^{15}N$) [27]. Other analyses yielded $T_0 = 43387.29 \pm 0.20$ [28], 43387.35 [22], and 43387.60 cm^{-1} (calibration against the Ba II emission line at 230.4 nm) [29]. Band head measurements led to $T_0 = 43384.9$ ($^{32}S^{14}N$) [13, 25], 43382.6 ($^{32}S^{15}N$) [25], and 43384.2 cm^{-1} ($^{34}S^{14}N$) [13]. An absorption spectrum gave $T_0 = 43386.4$ cm^{-1} (Q_1 head) [19].

The Rydberg nature of $C^2\Sigma^+$ seems to have been first discussed in [5] in view of the small internuclear distance. For the Rydberg orbital (S4s)σ (see Table 1, p. 9), a quantum defect $\alpha = 2.26$ was derived from the formula $T_e = T_\infty - R/(n-\alpha)^2$ employing $T_\infty \approx 79500$ cm^{-1} (see "Ionization Potential", p. 17) and n = 4 (R = Rydberg constant). The reasonable agreement with α(S4s) = 2.04 of sulfur confirmed the identification of $C^2\Sigma^+$ as a Rydberg state [27].

A spin-rotation splitting constant of $\gamma_0 = 0.0055$ cm^{-1} was given [30].

Predissociation of the $C^2\Sigma^+$ state (in its vibrational levels v=1 and 2, possibly also in v=0) was inferred from intensities measured in emission, absorption, and fluorescence spectra of the $C^2\Sigma^+ \leftrightarrow X^2\Pi$ system. Two mechanisms for predissociation were discussed [26]: (1) That by a low-lying $^4\Pi$ state, dissociating to the ground-state atoms, was consistent with MRD-CI calculations [18]. (2) That of an accidental predissociation through a dissociative $^2\Sigma^+$ state, crossed by $H^2\Pi$ and $I^2\Sigma^+$ [26], was ruled out, since such a $^2\Sigma^+$ state did not exist [18]. Predissociation could also be caused by states $^2\Phi$, $A^2\Delta$, and $G^2\Sigma^-$ (via a repulsive $^4\Sigma^+$ state) [18]. Compare also the diabatic and adiabatic potential curves from similar MRD-CI calculations in [4].

Doublet $G^2\Sigma^-$

The state was identified by three emission bands resulting from transitions to the $X^2\Pi$ ground state [12]. Its existence had previously been inferred from a positive sign of the Λ-doubling constant p_0 of the ground state (see p. 21) [31]. The term energy was given as $T_0 = 43178.8$ cm^{-1} [12]. $T_e \approx 43000$ cm^{-1} had been predicted earlier [5]. An approximate term energy of 51000 cm^{-1} was derived [32] from the measured p_0 value.

For the electronic configuration, see remark in the section on $A^2\Delta$.

A spin-rotation splitting constant of $\gamma = +0.034$ cm^{-1} was given [12].

Inverted Doublet $H^2\Pi_i$

The state was detected by two emission bands resulting from transitions to vibrationally excited levels of the ground state [33]. A rotational analysis of these two and six further emission bands led to $T_0 = 43651.8$ cm^{-1} for $H^2\Pi_{3/2}$ and 43827.6 cm^{-1} for $H^2\Pi_{1/2}$ [12]. When referred to the average of the ground-state minima, these energies were given as 44148.09 and 44323.78 cm^{-1} (original value 44328.78 corrected in view of the A_0 given, see below) [27]. $T_e = 43941 \pm 5$ cm^{-1} was reported earlier [33] and $T_e \approx 43000$ cm^{-1} was roughly estimated [5].

For the electronic configuration, see remark in the section on $B^2\Pi_r$.

The spin-orbit splitting constant (for the vibrational level v=0) was given as $A_0 = -175.69$ [27], -175.8 [12], -172.4 [33], or -172.3 cm^{-1} [24].

Doublet $I^2\Sigma^+$

The state was first fully characterized by a rotational analysis of two emission bands (ending in vibrational levels v = 1 and 2 of the ground state) [12]. The $^2\Sigma$ nature of the state had been earlier based on the vibrational analysis of bands of the isotopic species $^{32}S^{14}N$ and $^{32}S^{15}N$ [34] or $^{32}S^{14}N$ and $^{34}S^{14}N$ [13], see also [8, 23]. The rotational analysis yielded $T_0 = 44271.9$ cm^{-1} [12]. When referred to the average of the ground-state minima, this energy was given as 44768.00 cm^{-1} [27].

For the electronic configuration, see remark in the section on $A^2\Delta$.

The spin-rotation splitting constant was given as $\gamma_0 \approx 0.1$ cm^{-1} [12, 24].

Rydberg Doublets $E^2\Pi_r$, $J^2\Sigma^+$, and $F^2\Delta$

The three states were definitely assigned in [27]. Two of them, E("$^2\Pi$ or $^2\Delta$") and F("$^2\Sigma$") had been detected in [23]. The lowest vibrational level (v = 0) of $E^2\Pi_r$ is strongly perturbed by $H^2\Pi_i$ (for deperturbed energies of the spin-orbit split components, see Table 1, p. 9). Quantum defects α of the three states were derived from the relation $T_e = T_\infty - R/(n-\alpha)^2$ (see explanations in the section on $C^2\Sigma^+$) [27]:

state	$E^2\Pi$	$J^2\Sigma^+$	$F^2\Delta$
Rydberg orbital	(S4p)π	(S4p)σ	(S3d)δ
α	2.05	1.86	0.84
α(S)	1.61	1.61	0.32

The spin-orbit splitting constant of $E^2\Pi$ was given as $A \approx 95$ cm^{-1} [27]. The deperturbed constant amounts to 45 cm^{-1} [24, 27].

References:

[1] Jenouvrier, A.; Pascat, B. (Can. J. Phys. **58** [1980] 1275/90, 1287/9).
[2] Jenouvrier, A. (Diss. Reims 1974 from [3, 4]).
[3] Bialski, M.; Grein, F. (J. Mol. Spectrosc. **61** [1976] 321/31, 324).
[4] Lie, G. C.; Peyerimhoff, S. D.; Buenker, R. J. (J. Chem. Phys. **82** [1985] 2672/8).
[5] Smith, J. J.; Meyer, B. (J. Mol. Spectrosc. **14** [1964] 160/72, 171).
[6] Narasimham, N. A.; Srikameswaran, K. (Nature **197** [1963] 370).
[7] Narasimham, N. A.; Srikameswaran, K. (Proc. Indian Acad. Sci. A **59** [1964] 227/40).
[8] Asundi, R. K. (J. Chem. Phys. **43** [1965] S 24).
[9] Pannetier, G.; Goudmand, P.; Dessaux, O.; Arditi, I. (Compt. Rend. Congr. Natl. Soc. Savantes Sect. Sci. **90** Pt. 1 [1965/66] 117/27; C.A. **67** [1967] No. 77697).
[10] Pannetier, G.; Dessaux, O.; Arditi, I.; Goudmand, P. (Bull. Soc. Chim. France **1966** 313/8).

[11] Goudmand, P.; Dessaux, O. (J. Chim. Phys. Physicochim. Biol. **64** [1967] 135/40).
[12] Jenouvrier, A.; Pascat, B. (Can. J. Phys. **51** [1973] 2143/61, 2152, 2160).
[13] Narasimham, N. A.; Subramanian, T. K. B. (J. Mol. Spectrosc. **29** [1969] 294/304).
[14] Narasimham, N. A.; Raghuveer, K.; Balasubramanian, T. K. (J. Mol. Spectrosc. **54** [1975] 160/1).
[15] Jenouvrier, A.; Daumont, D. (J. Mol. Spectrosc. **61** [1976] 313/5).
[16] Raghuveer, A.; Narasimham, N. A. (J. Mol. Spectrosc. **70** [1978] 323/5).
[17] Karna, S. P.; Grein, F. (Intern. J. Quantum Chem. **29** [1986] 755/66, 757).
[18] Karna, S. P.; Grein, F. (J. Mol. Spectrosc. **120** [1986] 284/91).
[19] Barrow, R. F.; Downie, A. R.; Laird, R. K. (Proc. Phys. Soc. [London] A **65** [1952] 70/1).
[20] Barrow, R. F.; Drummond, G.; Zeeman, P. B. (Proc. Phys. Soc. [London] A **67** [1954] 365/77, 375/6).

[21] Fowler, A.; Bakker, C. J. (Proc. Roy. Soc. [London] A **136** [1932] 28/36).
[22] Zeeman, P. B. (Can. J. Phys. **29** [1951] 174/85, 185).
[23] Joshi, K. C. (Z. Physik **191** [1966] 126/36).
[24] Huber, K. P.; Herzberg, G. (Molecular Spectra and Molecular Structure, Vol. 4, Constants of Diatomic Molecules, Van Nostrand Reinhold, New York 1979, pp. 484/7).
[25] Narasimham, N. A.; Srikameswaran, K. (Proc. Indian Acad. Sci. A **56** [1962] 316/24).
[26] Chiu, C. L.; Silvers, S. J. (J. Chem. Phys. **63** [1975] 1095/8).
[27] Vervloet, M.; Jenouvrier, A. (Can. J. Phys. **54** [1976] 1909/23, 1913, 1916, 1920, 1923).
[28] Balasubramanian, T. K.; Narasimham, N. A. (private communication to [29]).
[29] Chiu, C. L.; Silvers, S. J. (J. Mol. Spectrosc. **61** [1976] 316/8).
[30] Balasubramanian, T. K.; Narasimham, N. A. (J. Mol. Spectrosc. **53** [1974] 128/9).

[31] Amano, T.; Saito, S.; Hirota, E.; Morino, Y. (J. Mol. Spectrosc. **32** [1969] 97/107, 102/3).
[32] Hirota, E. (J. Phys. Chem. **87** [1983] 3375/83).
[33] Narasimham, N. A.; Balasubramanian, T. K. (J. Mol. Spectrosc. **40** [1971] 511/8).
[34] Narasimham, N. A.; Srikameswaran, K. (Proc. Indian Acad. Sci. A **56** [1962] 325/8).

1.1.3.2 **Ionization Potential** E_i in eV

A vertical ionization potential $E_i(vert) = 8.87 \pm 0.01$ (71500 cm^{-1}) was measured by photoelectron (PE) spectroscopy using HeI and NeI radiation [1]. This value is in reasonable agreement with an estimation $E_i \approx 70000\ cm^{-1}$ (8.7 eV), which was based on a comparison of quantum defects α of Rydberg states with those of S atomic orbitals (see p. 16) [2]. An older and higher estimate, $E_i = 9.85 \pm 0.28$ (79500 cm^{-1}), from which the α in [2] had been calculated, was obtained using the formation enthalpies of SN (2.91 ± 0.26 eV, see p. 38) and of SN^+ (outdated value 12.76 ± 0.10 eV, see p. 41) [3].

Theoretical values for E_i from quantum-chemical calculations (ad = adiabatic): Ab initio calculations of total (molecular and ionic) energies yielded $E_i(ad) = 8.31$ and $E_i(vert) = 8.48$ at an MRD–CI level [4], $E_i(ad) = 8.82$, $E_i(vert) = 9.14$ [5], and 8.74 [1] at the SCF level. An earlier SCF calculation at the experimental internuclear distance yielded $E_i = 9.75$ [3]. E_i was also derived using Koopmans' theorem [3, 5]. For ionization from a 2π orbital, see [1]. SCF–Xα–SW calculations were applied to ionization from the orbitals 5σ through 3π [6].

References:

[1] Dyke, J. M.; Morris, A.; Trickle, I. R. (J. Chem. Soc. Faraday Trans. II **73** [1977] 147/51).
[2] Vervloet, M.; Jenouvrier, A. (Can. J. Phys. **54** [1976] 1909/23, 1923).
[3] O'Hare, P. A. G. (J. Chem. Phys. **52** [1970] 2992/6).
[4] Karna, S. P.; Grein, F. (Chem. Phys. **109** [1986] 35/8).
[5] Karpfen, A.; Schuster, P.; Petkov, J.; Lischka, H. (J. Chem. Phys. **68** [1978] 3884/90).
[6] Ringers, D. A. (AD-A 037802 [1977] 1/18, 12; C.A. **87** [1977] No. 90996).

1.1.3.3 **Electron Affinity** A in eV

A = 1.194 ± 0.011 was derived from the photoelectron spectrum of the SN^- ion, measured with Ar II laser radiation (488 nm, 2.54 eV). A rotational correction (accounting for the participation of unresolved rotational states of SN^- and SN, more fully discussed in [1]) and a spin-orbit correction (due to unresolved spin-orbit splitting of the SN ground state) were subtracted from the observed SN($X^2\Pi$, v=0)←SN^-($X^3\Sigma$, v=0) transition energy [2]. A similar value, A=1.3±0.3, was derived from the relation $D(SN^-)=A(SN)+D(SN)-A(S)$, with a dissociation energy D(SN)=4.8 eV (see p. 29) and A(S)=2.07±0.07 eV (due to [3]). $D(SN^-)$ (see p. 49) was calculated theoretically by an SCF (near HF) method (at the experimental SN internuclear distance) and was corrected by a correlation energy term taken from the related, and better known, NO^- case [4]. Semiempirical calculations, starting from thermochemical relations similar to that given above, yielded A=1.1 and 1.5 [5].

References:

[1] Engelking, P. C.; Lineberger, W. C. (J. Chem. Phys. **66** [1977] 5054/8).
[2] Burnett, S. M.; Feigerle, C. S.; Stevens, A. E.; Lineberger, W. C. (J. Phys. Chem. **86** [1982] 4486/8).
[3] Berry, R. S. (Chem. Rev. **69** [1969] 533/42).
[4] O'Hare, P. A. G. (J. Chem. Phys. **54** [1971] 4124/6).
[5] Karachevtsev, G. V. (Zh. Fiz. Khim. **61** [1987] 2070/4; Russ. J. Phys. Chem. **61** [1987] 1078/81).

1.1.3.4 ^{14}N Magnetic Hyperfine Coupling Constants a, b, c, d in MHz

The magnetic hyperfine (hf) interaction between the unpaired electron and the nucleus ^{14}N (nuclear spin I=1) comprises an electronic orbital angular momentum-I dipolar interaction (constant a) and interactions involving the electron spin (constants b, c, and d, where b also contains the Fermi contact interaction b_F in addition to the electron spin-I dipolar interaction). The full definitions of constants a through d (see original paper) make use of different types of averages, i.e., either over the orbital distribution of the unpaired electron (for a) or over the spin distribution (for b, c, d). One has $b_F=b+c/3$ [1]. If the two types of averaging are not distinguished (see, e.g. [2]), one has $a=c/3+d$ [1, 3].

Individual constants or certain combinations of them were first obtained from electron paramagnetic resonance (EPR) measurements (see p. 30), performed on the "magnetic" $X^2\Pi_{3/2}$ component of the ground state (J=3/2) and yielding the hf component $h=a+(b+c)/2$ along the molecular axis (see [4], with component given as $a+b/2+c$). Microwave (MW) spectra were measured for both ground-state components (rotational transitions J″+1←J″) and yielded $a-(b+c)/2$ and d (from $X^2\Pi_{1/2}$), $a+(b+c)/2$ (from $X^2\Pi_{3/2}$), and b (see [5]). An hf splitting was also observed in a far-IR laser magnetic resonance (LMR) spectrum taken in the $X^2\Pi_{3/2}$ state. It was consistent with the more precise EPR and MW measurements [6].

Constants measured for the lowest vibrational level (v=0) of the $X^2\Pi$ ground state of $^{32}S^{14}N$ are shown below (for v>0 and other isotopic species, see the remarks):

a+(b+c)/2 in MHz	a−(b+c)/2 in MHz	b in MHz	d in MHz	method	J	Ref.	remark
56.872(246)	67.645(99)	39.8(57)	86.953(81)	MW	J″≤13/2	[7]	a)
57.15(20)	67.60(12)	36.3(74)	86.972(90)	MW	J″≤5/2	[5]	b)
56.35(83)	67.45(50)	—	87.03(42)	MW	J″≤3/2	[3]	c)
56.99(23)	—	(0)	—	EPR	3/2	[8]	d)
56.8(5)	—	—	—	EPR	3/2	[9]	e)
(57)	—	—	—	LMR	J″=17/2	[6]	f)

a) Transitions in $X^2\Pi_{3/2}$ only up to J″=11/2. Data for v=1, see original paper. Measurements extended up to v=5 (see below). – b) Constants for $^{34}S^{14}N$ were 57.23(52), 67.49(18), 32.5(114), and 87.08(15). – c) Only J″ = 1/2 in $X^2\Pi_{1/2}$, 3/2 in $X^2\Pi_{3/2}$. The constant c was estimated as −75.4 from the expression 3(a−d) (see introduction) or as −43.5 (from another single-configuration approach, c = −d/2). For $^{34}S^{14}N$: a−(b+c)/2 = 67.02(83), d = 87.22(102) [3]; see also [10]. – d) Constant b fixed to zero. a+(b+c)/2 = 57.61(23) was obtained for a "theoretical" value b = 63.66, estimated by using the relation c = 3(a−d) and MW data for a−(b+c)/2 and d from the foregoing row. – e) ^{14}N hf splitting reported as 22 ± 1 G in [11], and ~20 G in [12], see also [13]. – f) ^{14}N hf splitting was observed in the LMR spectrum taken with the 680 μm line of the vinyl bromide (C_2H_3Br) laser, corresponding to the transition J′ = 19/2, M_J' = 5/2 ← J″ = 17/2, M_J'' = 7/2 (perpendicular polarization). The splitting was consistent with a predicted value of 20.1 G (using a+(b+c)/2 = 57 from the literature) [6]. An LMR spectrum taken with the 514 μm line of the DCOF laser exhibited no resolved hf structure [6, 14].

The vibrational dependence of the constants a, b_F, c/3, and d was obtained from MW spectra measured in vibrational levels up to v = 5 for $X^2\Pi_{1/2}$ and up to v = 4 for $X^2\Pi_{3/2}$. The following coefficients $Y^{(n)}$ of a Dunham-type series (for each constant), see p. 12, were given [7]:

constant	a	b_F	c/3	d
$Y^{(0)}$ in MHz	62.356(181)	23.13(267)	−16.97(129)	86.837(138)
$Y^{(1)}$ in MHz	−0.228(243)	0	0	0.230(204)

References:

[1] Hirota, E. (J. Phys. Chem. **87** [1983] 3375/83).
[2] Morton, J. R.; Preston, K. F. (Landolt-Börnstein New Ser. Group II **17a** [1987] 5/198, 5).
[3] Amano, T.; Saito, S.; Hirota, E.; Morino, Y. (J. Mol. Spectrosc. **32** [1969] 97/107, 102, 106).
[4] Carrington, A. (Proc. Colloq. AMPERE **15** [1968/69] 23/41, 37).
[5] Lovas, F. J.; Suenram, R. D. (J. Mol. Spectrosc. **93** [1982] 416/22).
[6] Anacona, J. R.; Davies, P. B. (Chem. Phys. Letters **108** [1984] 128/31).
[7] Anacona, J. R.; Bogey, M.; Davies, P. B.; Demuynck, C.; Destombes, J. L. (Mol. Phys. **59** [1986] 81/8).
[8] Uehara, H.; Morino, Y. (Mol. Phys. **17** [1969] 239/48, 245).
[9] Carrington, A.; Howard, B. J.; Levy, D. H.; Robertson, J. C. (Mol. Phys. **15** [1968] 187/200, 199).
[10] Lovas, F. J.; Tiemann, E. (J. Phys. Chem. Ref. Data **3** [1974] 609/769, 739).

[11] Carrington, A.; Levy, D. H. (J. Chem. Phys. **44** [1966] 1298/9).
[12] Carrington, A.; Levy, D. H. (J. Chem. Phys. **71** [1967] 2/11).
[13] Carrington, A. (Proc. Roy. Soc. [London] A **302** [1968] 291/304, 300).
[14] Anacona, J. R.; Davies, P. B. (Infrared Phys. **25** [1985] 233/7).

1.1.3.5 ^{14}N Quadrupole Coupling Constants eq_1Q, eq_2Q in MHz

In addition to the usual (axial) coupling constant eq_1Q, which is proportional to the electric field gradient $\partial^2V/\partial z^2$, a nonaxial constant eq_2Q may be determined in cases with $|\Lambda| = 1$, see [1, 2]. Both constants were obtained from the SN microwave spectrum measured in the vibrational levels v ≤ 5 for $X^2\Pi_{1/2}$ (rotational transitions up to J = 15/2 ← 13/2) and v ≤ 4 for $X^2\Pi_{3/2}$ (up to J = 13/2 ← 11/2). Equilibrium values (first coefficients $Y^{(0)}$ of the Dunham-type series; second coefficients $Y^{(1)}$ vanished) were then given as eq_1Q = −0.6752(282) and eq_2Q = −5.84(228) [3].

Older microwave (MW) and electron paramagnetic resonance (EPR) measurements yielded only the axial component eq_1Q:

$^{32}S^{14}N$	$^{34}S^{14}N$	measurement	Ref.
−2.696(196)	−2.80(22)	MW, v=0, up to $J=7/2 \leftarrow 5/2$	[4]
−2.62(73)	−1.74(156)	MW, v=0, up to $J=5/2 \leftarrow 3/2$	[5, 6]
−2.86(31)	—	EPR, $^2\Pi_{3/2}$, v=0, J=3/2	[7]
−3.1(5)	—	EPR, $^2\Pi_{3/2}$, v=0, J=3/2	[8]

References:

[1] Lin, C. C. (Phys. Rev. [2] **119** [1960] 1027/8).
[2] Hirota, E. (J. Phys. Chem. **87** [1983] 3375/83).
[3] Anacona, J. R.; Bogey, M.; Davies, P. B.; Demuynck, C.; Destombes, J. L. (Mol. Phys. **59** [1986] 81/8).
[4] Lovas, F. J.; Suenram, R. D. (J. Mol. Spectrosc. **93** [1982] 416/22).
[5] Amano, T.; Saito, S.; Hirota, E.; Morino, Y. (J. Mol. Spectrosc. **32** [1969] 97/107, 102).
[6] Lovas, F. J.; Tiemann, E. (J. Phys. Chem. Ref. Data **3** [1974] 609/769, 739).
[7] Uehara, H.; Morino, Y. (Mol. Phys. **17** [1969] 239/48).
[8] Carrington, A.; Howard, B. J.; Levy, D. H.; Robertson, J. C. (Mol. Phys. **15** [1968] 187/200, 199).

1.1.3.6 Dipole Moment μ in D. Quadrupole Moment

$\mu=1.81\pm0.02$ for the ground-state component $X^2\Pi_{1/2}$ was measured by the Stark effect in the microwave spectrum (transition $J=3/2 \leftarrow 1/2$ in vibrational level v=0) [1], see also [2]. $\mu=1.86\pm0.03$ for the component $X^2\Pi_{3/2}$ was measured by the Stark splitting in the electron paramagnetic resonance spectrum (v=0, J=3/2) [3], thus correcting an older value $\mu=1.35\pm0.10$ from [4]. The dipole moments of both ground-state components are expected to be equal within the experimental uncertainty [1, 5].

According to several quantum-chemical calculations (ab initio [7, 8], SCF–Xα–SW [9]; see also p. 11), the negative end of the dipole seems to be at the N atom. Two MRD–CI calculations yielded $\mu=1.97$ [7] and 1.951 [10]. SCF results were strongly basis-set dependent, with the restricted HF method often giving better values than the unrestricted one, see RHF results [6, 11] and UHF results [8, 12, 13].

For the $B'^2\Sigma^+$ state, $\mu=2.95$ was calculated by an MRD–CI method [7].

The quadrupole moment was calculated by an RHF method to be -0.6734×10^{-26} esu·cm² at the experimental r (and reckoned from the center of mass of the radical) [6].

References:

[1] Amano, T.; Saito, S.; Hirota, E.; Morino, Y. (J. Mol. Spectrosc. **32** [1969] 97/107, 105).
[2] Lovas, F. J.; Tiemann, E. (J. Phys. Chem. Ref. Data **3** [1974] 609/769, 739).
[3] Byfleet, C. R.; Carrington, A.; Russell, D. K. (Mol. Phys. **20** [1971] 271/7).
[4] Carrington, A.; Howard, B. J.; Levy, D. H.; Robertson, J. C. (Mol. Phys. **15** [1968] 187/200, 199).
[5] Morino, Y. (private communication to [6]).
[6] O'Hare, P. A. G. (J. Chem. Phys. **52** [1970] 2992/6).
[7] Karna, S. P.; Grein, F. (J. Mol. Spectrosc. **120** [1986] 284/91).

[8] Deutsch, P. W.; Curtiss, L. A. (Chem. Phys. Letters **51** [1977] 125/31).
[9] Salahub, D. R.; Messmer, R. P. (Phys. Rev. [3] B **14** [1976] 2592/602, 2599; J. Chem. Phys. **64** [1976] 2039/47).
[10] Lie, G. C.; Peyerimhoff, S. D.; Buenker, R. J. (J. Chem. Phys. **82** [1985] 2672/8).

[11] Karpfen, A.; Schuster, P.; Petkov, J.; Lischka, H. (J. Chem. Phys. **68** [1978] 3884/90).
[12] Findlay, R. H.; Palmer, M. H.; Downs, A. J.; Egdell, R. G.; Evans, R. (Inorg. Chem. **19** [1980] 1307/14).
[13] Török, F.; Pulay, P.; Szondy, T.; Nagy, P. (Acta Chim. [Budapest] **80** [1974] 139/45).

1.1.3.7 Λ-Doubling Constants

Electronic Ground State $X^2\Pi$

Two constants, p and q, describe the Λ-type splitting of rotational levels of a $^2\Pi$ state, arising from interaction with neighboring $^2\Sigma^+$ states, as $p(J+1/2)$ for the $^2\Pi_{1/2}$ component and as $(2q\cdot B/A+p\cdot B^2/A^2)(J-1/2)(J+1/2)(J+3/2)$ for the $^2\Pi_{3/2}$ component (A = spin-orbit splitting, B = rotational constant), see [1, 2] and (with p term in the $^2\Pi_{3/2}$ expression omitted) [3]. For the full definitions of p and q and also of the corresponding "parity independent" constants p* and q* ("L-uncoupling constants"), see [1, 4]. Simplified expressions arise if only one $^2\Sigma$ state interacts, lying at an energy ΔE above $^2\Pi$ and related to $^2\Pi$ by "pure precession", namely $p=\pm 4AB/\Delta E$ and $q=\pm 4B^2/\Delta E$, yielding the relation $q=p\cdot B/A$. The positive signs are valid for interaction with a $^2\Sigma^-$ state (with $\Delta E>0$, $A>0$), see [1] and (for p) also [3]. Λ-type splitting was also described (see [5]) by two constants, α_P and β_P, defined in [6]. These are connected by $\alpha_P-2\beta_P=\beta_P\cdot A/B$ [5] and are related to p by $p=-2\alpha_P-(\alpha_P-2\beta_P)\cdot A/\Delta E$ [7, 8]. Then, $\beta_P=-q/2$.

The following table shows constants p_v (for two vibrational levels v = 0 and 1), derived from microwave, tunable laser IR, and electronic emission spectra:

p_0 in MHz	p_1 in MHz	spectrum	Ref.	remark
396.95(72)	396.77(102)	microwave, up to $J=15/2\leftarrow 13/2$	[9]	a)
397.409(76)	—	microwave, up to $J=7/2\leftarrow 5/2$	[10]	b)
397.32(40)	—	microwave, up to $J=5/2\leftarrow 3/2$	[8]	c)
(397.32)	397.5(39)	laser IR, up to P(23/2), R(47/2)	[11]	d)
37_5	36_0	emission $C^2\Sigma^+\rightarrow X^2\Pi$	[12]	e)
360(60)	—	emission $A^2\Delta\rightarrow X^2\Pi$	[13]	f)
330	390	emission $A^2\Delta\rightarrow X^2\Pi$	[14]	g)

a) Spectrum measured up to v = 5 (see below). The tabulated p_0 and p_1 are from separate fittings for v = 0 and 1. – b) For $^{34}S^{14}N$: p_0 = 390.384(102) [10]. Compare also [15] for discussion of the sign of p_0 and derivation of ΔE. – c) Given as 0.013253(13) cm^{-1} in [11, 16]. $\alpha_P=-198.24(38)$ MHz (for $^{32}S^{14}N$). For $^{34}S^{14}N$: p_0 = 389.61(59) [8]. See also [17]. – d) $v=1\leftarrow 0$ vibration-rotation band; p_0 was taken from [8]. Original value p_1 = 0.01326(13) cm^{-1} [11]. – e) 0.012_5 and 0.012_0 cm^{-1}. Rotational analysis of $v=0\rightarrow 1$, 0 bands [12]. Previously, from the Zeeman [18] measurement of the same emission system, p_0 = 330, p_1 = 270 [14]. – f) 0.012(2) cm^{-1} [13]. See also [19]. – g) 0.011 and 0.013 cm^{-1} [14].

For p_v, the coefficients of a Dunham-type series, $p_v=\Sigma\, p^{(n)}\,(v+1/2)^n$, summed over n = 0 to 3, were obtained from the microwave spectrum measured up to v = 5: $p^{(0)}$ = 396.56(54) MHz, $p^{(1)}$ = 593(129) kHz, $p^{(2)}$ = – 74.8(256) kHz, and $p^{(3)}$ = 0 [9].

The second constant, q_v, could be obtained from the microwave spectrum referred to above [9], although Λ doubling in the $X^2\Pi_{3/2}$ component (measured up to v = 4) was not resolved. A mixing of $^2\Pi_{1/2}$ and $^2\Pi_{3/2}$ levels occurred at the high J values used, leading to finite magnitudes of off-diagonal terms in the rotational Hamiltonian. Thus, $q_0 = 0.30(39)$ and $q_1 = 0.67(63)$ MHz were derived. Dunham coefficients (see above) were given as $q^{(0)} = 366(285)$ kHz, $q^{(n)} = 0$ for n = 1 to 3 [9]. $\beta_P \approx -0.34$ MHz (or $q_0 = 0.68$ MHz) was obtained from far-IR laser magnetic resonance spectra of the radical in its $X^2\Pi_{3/2}$ (v = 0) state, taken at 680 μm (J = 19/2 ← 17/2, M_J = 5/2 ← 7/2) and 514 μm (J = 25/2 ← 23/2, M_J = 21/2 ← 21/2) and showing Λ doubling in $X^2\Pi_{3/2}$ for the first time [20], see also [21]. $\beta_P = -0.682(2)$ MHz had been derived from α_P (see remark c) of the table above) by the relation given at the end of the first section [8].

Electronically Excited States

For $B^2\Pi_{1/2}$, $p_0 = -0.012\ cm^{-1}$ was obtained from an analysis of the electronic emission system $B^2\Pi \rightarrow X^2\Pi$ [13, 19]. For $A^2\Delta$, no Λ-doubling constants could be derived from the respective emission spectrum [13, 14].

References:

[1] Mulliken, R. S.; Christy, A. (Phys. Rev. [2] **38** [1931] 87/119, 92/4).

[2] Favero, P. G.; Mirri, A. M.; Gordy, W. (Phys. Rev. [2] **114** [1959] 1534/7).

[3] Gordy, W.; Cook, R. L. (Microwave Molecular Spectra, Techniques of Organic Chemistry, Vol. IX, 2nd Ed., Pt. II, Wiley, New York 1970, p. 103).

[4] Gallagher, J. J.; Johnson, C. M. (Phys. Rev. [2] **103** [1956] 1727/37, 1731).

[5] Amano, T.; Hirota, E.; Morino, Y. (J. Mol. Spectrosc. **27** [1968] 257/65).

[6] Dousmanis, G. C.; Sanders, T. M., Jr.; Townes, C. H. (Phys. Rev. [2] **100** [1955] 1735/54).

[7] Amano, T.; Saito, S.; Hirota, E.; Morino, Y.; Johnson, D. R.; Powell, F. X. (J. Mol. Spectrosc. **30** [1969] 275/89, 280).

[8] Amano, T.; Saito, S.; Hirota, E.; Morino, Y. (J. Mol. Spectrosc. **32** [1969] 97/107, 102).

[9] Anacona, J. R.; Bogey, M.; Davies, P. B.; Demuynck, C.; Destombes, J. L. (Mol. Phys. **59** [1986] 81/8).

[10] Lovas, F. J.; Suenram, R. D. (J. Mol. Spectrosc. **93** [1982] 416/22).

[11] Matsumura, K.; Kawaguchi, K.; Nagai, K.; Yamada, C.; Hirota, E. (J. Mol. Spectrosc. **84** [1980] 68/73).

[12] Balasubramanian, T. K.; Narasimham, N. A. (J. Mol. Spectrosc. **53** [1974] 128/9).

[13] Jenouvrier, A.; Pascat, B. (Can. J. Phys. **51** [1973] 2143/61, 2144, 2160).

[14] Barrow, R. F.; Drummond, G.; Zeeman, P. B. (Proc. Phys. Soc. [London] A **67** [1954] 365/77, 374/5).

[15] Hirota, E. (J. Phys. Chem. **87** [1983] 3375/83).

[16] Huber, K. P.; Herzberg, G. (Molecular Spectra and Molecular Structure, Vol. 4, Constants of Diatomic Molecules, Van Nostrand Reinhold, New York 1979, pp. 486/7).

[17] Lovas, F. J.; Tiemann, E. (J. Phys. Chem. Ref. Data **3** [1974] 609/769, 739).

[18] Zeeman, P. B. (Can. J. Phys. **29** [1951] 174/85).

[19] Jenouvrier, A.; Pascat, B. (Can. J. Phys. **58** [1980] 1275/90, 1289).

[20] Anacona, J. R.; Davies, P. B. (Chem. Phys. Letters **108** [1984] 128/31).

[21] Anacona, J. R.; Davies, P. B. (Infrared Phys. **25** [1985] 233/7).

1.1.3.8 Rotational Constants

Electronic Ground State $X^2\Pi$

The table below lists the equilibrium constant B_e (together with α_e and γ_e) and the constants B_0 and D for the lowest vibrational level (v = 0). The centrifugal distortion constant D exhibited no definite v dependence, and thus any index was omitted; see, however, the remarks to the table. "Effective" values (eff), which are slightly different for both components of a spin-orbit splitted state (see [1]) are given in another table below. Spectroscopic methods used are abbreviated as MW (microwave spectrum), laser IR (IR absorption with a tunable diode laser), UV (electronic emission), and EPR (electron paramagnetic resonance).

B_e in MHz	23249.6573(120)	(23247.7_6)	23250.4(3)	—	—
α_e in MHz	188.3874(189)	(183.$_5$)	188.7(5)	—	—
γ_e in kHz	−167.1(84)	—	—	—	—
B_0 in MHz	23155.4226(108)	23156.01(16)	—	23153	23161(10)
D in kHz	37.94(21)	28(12)	38.7(99)	37.5	—
method	MW	MW	laser IR	UV	EPR
Ref.	[2]	[3]	[4]	[5]	[6]
remark	a)	b)	c)	d)	e)

a) Equilibrium constants from a common analysis (see [7]) of all rotational transitions measured in vibrational levels $v \leq 5$. This fitting yielded also a fourth coefficient in the Dunham-type series expansion of B_v (−12.26(105) kHz) and D_e = 37.901(153) kHz (all higher coefficients in an expansion of D_v were zero). B_0 and D in the table are from an independent fitting for v = 0. – b) B_e (not explicitly given in the paper [3]) was determined (in order to get an equilibrium distance, see p. 25) from B_0, using an α_e value due to Zeeman [8]. See also [9]. – c) The vibration-rotation band $v = 1 \leftarrow 0$ was observed and yielded a B_1 value. α_e (0.006296(16) cm^{-1}) and B_e (0.775549(10) cm^{-1}) were then obtained using the B_0 value (0.772401(5) cm^{-1}) from [3]. D (1.29(33) $\times 10^{-6}$ cm^{-1}, termed D_e) was found to be in good agreement with another MW result, D = 1.27 $\times 10^{-6}$ cm^{-1} [10], and with a calculated value $D_e = 4B_e^3/\omega_e^2 = 1.26 \times 10^{-6}$ cm^{-1} [4]. – d) The $v = 0 \rightarrow 1,0$ bands of the $C^2\Sigma^+ \rightarrow X^2\Pi$ emission system were reanalyzed, yielding B_0 (0.7723_0 cm^{-1}), D (termed D_0, 1.25 $\times 10^{-6}$ cm^{-1}), B_1, and D_1. As a final B_0, the MW result [3] was adopted [5]. The first rotational analysis of these bands had yielded B_e = 23193 (0.77364 cm^{-1}), α_e = 184 (0.00612 cm^{-1}), and B_0 = 23101 MHz (0.77057 cm^{-1}) [8]; for B_e and α_e, see also [11]. – e) EPR observed in the J = 3/2 (v = 0) state of the $X^2\Pi_{3/2}$ component [6]. An earlier EPR study [12] had first shown that the Zeeman [8] B_0 value (see remark d)) was probably too low and should be changed into B_0 = 23150 MHz (0.7722 cm^{-1}).

Effective values of the rotational constants B_0, D_0, B_e, and α_e for both ground-state components $X^2\Pi_{1/2}$ and $X^2\Pi_{3/2}$ are shown in the next table:

	B_0^{eff} in MHz	D_0^{eff} in kHz	B_e^{eff} in MHz	α_e^{eff} in MHz
$X^2\Pi_{1/2}$	23072.229(20)	35.7(72)	2317_4	18_9
$X^2\Pi_{3/2}$	23238.697(20)	42.0(105)	2331_5	19_2
Ref.	[13]	[4]	[14, 15]	
remark	a)	b)	c)	

a) From an MW spectrum of $^{32}S^{14}N$. For $^{34}S^{14}N$: B_0^{eff} = 22661.682(42) and 22822.292(54) MHz [13]. Earlier MW spectra of both isotopic species yielded almost identical values (no result for $^{34}S^{14}N$ in $X^2\Pi_{3/2}$) [3], see also [9, 16]. B_0^{eff} was also obtained from UV spectra of the $B^2\Pi \rightarrow X^2\Pi$ [17] and $C^2\Sigma^+ \rightarrow X^2\Pi$ [8] systems. – b) From an IR absorption ($v = 1 \leftarrow 0$) study using a tunable diode laser. B_0^{eff} was adopted from [3]. No change of D^{eff} with v was found [4]. The more recent MW spectrum [13] was analyzed assuming D_0^{eff} to be identical for both components, thus 36.3(9) for $^{32}S^{14}N$ and 36.8(35) kHz for $^{34}S^{14}N$ [13]. – c) From UV spectra of several emission systems [14, 15]. The system $H^2\Pi \rightarrow X^2\Pi$ yielded B_e^{eff}, α_e^{eff}, and γ_e^{eff} for the $X^2\Pi_{1/2}$ component [18].

Electronically Excited States

Constants B and D in the following table were obtained from UV spectra of the respective emission systems (or from perturbations executed by certain states upon such systems). B is usually valid for v = 0, equilibrium values are marked by an asterisk.

state	$a^4\Pi$	$b^4\Sigma^-$	$B^2\Pi_{1/2}$	$B^2\Pi_{3/2}$	$B'\,^2\Sigma^+$	$A^2\Delta_{3/2}$	$C^2\Sigma^+$
B in cm^{-1}	~0.56*	0.678*	0.5962*	0.6013*	0.7945*	0.6850	0.8275
D in 10^{-6} cm^{-1}	—	—	1.3	1.3	~1.2	—	1.2
Ref.	[15]		[14, 15]		[15]	[14]	[19]
remark	a)		b)		c)	d)	e)

state	$G^2\Sigma^-$	$H^2\Pi_{1/2}$	$H^2\Pi_{3/2}$	$I^2\Sigma^+$	$E^2\Pi_r$	$J^2\Sigma^+$	$F^2\Delta$
B in cm^{-1}	0.6905(2)	0.5972(2)	0.5915(5)	0.6940(2)	0.821	~0.825	0.8367
D in 10^{-6} cm^{-1}	2.5	1.75	(1.8)	5	—	—	1.05
Ref.	[14]	[14]		[14, 19]	[19]		
remark	f)	g)		h)	i)		

a) From perturbations observed in the $B^2\Pi \rightarrow X^2\Pi$ emission system. $\alpha_e = 0.007$ cm^{-1} for $b^4\Sigma^-$ [15]. – b) $\alpha_e = 0.0048$ ($B^2\Pi_{1/2}$) and 0.0046 cm^{-1} ($B^2\Pi_{3/2}$) [14]. $B_e = 0.5962$ and 0.5959, $\alpha_e = 0.0048$ and 0.0041 cm^{-1} [20]. For older data, see [11, 17, 21]. – c) D is a "deperturbed" constant, as well as $B_0 = 0.78615(13)$. $\alpha_e = 0.0165$ cm^{-1} [15]. $B_e \approx 0.78$ cm^{-1} in [22], where perturbations between $B'\,^2\Sigma^+$ and $B^2\Pi$ were revealed and earlier data [23] were corrected. – d) For $A^2\Delta$: $B_0 = 0.6964_2$, $D_0 = 1.45 \times 10^{-6}$ cm^{-1} [5]. – e) For $^{32}S^{15}N$: $B_0 = 0.7895$, $D_0 = 1.5 \times 10^{-6}$ cm^{-1} [19]. Previously (for $^{32}S^{14}N$): $B_0 = 0.8287_6$ [5], 0.8283 [24], 0.8267_0 [8, 11]; $D = 1.404 \times 10^{-6}$ cm^{-1} [5]. The early result $D = 0.77 \times 10^{-6}$ cm^{-1} [8] was corrected [11, 24] using the expression $4B^3/\omega^2$ for D.

f) $D = 2.6 \times 10^{-6}$ cm^{-1} [19]. – g) D in the $H^2\Pi_{3/2}$ component for v = 2 [14]. For this (lower lying) component also given: $B_0 = 0.5925$ [19], $B_e = 0.5944$, $\alpha_e = 0.0058$ [14], 0.0059 cm^{-1} [16]. For $H^2\Pi_{1/2}$: $B_0 = 0.5971_3$ and 0.5864_0, $D = 1.7_0$ and $1.6_4 \times 10^{-6}$ cm^{-1} for $^{32}S^{14}N$ and $^{34}S^{14}N$, respectively [18]. For $H^2\Pi$: $D = 1.7 \times 10^{-6}$ cm^{-1} [19]. – h) D not given in the final tables in [14, 19]. – i) "Deperturbed" B_0 given for $E^2\Pi$ (as 0.8211(40) cm^{-1} in the text) [19].

References:

[1] Herzberg, G. (Molecular Spectra and Molecular Structure, Vol. 1, Spectra of Diatomic Molecules, 2nd Ed., Van Nostrand Reinhold, New York 1950, p. 233).

[2] Anacona, J. R.; Bogey, M.; Davies, P. B.; Demuynck, C.; Destombes, J. L. (Mol. Phys. **59** [1986] 81/8).

[3] Amano, T.; Saito, S.; Hirota, E.; Morino, Y. (J. Mol. Spectrosc. **32** [1969] 97/107, 102).

[4] Matsumura, K.; Kawaguchi, K.; Nagai, K.; Yamada, C.; Hirota, E. (J. Mol. Spectrosc. **84** [1980] 68/73).

[5] Balasubramanian, T. K.; Narasimham, N. A. (J. Mol. Spectrosc. **53** [1974] 128/9).

[6] Uehara, H.; Morino, Y. (Mol. Phys. **17** [1969] 239/48).

[7] Endo, Y.; Saito, S.; Hirota, E. (J. Mol. Spectrosc. **92** [1982] 443/50).

[8] Zeeman, P. B. (Can. J. Phys. **29** [1951] 174/85, 184/5).

[9] Lovas, F. J.; Tiemann, E. (J. Phys. Chem. Ref. Data **3** [1974] 609/769, 739).
[10] Lovas, F. J. (private communication to [4]).

[11] Barrow, R. F.; Drummond, G.; Zeeman, P. B. (Proc. Phys. Soc. [London] A **67** [1954] 365/77, 373/5).
[12] Carrington, A.; Howard, B. J.; Levy, D. H.; Robertson, J. C. (Mol. Phys. **15** [1968] 187/200, 197).
[13] Lovas, F. J.; Suenram, R. D. (J. Mol. Spectrosc. **93** [1982] 416/22).
[14] Jenouvrier, A.; Pascat, B. (Can. J. Phys. **51** [1973] 2143/61, 2151/4, 2160).
[15] Jenouvrier, A.; Pascat, B. (Can. J. Phys. **58** [1980] 1275/90, 1289).
[16] Huber, K. P.; Herzberg, G. (Molecular Spectra and Molecular Structure, Vol. 4, Constants of Diatomic Molecules, Van Nostrand Reinhold, New York 1979, pp. 484/7).
[17] Narasimham, N. A.; Srikameswaran, K. (Nature **197** [1963] 370; Proc. Indian Acad. Sci. A **59** [1964] 227/40, 229, 239).
[18] Narasimham, N. A.; Balasubramanian, T. K. (J. Mol. Spectrosc. **40** [1971] 511/8).
[19] Vervloet, M.; Jenouvrier, A. (Can. J. Phys. **54** [1976] 1909/23, 1913, 1916, 1920).
[20] Narasimham, N. A.; Subramanian, T. K. B. (J. Mol. Spectrosc. **29** [1969] 294/304, 297).

[21] Smith, J. J.; Meyer, B. (J. Mol. Spectrosc. **14** [1964] 160/72, 168).
[22] Jenouvrier, A.; Daumont, D. (J. Mol. Spectrosc. **61** [1976] 313/5).
[23] Narasimham, N. A.; Raghuveer, K.; Balasubramanian, T. K. (J. Mol. Spectrosc. **54** [1975] 160/1).
[24] Barrow, R. F.; Downie, A. R.; Laird, R. K. (Proc. Phys. Soc. [London] A **65** [1952] 70/1).

1.1.3.9 Internuclear Distance r

Equilibrium internuclear distances r_e are reported for the electronic ground state (first table below) and for part of the electronically excited states (second table, also containing r_0 values). All data were derived from the respective rotational constants B_e or B_0 (see pp. 23/4).

Electronic Ground State $X^2\Pi$

r_e in Å	1.49403(4)	1.4940_2	1.4938(2)
Ref.	[1]	[2]	[3]
remark	a)	b)	c)

a) B_e from a tunable laser IR spectrum [1]. – b) B_e apparently based on the microwave work [3] (see last column), but converted to r_e using a more recent value of Planck's constant h (than in [3]). – c) B_e based on a microwave spectrum for v=0 and an α_e value from the literature. An uncertainty of $\pm 0.0001\ cm^{-1}$ was assumed for α_e, and the uncertainty arising from h was also included.

The early value $r_e = 1.495_7$ Å [4, 5] was based on a rotational constant that was too small.

Two effective B_e values for the spin-orbit split ground-state components led to $r_e(X^2\Pi_{1/2}) = 1.4955$ and $r_e(X^2\Pi_{3/2}) = 1.4901$ Å [6, 7].

Electronically Excited States

state	$a^4\Pi$	$B^2\Pi$	$B'^2\Sigma^+$	$A^2\Delta_{3/2}$	$C^2\Sigma^+$	$G^2\Sigma^-$	$H^2\Pi$	$I^2\Sigma^+$	$E^2\Pi$	$J^2\Sigma^+$	$F^2\Delta$
type of r ..	r_e	r_e	r_e	r_0	r_0	r_0	r_e	r_0	r_0	r_0	r_0
r in Å	1.75	1.697	1.4747	1.5897	1.446_4	1.5834	1.702	1.5793	1.45_2	(1.44_9)	1.438_4
Ref.	[7]	[2]	[7]	[2, 6]	[2]	[2, 6]	[2]	[2, 6]	[2]	[2]	[2]
remark		a)	b)	c)	d)		e)		f)	f), g)	f)

a) Apparently based on effective B_e values from [6]. These had been converted to $r_e(B^2\Pi_{1/2})=1.7064$ and $r_e(B^2\Pi_{3/2})=1.6961$ Å [6], see also [7]. A previous value is $r_e=1.703$ [8]. For older (effective) r_e values, see [9, 10], for a preliminary r_0 value, see [5]. – b) $r_e=1.49$ (uncertain) [2] was based on an earlier B_e value from [11]. $r_0=1.498$ had been given in [12]. – c) $r_e=1.576_8$ for $A^2\Delta$ was given in [5]. – d) Based on B_0 from [13]. Previously, $r_0=1.446_9$ [5]. – e) Apparently based on effective B_0 values from [6]. These had been converted to $r_0(H^2\Pi_{1/2})=1.7025$ and $r_0(H^2\Pi_{3/2})=1.7134$ Å [6]. An older value was $r_0=1.705$ [8]. – f) Apparently based on B_0 values from [13]. – g) Uncertain value.

Equilibrium internuclear distances for many electronic states (including the ground state) were also calculated by ab initio MRD–CI methods, see [14, 15], and by full-valence CI methods, see [16]. For the ground state, see also SCF work cited on p. 12 ("Total energies") and a calculation by the δ-function model of chemical binding [17].

References:

[1] Matsumura, K.; Kawaguchi, K.; Nagai, K.; Yamada, C.; Hirota, E. (J. Mol. Spectrosc. **84** [1980] 68/73).
[2] Huber, K. P.; Herzberg, G. (Molecular Spectra and Molecular Structure, Vol. 4, Constants of Diatomic Molecules, Van Nostrand Reinhold, New York 1979, pp. 484/7).
[3] Amano, T.; Saito, S.; Hirota, E.; Morino, Y. (J. Mol. Spectrosc. **32** [1969] 97/107, 102).
[4] Zeeman, P. B. (Can. J. Phys. **29** [1951] 174/85, 185).
[5] Barrow, R. F.; Drummond, G.; Zeeman, P. B. (Proc. Phys. Soc. [London] A **67** [1954] 365/77, 375).
[6] Jenouvrier, A.; Pascat, B. (Can. J. Phys. **51** [1973] 2143/61, 2160).
[7] Jenouvrier, A.; Pascat, B. (Can. J. Phys. **58** [1980] 1275/90, 1289).
[8] Narasimham, N. A.; Balasubramanian, T. K. (J. Mol. Spectrosc. **40** [1971] 511/8).
[9] Narasimham, N. A.; Subramanian, T. K. B. (J. Mol. Spectrosc. **29** [1969] 294/304, 297).
[10] Narasimham, N. A.; Srikameswaran, K. (Proc. Indian Acad. Sci. A **59** [1964] 227/40, 239).

[11] Jenouvrier, A.; Daumont, D. (J. Mol. Spectrosc. **61** [1976] 313/5).
[12] Narasimham, N. A.; Raghuveer, K.; Balasubramanian, T. K. (J. Mol. Spectrosc. **54** [1975] 160/1).
[13] Vervloet, M.; Jenouvrier, A. (Can. J. Phys. **54** [1976] 1909/23, 1920).
[14] Lie, G. C.; Peyerimhoff, S. D.; Buenker, R. J. (J. Chem. Phys. **82** [1985] 2672/8).
[15] Karna, S. P.; Grein, F. (J. Mol. Spectrosc. **120** [1986] 284/91).
[16] Bialski, M.; Grein, F. (J. Mol. Spectrosc. **61** [1976] 321/31, 324).
[17] Lippincott, E. R.; Dayhoff, M. O. (Spectrochim. Acta **16** [1960] 807/34, 824).

1.1.3.10 Vibrational Constants ω_e, $\omega_e x_e$, and $\omega_e y_e$ in cm^{-1}

Electronic Ground State $X^2\Pi$

The constants in the following table were based upon band origins, which were obtained from rotational analyses of laser-induced fluorescence (LIF) spectra, of a tunable laser infrared absorption spectrum (IR), and of electronic emission bands (UV).

ω_e in cm^{-1}	1219.90(110)	1218.50(250)	1218.80	1218.97	1218.90	1218.7
$\omega_e x_e$ in cm^{-1}	7.45(6)	7.24(34)	(7.31)	7.26	7.30	7.28
$\omega_e y_e$ in cm^{-1}	—	−0.007(11)	—	−0.0053		—
spectrum	LIF		IR	UV		UV
Ref.	[1]		[2]	[3]		[4]
remark	a)		b)	c)		d)

a) Fluorescence $B^2\Pi \rightarrow X^2\Pi$ measured up to v″ = 26 (also $A^2\Delta \rightarrow X^2\Pi$ up to v″ = 6 and $C^2\Sigma^+ \rightarrow X^2\Pi$ up to v″ = 5). The quadratic fit (left column) was preferable to the cubic fit (right column). – b) Vibration-rotation transitions in the v = 1 ← 0 fundamental band, yielding $\omega_e - 2\omega_e x_e = 1204.1824(11)\ cm^{-1}$. For $\omega_e x_e$, two values (for $X^2\Pi_{1/2}$ and $X^2\Pi_{3/2}$) from [5] were averaged (see next remark). – c) Values to the left are for $X^2\Pi_{1/2}$, to the right for $X^2\Pi_{3/2}$, derived from the $B^2\Pi \rightarrow X^2\Pi$ and $B'^2\Sigma^+ \rightarrow X^2\Pi$ emission systems [3]. Previous values are $\omega_e = 1219.14$ and 1218.90, $\omega_e x_e = 7.28$ and 7.34 [5]. – d) From the literature.

$\omega_e = 1218.60$ and $\omega_e x_e = 7.25$ (for $X^2\Pi_{1/2}$) were based [6] upon a $\Delta G_{9/2}$ value measured in the $H^2\Pi_{1/2} \rightarrow X^2\Pi_{1/2}$ system [6] and a $\Delta G_{1/2}$ value measured in $C^2\Sigma^+ \rightarrow X^2\Pi_{1/2}$ [7].

ω_e and $\omega_e x_e$ have also been based upon the band heads observed in the longer known emission systems $A^2\Delta \rightarrow X^2\Pi$ and $C^2\Sigma^+ \rightarrow X^2\Pi$ [7 to 12], in $B^2\Pi \rightarrow X^2\Pi$ [13, 14, 15], and in $H^2\Pi \rightarrow X^2\Pi$ [16]. From a survey of previous work came $\omega_e = 1219 \pm 1$ and $\omega_e x_e = 7.5 \pm 0.2$ [17], earlier given (without error limits) in [21].

Electronically Excited States

state	$a^4\Pi$	$b^4\Sigma^-$	$B^2\Pi$	$B'^2\Sigma^+$	$A^2\Delta_{3/2}$		$C^2\Sigma^+$	$G^2\Sigma^-$	$H^2\Pi_{3/2}$	$I^2\Sigma^+$
ω_e	~760	935	794.9(9)	1091.5	960.3	958.0	1414	(879.8)	767.6	(~1008)
$\omega_e x_e$	~4	7.2	3.72(12)	24.5	8.4	7.5	5.25	—	5.0	—
Ref.	[3]		[1]	[3]	[10]	[11, 12]	[1]	[4]	[4, 18]	[4, 18]
remark	a)		b)	c)	d)		e)	f)	g)	h)

a) From perturbations observed for the $B^2\Pi \rightarrow X^2\Pi$ emission system. – b) Laser-induced fluorescence. Levels v = 0 through 12 were excited in $B^2\Pi$ [1]. $\omega_e = 797.35$ and 798.70, $\omega_e x_e = 3.72$ and 3.58 for $B^2\Pi_{1/2}$ and $B^2\Pi_{3/2}$, respectively, were derived from band origins of $B^2\Pi \rightarrow X^2\Pi$ [3]; for similar data, see [4, 5]. Older values, due to band origins, see [19, 20], due to band heads, see [13, 14, 15, 21]. – c) $\omega_e y_e = -0.7$ [3]. Previously, $\omega_e \approx 1060$ [22], supplemented by $\omega_e x_e \approx 15$ [4], had been given. – d) From $^{32}S^{14}N$ band heads. For $^{32}S^{15}N$, see [10]. For the $A^2\Delta_{5/2}$ component, see original papers [10, 11, 12]. The $A^2\Delta$ state, especially its $^2\Delta_{5/2}$ component, is strongly perturbed, and thus only $\Delta G_{1/2}$ values from [9] are given in [4] (for both components). $\omega_e \approx 950$ for $A^2\Delta_{3/2}$ [5]. – e) Values assumed [1], since the observed $\Delta G_{v+1/2}$ did not behave regularly: $\Delta G_{1/2} = 1389.2$, $\Delta G_{3/2} = 1413.7$ (from origins) [18], $\Delta G_{5/2} = 1372$, $\Delta G_{7/2} = 1378\ cm^{-1}$ (from heads) [11], see also [4]. Previously given values are: $\omega_e = 1401.1$, $\omega_e x_e = 6.7$ (from $^{32}S^{14}N$ band heads; for $^{32}S^{15}N$, see original paper) [10], $\omega_e = 1403.0$, $\omega_e x_e = 8.7$ [11]. From an early absorption study (assuming for x_e the ground-state value): $\omega_e \approx 1403._5$, $\omega_e x_e \approx 8.7_5$ [23]. –

f) $\Delta G_{1/2}$ value [4]. Originally given were $\Delta G_{1/2}=878.97$ [18] and 879.31 cm^{-1} [5]. – g) $\omega_e=765.4$ for $H^2\Pi_{3/2}$ was given in a table [18]. Previously, $\omega_e\approx775$ (for both components) [5], $\omega_e=761.04$, and $\omega_ex_e=5.10$ [16] were given. $\omega_e=813\pm10$ for $H^2\Pi_{1/2}$ was obtained from isotopic shifts of band origins [6]. – h) $\Delta G_{1/2}$ value [4, 18]. $\omega_e\approx1160$ was estimated from isotopic shifts of band heads [24].

For many electronic states, including the ground state, ω_e was obtained from ab initio MRD–CI calculations [25, 26]. See also the earlier full-valence CI calculations for valence states [27]. An SCF calculation for the ground state is in [28]. ω_e and ω_ex_e were derived from the δ-function model of chemical binding (applied to the ground state) [29].

References:

[1] Jeffries, J. B.; Crosley, D. R.; Smith, G. P. (J. Phys. Chem. **93** [1989] 1082/90).
[2] Matsumura, K.; Kawaguchi, K.; Nagai, K.; Yamada, C.; Hirota, E. (J. Mol. Spectrosc. **84** [1980] 68/73).
[3] Jenouvrier, A.; Pascat, B. (Can. J. Phys. **58** [1980] 1275/90, 1289).
[4] Huber, K. P.; Herzberg, G. (Molecular Spectra and Molecular Structure, Vol. 4, Constants of Diatomic Molecules, Van Nostrand Reinhold, New York 1979, pp. 484/7).
[5] Jenouvrier, A.; Pascat, B. (Can. J. Phys. **51** [1973] 2143/61, 2157, 2160).
[6] Narasimham, N. A.; Balasubramanian, T. K. (J. Mol. Spectrosc. **40** [1971] 511/8).
[7] Zeeman, P. B. (Can. J. Phys. **29** [1951] 174/85, 175, 185).
[8] Fowler, A.; Bakker, C. J. (Proc. Roy. Soc. [London] A **136** [1932] 28/36).
[9] Barrow, R. F.; Drummond, G.; Zeeman, P. B. (Proc. Phys. Soc. [London] A **67** [1954] 365/77, 375).
[10] Narasimham, N. A.; Srikameswaran, K. (Proc. Indian Acad. Sci. A **56** [1962] 316/24).

[11] Joshi, K. C. (Z. Physik **191** [1966] 126/36, 129/30).
[12] Joshi, K. C. (Z. Physik. Chem. [N.F.] **55** [1967] 173/8).
[13] Pannetier, G.; Goudmand, P.; Dessaux, O.; Arditi, I. (Compt. Rend. Congr. Natl. Soc. Savantes Sect. Sci. **90** Pt. 1 [1965/66] 117/27; C.A. **67** [1967] No. 77697).
[14] Pannetier, G.; Dessaux, O.; Arditi, I.; Goudmand, P. (Bull. Soc. Chim. France **1966** 313/8).
[15] Goudmand, P.; Dessaux, O. (J. Chim. Phys. Physicochim. Biol. **64** [1967] 135/40).
[16] Patel, M. M. (Z. Physik **173** [1963] 347/51).
[17] O'Hare, P. A. G. (J. Chem. Phys. **52** [1970] 2992/6).
[18] Vervloet, M.; Jenouvrier, A. (Can. J. Phys. **54** [1976] 1909/23, 1917, 1920).
[19] Narasimham, N. A.; Balasubramanian, T. K. (J. Mol. Spectrosc. **29** [1969] 294/304, 297).
[20] Narasimham, N. A.; Srikameswaran, K. (Proc. Indian Acad. Sci. A **59** [1964] 227/40, 239).

[21] Smith, J. J.; Meyer, B. (J. Mol. Spectrosc. **14** [1964] 160/72, 163, 168).
[22] Jenouvrier, A.; Daumont, D. (J. Mol. Spectrosc. **61** [1976] 313/5).
[23] Barrow, R. F.; Downie, A. R.; Laird, R. K. (Proc. Phys. Soc. [London] A **65** [1952] 70/1).
[24] Narasimham, N. A.; Srikameswaran, K. (Proc. Indian Acad. Sci. A **56** [1962] 325/8).
[25] Lie, G. C.; Peyerimhoff, S. D.; Buenker, R. J. (J. Chem. Phys. **82** [1985] 2672/8).
[26] Karna, S. P.; Grein, F. (J. Mol. Spectrosc. **120** [1986] 284/91).
[27] Bialski, M.; Grein, F. (J. Mol. Spectrosc. **61** [1976] 321/31, 324).
[28] Dyke, J. M.; Morris, A.; Trickle, I. R. (J. Chem. Soc. Faraday Trans. II **73** [1977] 147/51).
[29] Lippincott, E. R.; Dayhoff, M. O. (Spectrochim. Acta **16** [1960] 807/34, 825).

1.1.3.11 Dissociation Energy D in eV

For the $X^2\Pi$ **ground state,** $D_e \geq 4.95$ was inferred from a reanalysis of the $B^2\Pi \rightarrow X^2\Pi$ emission system: No rovibronic levels of $B^2\Pi$ were found above 39935 cm^{-1} (corresponding to $v'=12$, $J'=51/2$ and referred to the average of the $X^2\Pi_{1/2}$ and $X^2\Pi_{3/2}$ potential minima), i.e., the $B^2\Pi$ state should be predissociated above that energy [1]. Previously, lower limits had been reported as 4.93 (from the $B^2\Pi \rightarrow X^2\Pi$ emission with $v'=12$ as highest observed vibrational level) [2] and 4.83 ($v'=11$ was observed in [3] and was located at 38984 cm^{-1}) [4]. An upper limit of 5.63 (45380 cm^{-1}) was obtained from an apparent predissociation of the level $v'=1$ of the $C^2\Sigma^+$ state. A short lifetime of the $v'=0$ level (see p. 36) might indicate predissociation of this level, too, yielding $D_0 < 5.45$ (43994 cm^{-1}) [4].

Approximate dissociation energies, which agree reasonably with the above values, follow from empirical corrections, lowering the results of linear Birge-Sponer extrapolations (LBX) of the ground-state vibrational levels, $D_0(LBX) = \omega_e^2/4\omega_e x_e - \omega_e/2 \approx \omega_0^2/4\omega_0 x_0$ or $D_e(LBX) = \omega_e^2/4\omega_e x_e$, see [5, 6]. Thus, $D_0 = 4.8 \pm 0.25$ (for the lowest vibrational level, $v=0$, and also for $T=0$ K) followed from $D_0(LBX) = 6.1$ (arising from $\omega_e = 1219$, $\omega_e x_e = 7.5$ cm^{-1}), if a correction factor $0.448 + 5.256/r_e \cdot (E_i(S) - A(N))$ from [7] was applied, with $r_e = 1.4957$ Å, $E_i(S) = 10.36$ eV, and $A(N) = 0.04$ eV [8]. The corresponding D_e value (referred to the potential minimum) was given as $D_e = 4.87 \pm 0.25$ (39300 ± 2000 cm^{-1}) [4, 9] or $D_e = 4.88 \pm 0.25$ [2]. $D_0 = 5.0 \pm 0.7$ was obtained by lowering $D_0(LBX) = 6.1$ or 5.7 (see below) by roughly 20% [6], $D_0 = 5.0 \pm 1$ by lowering $D_0(LBX) = 6.0$ [5]. The LBX data 6.1 and 5.7 eV [6] were based on vibrational constants in [10, 11] and [12], respectively, 6.0 eV [5] and 5.9 eV [13] on constants in [14].

An empirical formula based by Gaydon [6] on a Lippincott [19] potential function and involving the rotational constant B_e, i.e., $D_0 = \omega_e^2/(5.33\omega_e x_e - 2B_e)$, yielded $D_0 = 4.79$ (with $B_e = 0.7736$ cm^{-1}; ω_e, $\omega_e x_e$ as above) [8].

A semiempirical δ-function model of chemical binding was also used to calculate D_e (130.8 kcal/mol or 5.67 eV) and D_0 (129.1 kcal/mol or 5.60 eV) [15].

A purely theoretical result ($D_e = 0.75$ eV) from an HF calculation was corrected by comparison with the better known NO radical to yield $D_0 = 5.2$ [8]. $D_e = 4.07$ followed from an MRD–CI calculation [16]. For an earlier full-valence CI calculation, see [17].

Electronically **excited states.** For $B^2\Pi$, a linear Birge-Sponer extrapolation was used [18]. MRD–CI calculations have been performed for many states [16] and full-valence CI calculations for the valence states [17].

References:

[1] Jenouvrier, A.; Pascat, B. (Can. J. Phys. **58** [1980] 1275/90, 1288/9).
[2] Murai, H.; Yagi, T.; Obi, K.; Tanaka, I. (Chem. Phys. Letters **61** [1979] 513/7).
[3] Peyron, M.; My, Lam Thanh (J. Chim. Phys. Physicochim. Biol. **64** [1967] 129/34).
[4] Chiu, C. L.; Silvers, S. J. (J. Chem. Phys. **63** [1975] 1095/8).
[5] Gaydon, A. G. (Dissociation Energies and Spectra of Diatomic Molecules, 2nd Ed., Chapman and Hall, London 1953, pp. 76, 99/100, 228).
[6] Gaydon, A. G. (Dissociation Energies and Spectra of Diatomic Molecules, 3rd Ed., Chapman and Hall, London 1968, pp. 252/3, 277).
[7] Hildenbrand, D. L.; Murad, E. (J. Chem. Phys. **51** [1969] 807/11).
[8] O'Hare, P. A. G. (J. Chem. Phys. **52** [1970] 2992/6).
[9] Dyke, J. M.; Morris, A.; Trickle, I. R. (J. Chem. Soc. Faraday Trans. II **73** [1977] 147/51).
[10] Smith, J. J.; Meyer, B. (J. Mol. Spectrosc. **14** [1964] 160/72, 168).

[11] Patel, M. M. (Z. Physik **173** [1963] 347/51).
[12] Joshi, K. C. (Z. Physik **191** [1966] 126/36).
[13] Herzberg, G. (Molecular Spectra and Molecular Structure, Vol. 1, Spectra of Diatomic Molecules, 2nd Ed., Van Nostrand Reinhold, New York 1950, p. 558).
[14] Fowler, A.; Bakker, C. J. (Proc. Roy. Soc. [London] A **136** [1932] 28/36).
[15] Lippincott, E. R.; Dayhoff, M. O. (Spectrochim. Acta **16** [1960] 807/34, 824).
[16] Lie, G. C.; Peyerimhoff, S. D.; Buenker, R. J. (J. Chem. Phys. **82** [1985] 2672/8).
[17] Bialski, M.; Grein, F. (J. Mol. Spectrosc. **61** [1976] 321/31, 324).
[18] Pannetier, G.; Dessaux, O.; Arditi, I.; Goudmand, P. (Bull. Soc. Chim. France **1966** 313/8).
[19] Lippincott, E. R.; Schroeder, R. (J. Chem. Phys. **23** [1955] 1131/41), Steele, D.; Lippincott, E. R. (J. Chem. Phys. **35** [1961] 2065/75).

1.1.3.12 **Electron Paramagnetic Resonance** (EPR) **Spectrum**

The spectrum was first observed by reacting N atoms (from a microwave discharge through N_2) with H_2S in a flow system (giving primarily the SH resonance). The spectrum consisted of three widely separated triplets and was centered at a spectroscopic splitting factor $g_J = 0.8$. Each triplet was ascribed to a ^{14}N (nuclear spin I=1) hyperfine (hf) splitting. The separations between them, due to second-order Zeeman splitting, were consistent with the rotational constant B_0 of the radical. The weak and rather broad lines did, however, not yet allow a determination of the ^{14}N quadrupole coupling constant. The spectrum could be ascribed to the lowest rotational level (J=3/2) of the upper component $X^2\Pi_{3/2}$ of the SN ground state (the lower component $X^2\Pi_{1/2}$ being "nonmagnetic") [1 to 3]: The pertaining expression for g_J of a linear molecule belonging to Hund's case (a), $g_J = (\Lambda+\Sigma)(\Lambda+2\Sigma)/J(J+1)$ (see [4]), yields 0.8 for $\Lambda=1$, $\Sigma=1/2$, $J=3/2$ and 0 for $\Lambda=1$, $\Sigma=-1/2$ $(J=1/2)$.

A stronger spectrum was obtained by reacting the N atoms with S_2Cl_2. It allowed a more detailed theoretical analysis, accounting first of all for the ^{14}N quadrupole coupling constant, but also for the dependence of the spectrum on the nuclear rotational g factor g_R, describing the interaction of the magnetic field with the magnetic moment of the rotating nuclei, and for the dependence of the center of the spectrum on the spin-orbit splitting constant A and on the exact value of the electronic g factor, $g_L + g_S/2$ (replacing $\Lambda+2\Sigma$ in the g_J expression above, see [5]). The spectrum was measured at 8589.65 MHz (field strengths see original paper) and was interpreted using A_0 and B_0 from the literature (see pp. 12, 23) and a theoretical value $g_R = 0.000272$ (rigid rotor model). It then yielded values for the hf splitting constant h (see p. 18) and the quadrupole coupling constant eqQ (see p. 20), and indicated a modification of the older B_0 value and slight corrections to $g_L + g_S/2$ and to g_R [6]. Compare also the earlier report [7], where, however, the use of OCS (instead of S_2Cl_2) was mentioned.

An even more intense spectrum resulted from passing a mixture of N_2 and SCl_2 through a microwave discharge (field strengths and frequencies, lying around 9.2 GHz, but measured separately for each line, see original paper). The interpretation made use of A_0 from [8] and of $g_R = (-)0.000272$ (sign convention different from [6]) and took also account of two "L-uncoupling constants", which were based on the Λ-doubling constant β_P given in [8], too (see p. 22). The spectrum then yielded values for B_0 (see p. 23), h (see p. 18), eqQ (see p. 20), and $g_L + g_S/2 = 2.00106 \pm 0.00004$. This latter value was consistent with g_S lying between the free-spin value (2.002319) and the values of SO (2.00199) or O_2 (2.001997) [9].

References:

[1] Carrington, A.; Levy, D. H. (J. Chem. Phys. **44** [1966] 1298/9).
[2] Carrington, A.; Levy, D. H. (J. Phys. Chem. **71** [1967] 2/11, 8/9).

[3] Carrington, A. (Proc. Roy. Soc. [London] A **302** [1968] 291/304, 300).
[4] Margenau, H.; Henry, A. (Phys. Rev. [2] **78** [1950] 587/92).
[5] Gordy, W.; Cook, R. L. (Microwave Molecular Spectra, Techniques of Organic Chemistry, Vol. IX, 2nd Ed., Pt. II, Wiley, New York 1970, p. 405).
[6] Carrington, A.; Howard, B. J.; Levy, D. H.; Robertson, J. C. (Mol. Phys. **15** [1968] 187/200, 188, 195, 197).
[7] Carrington, A. (Proc. Colloq. AMPERE **15** [1968/69] 23/41, 27).
[8] Amano, T.; Saito, S.; Hirota, E.; Morino, Y. (J. Mol. Spectrosc. **32** [1969] 97/107, 102/3).
[9] Uehara, H.; Morino, Y. (Mol. Phys. **17** [1969] 239/48, 241/2, 245/6).

1.1.3.13 Microwave Spectrum

All transitions observed for SN ($X^2\Pi$) show a resolved hyperfine (hf) structure (due to ^{14}N), but Λ doubling was resolved only for the lower-lying ground-state component $X^2\Pi_{1/2}$ [1 to 4].

Formation of $^{32}S^{14}N$ within the absorption cell by discharging first H_2S and then N_2, allowed the observation of rotational transitions in several vibrational levels, v = 0 through 5 in $X^2\Pi_{1/2}$ and v = 0 through 4 in $X^2\Pi_{3/2}$. Altogether 145 individual lines, i.e., 106 for $X^2\Pi_{1/2}$, extending from 69002.990(50) (v = 0, J = 3/2←1/2, F = 5/2←3/2) to 346220.775(250) MHz (v = 0, J = F = 15/2←13/2), and 39 for $X^2\Pi_{3/2}$, extending from 116147.270(80) (v = 0, J = 5/2←3/2, F = 5/2←5/2) to 302062.526(150) MHz (v = 0, J = 13/2←11/2, F = 11/2←9/2), were listed. They were used for the determination of fine structure (p. 12), hf structure (pp. 18/9), Λ doubling (p. 21), and rotational constants (p. 23) [1]. Earlier work was done for both isotopic species, $^{32}S^{14}N$ and $^{34}S^{14}N$, formed from discharges (outside the absorption cell) of flowing N_2–H_2S [2], N_2–SCl_2 [2, 3], or N_2–S_2Cl_2 [3] mixtures, but were restricted to the lowest vibrational level v = 0. Three rotational transitions (J = 3/2←1/2 to 7/2←5/2) in $X^2\Pi_{1/2}$ and one (J = 5/2←3/2) in $X^2\Pi_{3/2}$ were observed for both isotopes in [2], two (J = 3/2←1/2, 5/2←3/2) in $X^2\Pi_{1/2}$ and one (J = 5/2←3/2) in $X^2\Pi_{3/2}$ for $^{32}S^{14}N$, one (J = 3/2←1/2) in $X^2\Pi_{1/2}$ for $^{34}S^{14}N$ in [3], see also [4]. The frequency range thus began at 67774.75(3) MHz [2] or 67775.33 MHz [3, 4], measured for $^{34}S^{14}N$ ($X^2\Pi_{1/2}$, J = 3/2←1/2, F = 5/2←3/2). The frequency 69016.92 MHz of $^{32}S^{14}N$ ($X^2\Pi_{1/2}$, J = F = 3/2←1/2) [3, 4] is too low by ~1 MHz [2].

References:

[1] Anacona, J. R.; Bogey, M.; Davies, P. B.; Demuynck, C.; Destombes, J. L. (Mol. Phys. **59** [1986] 81/8).
[2] Lovas, F. J.; Suenram, R. D. (J. Mol. Spectrosc. **93** [1982] 416/22).
[3] Amano, T.; Saito, S.; Hirota, E.; Morino, Y. (J. Mol. Spectrosc. **32** [1969] 97/107, 104/5).
[4] Lovas, F. J.; Tiemann, E. (J. Phys. Chem. Ref. Data **3** [1974] 609/769, 740).

1.1.3.14 Far-Infrared Laser Magnetic Resonance (LMR) Spectrum

Resonance absorption of SN radicals, formed from a microwave discharge through an N_2–SCl_2 mixture, was detected in magnetic fields of 10 to 20 kG for the 680 μm line (440520.5 MHz) of the vinyl bromide (C_2H_3Br) laser and for the 514 μm line (~581323 MHz) of the DCOF laser. The respective rotational transitions were J = 19/2←17/2 and J = 25/2←23/2 in the lowest vibrational level (v = 0) of the "magnetic" component $X^2\Pi_{3/2}$ of the ground state [1, 2]. The 680 μm spectrum in parallel polarization ($\Delta M_J = 0$) consisted of component lines with M_J = 11/2 through 17/2, in perpendicular polarization ($\Delta M_J = -1$) of lines with M_J = 3/2 through 17/2 (and a single line with $\Delta M_J = +1$; see graphs in original paper) [1]. For the 514 μm spectra, see [2]. Each M_J component showed a structure due to Λ doubling (which had never before been resolved in the $X^2\Pi_{3/2}$ state) [1, 2]. The ^{14}N hyperfine structure was resolved only in the 680 μm spectrum [1].

References:

[1] Anacona, J. R.; Davies, P. B. (Chem. Phys. Letters **108** [1984] 128/31).
[2] Anacona, J. R.; Davies, P. B. (Infrared Phys. **25** [1985] 233/7).

1.1.3.15 Infrared Absorption

The $v=1 \leftarrow 0$ vibration-rotation band was observed between 1180 and 1240 cm^{-1} with ground-state radicals, formed from a microwave discharge through an N_2–S_2Cl_2 mixture, by using a tunable diode laser. Twelve lines (7 from the P and 5 from the R branch), each split by Λ doubling, were found for the $X^2\Pi_{1/2}$ state, and eleven unsplit lines (6 from the P and 5 from the R branch) for the $X^2\Pi_{3/2}$ state. Band origins were 1204.2755(12) cm^{-1} for $^2\Pi_{1/2}$ and 1204.0892(19) cm^{-1} for $^2\Pi_{3/2}$ [1].

A very weak IR absorption band at 1225 cm^{-1}, observed in a matrix-isolation study of the products of $(SN)_x$ volatilization, could have been due to the SN radical [2].

References:

[1] Matsumura, K.; Kawaguchi, K.; Nagai, K.; Yamada, C.; Hirota, E. (J. Mol. Spectrosc. **84** [1980] 68/73).
[2] Teichman, R. A., III; Nixon, E. R. (Inorg. Chem. **15** [1976] 1993/4).

1.1.3.16 Visible and Ultraviolet Emission and Absorption

All observed electronic transition systems involve the regular ground state $X^2\Pi_{3/2,1/2}$ as the lower state. They are dealt with below in several subsections, arranged according to increasing upper-state energy. Work on the respective emission bands is always mentioned first; a short table lists wavelength ranges, read from figures or tables (λ), vibrational levels involved (v′, v″), other isotopic species studied, and the main concern, such as rotational (rot.) or vibrational (vib.) analyses or band intensities (often including a discussion of the electronic transition moment R_e, of Franck-Condon factors $q_{v'v''}$, and of r centroids $\bar{r}_{v',v''}$). Lifetimes (τ) or quenching rates (k_q), extracted from the temporal behavior of the emission, are referred to subsequently. Absorption studies (including excitation scans from laser-induced fluorescence, LIF) are mentioned at the end.

1.1.3.16.1 $B^2\Pi_{3/2,1/2} \leftrightarrow X^2\Pi_{3/2,1/2}$ System

Emission bands are shaded to the red [1 to 4].

λ in nm	v′	v″	isotopes	concern	Ref.	remark
250 to 565	0 to 12	(0 to 11)	$^{32}S^{15}N$	rot. analysis	[5]	a)
250 to 365	0 to 12	0 to 4	—	rot. analysis	[4]	b)
265 to 290	6 to 10	0	$^{32}S^{15}N$, $^{34}S^{14}N$	rot., vib. analysis	[1, 6]	c)

λ in nm	v′	v″	isotopes	concern	Ref.	remark
280 to 500	0 to 7	0 to 9	$^{32}S^{15}N$, $^{34}S^{14}N$	rot., vib. analysis	[7]	d)
276.1	8	0	—	rot. analysis	[8]	e)
394 to 496	0 to 2	4 to 10	—	vib. analysis	[2, 3]	f)
255 to 491	0 to 12	0 to 12	—	vib. analysis	[9]	g)
260 to 520	0 to 11	0 to 10	—	intensities	[10]	h)
300 to 450	0 to 4	0 to 7	—	intensities	[11]	i)
300 to 700	0 to 11	0 to 26	—	intensities, LIF	[12]	j)
250 to 500	0 to 12	(0 to 9)	—	kinetics	[13]	k)
220 to 520	0 to 12	0 to 10	—	kinetics	[14]	

a) Various perturbations by $a^4\Pi$, $b^4\Sigma^-$, and $B'^2\Sigma^+$ were studied; v″ not reported. – b) Altogether 25 "subbands" (11 with $\Omega=1/2$, 14 with $\Omega=3/2$). – c) $^{34}S^{14}N$ vibrational analysis (v′=6 to 9) in [6]. – d) Rotational lines given for 2 subbands ($\Omega=1/2$, $v=0\rightarrow4$, 5). – e) Perturbations by $B'^2\Sigma^+$ studied. – f) For v″=4 to 9, see also [15]. For wavelengths (in a table comprising also other systems), see also [16], correcting [17]. Preliminary data, see [18]. For relative intensities (v′=0 to 9, v″=0 to 10), Franck-Condon factors $q_{v'v''}$, and R_e, see [19], for $q_{v'v''}$ (v′=0 to 9, v″=0 to 4), see also [20]. – g) Forty-two new subband heads (22 for $\Omega=1/2$, 20 for $\Omega=3/2$). – h) For R_e based upon these intensities, see [19]. – i) For $q_{v'v''}$ (v′=0 to 6, v″=0 to 9) and $R_e(\bar{r}_{v'v''})$, see original paper [11]. – j) Comparisons with calculated $q_{v'v''}$ from RKR potential curves (v′=0 through 11, v″≤18, using calculations [21] for the whole range) and from Morse curves (v′=0 to 2, v″≤14) showed R_e to be constant with r [12]. – k) v″ not stated. Plots of intensities vs. partial pressures (of active nitrogen, sulfur, or S_2Cl_2) yielded k_r/k_q.

Quenching (k_q) and vibrational relaxation (k_v) rate constants as well as zero-pressure lifetimes (τ_0) could be obtained from the time behavior of the LIF from certain vibrational (v′) and rotational (J′) levels. τ_0 could be identified with the radiative lifetime (τ_r), since all excited levels were below the first dissociation limit. The observed time behavior reflected perturbation of many levels (already known from purely spectroscopic work). Thus, multi-exponential decay curves were observed for v′=0 (J′=17/2) [22] and v′=8, 10, and 11 [23]. Except for v′=0, the "quenching rate constants" given below for N_2 as a quencher are really the sum k_q+k_v (in 10^{-11} $cm^3\cdot molecule^{-1}\cdot s^{-1}$). For v′=1, k_v could be determined (see the remarks):

v′	0	1	6	7	8	9	12
k_q+k_v	0.93(12)	2.01(12)	0.25(6)	≤0.5	1.3 to 1.9	1.5, 2.0	7.2 to 12.1
Ref.	[22]	[22]	[22]	[23]	[23]	[23]	[23]
remark	a)	b)	c)		d), e)	d)	d), f)

a) k_q value. For the series of further quenchers, NH_3, N_2O, CH_4, O_2, SF_6, and He, k_q was decreasing from 33.6(25) to 0.037(12)×10^{-11} $cm^3\cdot molecule^{-1}\cdot s^{-1}$. – b) $k_v(v'=1\rightarrow0)$= 0.49(12)×10^{-11} $cm^3\cdot molecule^{-1}\cdot s^{-1}$. k_q+k_v for seven further quenchers decreased from 9.11 (CO_2) to 0.13 (He). k_v was given for N_2O, O_2, and SF_6. – c) $k_q+k_v\leq0.5\times10^{-11}$ $cm^3\cdot molecule^{-1}\cdot s^{-1}$ [23]. Values for six further quenchers decreased from 2.32 (CO_2) to 0.17 (He) [22]. – d) Depending on Ω and J′. – e) The increased values (compared to v′=6) were attributed to the influence of the state $B'^2\Sigma^+$. – f) The highly increased values (compared to those at lower v′) were ascribed to collision-induced transfer to high rotational levels, which predissociate.

The radiative lifetime τ_r was found near ~1 μs for levels v′ = 0 through 11 [22]. Explicit data, including the somewhat exceptional case v′ = 12, are shown below:

v′	0	1	6	6	7
τ_r in μs	1.04(10)	1.00(10)	0.98(10)	1.01 to 1.05*)	1.03, 1.04*)
Ref.	[22]	[22]	[22]	[23]	[23]
v′	8	9	10	11	12
τ_r in μs	1.02 to 1.12*)	1.08, 1.11*)	1.01	1.05	1.00 to 1.29*)
Ref.	[23]	[23]	[23]	[23]	[23]

*) Depending on Ω and J′.

Absorption was monitored by LIF excitation scans, showing individual rotational lines; see spectra for v = 1←0 (324.0 to <327.0 nm), 8←0 (276.25 to 276.75 nm, showing Λ-type doubling) in [12], 9←0 (270.8 to 272.2 nm), 11←0 (260.6 to 261.5 nm; only Ω = 1/2) in [23], and 12←0 (256.5 to 257.2 nm, with onset of predissociation) in [12].

References:

[1] Narasimham, N. A.; Srikameswaran, K. (Nature **197** [1963] 370; Proc. Indian Acad. Sci. A **59** [1964] 227/40).
[2] Pannetier, G.; Dessaux, O.; Arditi, I.; Goudmand, P. (Bull. Soc. Chim. France **1966** 313/8).
[3] Goudmand, P.; Dessaux, O. (J. Chim. Phys. Physicochim. Biol. **64** [1967] 135/40).
[4] Jenouvrier, A.; Pascat, B. (Can. J. Phys. **51** [1973] 2143/61, 2154/7).
[5] Jenouvrier, A.; Pascat, B. (Can. J. Phys. **58** [1980] 1275/90).
[6] Narasimham, N. A.; Subramanian, T. K. B. (J. Mol. Spectrosc. **29** [1969] 294/304, 296).
[7] Smith, J. J.; Meyer, B. (J. Mol. Spectrosc. **14** [1964] 160/72).
[8] Raghuveer, K.; Narasimham, N. A. (J. Mol. Spectrosc. **70** [1978] 323/5).
[9] Vidal, B.; Dessaux, O.; Marteel, J.-P.; Goudmand, P. (Compt. Rend. C **268** [1969] 2140/1).
[10] Peyron, M.; My, Lam Thanh (J. Chim. Phys. Physicochim. Biol. **64** [1967] 129/34).

[11] Obase, H.; Nagano, I.; Tsuji, M.; Nishimura, Y. (J. Chem. Phys. **89** [1988] 257/61).
[12] Jeffries, J. B.; Crosley, D. R.; Smith, G. P. (J. Phys. Chem. **93** [1989] 1082/90).
[13] Murai, H.; Yagi, T.; Obi, K.; Tanaka, I. (Chem. Phys. Letters **61** [1979] 513/7).
[14] Tabayashi, K.; Shobatake, K. (J. Chem. Phys. **87** [1987] 7344/5).
[15] Pannetier, G.; Goudmand, P.; Dessaux, O.; Arditi, I. (Compt. Rend. Congr. Natl. Soc. Savantes Sect. Sci. **90** Pt. 1 [1965/66] 117/27; C.A. **67** [1967] No. 77697).
[16] Pannetier, G.; Dessaux, O.; Arditi, I.; Goudmand, P. (Compt. Rend. **259** [1964] 2198/9).
[17] Pannetier, G.; Goudmand, P.; Dessaux, O.; Tavernier, N. (J. Chim. Phys. Physicochim. Biol. **61** [1964] 395/406).
[18] Pannetier, G.; Goudmand, P.; Dessaux, O.; Tavernier, N. (Compt. Rend. **255** [1962] 91/3).
[19] Petropoulos, B.; Dessaux, O.; Chaffiol, D.; Goudmand, P. (Compt. Rend. B **265** [1967] 355/8).
[20] Petropoulos, B.; Dessaux, O.; Goudmand, P. (Compt. Rend. C **270** [1970] 1223/6).

[21] Raghuveer, K.; Narasimham, N. A. (J. Astrophys. Astron. **3** [1982] 13/25, 20/1).
[22] Jeffries, J. B.; Crosley, D. R. (J. Chem. Phys. **86** [1987] 6839/46).
[23] Matsumi, Y.; Munakata, T.; Kasuya, T. (J. Phys. Chem. **88** [1984] 264/9).

1.1.3.16.2 $B'^2\Sigma^+ \rightarrow X^2\Pi_{3/2,1/2}$ Emission

λ in nm	v′	v″	isotopes	concern	Ref.
268.7, 270.3*)	1	0	$^{32}S^{15}N$**)	rot. analysis	[1]

*) Both subbands, showing an open structure, were detected (but erroneously assigned to v′=0) in [2]. – **) For $^{32}S^{15}N$, the two subbands were observed at 268.87 and 270.77 nm, and two violet degraded bands at 276.5 and 278.25 nm were found and assigned to $v=0 \rightarrow 0$ [1].

References:

[1] Jenouvrier, A.; Daumont, D. (J. Mol. Spectrosc. **61** [1976] 313/5).

[2] Narasimham, N. A.; Raghuveer, K.; Balasubramanian, T. K. (J. Mol. Spectrosc. **54** [1975] 160/1).

1.1.3.16.3 $A^2\Delta_{5/2,3/2} \leftrightarrow X^2\Pi_{3/2,1/2}$ System (β Bands)

Emission bands are shaded to the red [1 to 5].

λ in nm	v′	v″	isotopes	concern	Ref.	remark
235 to 270	0 to 3	0 to 3	—	rot. analysis	[5]	a)
245 to 252	0, 1	0, 1	—	rot. analysis	[3]	b)
240 to 278	0 to 3	0 to 4	$^{32}S^{15}N$	vib. analysis	[6]	
240 to 270	0 to 2	0 to 4	$^{34}S^{14}N$	vib. analysis	[7]	
240 to 270	0 to 2	0 to 3	—	vib. analysis	[1, 2]	c)
235 to 280	0 to 3	0 to 5	—	vib. analysis, intensities	[4]	d)
245 to 290	0, 1	0 to 6	—	intensities, time behavior	[8]	e)

a) Altogether 17 subbands (seven with $\Omega=3/2 \rightarrow 1/2$, nine with $\Omega=5/2 \rightarrow 3/2$, one with $\Omega=3/2 \rightarrow 3/2$). Perturbations by $^2\Phi$ [5]. For a rotational analysis of the $v=0 \rightarrow 0$ band, see also [9]. – b) Upper state first identified as $^2\Delta$. Six subbands (three for both, $\Omega=5/2 \rightarrow 3/2$ and $3/2 \rightarrow 1/2$) [3]. See also [10]. – c) Upper state considered as $^2\Pi$. – d) Vibrational band strengths were calculated from Morse potential curves [11]. – e) LIF study. A comparison with Franck-Condon factors, calculated from Morse potential curves, revealed discrepancies, attributed to the $r_e(A^2\Delta)$ value used.

Fluorescence lifetimes were measured by LIF in the presence of 1 Torr N_2 and 40 mTorr SF_6. They depended irregularly on Ω′ and v′ (in ns): 100 (Ω′=3/2) and 140 (Ω′=5/2) for v′=0, 290 (Ω′=3/2 and 5/2) for v′=1 [8].

Absorption was monitored by LIF excitation scans of the two subbands $v=0 \leftarrow 0$ (250.8 to 251.2 nm, $\Omega=3/2 \leftarrow 1/2$, and 252.0 to 252.4 nm, $\Omega=5/2 \leftarrow 3/2$) [8]. $A^2\Delta \leftarrow X^2\Pi$ absorption around 250 nm was also observed by flash photolysis of OCS in the presence of N_2O [12].

References:

[1] Fowler, A.; Bakker, C. J. (Proc. Roy. Soc. [London] A **136** [1932] 28/36).

[2] Zeeman, P. B. (Can. J. Phys. **29** [1951] 174/85).

[3] Barrow, R. F.; Drummond, G.; Zeeman, P. B. (Proc. Phys. Soc. [London] A **67** [1954] 365/77).

[4] Joshi, K. C. (Z. Physik **191** [1966] 126/36).

[5] Jenouvrier, A.; Pascat, B. (Can. J. Phys. **51** [1973] 2143/61, 2144).
[6] Narasimham, N. A.; Srikameswaran, K. (Proc. Indian Acad. Sci. A **56** [1962] 316/24).
[7] Narasimham, N. A.; Subramanian, T. K. B. (J. Mol. Spectrosc. **29** [1969] 294/304, 298).
[8] Jeffries, J. B.; Crosley, D. R.; Smith, G. P. (J. Phys. Chem. **93** [1989] 1082/90).
[9] Balasubramanian, T. K.; Narasimham, N. A. (J. Mol. Spectrosc. **53** [1974] 128/9).
[10] Barrow, R. F.; Downie, A. R.; Laird, R. K. (Proc. Phys. Soc. [London] A **65** [1952] 70/1).

[11] Joshi, K. C. (Z. Physik. Chem. [N.F.] **55** [1967] 173/8).
[12] Donovan, R. J.; Breckenridge, W. H. (Chem. Phys. Letters **11** [1971] 520/2).

1.1.3.16.4 $C^2\Sigma^+ \leftrightarrow X^2\Pi_{3/2,1/2}$ System (γ Bands)

Emission bands are shaded to the violet [1 to 3]. For an empirical estimate of the electronic transition strength, see [4].

λ in nm	v′	v″	isotopes	concern	Ref.	remark
220 to 240	0 to 2	0 to 4	$^{32}S^{15}N$	rot. analysis	[3]	a)
230.4 to 238.4	0	0, 1	—	rot. analysis	[2]	b)
216 to 251	0 to 4	0 to 5	$^{32}S^{15}N$	vib. analysis	[5]	
216 to 246	0 to 2	0 to 2	$^{34}S^{14}N$	vib. analysis	[6]	
230 to 260	0	0 to 4	—	vib. analysis	[1]	c)
215 to 257	0 to 4	0 to 6	—	vib. analysis, intensities	[7]	d)
230 to 270	0	0 to 5	—	intensities, LIF	[8]	e)
216.4 to 251.1	0 to 2	0 to 4	—	intensities	[9]	f)
228.0 to 262.5	0	0 to 3	—	Hanle effect	[10]	g)

a) Perturbations in the rotational structures for v′=1, 2. – b) Same bands independently analyzed [11]. For a reanalysis, see [12]. – c) Band heads were remeasured in [2]. – d) Vibrational band strengths were calculated from Morse potential curves [13]. – e) A comparison with Franck-Condon factors, calculated from Morse potential curves, and r centroids showed a decrease of R_e with increasing r (in agreement with an ab initio calculation [14]). For lifetimes, see below. – f) Emission and fluorescence spectra were combined with an absorption spectrum to obtain populations of levels v′=0, 1, 2. Franck-Condon factors were calculated from Morse potentials. – g) Excitation by the Ba II 230.4 nm line. For τ_r, see below [10]. Rotational levels (in v′=0), excited by that line, were studied in [15].

A low-pressure lifetime τ=6.5 ns of the v′=0 level of $C^2\Sigma^+$ was derived from the Hanle effect [10]. This low value (if compared to that of the lowest Rydberg level of NO) suggested that the lifetime was shortened by predissociation [9, 10]. The value was consistent with an LIF study where times shorter than 15 ns could not be measured [8]. The comparison with a calculated radiative lifetime τ_r=28 to 33 ns [14] also showed that τ was due mainly to predissociation [8, 14]. For an estimation of quenching rate constants, see [16].

Absorption of the v=0←0 and 1←0 bands was first observed by Laird [17]; for wavelengths of certain band heads, see [11]. These two bands, together with v=0←1, were observed by flash photolysis of OCS in the presence of N_2O [18]. Bands v=2, 1, 0←0 were studied using a deuterium lamp [9]. The absorption spectrum was also monitored by an excitation scan of the 0←0 band between 230 and 232 nm [8, 16].

References:

[1] Fowler, A.; Bakker, C. J. (Proc. Roy. Soc. [London] A **136** [1932] 28/36).
[2] Zeeman, P. B. (Can. J. Phys. **29** [1951] 174/85).
[3] Vervloet, M.; Jenouvrier, A. (Can. J. Phys. **54** [1976] 1909/23, 1910/3).
[4] Monyakin, A. P.; Kuznetsova, L. A.; Kuz'menko, N. E.; Kuzyakov, Yu. Ya.; Plastinin, Yu. A. (Opt. Spektroskopiya **48** [1980] 443/6; Opt. Spectrosc. [USSR] **48** [1980] 243/5).
[5] Narasimham, N. A.; Srikameswaran, K. (Proc. Indian Acad. Sci. A **56** [1962] 316/24).
[6] Narasimham, N. A.; Subramanian, T. K. B. (J. Mol. Spectrosc. **29** [1969] 294/304).
[7] Joshi, K. C. (Z. Physik **191** [1966] 126/36).
[8] Jeffries, J. B.; Crosley, D. R.; Smith, G. P. (J. Phys. Chem. **93** [1989] 1082/90).
[9] Chiu, C. L.; Silvers, S. J. (J. Chem. Phys. **63** [1975] 1095/8).
[10] Silvers, S. J.; Chiu, C. L. (J. Chem. Phys. **61** [1974] 1475/9).

[11] Barrow, R. F.; Downie, A. R.; Laird, R. K. (Proc. Phys. Soc. [London] A **65** [1952] 70/1).
[12] Balasubramanian, T. K.; Narasimham, N. A. (J. Mol. Spectrosc. **53** [1974] 128/9).
[13] Joshi, K. C. (Z. Physik. Chem. [N.F.] **55** [1967] 173/8).
[14] Lie, G. C.; Peyerimhoff, S. D.; Buenker, R. J. (J. Chem. Phys. **82** [1985] 2672/8).
[15] Chiu, C. L.; Silvers, S. J. (J. Mol. Spectrosc. **61** [1976] 316/8).
[16] Jeffries, J. B.; Crosley, D. R. (Combust. Flame **64** [1986] 55/64).
[17] Laird, R. K. (from Barrow, R. F., Discussions Faraday Soc. No. 9 [1950] 81).
[18] Donovan, R. J.; Breckenridge, W. H. (Chem. Phys. Letters **11** [1971] 520/2).

1.1.3.16.5 Further Emission Systems

system	λ in nm	v′	v″	shading	isotopes	Ref.	remark
$G^2\Sigma^- \rightarrow X$	226 to 240	0, 1	0, 1	red	$^{32}S^{15}N$	[1]	a)
$H^2\Pi_{1/2,3/2} \rightarrow X$	225 to 268	0, 2	1 to 6	red	—	[1]	b)
	255.0 to 266.5	0	4, 5	red	$^{34}S^{14}N$	[2]	b)
$I^2\Sigma^+ \rightarrow X$	220 to 240	0	0 to 2	red	—	[1, 3]	c)
	232 to 240	0	1, 2	red	$^{32}S^{15}N$	[4, 5]	d)
$E^2\Pi_{3/2,1/2} \rightarrow X$	190 to 205	0	0 to 2	red	—	[3]	e)
$J^2\Sigma^+ \rightarrow X$	178.5 to 184.0	0	0, 1	violet	—	[3]	
$F^2\Delta_{5/2,3/2} \rightarrow X$	175 to 190	0	0 to 2	violet	—	[3, 6]	f)

a) Rotational analysis. The earlier attribution of part of these bands to the SN^+ ion [7] was thus corrected [1]. For an excitation scan of the 0–0 band at 232.7 nm ($\Omega = 3/2$), see [8]. – b) Rotational analyses [1, 2]. For a vibrational analysis, see [9]. – c) Rotational analysis. – d) Vibrational analysis [4, 5]. A few bands also in [6]. – e) Rotational analysis showing perturbations in the v′ = 0 level [3]. For an earlier observation, see [6]. – f) Rotational analysis for v″ = 0, 1 in [3], vibrational analysis in [6].

References:

[1] Jenouvrier, A.; Pascat, B. (Can. J. Phys. **51** [1973] 2143/61).
[2] Narasimham, N. A.; Balasubramanian, T. K. (J. Mol. Spectrosc. **40** [1971] 511/8).
[3] Vervloet, M.; Jenouvrier, A. (Can. J. Phys. **54** [1976] 1909/23).
[4] Narasimham, N. A.; Srikameswaran, K. (Proc. Indian Acad. Sci. A **56** [1962] 325/8).
[5] Narasimham, N. A.; Subramanian, T. K. B. (J. Mol. Spectrosc. **29** [1969] 294/304).

[6] Joshi, K. C. (Z. Physik **191** [1966] 126/36).
[7] Dressler, K. (Helv. Phys. Acta **28** [1955] 563/90, 570).
[8] Jeffries, J. B.; Crosley, D. R.; Smith, G. P. (J. Phys. Chem. **93** [1989] 1082/90).
[9] Patel, M. M. (Z. Physik **173** [1963] 347/51).

1.1.4 Heat of Formation. Equilibrium Constants. Heat Capacity and Thermodynamic Functions

The heat of formation value for the electronic ground state, $\Delta H^\circ_{f,0} = 2.91 \pm 0.26$ eV (67.10 ± 6.00 kcal/mol), was derived from the dissociation energy D°_0 (SN) = 4.8 eV with the aid of the equation $\Delta H^\circ_{f,0}$ (SN) = 1/2 D°_0 (N_2) + $\Delta H^\circ_{f,0}$ (S) − D°_0 (SN). For D°_0 (N_2) and $\Delta H^\circ_{f,0}$ (S) the values 9.76 ± 0.1 and 2.83 ± 0.05 eV, respectively, are taken [1].

$H^\circ_{f,298} = 63 \pm 25$ kcal/mol was calculated [2] from D°_0(SN) = 5.0 ± 0.7 eV (115 ± 25 kcal/mol) [3]. A tabulation of $\Delta H^\circ_{f,T}$ and $\Delta G^\circ_{f,T}$ at T = 0 to 6000 K (100 K intervals) was published [2].

Equilibrium constants for $S_2 + N_2 \rightleftharpoons 2$ SN, heat capacity, and thermodynamic function for the ideal gas SN were calculated and tabulated for T = 0 to 6000 K at 100 K intervals from partition functions [2] which were based on the spectroscopic constants $\omega_e = 1220.0$ cm^{-1}, $\omega_e x_e = 7.75$ cm^{-1} [4, p. 558], $B_e = 0.7762$ cm^{-1}, $\alpha_e = 0.0064$ cm^{-1} (Pekeris relation, cf. [4, p. 108]), and $r_e = 1.495(7)$ Å [5].

Selected values from [2] are given in Table 2.

Table 2
Equilibrium Constants log K_p, for $S_2 + N_2 \rightleftharpoons 2SN$, Heat Capacity C°_p, Entropy S°, Gibbs Free Energy Function $-(G^\circ - H^\circ_{298})/T$ and Enthalpy $H^\circ - H^\circ_{298}$ of SN in the Ideal Gas State [2].

T in K	log K_p	C°_p in cal·mol^{-1}·K^{-1}	S° in cal·mol^{-1}·K^{-1}	$-(G^\circ - H^\circ_{298})/T$ in cal·mol^{-1}·K^{-1}	$H^\circ - H^\circ_{298}$ in kcal/mol
0	∞	0.000	0.000	∞	−2.246
100	−132.861	7.720	44.614	59.878	−1.526
200	−63.933	7.697	50.011	53.751	−0.748
298	−41.251	7.599	53.055	53.055	0.000
400	−29.517	7.742	55.305	53.355	0.780
600	−18.245	8.146	58.522	54.573	2.370
800	−12.343	8.433	60.908	55.871	4.030
1000	−9.735	8.614	62.811	57.075	5.736
1500	−6.251	8.843	66.354	59.615	10.108
2000	−4.506	8.952	68.914	61.634	14.560
3000	−2.757	9.072	72.569	64.710	23.578
4000	−1.881	9.153	75.190	67.017	32.691
5000	−1.354	9.222	77.240	68.864	41.879
6000	−1.002	9.288	78.926	70.405	51.135

The spectroscopic constants for the lowest five electronic states of SN [6] were used to give polynomial expansions of the partition function Q and the equilibrium constant K_p (for $SN \rightleftharpoons S + N$) for the temperature range T = 1000 to 9000 K: $\log Q = a_0 + a_1 \cdot \log \Theta + a_2 \cdot (\log \Theta)^2$

with $\Theta = 5040/T$ and $a_0 = 4.8052$, $a_1 = -1.9619$, $a_2 = 0.3140$; $\log K_p = b_0 + b_1 \cdot \log \Theta + b_2 (\log \Theta)^2 + b_3 \cdot (\log \Theta)^3 - \Theta D_0^\circ(SN)$ with p in Pa, $b_0 = 10.9036$, $b_1 = -0.7786$, $b_2 = 0.0941$, $b_3 = -0.3808$, and D_0° (SN) = 4.8 eV [7]. The following polynomial expansion of the partition function Q for the temperature range T = 1000 to 5000 K was also derived from available molecular data [6]: $\ln Q = a_0 \ln Z + a_1/2 \cdot Z^1 + a_2/6 \cdot Z^2 + a_3/12 \cdot Z^3 + a_4/20 \cdot Z^4 - a_5/Z + a_6$ with $Z = T/1000$, $a_0 = 1.39727$, $a_1 = 6.53608$, $a_2 = -2.59313$, $a_3 = 4.77290$, $a_4 = -3.14540$, $a_5 = -1.35567$, $a_6 = 8.51400$ [8].

References:

[1] O'Hare, P. A. G. (J. Chem. Phys. **52** [1970] 2992/6).
[2] Stull, D. R.; Prophet, H. (NSRDS-NBS-37 [1971]).
[3] Gaydon, A. (Dissociation Energies and Spectra of Diatomic Molecules, 3rd Ed., Chapman & Hall, London 1968, p. 277).
[4] Herzberg, G. (Molecular Spectra and Molecular Structure, Spectra of Diatomic Molecules, Vol. 1, Van Nostrand, Princeton, N.J., 1961).
[5] Tables of Interatomic Distances and Configuration in Molecules and Ions, The Chemical Society, London 1958.
[6] Huber, K. P.; Herzberg, G. (Molecular Spectra and Molecular Structure, Constants of Diatomic Molecules, Vol. 4, Van Nostrand Reinhold, New York 1979, pp. 484/7).
[7] Sauval, A.J.; Tatum, J. B. (Astrophys. J. Suppl. Ser. **56** [1984] 193/209, 201).
[8] Rossi, S. C. F.; Maciel, W. J.; Benevides-Soares, P. (Astron. Astrophys. **148** [1985] 93/6).

1.1.5 Chemical Reactions

SN radicals produced in the gas phase, e.g. by reaction of active nitrogen with sulfur vapor [1] or H_2S [2] and by pyrolysis of S_7NH or microwave discharge of ClS≡N vapor [1] in a fast flow apparatus at low pressure, polymerize to give deposits of black $(SN)_x$ on the walls of the reaction vessel or the cooler connecting tubes. The SN radical generated by the reaction of active nitrogen and S_2 vapor forms S_4N_4 and at least two other sulfur nitrides in the gas phase and at the surface of the reaction vessel when the N atom flow rate is approximately twice that of S_2. In the presence of excess N atoms, the SN intermediate is rapidly destroyed, probably by the reaction $N + SN \rightarrow N_2 + S$ [3].

From the presence of cyclic S_2N_2 and S_4N_4 in the condensate and S_3N_3 in the vapor of pyrolyzed $(SN)_x$ and the similar mass spectra of S_4N_4 and $(SN)_x$, oligomerization of SN in the gas phase to give cyclic molecules is proposed: $SN \overset{+SN}{\rightleftharpoons} S_2N_2 \overset{+SN}{\rightleftharpoons} S_3N_3 \overset{+SN}{\rightleftharpoons} S_4N_4$ [4, 5].

For the repolymerization process of $(SN)_x$ vapor passed through silver wool at 155 and 160°C, a mechanism is proposed in which SN radicals and acyclic $(SN)_x$, $x \geqq 2$, are involved: $(SN)_x + SN \rightarrow (SN)_{x+1}$ [4, 5]. See also vaporization of $(SN)_x$, pp. 136/41.

The reactions $N + SN \rightarrow N_2 + S$ and $O + SN \rightarrow SO + N$ are also discussed to occur in rich, moist, CO–Ar–O_2 flames with fuel nitrogen as NO, C_2N_2, and NH_3 and fuel sulfur as SO_2 [6] and in CH_4 flames doped with H_2S and NH_3 and burned with O_2, N_2O, or air at atmospheric pressure. The SN radical is a direct interconnection between the NO_x and SO_x production mechanisms [7].

For the high temperatures of combustion reactions, estimates of the rate constants of SN reactions $X + SN \rightarrow (X \cdots S \cdots N)^{\ddagger} \rightarrow XS + N$ and $X + NS \rightarrow (X \cdots N \cdots S)^{\ddagger} \rightarrow XN + S$, with X = H, O, N, S, C were made. From transition state theory and estimates of the heat capacity of activation for atom transfer reactions between atoms and diatomic molecules it could be shown that the form of the rate constant is $k = AT^{0.5} \exp(-C/RT)$ where $A = 10^{16.6} \exp(\Delta S^{\circ}_{300}{}^{\ddagger}/R)$ in $cm^3 \cdot mol^{-1} \cdot s^{-1}$ ($\Delta S^{\circ}_{300}{}^{\ddagger}$ = the entropy of activation at 300 K); $C = -(\Delta H^{\circ}_{300}{}^{\ddagger} + 1)$ in kcal/mol ($\Delta H^{\circ}_{300}{}^{\ddagger}$ = the heat of activation of 300 K). The triatomic intermediate, the reaction, and the values of log A, C, and log k at 1500 K for these reactions are given below [8]:

intermediate	reaction	log (A/ $cm^3 \cdot mol^{-1} \cdot s^{-1}$)	C in kcal/mol	log ($k/cm^3 \cdot mol^{-1} \cdot s^{-1}$) at 1500 K
HNS	$H + SN \rightarrow HS + N$	12.4	31.2	9.5
HNS	$H + NS \rightarrow HN + S$	12.4	41.2	8.0
NOS	$O + SN \rightarrow OS + N$	11.8	8.0	12.2
NOS	$O + NS \rightarrow ON + S$	11.8	0	13.4
N_2S	$N + SN \rightarrow NS + N$	11.8	8.0	12.2
N_2S	$N + NS \rightarrow N_2 + S$	11.8	0	13.4
NS_2	$S + SN \rightarrow S_2 + N$	11.3	21.6	9.8
NS_2	$S + NS \rightarrow SN + S$	11.8	0	13.4
CNS	$C + SN \rightarrow CS + N$	11.8	8.0	12.2
CNS	$C + NS \rightarrow CN + S$	11.8	0	13.4

References:

[1] Dyke, J. M.; Morris, A.; Trickle, I. R. (J. Chem. Soc. Faraday Trans. II **73** [1977] 147/51).
[2] Westbury, R. A.; Winkler, C. A. (Can. J. Chem. **38** [1960] 334/42).
[3] Bett, J. A. S.; Winkler, C. A. (J. Phys. Chem. **68** [1964] 2501/8).
[4] Besenyei, E.; Eigendorf, G. K.; Frost, D. C. (J. Electron Spectrosc. Relat. Phenom. **41** [1986] 419/26).
[5] Besenyei, E.; Eigendorf, G. K.; Frost, D. C. (Inorg. Chem. **25** [1986] 4404/8).
[6] Wendt, J. O. L.; Wootan, E. C.; Corley, T. L. (Combust. Flame **49** [1983] 261/74).
[7] Jeffries, J. B.; Crosley, D. R. (Combust. Flame **64** [1986] 55/64).
[8] Benson, S. W.; Golden, D. M.; Lawrence, R. W.; Shaw, R.; Woolfolk, R. W. (EPA-600-2-75-019 [1975]).

1.2 Sulfur Nitride Ion (1+), Thiazyl Ion (1+), Thionitrosyl Ion (1+), SN^+

1.2.1 Sulfur Nitride Ion (1+), SN^+, in the Gas Phase

The ion in the $X\,^1\Sigma^+$ ground state was produced by photoionization of SN radicals using He I and Ne I radiation. The vertical ionization potential of the process $SN^+ (X\,^1\Sigma^+) \leftarrow SN (X\,^2\Pi)$ was measured as $E_i = 8.87 \pm 0.01$ eV [1]. SN^+ forms with high relative abundance by electron impact on compounds containing the SN group. SN^+ is produced from S_2N_2 and S_4N_4 at an electron energy of 25 eV with an appearance potential of a.p. $= 13.0 \pm 0.3$ and 14.1 eV, respectively [2, 3]. The appearance potential of SN^+ from FS≡N, a.p. $= 11.80 \pm 0.05$ eV, was attributed to $FS{\equiv}N + e^- \rightarrow SN^+ + F + 2e^-$ [4]. From an electron-impact (70 eV) investigation of $CH_3N{=}C{=}S$ the value a.p. $(SN^+) = 12.5 \pm 0.2$ eV was proposed for $CH_3N{=}C{=}S + e^- \rightarrow SN^+ + C_2H_3 + 2e^-$ [5]. The observation of the $A\,^1\Pi - X\,^1\Sigma^+$ system of SN^+ in emission in a He gas discharge tube study of SN was reported in 1955 [6, 7]. However, later these bands were rephotographed and assigned to the $G\,^2\Sigma^- \rightarrow X\,^2\Pi$ system of SN [8].

The formation enthalpy $\Delta H^{\circ}_{f,298} = 251$ kcal/mol (10.9 eV) was calculated from the SN^+ appearance potential a.p. $= 12.5 \pm 0.2$ eV for the assumed process $CH_3N{=}C{=}S + e^- \rightarrow SN^+ +$

$C_2H_3 + 2e^-$ [5]. The obsolete value of the formation enthalpy of SN^+, $\Delta H^\circ_{f,0} = 12.76 \pm 0.10$ eV [9], was derived from the SN^+ appearance potential from FS≡N a.p. $= 11.80 \pm 0.05$ eV [4] and the formation enthalpy $\Delta H^\circ_{f,298}$ (FS≡N) $= 1.78 \pm 0.09$ eV which is out-of-date.

The equilibrium bond length in the $X\,^1\Sigma^+$ ground state is estimated to be $r_e = 1.440 \pm 0.005$ Å from comparison of the intensity of the three observed vibrational components of the $SN^+(X\,^1\Sigma^+) \leftarrow SN(X\,^2\Pi)$ photoelectron band with calculated Franck-Condon factors. The vibrational constant $\omega_e = 1415 \pm 20$ cm^{-1} and the anharmonicity constant $\omega_e x_e = 15 \pm 10$ cm^{-1} for the $X\,^1\Sigma^+$ state were obtained from the position of the three vibrational components of that band [1]. Another experimental value r = 1.42(1) Å is obtained from a solid state X-ray diffraction study of $SN^+SbF_6^-$ and $SN^+Sb_2F_{11}^-$ [10]. The experimental r_e value is in accord with theoretical calculations, e.g. [11 to 15]. The very short equilibrium bond distance of $r_e = 1.25$ Å derived from an SN emission spectrum in which lines were tentatively assigned to the $A^1\Pi$–$X^1\Sigma^+$ system of SN^+ [6, 7] seems to be erroneous (see above).

The dissociation energy for the process SN^+ $(X\,^1\Sigma^+) \rightarrow N(^4S) + S^+(^4S)$ is estimated to be $D_e = 6.37 \pm 0.25$ eV from D_e (SN) $= 4.87 \pm 0.25$ eV, E_i (S) $= 10.36$ eV, and the first vertical ionization potential of SN, $E_i = 8.87 \pm 0.01$ eV [1].

The molecular and electronic structure of the ground state $X\,^1\Sigma^+$ was the subject of several quantum chemical calculations. Table 3 gives a survey of publications and calculated molecular properties. The following abbreviations and symbols are used:

MO	molecular orbital	ε_i	orbital energy
LCAO	linear combination of atomic orbitals	r_e	equilibrium bond length
SCF	self-consistent field	ω_e	vibrational constant
MRD	multi-reference double-excitation	B_e	rotational constant
CI	configuration interaction	f	force constant
CEPA	coupled electron pair approximation	μ	dipole moment
PNO	pair natural orbital	Q	quadrupole moment
M	method of moments	pop	population analysis
E_t	total energy		

Table 3
Ab initio Quantum Chemical Calculations for $SN^+(X\,^1\Sigma^+)$.

method	calculated properties	Ref.
MRD–CI	E_t, r_e, ω_e, B_e, μ	[11]
LCAO–MO SCF + CEPA	r_e, f, ω_e, pop	[12]
LCAO–MO SCF + PNO–CI + CEPA	E_t, r_e, f	[14]
LCAO–MO SCF	E_t, r_e	[13]
LCAO–MO SCF + M	E_t, ε_i, r_e, f	[15]
LCAO–MO SCF	E_t	[16]
LCAO–MO SCF	E_t, ε_i, μ, Q based on erroneous $r_e = 1.25$ Å [6, 7]	[9]

The SN^+ $(X\,^1\Sigma^+)$ ion has the electron configuration $1\sigma^2\,2\sigma^2\,3\sigma^2\,4\sigma^2\,1\pi^4\,5\sigma^2\,6\sigma^2\,2\pi^4\,7\sigma^2$ [11]. The population analysis based on an ab initio LCAO–MO SCF method yields the atomic net charges +0.99 at S and +0.01 at N, which clearly indicate an electronic structure $^+$S≡N although the shared electron number of 2.59 points to a weak triple bond [12].

The dipole moment of the ground state at r_e relative to the center of mass was calculated to be $\mu = 2.291$ D at the MO–SCF and $\mu = 2.071$ D at the CI level, with the polarity $S^{2+}N^-$. Spectroscopic constants for the ground state $^1\Sigma^+$ and other nine low-lying valence states of SN^+ calculated using ab initio MRD–CI methods are compiled in Table 4 [11].

Table 4

Calculated Spectroscopic Constants for SN^+ Using ab initio MRD–CI Methods [11].
Total energy, E_t; electronic excitation energy, T_e; equilibrium bond length, r_e; vibrational constant, ω_e; rotational constant, B_e.

molecular state	E_t in a.u.	T_e in eV	r_e in Å	ω_e in cm^{-1}	B_e in cm^{-1}
$^1\Sigma^+$	−451.9014	0.0	1.443	1405.1	0.8319
$^3\Sigma^+$	−451.7713	3.54	1.609	952.6	0.6686
$^3\Pi$	−451.7310	4.64	1.475	1247.7	0.7959
$^3\Delta$	−451.7235	4.84	1.641	1002.8	0.6426
$^5\Sigma^+$	−451.700	5.48	1.90		
$^3\Sigma^-$	−451.6948	5.62	1.644	1024.1	0.6401
$^1\Sigma^-$	−451.6910	5.72	1.620	972.8	0.6595
$^1\Pi$	−451.6837	5.92	1.473	1325.9	0.7973
$^1\Delta$	−451.6750	6.16	1.620	932.8	0.6597
$^5\Pi$	−451.652	6.79	1.773		

The spectroscopic constants for the electronic ground state of SN^+ [18, pp. 486/7] were used to give polynominal expansions of the partition function Q and the equilibrium constant K_p (for $SN^+ \rightleftharpoons S + N^+$) for the temperature range T = 1000 to 9000 K: $\log Q = a_0 + a_1 \cdot \log \Theta + a_2 \cdot (\log \Theta)^2$ with $\Theta = 5040/T$ and $a_0 = 4.1450$, $a_1 = -1.8975$, $a_2 = 0.3832$; $\log K_p = b_0 + b_1 \cdot \log \Theta + b_2 \cdot (\log \Theta)^2 + b_3 \cdot (\log \Theta)^3 + b_4 \cdot (\log \Theta)^4 - \Theta\, D_0^\circ\, (SN^+)$ with p in Pa, $b_0 = 11.2307$, $b_1 = -0.8277$, $b_2 = 0.5112$, $b_3 = -1.3707$, $b_4 = 0.7033$, and $D_0^\circ\, (SN^+) = 6.3$ eV [17].

References:

[1] Dyke, J. M.; Morris, A.; Trickle, I. R. (J. Chem. Soc. Faraday Trans. II **73** [1977] 147/51).
[2] Smith, R. D.; Wyatt, J. R.; DeCorpo, J. J.; Saalfeld, F. E.; Moran, M. J.; MacDiarmid, A. G. (Chem. Phys. Letters **41** [1976] 362/4).
[3] Smith, R. D.; Wyatt, J. R.; DeCorpo, J. J.; Saalfeld, F. E.; Moran, M. J.; MacDiarmid, A. G. (J. Am. Chem. Soc. **99** [1977] 1726/30).
[4] Glemser, O.; Müller, A.; Böhler, D.; Krebs, B. (Z. Anorg. Allgem. Chem. **357** [1968] 184/9).
[5] Hobrock, B. G.; Shenkel, R. C.; Kiser, R. W. (J. Phys. Chem. **67** [1963] 1684/6).
[6] Dressler, K. (Helv. Phys. Acta **28** [1955] 317/8).
[7] Dressler, K. (Helv. Phys. Acta **28** [1955] 563/90).
[8] Jenouvrier, A.; Pascat, B. (Can. J. Phys. **51** [1973] 2143/61, 2148).
[9] O'Hare, P. A. G. (J. Chem. Phys. **52** [1970] 2992/6).
[10] Clegg, W.; Glemser, O.; Harms, K.; Hartmann, G.; Mews, R.; Noltemeyer, M.; Sheldrick, G. M. (Acta Cryst. B **37** [1981] 548/52).

[11] Karna, S. P.; Grein, F. (Chem. Phys. **109** [1986] 35/8).
[12] Zirz, C.; Ahlrichs, R. (Inorg. Chem. **23** [1984] 26/31).

[13] Seeger, R.; Seeger, U.; Bartetzko, R.; Gleiter, R. (Inorg. Chem. **21** [1982] 3473/7).
[14] Karpfen, A.; Schuster, P.; Petkov, J.; Lischka, H. (J. Chem. Phys. **68** [1978] 3884/90).
[15] Török, F.; Pulay, P.; Szondy, F.; Nagy, P. (Acta Chim. [Budapest] **80** [1974] 139/45).
[16] Findlay, R. H.; Palmer, M. H.; Downs, A. J.; Egdell, R. G.; Evans, R. (Inorg. Chem. **19** [1980] 1307/14).
[17] Sauval, A. J.; Tatum, J. B. (Astrophys. J. Suppl. Ser. **56** [1984] 193/209, 201).
[18] Huber, K. P.; Herzberg, G. (Molecular Spectra and Molecular Structure, Vol. 4, Constants of Diatomic Molecules, Van Nostrand Reinhold, New York 1979).

1.2.2 Salts of the Sulfur Nitride Ion (1+), Thiazyl Ion (1+), Thionitrosyl Ion (1+), SN^+

$SN^+SO_3F^-$

The title compound was obtained when 2.14 mmol of SO_3 was sublimed onto 0.715 mmol of $S_3N_3F_3$ at 77 K, followed by 10 mL of SO_2. The solution was allowed to warm to room temperature. After 3 h large cream-colored needles appeared in a green solution. The crystals were washed with SO_2 until they became white.

In the Raman spectrum 30 lines in the range 1438 to 52 cm^{-1} were observed: 1438 s, ν(S≡N); 1425 cm^{-1} w, ν(^{34}S≡N). Six lines can be assigned to SO_3F^- [1].

$SN^+SO_3CF_3^-$

The salt was formed by reaction of $SN^+AsF_6^-$ with a slight excess of $NO^+SO_3CF_3^-$ in liquid SO_2 at room temperature for 24 h.

The colorless solid melts at 120°C and sublimes in vacuum. It is extremely sensitive to humidity. In the Raman spectrum the line at 1443 cm^{-1} is assigned to ν(S≡N) [2].

$SN^+BF_4^-$

The off white solid was produced by mixing gaseous FS≡N with BF_3 in a glass-metal vacuum system at −23°C. A molecular weight of 66.50 (calculated 66.44) was determined from vapor density measurements of the volatiles above a sample of the solid at room temperature.

IR (solid film) at low temperature: ν (in cm^{-1}) = 1443 m, ν(S≡N); 1060 vs, $\nu_3(BF_4^-)$; 763 m, $\nu_1(BF_4^-)$; 528 s, $\nu_4(^{11}BF_4^-)$; 520 s, $\nu_4(^{10}BF_4^-)$.

The volatiles above the solid dissociate at room temperature at low pressure to give FS≡N and BF_3 [11].

$SN^+BF_4^-$ reacts with equimolar amounts of the arenes hexamethylbenzene, 1,3,5-trimethylbenzene, naphthalene, and 9,10-dimethylanthracene in CH_2Cl_2 or SO_2 at room temperature for some minutes to give the corresponding 1:1 arene complexes $SN^+BF_4^- \cdot$arene [18].

$SN^+PF_6^-$

The yellow crystalline solid formed by mixing gaseous FS≡N with PF_5 in a glass-metal vacuum system at −23°C [11]. A solution of $SN^+PF_6^-$ in nitromethane was previously prepared from $S_3N_3Cl_3$ and $Ag^+PF_6^-$, mole ratio 1:3. It was used in situ without isolation for the preparation of SN complex compounds [15].

IR (solid film) at low temperature: ν (in cm^{-1}) = 1443 m, ν(S≡N); 820 vs, $\nu_2(PF_6^-)$; 555 s, $\nu_3(PF_6^-)$.

The volatiles, which are in equilibrium with the solid state, dissociate at room temperature at low pressure to give PF_5 [11].

$SN^+PF_6^-$ reacts analogously as $SN^+BF_4^-$ with arenes to give 1:1 arene complexes $SN^+PF_6^- \cdot$arene [18]; see above.

$SN^+AsF_6^-$

A convenient simple synthesis of solid, pure $SN^+AsF_6^-$ in approximately 75% yield according to the equation $S_3N_3Cl_3 + 3Ag^+AsF_6^- \rightarrow 3AgCl + 3SN^+AsF_6^-$ was reported by [12]. A mixture of 3.45 mmol of $S_3N_3Cl_3$ and 23.48 mmol of $Ag^+AsF_6^-$ in 81.53 mmol of SO_2, which was condensed onto this mixture at −196°C, reacted at room temperature. After 20 h stirring the soluble product was sublimed at 120°C/10^{-3} Torr over a period of 3 d [12].

The salt formed by simultaneous introduction of FS≡N (p = 1 Torr) and AsF_5 (p = 1 Torr) into a quartz flask at room temperature [3, 4] or by reaction of condensed FS≡N and AsF_5 at −23°C [11]. For preparative purposes the reaction was carried out in liquid SO_2. $SN^+AsF_6^-$ was isolated by sublimation at 40°C in an oil-pump vacuum [2].

$SN^+AsF_6^-$ was obtained when $S_3N_3F_3$ reacted with AsF_5 in the mole ratio 1:6 in liquid SO_2 at room temperature [1]. The compound sublimed on heating $S_3N_3F_2^+AsF_6^-$ at 85°C in vacuum [5]. A mixture of $SN^+AsF_6^-$ and $S_3N_3F_2^+AsF_6^-$ in the ratio ~3:1 formed when 103 mmol of AsF_5 was passed into a suspension of 7.5 mmol of $S_4N_4F_4$ in 50 mL of liquid SO_2 at −50°C and the mixture was stirred at −40°C for 48 h. $SN^+AsF_6^-$ can be separated by fractional sublimation at room temperature in the vacuum [5].

The formation of $SN^+AsF_6^-$ was postulated from the decomposition of $S_3N_3F_2^+$ AsF_6^- in saturated SO_2 solution at room temperature during several months. The reaction proceeds according to the equation 2 $S_3N_3F_2^+$ $AsF_6^- + SO_2 \rightarrow S_4N_5^+$ $AsF_6^- + SN^+AsF_6^- + 2$ OSF_2 [6].

^{14}N NMR spectrum (1.05 M in SO_2/CH_3NO_2 external) at 22°C: δ = +202 ppm [17]. A ^{19}F NMR signal for SF in AsF_3 solution was not observed [3, 4]. Raman spectra show the ionic structure of the compound: ν (in cm^{-1}) (rel. int.) = 1438 (100), ν(S≡N); 1428 (4), ν(^{34}S≡N); 680 (88), $\nu_1(AsF_6^-)$; 572 (18), $\nu_2(AsF_6^-)$; 360 (14), $\nu_5(AsF_6^-)$ [1]. Similar frequency values are given in [3, 4, 11]. The SN valence force constant f(SN) = 11.85 mdyn/Å and the bond order N(SN) = 2.6 were calculated from the value ν(S≡N) = 1437 cm^{-1} [3, 4].

The reaction of $SN^+AsF_6^-$ and S_8 in liquid SO_2 at room temperature leads to $[S{=}N{=}S]^+AsF_6^-$ according to the equation $8SN^+AsF_6^- + S_8 \rightarrow 8[S{=}N{=}S]^+AsF_6^-$ [12, 13].

The reaction with a slight excess of $NO^+SO_3CF_3^-$ in liquid SO_2 at room temperature for 24 h gave $SN^+SO_3CF_3^-$. When $SN^+AsF_6^-$ was reacted with S_4N_4 in the mole ratio 1:1 in $SOCl_2$ at room temperature, $S_5N_5^+AsF_6^-$ formed in 80% yield. The analogous reaction in liquid SO_2 at room temperature for several days gave a black solid of polymeric structure $(S_5N_5^+AsF_6^-)_x$ in 60% yield and other compounds not identified. The ratio of (SN) : AsF_6^- in the polymer can be varied by alteration of the reaction stoichiometry. $SN^+AsF_6^-$ reacted with a slight excess of SCl_2 in liquid SO_2 at room temperature for ~10 h to give $(ClS)_2N^+AsF_6^-$ in quantitative yield. The analogous reaction with CF_3SCl and $(CF_3)_2C{=}NSCl$ yielded $[CF_3SNSCl]^+AsF_6^-$ and $[(CF_3)_2C{=}NSNSCl]^+AsF_6^-$, respectively [2].

The cation-exchange reaction $SN^+AsF_6^- + Cs^+F^- \rightarrow FS{\equiv}N + Cs^+AsF_6^-$ at 65 to 110°C is a route for the preparation of FSN (see p. 208) [12].

Stirring a mixture of equimolar amounts of $Re(CO)_5Cl$ and $SN^+AsF_6^-$ in liquid SO_2 at −15°C for 5 h gave $[Re(CO)_5(N{\equiv}SCl)]^+AsF_6^-$ in almost quantitative yield. The analogous reaction with $Re(CO)_5Br$ in SO_2 at −40°C for 3 h gave $[Re(CO)_5(N{\equiv}SBr)]^+AsF_6^-$ in 47.6% yield [7]. $SN^+AsF_6^-$ reacted with equimolar amounts of $[Re(CO)_5(SO_2)]^+AsF_6^-$ in liquid SO_2 at room temperature for

3 d to give $[Re(CO)_5(NS)]^{2+}(AsF_6^-)_2$ in almost quantitative yield. Treatment with equimolar amounts of $(\eta^5\text{-}C_5H_5)Mn(CO)_3$ in SO_2 at −60°C yielded $[(\eta^5\text{-}C_5H_5)Mn(CO)_2(NS)]^+AsF_6^-$ almost quantitatively. Stirring a mixture of equimolar amounts of $SN^+AsF_6^-$ and $[(\eta^5\text{-}C_5H_5)Fe(CO)_2(SO_2)]^+AsF_6^-$ in SO_2 at room temperature for 12 h gave $[(\eta^5\text{-}C_5H_5)Fe(CO)_2(NS)]^{2+}(AsF_6^-)_2$ in almost quantitative yield [8].

$SN^+AsF_6^-$ reacts with equimolar amounts of the arenes hexamethylbenzene, 1,3,5-trimethylbenzene, naphthalene, and 9,10-dimethylanthracene in CH_2Cl_2 or SO_2 at room temperature for some minutes to give the corresponding 1:1 arene complexes $SN^+BF_4^-\cdot$arene. Treatment of highly oriented pyrolytic graphite with a solution of $SN^+AsF_6^-$ in SO_2 for 8 d at room temperature gave a product in which only AsF_6^- is intercalated into the graphite [18].

$SN^+F_5TeOAsF_5^-$

The salt was prepared by reaction of $F_5TeOS{\equiv}N$ with excess AsF_5 at −60°C for 24 h.

^{19}F NMR ($SO_2/CFCl_3$): δ (in ppm) = −39.2 (F(A)), −42.6 ($F(B_4)$); $^2J(F(A), F(B_4))$ = 186.7 Hz(?). IR (Kel-F trituration): 1436 cm^{-1}, ν(S≡N).

The colorless solid decomposes slowly at room temperature [9].

$SN^+SbF_6^-$

The salt was obtained by simultaneous introduction of FS≡N (p = 1 Torr) and SbF_5 (p = 1 Torr) into a quartz flask at room temperature [3, 4]. On a preparative scale, the reaction was carried out in liquid SO_2. The product was purified by heating to 100°C followed by recrystallization of the residue from a solution of SO_2–SO_2ClF [2].

The crystal structure of $SN^+SbF_6^-$ was determined by X-ray diffraction at 293 (2) K. The crystals are orthorhombic, space group Pbca–D_{2h}^{15} (No. 61), with a = 13.999(6), b = 8.363(4), c = 10.468(4) Å; V = 1225.5 Å^3; Z = 8; D_x = 3.055 g/cm^3. R = 0.032 for 2025 independent reflexions. The structure consists of discrete ions. The observed SN distance is 1.361 (6) Å at 293 K. A librational analysis including the structures of $SN^+Sb_2F_{11}^-$ at 121.5 and 293 K led to the corrected value r(SN) = 1.42 (1) Å. The coordination of Sb by F is octahedral, see **Fig. 2**. The Sb–F bond lengths have values of 1.85 to 1.88 Å. The S atom is surrounded by five F atoms at distances in the range 2.6 to 2.8 Å. The five short S···F interactions are associated with the five longest Sb–F bonds; the interaction with S lengthens the Sb–F bonds by about 0.02 Å. Atomic coordinates and isotropic thermal parameters are given in the paper [10].

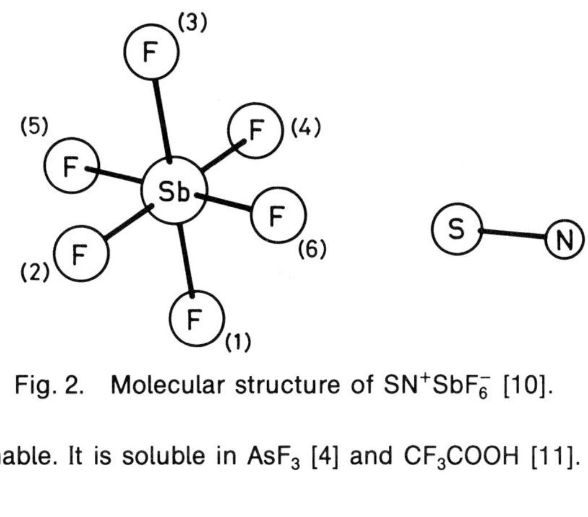

Fig. 2. Molecular structure of $SN^+SbF_6^-$ [10].

$SN^+SbF_6^-$ is sublimable. It is soluble in AsF_3 [4] and CF_3COOH [11].

^{19}F NMR ($AsF_3/CFCl_3$?): $\delta=-110$ ppm (SbF_6^-) [4]. Raman (liquid?): ν (in cm^{-1})=1448, ν(S≡N); 657, $\nu_1(SbF_6^-)$; 563, $\nu_2(SbF_6^-)$; 284, $\nu_5(SbF_6^-)$ [2]. Similar frequency values are given by [4].

The redox behavior of $SN^+SbF_6^-$ was investigated by electrochemical and ESR spectroscopic studies. The cyclic voltammogram of 5 mM $SN^+SbF_6^-$ in CF_3COOH–$(CF_3CO)_2O$ with 0.1 M $(n\text{-}C_4H_9)_4N^+$ BF_4^- as supporting electrolyte was conducted on a Pt electrode vs. the saturated calomel electrode. At the half-wave potential $E_{1/2}=+0.76$ V an irreversible reduction peak appears. From the irreversible reduction follows a short lifetime of the species formed at the electrode in the primary reduction process. The ESR spectrum of this solution shows the quintet of $S_3N_2^{+\bullet}$ and a 1:2:3:2:1 quintet ($a_N=0.63$ mT) of 1:1:1 triplets ($a_N=0.20$ mT) of an unknown radical [16]. $SN^+SbF_6^-$ reacted with S_4N_4 in the mole ratio 1:1 in $SOCl_2$ at room temperature to give $S_5N_5^+SbF_6^-$ in 80% yield. The reaction with a slight excess of SCl_2 in liquid SO_2 at room temperature for about 10 h gave $(ClS)_2N^+SbF_6^-$ in almost quantitative yield. The analogous reaction with CF_3SCl and $(CF_3)_2C{=}NSCl$ yielded $[CF_3SNSCl]^+SbF_6^-$ and $[(CF_3)_2C{=}NSNSCl]^+SbF_6^-$, respectively [2]. Treatment with SbF_5 in liquid SO_2ClF gave $SN^+Sb_2F_{11}^-$ [10]. $SN^+SbF_6^-$ reacted with equimolar amounts of $Cs^+OTeF_5^-$ in SO_2 at −35°C for 0.75 h to give $F_5TeOS{\equiv}N$ in 87% yield [9].

The reaction of $SN^+SbF_6^-$ with equimolar amounts of the arenes hexamethylbenzene, 1,3,5-trimethylbenzene, naphthalene, and 9,10-dimethylanthracene in CH_2Cl_2 or SO_2 at room temperature for some minutes gave the corresponding 1:1 arene complexes $SN^+SbF_6^- \cdot$arene. Treatment of highly oriented pyrolytic graphite with a solution of $SN^+SbF_6^-$ in SO_2 for 8 d at room temperature gave a product in which only SbF_6^- is intercalated into the graphite [18].

$SN^+Sb_2F_{11}^-$

The pale yellow compound was obtained when $S_3N_3F_3$ reacted with SbF_5 (mole ratio 1:6) in liquid SO_2 at room temperature [1]. Previously it was synthesized from $SN^+SbF_6^-$ and SbF_5 in liquid SO_2ClF. It is extremely moisture-sensitive [10].

The crystal structure was determined by X-ray diffraction at 293 (2) K and 121.5 (2.0) K. The crystals are monoclinic, space group $P2_1/c$–C_{2h}^5 (No. 14), with a=8.374(2), b=11.792(3), c=10.108(3) Å, β=91.89(4)°; V=997.6 Å^3; Z=4; D_x=3.328 g/cm^3 at 293(2) K and a=8.191(8), b=11.760(8), c=9.935(7) Å, β=92.88(2)°; V=955.8 Å^3; Z=4; D_x=3.474 g/cm^3 at 121.5(2.0) K. The structure was refined to R=0.036 for 2873 independent reflexions at 293 K and 0.037 for 1691 independent reflexions at 121.5 K. The structure consists of discrete ions. The observed SN distance is 1.354(6) Å at 293 K but 1.404(6) Å at 121.5 K. A librational analysis including the structure of $SN^+SbF_6^-$ at 293 K led to the corrected value r(SN)=1.42(1) Å. The molecular structure at 121.5 K is shown in **Fig. 3**. The coordination of Sb by F is octahedral, with angles of 151.5(2)° for 293 K and 151.2(2)° for 121.5 K at the bridging F atom. The bridging Sb–F bonds with r(Sb–F)=2.03 Å are both about 0.18 Å longer than the terminal bonds with r(Sb–F)=1.85 Å. The S atom is surrounded by five F atoms at distances in the range 2.6 to 2.8 Å. The five short S···F interactions are associated with the five longest Sb–F bonds with r(Sb–F)=1.86 to 1.87 Å. The interaction with S lengthens the Sb–F bonds by about 0.02 Å. Atomic coordinates and isotropic thermal parameters are given in the paper [10].

The Raman spectrum shows a strong band at 1436 cm^{-1}, attributed to the SN^+ ion, and nine other unassigned bands [1].

$SN^+SbCl_6^-$

The compound was prepared from a gas-phase reaction of ClSN and $SbCl_5$. No details are given [14].

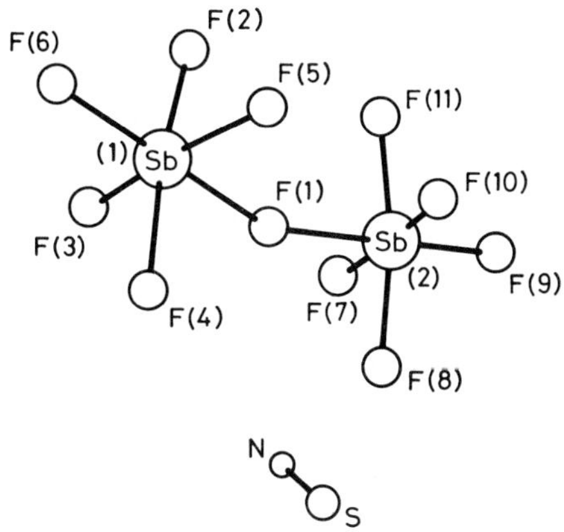

Fig. 3. Molecular structure of $SN^{+}Sb_2F_{11}^{-}$ [10].

References:

[1] Gillespie, R. J.; Sawyer, J. F.; Slim, D. R.; Tyrer, J. D. (Inorg. Chem. **21** [1982] 1296/302).
[2] Mews, R. (Angew. Chem. **88** [1976] 757/8; Angew. Chem. Intern. Ed. Engl. **15** [1976] 691/2).
[3] Glemser, O.; Koch, W. (Angew. Chem. **83** [1971] 145/6; Angew. Chem. Intern. Ed. Engl. **10** [1971] 127).
[4] Glemser, O.; Koch, W. (Anales Asoc. Quim. Arg. **59** [1971] 143/8).
[5] Mews, R.; Wagner, D. L.; Glemser, O. (Z. Anorg. Allgem. Chem. **412** [1975] 148/54).
[6] Isenberg, W.; Mews, R. (Z. Naturforsch. **37b** [1982] 1388/92).
[7] Hartmann, G.; Mews, R. (Z. Naturforsch. **40b** [1985] 343/6).
[8] Hartmann, G.; Mews, R. (Angew. Chem. **97** [1985] 218/9; Angew. Chem. Intern. Ed. Engl. **24** [1985] 202/3).
[9] Hoppenheit, R.; Mews, R. (Chem. Ber. **118** [1985] 4276/80).
[10] Clegg, W.; Glemser, O.; Harms, K.; Hartmann, G.; Mews, R.; Noltemeyer, M.; Sheldrick, G. M. (Acta Cryst. B **37** [1981] 548/52).

[11] Miller, K. V.; Emken, W. C.; Duncan, L. C. (J. Fluorine Chem. **26** [1984] 125/32).
[12] Apblett, A.; Banister, A. J.; Biron, D.; Kendrick, A. G.; Passmore, J.; Schriver, M.; Stojanac, M. (Inorg. Chem. **25** [1986] 4451/2).
[13] Mews, R. (private communication from [12]).
[14] Hey, R. G. (Diss. Durham, U. K., 1980 from [12]).
[15] Herberhold, M.; Haumaier, L. (Z. Naturforsch. **35b** [1980] 1277/80).
[16] Fritz, H. P.; Bruchhaus, R.; Mews, R.; Höfs, H. U. (Z. Anorg. Allgem. Chem. **525** [1985] 214/20).
[17] Passmore, J.; Schriver, M. J. (Inorg. Chem. **27** [1988] 2749/51).
[18] Brownstein, S.; Louie, B. (Can. J. Chem. **65** [1987] 1361/3).

1.2.3 Arene Complexes of Salts of the Sulfur Nitride Ion (1+), $SN^{+}BF_4^{-}$ · arene, $SN^{+}MF_6^{-}$ · arene (M = P, As, Sb); arene = 1,3,5-$(CH_3)_3C_6H_3$, $(CH_3)_6C_6$, $C_{10}H_8$, 9,10-$(CH_3)_2C_{14}H_8$ 9,10-$(CH_3)_2C_{14}H_8$

The arene complexes of the sulfur nitride salts compiled in Table 5, p. 48, were prepared by the methods described for $SN^{+}AsF_6^{-}\cdot(CH_3)_6C_6$:

Table 5
Spectral Properties of Arene Complexes of Salts of the Sulfur Nitride Ion (1+).

complex	^{1}H NMR(CH_2Cl_2/TMS) at −40°C, δ in ppm CH$_3$	CH	^{13}C NMR(CH_2Cl_2/TMS) at −40°C, δ in ppm CH$_3$	C(arom)	^{19}F NMR(CH_2Cl_2/TMS) at −40°C, δ in ppm	IR(KBr) ν in cm^{-1} ν(SN)	ν(MF)	UV-visible (CH_2Cl_2) λ_{CT} in nm
$SN^+BF_4^- \cdot 1,3,5\text{-}(CH_3)_3C_6H_3$	—	—	—	—	−140.8	1386	1085	324
$SN^+BF_4^- \cdot (CH_3)_6C_6$	2.40	—	16.8	147.0	−139.3	1400	1087	310
$SN^+BF_4^- \cdot C_{10}H_8$	—	—	—	—	−147.8	1380	1089	304
$SN^+BF_4^- \cdot 9,10\text{-}(CH_3)_2C_{14}H_8$	—	—	—	—	−148	1400	1089	—
$SN^+PF_6^- \cdot 1,3,5\text{-}(CH_3)_3C_6H_3$	—	—	—	—	−70.6 (J = 720 Hz)	1400	840 560	315
$SN^+PF_6^- \cdot (CH_3)_6C_6$	2.39	—	—	—	−69.8 (J = 710 Hz)	1408	838 563	361
$SN^+PF_6^- \cdot C_{10}H_8$	—	—	—	—	−66.8	1400	834 561	330
$SN^+PF_6^- \cdot 9,10\text{-}(CH_3)_2C_{14}H_8$	—	—	—	—	−70.7 (J = 700 Hz)	1389	850 564	—
$SN^+AsF_6^- \cdot 1,3,5\text{-}(CH_3)_3C_6H_3$	2.56	7.30	—	—	−64.0	1384	700	400
$SN^+AsF_6^- \cdot (CH_3)_6C_6$	2.40	—	17.1	147.2	−57.3 (J = 920 Hz)	1400	702	395
$SN^+AsF_6^- \cdot C_{10}H_8$	—	—	—	—	−63.1	1404	703	325
$SN^+AsF_6^- \cdot 9,10\text{-}(CH_3)_2C_{14}H_8$	—	—	—	—	−63.4	1383	705	—
$SN^+SbF_6^- \cdot 1,3,5\text{-}(CH_3)_3C_6H_3$	2.43	7.2	—	—	−120.8	1400	665	310
$SN^+SbF_6^- \cdot (CH_3)_6C_6$	2.40	—	17.1	147.3	−113.5	1406	668	406
$SN^+SbF_6^- \cdot C_{10}H_8$	—	—	—	—	−125	1401	667	358
$SN^+SbF_6^- \cdot 9,10\text{-}(CH_3)_2C_{14}H_8$	—	—	—	—	—	1415	667	—

Treatment of $SN^+AsF_6^-$ with equivalent amounts of hexamethylbenzene in CH_2Cl_2 or SO_2 for several minutes resulted in a dark red-brown solid product after removing all volatile material in vacuum. The preparation was also performed by condensing one equivalent each of AsF_5 and FSN onto a solution of hexamethylbenzene in CH_2Cl_2 or SO_2. The reaction mixture was worked up as described above.

All of these compounds are thermally unstable in solution at room temperature. They decompose fairly rapidly in air but can be stored in glass containers at room temperature in an inert atmosphere.

The intensely colored charge-transfer complexes were identified by the spectral properties listed in Table 5, p. 48.

The ^1H and ^{13}C NMR spectra of the complexes reflect the deshielding of the aromatic hydrocarbon by exhibiting resonances displayed downfield from those of the uncomplexed arene. In the ^{19}F NMR spectra, the observed chemical shifts are those expected for BF_4^- or MF_6^- (M = P, As, Sb) anions. From the IR spectra, it follows that the effect of an aromatic hydrocarbon on SN^+ salts is to consistently lower the S–N stretching frequency by about 40 cm^{-1}. The UV-visible spectra of the complexes, where arene is 1,3,5-$(CH_3)_3C_6H_3$, $(CH_3)_6C_6$, and $C_{10}H_8$, display strong absorptions owing to the aromatic hydrocarbon in the region of 250 to 300 nm. The charge-transfer band is present as a weaker absorption at slightly longer wavelength. The charge-transfer peaks for the 9,10-$(CH_3)_2C_{14}H_8$ derivatives were obscured by broadened bands due to the 9,10-dimethylanthracene and thus the position of the maxima for the charge-transfer peaks could not be measured.

The conductivity of $SN^+AsF_6^- \cdot (CH_3)_6C_6$ at room temperature was measured as 3.3×10^{-11} $\Omega^{-1} \cdot cm^{-1}$.

Reference:

Brownstein, S.; Louie, B. (Can. J. Chem. **65** [1987] 1361/3).

1.3 Sulfur Nitride Ion (1−), Thiazyl Ion (1−), SN^-

SN^- ions are produced in a gas discharge source by use of a thoriated iridium filament with N_2O and a trace of CS_2 at a total pressure of about 0.1 Torr. The negative-ion beam is accelerated to 680 eV and mass analyzed by use of crossed electric and magnetic fields. Intersection of the SN^- ion beam by the 488 nm intracavity radiation of an Ar II laser allows the observation of the photoelectron spectrum. Photodetachment occurs from several vibrational levels, $SN^-(^3\Sigma, v''=0) \rightarrow SN(X^2\Pi, v'=0, 1, 2, 3)$ and $SN^-(^3\Sigma, v''=1, 2, 3) \rightarrow SN(X^2\Pi, v'=0)$. From a Franck-Condon analysis of the relative intensities of the observed transitions the equilibrium bond length $r_e = 1.58_9 \pm 0.02$ Å is obtained. Analysis of the $SN^-(^3\Sigma, v''=1, 2, 3) \rightarrow SN(X^2\Pi, v'=0)$ system gives the vibrational constant $\omega_e = 880 \pm 70$ cm^{-1}. From the difference of the laser photon energy (2.54 eV) and the observed $SN^-(^3\Sigma, v''=0) \rightarrow SN(X^2\Pi, v'=0)$ transition energy the electron affinity of SN, $A = 1.194 \pm 0.011$ eV, is obtained [1].

Following Mulliken's scheme, the electronic structure of SN^- is denoted by KKL $(z\sigma)^2 (y\sigma)^2 (x\sigma)^2 (w\pi)^4 (v\pi)^2$, $^3\Sigma$. The following molecular properties were calculated by an LCAO-MO SCF method based on an SN equilibrium bond length $r_e = 2.8265$ a.u. (1.496 Å): total energy $E_t = -451.95446$ a.u., dipole moment $(S^+–N^-)$ $\mu = 0.09$ D (from center of molecule), quadrupole moment $Q = -3.89 \times 10^{-26}$ esu·cm^2 (from center of mass); dissociation energy for $SN^-(^3\Sigma) \rightarrow S^-(^2P) + N(^4S)$, D = 0.1487 a.u. or $4.0_5 \pm 0.2$ eV (corrected by a correlation energy, which was taken from the better known NO^- case) [2].

References:

[1] Burnett, S. M.; Feigerle, C. S.; Stevens, A. E.; Lineberger, W. C. (J. Phys. Chem. **86** [1982] 4486/8).
[2] O'Hare, P. A. G. (J. Chem. Phys. **54** [1971] 4124/6).

1.4 Metal Complexes of Sulfur Nitride, Thionitrosyl Complexes

The range of transition-metal-thionitrosyl complexes was surveyed up to 1981 by Roesky, H. W.; Pandey, K. K. (Advan. Inorg. Chem. Radiochem. **26** [1983] 337/56) and Herberhold, M. (Nachr. Chem. Tech. Lab. **29** [1981] 365/9). Reviews which supplement the literature until 1985 are given by Kelly, P. F.; Woollins, J. D. (Polyhedron **5** [1986] 607/32) and Chivers, T.; Edelmann, F. (Polyhedron **5** [1986] 1661/99).

A short survey obtained with the aid of representative examples precedes the detailed description.

1.4.1 Survey

1.4.1.1 Preparation

By Reactions of Nitrido Complexes with S_8, CH_3CHCH_2S, and S_2Cl_2

The reaction of $Mo(N)(S_2CN(CH_3)_2)_3$ with S_8 or CH_3CHCH_2S gave $Mo(NS)(S_2CN(CH_3)_2)_3$; see p. 56. $M(N)N_3((C_6H_5)_2PCH_2CH_2P(C_6H_5)_2)$ with M = Mo or W reacted analogously with CH_3CHCH_2S; see pp. 58, 59.

Treatment of $M(N)Cl_2(P(CH_3)_2C_6H_5)_3$ with M = Tc, Re and $Os(N)Cl_3(P(CH_3)_2C_6H_5)_2$ with S_2Cl_2 yielded $M(NS)Cl_2(P(CH_3)_2C_6H_5)_3$ and $Os(NS)Cl_3(P(CH_3)_2C_6H_5)_2$, respectively; see pp. 60, 63, 82.

By Reactions of SN^+ Salts with Transition-Metal Complexes

The few examples of this preparation route are given by the following equations:

$(\eta^6\text{-}C_6H_6)Cr(CO)_3 + SN^+PF_6^- + CH_3CN \rightarrow [Cr(NS)(N{\equiv}CCH_3)_5]^{2+}\ (PF_6^-)_2$; see p. 55.

$(\eta^5\text{-}C_5H_5)Mn(CO)_3 + SN^+AsF_6^- \xrightarrow{SO_2} [(\eta^5\text{-}C_5H_5)Mn(NS)(CO)_2]^+AsF_6^- + CO$; see p. 60.

$[Re(CO)_5(SO_2)]^+AsF_6^- + SN^+AsF_6^- \xrightarrow{SO_2} [Re(NS)(CO)_5]^{2+}(AsF_6^-)_2 + SO_2$; see p. 70.

$Re(CO)_5Br + SN^+SbF_6^- \xrightarrow{SO_2} [Re(NS)(CO)_5]^{2+}(SbF_6^-)_2$; see p. 70.

$[(\eta^5\text{-}C_5H_5)Fe(CO)_2(SO_2)]^+AsF_6^- + SN^+AsF_6^- \xrightarrow{SO_2} [(\eta^5\text{-}C_5H_5)Fe(NS)(CO)_2]^{2+}(AsF_6^-)_2 + SO_2$; see p. 70.

By Reactions of $S_3N_3Cl_3$ with Transition-Metal Halides

This method was applied to the synthesis of Ru, Os, and Ir complexes:

$MCl_2(P(C_6H_5)_3)_3 + 1/3\ S_3N_3Cl_3 \xrightarrow{THF} M(NS)Cl_3(P(C_6H_5)_3)_2 + P(C_6H_5)_3$, M = Ru, Os; see pp. 73, 79.

$IrHCl_2(P(C_6H_5)_3)_3 + 1/3\ S_3N_3Cl_3 \xrightarrow{THF} Ir(NS)Cl_2(P(C_6H_5)_3)_2$; see p. 86.

The existence of Rh complexes, of which the analogous preparation method is described, is questionable, see pp. 76/9.

By Reactions of $S_3N_3Cl_3$ with Anionic Metal-Carbonyl Complexes

Organometallic-thionitrosyl complexes were prepared by addition of $S_3N_3Cl_3$ in THF to the anion: $Na^+[(\eta^5\text{-}C_5H_5)Cr(CO)_3]^- + \frac{1}{3}\,S_3N_3Cl_3 \xrightarrow{THF} (\eta^5\text{-}C_5H_5)Cr(NS)(CO)_2 + CO + NaCl$; see p. 52. The analogous Mo or W complexes could not be prepared in this way, however, due to preferential oxidation of the $[(\eta^5\text{-}C_5H_5)M(CO)_3]^-$ anion to the dimer $((\eta^5\text{-}C_5H_5)M(CO)_3)_2$ by $S_3N_3Cl_3$. Replacement of the cyclopentadienyl by the hydridotris(3,5-dimethylpyrazolyl)-borate ligand overcame this problem: $K^+[HB(3,5\text{-}(CH_3)_2C_3HN_2)_3M(CO)_3]^- + S_3N_3Cl_3 \xrightarrow{THF} HB(3,5\text{-}(CH_3)_2C_3HN_2)_3M(CO)_2(NS) + (HB(3,5\text{-}(CH_3)_2C_3HN_2)_3M(CO)_2)_2\ (\mu\text{-}S)$ with M = Mo, W; see pp. 116/7.

By Reactions of S_4N_4 with Transition-Metal Nitrides or Halides

The reaction of $Re(N)Cl_4$ and Re_2Cl_{10} with S_4N_4 and subsequent treatment of the crude products with $(C_6H_5)_4As^+Cl^-$ gave $((C_6H_5)_4As^+)_2[Re(NS)(NSCl)Cl_4]^{2-} \cdot CH_2Cl_2$ and $(C_6H_5)_4As^+$-$[Re(NS)Cl_5]^-$, respectively; see pp. 67, 65.

By Halide Abstraction from NSF- or NSCl-Ligands in Metal Complexes

The method is described by following equations:

$[Re(CO)_5(N{\equiv}SF)]^+AsF_6^- + AsF_5 \rightarrow [Re(NS)(CO)_5]^{2+}(AsF_6^-)_2$; see p. 70.

$OsCl_4(N{\equiv}SCl)_2 + (C_6H_5)_4As^+Cl^- \rightarrow (C_6H_5)_4As^+\ [Os((NS)_2Cl)Cl_4]^-$; see p. 84.

$ReCl_3(NSCl)_2(POCl_3) + 2\,C_5H_5N \rightarrow ReCl_2(NS)(NSCl)(NC_5H_5)_2 + 2\,Cl^- + POCl_3$; see p. 66.

1.4.1.2 Structure and Bonding

X-ray structural data for thionitrosyl complexes of Cr (see p. 52), Mo (see p. 57), Tc (see p. 60), Re (see pp. 66/9), Ru (see pp. 71/3), and Os (see pp. 80, 81, 84, 85) have been determined. The M–N≡S unit is essentially linear in all compounds and, unlike metal nitrosyls, there are no confirmed examples of complexes containing either bent – or bridging – thionitrosyl ligands. The NS bond lengths have values from 1.46 to 1.59 Å. The IR bands attributed to the ν(NS) stretching mode of metal-thionitrosyl complexes fall within the broad range 1065 to 1388 cm^{-1}. On the basis of these experimental results, the binding can be described by the following resonance structures $M^{\ominus}{=}N^{\oplus}{=}S\rangle \leftrightarrow \overset{2\ominus}{M}{-}\overset{\oplus}{N}{\equiv}\overset{\oplus}{S}|$.

1.4.1.3 Reactions

Nucleophilic halide addition was observed for Re and Os complexes:

$[Re(NS)(CO)_5]^{2+}(AsF_6^-)_2 + CsX \xrightarrow{SO_2} [Re(CO)_5(N{\equiv}SX)]^+\ AsF_6^- + Cs^+AsF_6^-$, where X = Cl, Br; see p. 70.

$Os(NS)_2Cl_4 + (C_6H_5)_4As^+Cl^- \rightarrow (C_6H_5)_4As^+[OsCl_4((NS)_2Cl)]^-$; see p. 79.

Alkylation occurs in the reaction

$Mo(NS)(S_2CN(CH_3)_2)_3 + (C_6H_5)_3C^+BF_4^- \rightarrow [Mo(NSC(C_6H_5)_3)(S_2CN(CH_3)_2)_3]^+\ BF_4^-$; see p. 57.

Substitution of NS by NO is known for several complexes. The reaction of $Os(NS)Cl_3(P(C_6H_5)_3)_2$ with NOCl or NOBr resulted in the formation of $Os(NO)XCl_2(P(C_6H_5)_3)_2$, where X = Cl, Br; see p. 81. In the reaction of $Ru(NS)X_3L_2$ (X = Cl, Br; L = $P(C_6H_5)_3$, $As(C_6H_5)_3$,

$Sb(C_6H_5)_3$, C_5H_5N) with NOCl, NOBr, and N_2O_3 in the presence of excess L, displacement of the NS group occurred and the products were nitrosyl and nitrosyl-nitrato (from N_2O_3) complexes of the general formulas $Ru(NO)X_3L_2$ and $Ru(NO)(NO_3)X_2L_2$; see pp. 74/5. Desulfurization of $Mo(NS)(S_2CN(CH_3)_2)_3$ was accomplished by reaction with $(n\text{-}C_4H_9)_3P$ to give Mo(N)-$(S_2CN(CH_3)_2)_3$, see p. 57.

1.4.2 Thionitrosyl Complexes of Cr

$(\eta^5\text{-}C_5H_5)Cr(NS)(CO)_2$

The complex was produced in 21% yield by dropping a solution of 3.35 mmol of $S_3N_3Cl_3$ in THF to a stirred THF solution of 10.0 mmol $Na^+[(\eta^5\text{-}C_5H_5)Cr(CO)_3]^-$ cooled to −78°C. The mixture was allowed to warm slowly to room temperature over a period of several hours. The complex was isolated by column chromatography on Forisil with $CH_3C_6H_5$ as eluent, followed by sublimation at 30°C/5 × 10^{-3} Torr [1, 2].

The red or red-violet crystals melt under N_2 at 68 to 69°C [1, 2].

The crystal structure was determined by X-ray diffraction. The complex crystallizes in space group $P2_1/m\text{–}C^2_{2h}$ (No. 11) of the monoclinic system with a = 6.528(1), b = 9.641(1), c = 7.096(1) Å, β = 99.22(1)°; V = 440.8 Å³. R = 0.025, R_W = 0.033 for 848 observed reflections. Z = 2. D_x = 1.65 g/cm³. The molecular structure is shown in **Fig. 4**. The crystal packing is illustrated in **Fig. 5**. Bond lengths and bond angles are compiled in Table 6. Atomic positional fractional parameters and anisotropic and isotropic thermal parameters are given in [1].

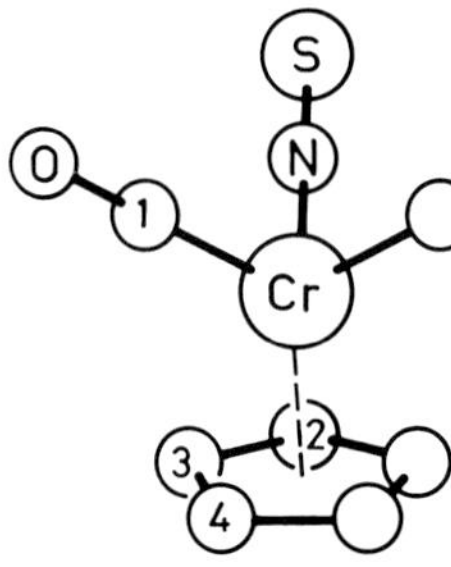

Fig. 4. Molecular structure of $(\eta^5\text{-}C_5H_5)Cr(NS)(CO)_2$ as viewed down the crystallographic mirror plane. Only the 2/3 cyclopentadienyl ring is shown and H atoms are omitted [1, 3].

The crystal structure consists of a well-separated array of discrete molecular units, the intermolecular distances corresponding to normal van der Waals contacts. Each molecule has crystallographic m symmetry and is in a "piano stool" configuration. The NS ligand coordinates essentially linearly to the metal via the N atom. The Cr–N bond length of 1.694(2) Å falls in the range of values reported for Cr–N bonds formed by linear terminal NO ligands. Hence, in a formal sense, the NS group (an overall three-electron donor in this bonding configuration) can be considered to be bonded as NS^+, a mode of coordination that involves synergistic coupling of σ- and π-bonding components. Consistent with the occurrence of back-donation from Cr π orbitals to NS π* orbitals is the fact that the N–S distance of 1.551(2) Å in the complex is longer than the $S{\equiv}N^+$ equilibrium bond length of 1.440(5) Å; see p. 41 [1, 3].

Table 6
Bond Lengths and Bond Angles of (η^5-C_5H_5)Cr(NS)$(CO)_2$ [1].

bond length	in Å	bond length	in Å	bond angle	in °
Cr–N	1.694(2)	N–S	1.551(2)	N–Cr–C(1)	94.78(7)
Cr–C(1)	1.883(2)	C(1)–O	1.131(2)	C(1)–Cr–C(1)[b)]	92.4(1)
Cr–C(2)	2.13(1)	Cr–C(2′)	2.23(2)	N–Cr–CP[a)]	130.1
Cr–C(3)	2.19(1)	Cr–C(3′)	2.18(2)	C(1)–Cr–CP[a)]	118.2
Cr–C(4)	2.22(1)	Cr–C(4′)	2.23(2)	Cr–N–S	176.8(1)
Cr–CP[a)]	1.85	C(2′)–C(3′)	1.42(2)	Cr–C(1)–O	178.1(2)
C(2)–C(3)	1.39(2)	C(3′)–C(4′)	1.32(3)	C(2)–C(3)–C(4)	104(1)
C(3)–C(4)	1.44(2)	C(4′)–C(4′)[b)]	1.36(5)	C(3)–C(2)–C(3)[b)]	114(2)
C(4)–C(4)[b)]	1.39(1)			C(3)–C(4)–C(4)[b)]	109(1)
				C(2′)–C(3′)–C(4′)	110(2)
				C(3′)–C(2′)–C(3′)[b)]	102(2)
				C(3′)–C(4′)–C(4′)[b)]	109(1)

[a)] CP represents the unweighted centroid of the disordered η^5-C_5H_5 ring.– [b)] Symmetry-related atom at x, 1/2 – y, z.

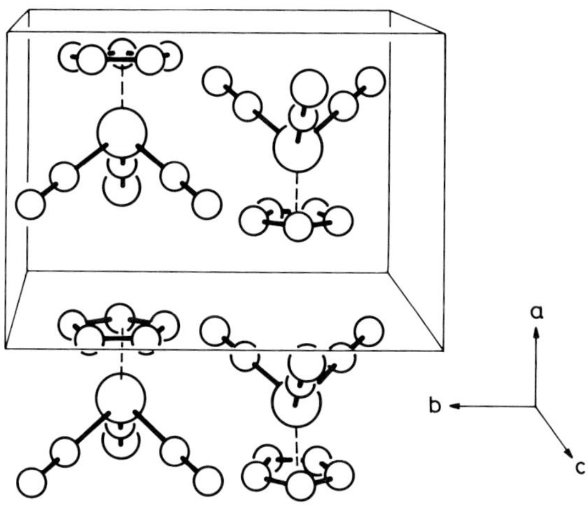

Fig. 5. Perspectivic view of the crystal packing of (η^5-C_5H_5)Cr(NS)$(CO)_2$ [1].

^{1}H NMR ($CDCl_3$/TMS): δ (in ppm) = 5.08 (s,C_5H_5) [1, 2]. (CH_3NO_2/TMS): δ (in ppm) = 5.21 (s,C_5H_5) [4].

^{13}C {^{1}H} NMR ($CDCl_3$/TMS): δ (in ppm) = 92.75 (C_5H_5), 239.43 (CO) [1, 2]; (CH_2Cl_2–$CDCl_3$/TMS): δ (in ppm) = 237 (CO) [5]. (CH_3NO_2/TMS): δ (in ppm) = 94.4 (C_5H_5), 241.2 (CO) [4].

^{14}N NMR (CH_2Cl_2/external neat CH_3NO_2): δ (in ppm) = 104 (NS), line width 90 Hz [5].

IR (hexane): ν (in cm^{-1}) = 2033, ν(CO); 1962, ν(CO); 1180, ν(NS) [1, 2, 5]. (CH_3NO_2): ν (in cm^{-1}) = 2033, ν(CO); 1963, ν(CO); 1154, ν(NS) [4].

UV-vis (CH_2Cl_2): λ_{max} = 470 nm, ε = 310 L·mol^{-1}·cm^{-1} [5].

References on p. 86

Valence level ionization energies determined by gas-phase He(I) and He(II) photoelectron spectroscopy and eigenvalues calculated by the Fenske-Hall MO method are given in Table 7 [6, 7].

Table 7
Valence Level Ionization Energies E_i and MO Energies ε_i of $(\eta^5\text{-}C_5H_5)Cr(NS)(CO)_2$ [7].

E_i in eV	$-\varepsilon_i$ in eV	MO	E_i in eV	$-\varepsilon_i$ in eV	MO
7.47	8.82	Cr $d_{x^2\text{-}y^2}$(2a′)	9.93	12.45	$\eta^5\text{-}C_5H_5$ e″(a″)
7.76	9.18	Cr d_{yx}(a″)	—	16.92	NS 7σ
7.96	9.29	Cr d_{xz}(1a′)	11.43	16.48	NS 2π (a′)
9.51	12.30	$\eta^5\text{-}C_5H_5$e″(a′)	11.42	16.69	NS 2π (a″)

From the electron impact mass spectrum (70 eV) a higher value of the first ionization energy $E_i = 7.83 \pm 0.1$ eV is obtained [1]. The orbital populations CO(5σ) 1.51, CO(2π) 0.52, NS (7σ) 1.61, NS (3π) 1.69 and the atomic charges Cr +0.88, C(CO) −0.09, O(CO) +0.05, $C(\eta^5\text{-}C_5H_5)$ −0.12, N −0.25, S +0.04 were obtained from Fenske-Hall MO calculations [6, 7].

The core level ionization energies E_i (in eV) = 581.66(4), Cr $2p_{3/2}$; 290.72(2), C 1s $(\eta^5\text{-}C_5H_5)$; 292.51(4), C 1s (CO); 539.10(1), O 1s; 404.89(5), N 1s; 168.51(3), S $2p_{3/2}$ were determined by gas-phase X-ray photoelectron spectroscopy [7].

From the difference between the appearance potential of the product ion and that of the reactant ion for the dissociation reactions

$(\eta^5\text{-}C_5H_5)Cr(NS)(CO)_2^+ \rightarrow (\eta^5\text{-}C_5H_5)Cr(NS)(CO)^+ + CO$ and
$(\eta^5\text{-}C_5H_5)Cr(NS)(CO)^+ \rightarrow (\eta^5\text{-}C_5H_5)Cr(NS)^+ + CO$ (see Table 8),

the average ionic bond dissociation energy $D(Cr^+\text{–}CO) = 0.6$ eV is obtained. The ionic bond dissociation energy $D(Cr^+\text{–}NS)$ for the process $(\eta^5\text{-}C_5H_5)Cr(NS)^+ \rightarrow (\eta^5\text{-}C_5H_5)Cr^+ + NS$ is 4.4 eV (see Table 8) [1].

The dark red-brown solid or red-violet crystals melt reversibly under N_2 at 68 to 69°C. They are quite volatile, subliming at about 30°C/5 × 10^{-3} Torr without attendant decomposition. The diamagnetic solid is freely soluble in common organic solvents to give blood-red solutions that eventually deposit some decomposed matter when exposed to air for several hours. The pure solid is reasonably stable in air but is best stored in an inert atmosphere.

Electron impact at 70 eV gives fragments which are given in the mass spectrum, see Table 8 [1, 2].

Table 8
EI Mass Spectrum (70 eV) of the Major Metal-Containing Ions from $(\eta^5\text{-}C_5H_5)Cr(NS)(CO)_2$ [1, 2].

appearance potential in eV	m/e	relative abundance in %	ion
7.83 ± 0.1	219	36	$(C_5H_5)Cr(NS)(CO)_2^+$
8.15 ± 0.1	191	8	$(C_5H_5)Cr(NS)(CO)^+$
9.07 ± 0.1	163	74	$(C_5H_5)Cr(NS)^+$
13.45 ± 0.1	117	100	$(C_5H_5)Cr^+$
—	52	63	Cr^+

References on p. 86

The reaction with excess NOCl in CH_2Cl_2 at ambient temperature for 3 h gave $(\eta^5\text{-}C_5H_5)Cr(NO)_2Cl$ in 27% yield. IR monitoring of the reaction indicated that it probably proceeded via $((\eta^5\text{-}C_5H_5)Cr(NO)Cl)_2$. Treatment of a stirred CH_2Cl_2–CH_3CN solution of $(\eta^5\text{-}C_5H_5)Cr(NS)(CO)_2$ with equimolar amounts of $NO^+PF_6^-$ at −78°C for 2 h gave $[(\eta^5\text{-}C_5H_5)Cr(NS)(CO)(NO)]^+PF_6^-$.

No reaction occurred when a mixture of $(\eta^5\text{-}C_5H_5)Cr(NS)(CO)_2$ and $P(C_6H_5)_3$ of the mole ratio 1:2 in $CH_3C_6H_5$ was held at reflux for 70 h. The complex was completely destroyed when a stirred benzene solution containing $(\eta^5\text{-}C_5H_5)Cr(NS)(CO)_2$ and $P(C_6H_5)_3$ in the mole ratio 1:2 was exposed to light of a medium-pressure Hg lamp for 6 h [1].

$(\eta^5\text{-}C_5(CH_3)_5)Cr(NS)(CO)_2$

The complex formed as a by-product in 10% yield by the reaction of $K^+[(\eta^5\text{-}C_5(CH_3)_5)Cr(CO)_3]^-$ with $S_3N_3Cl_3$ in the mole ratio 5:1 in THF at −78°C for 1 h. It was isolated by column chromatography on silica gel in $n\text{-}C_6H_{14}$ with $n\text{-}C_6H_{14}$–$C_6H_5CH_3$ (2:1) as eluent at −30°C and subsequent sublimation at 45 to 50°C/<1 Torr.

The dark red crystals melt at 64 to 65°C. IR (KBr): ν (in cm^{-1}) = 2000 vs, 1942 vs, all ν(CO); 1150 s, ν(NS). IR ($n\text{-}C_6H_{14}$): ν (in cm^{-1}) = 2005 s, 1950 s, all ν(CO).

The compound is soluble in all common organic solvents. The solutions are air-sensitive [8].

$[(\eta^5\text{-}C_5H_5)Cr(NS)(CO)(NO)]^+PF_6^-$

The complex salt was produced by reaction of $(\eta^5\text{-}C_5H_5)Cr(NS)(CO)_2$ with equimolar amounts of $NO^+PF_6^-$ in CH_2Cl_2–CH_3CN at −78°C for 2 h.

The blue-green or green-black solid is water-sensitive. It is insoluble in all but those solvents with which it reacts and could not be purified.

IR (Nujol): ν (in cm^{-1}) = 2122, ν(CO); 1790, ν(NO); 1243, ν(NS).

Treatment with an acetone solution of NaI at room temperature for 2 h gives $(\eta^5\text{-}C_5H_5)Cr(NO)_2I$ in 25% yield. This conversion involves the initial formation of $((\eta^5\text{-}C_5H_5)Cr(NO)I)_2$, which then slowly disproportionates to the isolated complex [1].

$[Cr(NS)(N{\equiv}CCH_3)_5]^{2+}(PF_6^-)_2$

The complex formed in 23% yield by dropwise addition of a CH_3NO_2 solution of $SN^+PF_6^-$ to a CH_3CN solution of $(\eta^6\text{-}C_6H_6)Cr(CO)_3$ in the mole ratio 2:1 at ambient temperature. The CH_3NO_2 solution of $SN^+PF_6^-$ was freshly prepared by reaction of $S_3N_3Cl_3$ with $Ag^+PF_6^-$ in the mole ratio 1:3 in CH_3NO_2. The residue of the reaction solution was reprecipitated twice from CH_3CN–$(C_2H_5)_2O$ (1:10) and extracted with CH_2Cl_2. The blue-green extraction solution was filtered and evaporated. The residue was recrystallized from CH_3CN–$(C_2H_5)_2O$ (1:3).

The conductivity in CH_3NO_2 shows the blue crystalline solid to be of the 1:2 electrolyte type.

^1H NMR (CD_3NO_2/TMS): δ = 1.97 ppm (br, CH_3). IR (KBr): ν (in cm^{-1}) = 2325, 2305, ν(CN); 1245, ν(NS).

Reaction with excess $C{\equiv}NC_4H_9\text{-}t$ in CH_3CN in the presence of Zn dust at ambient temperature gave $[Cr(NS)(C{\equiv}NC_4H_9\text{-}t)_5]^+PF_6^-$ described below [9].

References on p. 86

[Cr(NS)(C≡NC$_4$H$_9$-t)$_5$]$^+$PF$_6^-$

The complex was produced in 57% yield when a mixture of 0.88 mmol of [Cr(NS)-(N≡CCH$_3$)$_5$]$^{2+}$(PF$_6^-$)$_2$, 1 mL (8.85 mmol) of t-C$_4$H$_9$N≡C and 0.1 g Zn dust in 25 mL CH$_3$CN was stirred at ambient temperature for 2 h. The reaction solution was filtered and cooled to −20°C after addition of 30 mL (C$_2$H$_5$)$_2$O and 30 mL C$_5$H$_{12}$. The separated green oil was washed with (C$_2$H$_5$)$_2$O and dried in the high vacuum. The solid was recrystallized from CH$_2$Cl$_2$–(C$_2$H$_5$)$_2$O (1:1) and (1:2) on addition of C$_5$H$_{12}$.

The green diamagnetic salt decomposes at 126°C under N$_2$. Conductivity in CH$_3$NO$_2$ shows it to be of the 1:1 electrolyte type.

^{1}H NMR (CD$_3$NO$_2$/TMS): δ = 1.49 ppm (s, CH$_3$). IR (KBr): ν (in cm^{-1}) = 2203, 2135, 2070 sh, all ν(CN); 1135, ν(NS).

The reaction of NO$^+$PF$_6^-$ with a solution of [Cr(NS)(C≡NC$_4$H$_9$-t)$_5$]$^+$PF$_6^-$ in CH$_3$CN gave [Cr(NS)(C≡NC$_4$H$_9$-t)$_5$]$^{2+}$(PF$_6^-$)$_2$ (see below) and NO [9].

[Cr(NS)(C≡NC$_4$H$_9$-t)$_5$]$^{2+}$(PF$_6^-$)$_2$

The complex salt was prepared in 77% yield by adding a CH$_3$CN solution of NO$^+$PF$_6^-$ dropwise to a solution of [Cr(NS)(C≡NC$_4$H$_9$-t)$_5$]$^+$PF$_6^-$ in CH$_3$CN until only the new IR absorption band at 2225 cm^{-1} of the product remained. Subsequent addition of (C$_2$H$_5$)$_2$O to the solution caused crystallization.

The blue crystals melt with decomposition at 118°C under N$_2$. Conductivity in CH$_3$NO$_2$ shows the paramagnetic complex to be of the 1:2 electrolyte type.

^{1}H NMR (CD$_3$NO$_2$/TMS): δ = 1.96 ppm (s, CH$_3$). IR (KBr): ν (in cm^{-1}) = 2225, ν(CN); ~1220, ν(NS) [9].

1.4.3 Thionitrosyl Complexes of Mo

Mo(NS)(S$_2$CN(CH$_3$)$_2$)$_3$

The complex was prepared in 74% and ca. 84% yield by heating a mixture of Mo(N)(S$_2$CN(CH$_3$)$_2$)$_3$ and excess S$_8$ or 1 to 2 equivalents of CH$_3$CHCH$_2$S, respectively under reflux in CH$_3$CN for 0.5 h. The resulting solution was filtered and evaporated to near dryness at 10^{-2} Torr. The residue was triturated with CH$_3$OH and recrystallized from CH$_2$Cl$_2$–CH$_3$OH.

The complex was also obtained in about 50% yield by refluxing 1.6 mmol of MoO$_2$(S$_2$CN(CH$_3$)$_2$)$_2$, 0.8 mmol of (CH$_3$)$_2$NC(=S)SSC(=S)N(CH$_3$)$_2$, and 1.7 mmol of (CH$_3$)$_3$SiN$_3$ in CH$_3$CN for 0.5 h. The complex precipitated after concentrating the solution. It was recrystallized from CH$_2$Cl$_2$–CH$_3$OH [10, 11].

The orange-yellow complex is crystalline, stable in air, diamagnetic, and nonconducting in CH$_2$ClCH$_2$Cl solution. The molecular weight M = 521 is determined osmometricly [10].

The crystal and molecular structures have been determined by X-ray analysis. The complex is triclinic, space group P$\bar{1}$–C$_i^1$ (No. 2), with a = 9.090(6), b = 9.327(6), c = 12.080(8) Å, α = 95.010(2)°, β = 91.950(2)°, γ = 96.160(2)°; V = 1013.72 Å^3; D$_x$ = 1.66 g/cm^3; Z = 2. R = 0.068 using 1907 observed reflections.

The molecular structure is shown in **Fig. 6.** Bond lengths and bond angles are compiled in Table 9. The complex has a pentagonal-bipyramidal molecular geometry with the NS ligand axial and one of the chelating ligands spanning one axial and one equatorial site. The

N–S distance (1.59 Å) is consistent with a considerable degree of Mo $\rightarrow$ NS π-back bonding, since it is greater than that in the free ligand or the value calculated for sums of covalent radii for an N=S double bond.

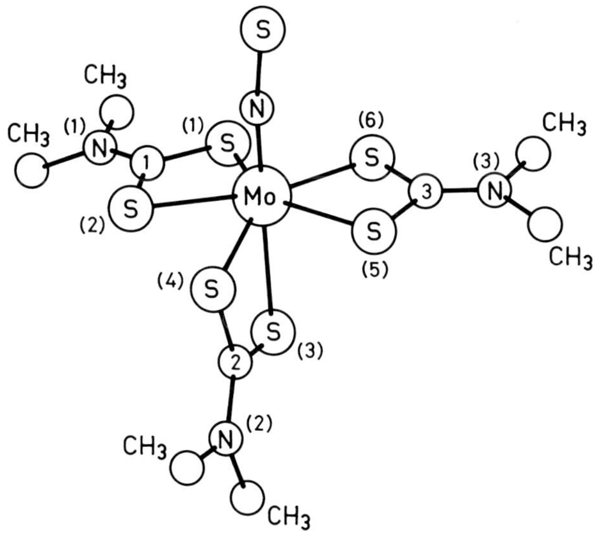

Fig. 6. Molecular structure of $Mo(NS)(S_2CN(CH_3)_2)_3$ [12].

Table 9

Selected Bond Lenghts and Bond Angles in $Mo(NS)(S_2CN(CH_3)_2)_3$ [12].

bond length	in Å	bond angle	in°	bond angle	in°
Mo–S(1)	2.515(4)	S(1)–Mo–S(2)	67.6(1)	N–Mo–S(1)	98.1(4)
Mo–S(2)	2.530(4)	S(3)–Mo–S(4)	69.8(1)	N–Mo–S(2)	98.2(3)
Mo–S(3)	2.602(4)	S(5)–Mo–S(6)	67.9(1)	N–Mo–S(3)	165.7(3)
Mo–S(4)	2.488(4)	S(1)–Mo–S(6)	73.3(1)	N–Mo–S(4)	98.0(3)
Mo–S(5)	2.515(4)	S(2)–Mo–S(4)	74.7(1)	N–Mo–S(5)	87.0(3)
Mo–S(6)	2.527(4)	S(4)–Mo–S(5)	75.3(1)	N–Mo–S(6)	93.8(3)
Mo–N	1.738(11)			S–N–Mo	172.0(7)
N–S	1.592(11)				

A figure of the crystal packing, the atomic positional parameters, and core atom deviations from the least-squares planes are given in [12].

^1H NMR ($NO_2C_6H_5$/TMS): $\delta = 3.16$ ppm (two overlapping ds 1:2:3; CH_3) [10, 11].

IR (Nujol?): ~1100 cm^{-1}, ν(NS) [11].

The mass spectrum shows strong peaks around m/e = 502 M^+, 456 $[Mo(S_2CN(CH_3)_2)_3]^+$, and 368 $[MoS(S_2CN(CH_3)_2)_2]^+$ [10, 11].

Reaction with $(n\text{-}C_4H_9)_3P$ in refluxing $CH_3C_6H_5$ or CH_3CN gives nearly quantitative yields of $Mo(N)(S_2CN(CH_3)_2)_3$ [10, 11]. Alkylation with $(C_6H_5)_3C^+BF_4^-$ gives $[Mo(NSC(C_6H_5)_3)(S_2CN(CH_3)_2)_3]^+BF_4^-$ [10].

References on p. 86

$Mo(NS)(S_2CN(C_2H_5)_2)_3$

The complex was prepared in 70 to 80% yield analogously to $Mo(NS)(S_2CN(CH_3)_2)_3$ from $Mo(N)(S_2CN(C_2H_5)_2)_3$ and S_8 or $CH_3\overline{CHCH_2S}$.

The complex is yellow, crystalline, stable in air, diamagnetic, and nonconducting in CH_2ClCH_2Cl solution [10, 11].

^1H NMR (CD_2Cl_2/TMS): δ (in ppm) = 1.32 (t, CH_3), 3.82 (q, CH_2) [10].

IR (Nujol?): 1100 cm^{-1}, ν(NS) [11].

The mass spectrum shows peaks attributable to the parent ion, together with ones assigned to $[Mo(S_2CN(C_2H_5)_2)_3]^+$ and $[MoS(S_2CN(C_2H_5)_2)_2]^+$ [11].

Desulfurization by $(n\text{-}C_4H_9)_3P$ in refluxing $CH_3C_6H_5$ or CH_3CN gives nearly quantitative $Mo(N)(S_2CN(C_2H_5)_2)_3$. Alkylation with $(C_6H_5)_3C^+BF_4^-$ produces $[Mo(NSC(C_6H_5)_3)(S_2CN(C_2H_5)_2)_3]^+$ BF_4^- [10].

$HB(3,5\text{-}(CH_3)_2C_3HN_2)_3Mo(NS)(CO)_2$

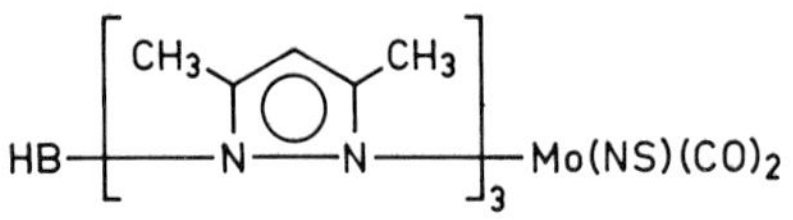

The complex was prepared in 24% yield by adding a THF solution of $S_3N_3Cl_3$ dropwise to an N_2-saturated THF solution of $K^+[HB(3,5\text{-}(CH_3)_2C_3HN_2)_3Mo(CO)_3]^-$ (mole ratio 1:3) at 0°C. The reaction mixture was allowed to warm to room temperature over a 1-h period. The complex was isolated and separated from $(HB(3,5\text{-}(CH_3)_2C_3HN_2)_3Mo(CO)_2)_2$ (μ-S) by column chromatography on silica gel with $C_6H_6-C_6H_{14}$ as eluent.

Recrystallization from $CH_2Cl_2-C_5H_{12}$ gave dark violet crystals which decomposed at > 240°C. The complex sublimes in vacuum at > 200°C.

The complex has a plane of symmetry as shown below, and the proton resonances appear as sharp signals with relative intensities of 2:1, representing the two 3,5-dimethylpyrazolyl groups cis to the thionitrosyl substituent and the one 3,5-dimethylpyrazolyl group trans to the thionitrosyl substituent.

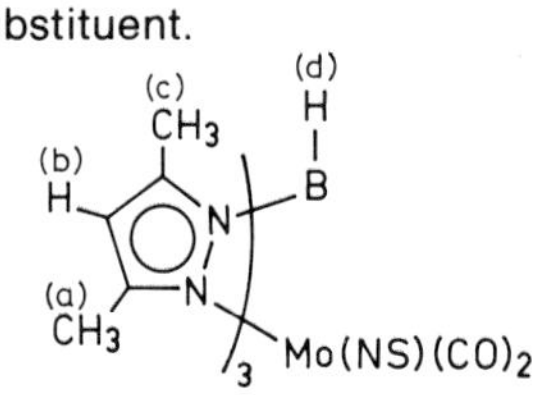

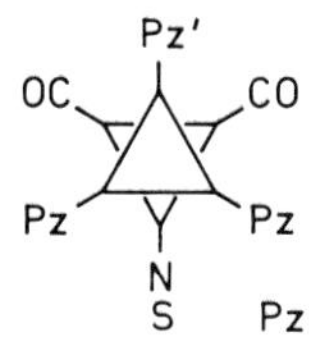

Pz = $3,5\text{-}(CH_3)_2C_3HN_2$

^1H NMR ($CDCl_3$/TMS): δ (in ppm) = 2.23 (s, CH_3, a'); 2.28 (s, CH_3, c'); 2.37 (s, 2 CH_3, a); 2.62 (s, 2 CH_3, c); 5.66 (s, CH, b'); 5.88 (s, 2 CH, b) [13].

^{13}C NMR ($CH_2Cl_2-CDCl_3$/TMS): δ = 225 ppm (CO).

^{14}N NMR (CH_2Cl_2/external neat CH_3NO_2): δ = 64 ppm (NS).

^{95}Mo NMR (CH_2Cl_2/external 2 M Na_2MoO_4 in H_2O at pH 11): δ = −192 ppm [5].

IR (KBr): ν (in cm^{-1}) = 2003, ν(CO); 1924, ν(CO); 1125, ν(NS); 1136, ν(NS) [5, 13].

UV-vis (CH_2Cl_2): λ_{max} (ε in $L \cdot mol^{-1} \cdot cm^{-1}$) = 515 nm (200), 445 nm (230) [5].

The complex dissolves in solvents such as $CHCl_3$, CH_2Cl_2, C_6H_6, and CH_3COCH_3 to give blood-red solutions that are stable for weeks in air [13].

References on p. 86

$Mo(NS)(N_3)((C_6H_5)_2PCH_2CH_2P(C_6H_5)_2)_2$

The complex formed in 65% yield by heating a mixture of 0.66 mmol of $Mo(N)(N_3)$-$((C_6H_5)_2PCH_2CH_2P(C_6H_5)_2)_2$ and 0.4 mL $CH_3\overline{CHCH_2S}$ in THF under reflux for 0.5 h. The filtrate was taken to dryness in vacuum, and the residue crystallized from CH_2Cl_2.

The orange crystals melt at 170 to 172°C with decomposition.

IR (KBr or Nujol): ν (in cm^{-1}) = 2065 vs, br, $\nu(N_3)$; 1065 s, ν(NS) [14].

$Mo(NS)Br((C_6H_5)_2PCH_2CH_2P(C_6H_5)_2)_2 \cdot 0.5\ CH_2Cl_2$

The complex was obtained in 60% yield by refluxing $Mo(N)(Br)((C_6H_5)_2PCH_2CH_2P(C_6H_5)_2)_2$ and $CH_3\overline{CHCH_2S}$ in THF.

The orange crystals, crystallized from CH_2Cl_2, melt at 167 to 171°C with decomposition.

IR (KBr or Nujol): 1065 cm^{-1} s, ν(NS) [14].

1.4.4 Thionitrosyl Complexes of W

$HB(3,5-(CH_3)_2C_3HN_2)_3W(NS)(CO)_2$

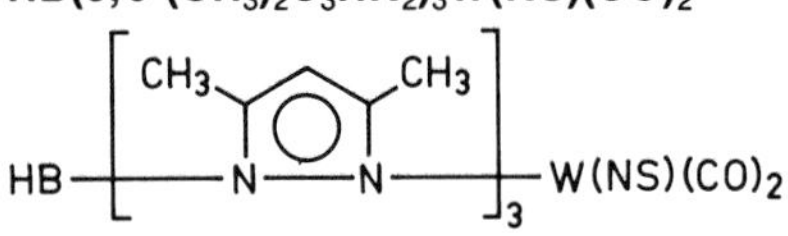

The compound was prepared in 10% yield by reaction of $K^+[HB(3,5-(CH_3)_2C_3HN_2)_3W(CO)_3]^-$ with $S_3N_3Cl_3$ in the mole ratio 3:1 in N_2-saturated THF solution at 0°C to room temperature over a 1-h period. The complex was separated from $(HB(3,5-(CH_3)_2C_3HN_2)_3W(CO)_2)_2$ (μ-S) by column chromatography on silica gel with $C_6H_6-C_6H_{14}$ as eluent.

The dark violet crystals decompose at >280°C and sublime in vacuum at >200°C. For the assignment of the ^1H NMR spectrum, see the scheme for the analogous Mo complex on p. 58.

^1H NMR ($CDCl_3$/TMS): δ (in ppm) = 2.26 (s, CH_3, a′), 2.29 (s, CH_3, c′), 2.38 (s, 2CH_3, a), 2.66 (s, 2CH_3, c), 5.70 (s, CH, b′), 5.92 (s, 2CH, b) [13].

^{13}C NMR ($CH_2Cl_2-CDCl_3$/TMS): δ = 227 ppm (CO).

^{14}N NMR (CH_2Cl_2/external neat CH_3NO_2): δ = 23 ppm (NS) [5].

IR (KBr): ν (in cm^{-1}) = 1988, ν(CO); 1902, ν(CO); 1127, ν(NS); 1140, ν(NS) [5, 13].

UV-vis (CH_2Cl_2): λ_{max}(ε in $L \cdot mol^{-1} \cdot cm^{-1}$) = 510 nm (290), 445 nm (350) [5].

The complex dissolves in $CHCl_3$, CH_2Cl_2, C_6H_6, and CH_3COCH_3 to give blood-red solutions that are stable for weeks in air [13].

$W(NS)(N_3)((C_6H_5)_2PCH_2CH_2P(C_6H_5)_2)_2$

The compound was obtained in 44% yield by heating 0.39 mmol of $W(N)(N_3)$-$((C_6H_5)_2PCH_2CH_2P(C_6H_5)_2)_2$ and 0.1 mL $CH_3\overline{CHCH_2S}$ under reflux for 2 h. After evaporating the solvent, the residue was crystallized from $CH_2Cl_2-C_6H_{14}$.

The red crystals melt at 144 to 146°C with decomposition.

IR (KBr or Nujol): ν (in cm^{-1}) = 2100 vs, $\nu(N_3)$; 1080 m, ν(NS) [14].

References on p. 86

1.4.5 Thionitrosyl Complex of Mn

$[(\eta^5\text{-}C_5H_5)Mn(NS)(CO)_2]^+AsF_6^-$

The compound was prepared almost quantitatively by reaction of $(\eta^5\text{-}C_5H_5)Mn(CO)_3$ with equimolar amounts of SN^+ AsF_6^- in SO_2 at −60°C. The solvent was evaporated at −40°C in vacuum. The light brown solid could not be isolated in analytically pure form.

IR (Nujol-Kel-F trituration): ν (in cm^{-1}) = 2108, ν(CO); 2070, ν(CO); 1284, ν(NS).

The compound quantitatively eliminates CO on standing in liquid SO_2 solution at higher temperature for a longer time to give a solid which is insoluble in liquid SO_2 [15].

1.4.6 Thionitrosyl Complexes of Tc

$Tc(NS)Cl_2(P(CH_3)_2C_6H_5)_3$

The complex was prepared in 40% yield by stirring a mixture of equimolar amounts of $Tc(N)Cl_2(P(CH_3)_2C_6H_5)_3$ and S_2Cl_2 in CH_2Cl_2 at room temperature for 2 h. After evaporation of the solvent, the residue was recrystallized twice from CH_2Cl_2−n-C_6H_{14} (1:4).

The light red crystals melt at 141 to 143°C. IR (KBr): 1177 cm^{-1}, ν(NS) [16].

$Tc(NS)Cl_3(P(CH_3)_2C_6H_5)_2$

The compound formed in 63% yield by refluxing $Tc(N)Cl_2(P(CH_3)_2C_6H_5)_3$ and S_2Cl_2 (mole ratio 1:2) in CH_2Cl_2 for 20 h. Dropwise addition of n-C_6H_{14} to the filtrate caused precipitation of the complex. The deep red crystals were recrystallized several times from CH_2Cl_2–n-C_6H_{14} (1:4).

The deep red crystals melt at 146 to 149°C. IR (KBr): 1240 cm^{-1}, ν(NS).

The ESR spectrum in CH_2Cl_2 at T = 295 K shows a 10-components-hyperfine structure caused by the nuclear spin of ^{99}Tc (I = 9/2). Following values of the components of the $\tilde{g}$- and ^{99}Tc-hyperfine structure tensor $\tilde{A}$(Tc) are calculated: $g_0 = 2.033 \pm 0.005$, a_0 (Tc) = $(148.6 \pm 3.0) \cdot 10^{-4}$ cm^{-1}; $g_{\parallel} = 2.027 \pm 0.005$, $A_{\parallel}$ (Tc) = $(237.1 \pm 3.0) \cdot 10^{-4}$ cm^{-1}. For the ESR spectrum in CH_2Cl_2 at T = 140 K the parameters of the vertical part can be estimated with the aid of the relations $g_{\perp} = (3g_0 - g_{\parallel})/2$ and $A_{\perp}$ (Tc) = $(3a_0(Tc) - A_{\parallel}(Tc))/2$ to be $g_{\perp} = 2.035$, $A_{\perp}$ (Tc) = 104.4 $\cdot 10^{-4}$ cm^{-1}. From these results the configuration $4t_{2g}^5$ (Tc^{2+}, S = 1/2) is confirmed [16].

$Tc(NS)Cl_2(S_2CN(C_2H_5)_2)_2$

The complex was prepared in 92% yield by addition of 0.43 mmol of S_2Cl_2 to a solution of 0.30 mmol of $Tc(N)(S_2CN(C_2H_5)_2)_2$ in 20 mL CH_2Cl_2. The dark brown solution was reduced in volume and applied to a silica-gel column. Unchanged nitrido complex was removed by elution with CH_2Cl_2. The orange fraction of the complex was eluted with $CH_3CO_2C_2H_5$−CH_2Cl_2 (1:3).

The complex was also obtained in 38% yield by reaction of 0.33 mmol of $Tc(N)(S_2CN(C_2H_5)_2)_2$ with 0.23 mmol of $SOCl_2$ in CH_2Cl_2.

The orange-brown crystals recrystallized from $CH_3CO_2C_2H_5$−CH_2Cl_2 (1:3) melt at 224 to 225°C with decomposition.

The crystal structure has been determined by single-crystal X-ray diffraction methods at 15°C. Single crystals were grown by slow evaporation at room temperature of either a $CH_3CO_2C_2H_5$−CH_2Cl_2 (1:3) solution (crystal 1) or a C_6H_6−$(CH_3)_2CO$ (1:1) solution (crystal 2). The crystals are orthorhombic, space group Pcmn−$D_{2h}^{16} = V_h^{16}$ (No. 62) with the following unit

References on p. 86

cell parameters. Crystal 1: a = 8.860(6), b = 15.690(2), c = 29.02(1) Å, V = 4034.2 $Å^3$. Z = 8. D_x = 1.687 g/cm^3. R = 0.202 for 1026 independent reflections. Crystal 2: a = 8.936(1), b = 15.681(1), c = 28.445(7) Å, V = 3985.9 $Å^3$. Z = 8. D_x = 1.708 g/cm^3. R = 0.078 for 2078 independent reflections.

The crystal lattice is disordered across a noncrystallographic mirror plane, the degree of disorder being 21.9(7)% for crystal 1 and 4.0(2)% for crystal 2. The crystal structure consists of two crystallographically independent molecules in the asymmetric unit. Each unique molecule contains a mirror plane on which are positioned the Tc and Cl atoms and the NS ligand. Both structures of crystal 1 and 2 are basically identical, with the exception of the degree of disorder, and the discussion below refers to the less disordered (crystal 2) structure. The two independent molecules are similar in gross features, and the Tc atoms have seven-coordinate pentagonal-bipyramidal environments. The molecular structure and atom numbering scheme are shown in **Fig. 7**, p. 62, for one molecule. Bond lengths and bond angles are given in Tables 10 and 11.

Table 10

Bond Lengths in the Two Independent Molecules of $Tc(NS)Cl_2(S_2CN(C_2H_5)_2)_2$ [17].

bond	molecule 1 in Å	molecule 2 in Å	bond	molecule 1 in Å	molecule 2 in Å
Tc–N(1)	1.80(2)	1.70(2)	C(1)–S(1)	1.69(2)	1.70(2)
Tc–S(1)	2.467(4)	2.461(4)	C(1)–S(2)	1.70(2)	1.71(2)
Tc–S(2)	2.461(5)	2.484(5)	C(1)–N(2)	1.33(2)	1.33(2)
Tc–Cl(1)	2.492(6)	2.414(5)	C(2)–N(2)	1.49(2)	1.47(2)
Tc–Cl(2)	2.401(7)	2.376(8)	C(2)–C(3)	1.48(2)	1.49(3)
Tc···S(3)	3.29	3.24	C(4)–N(2)	1.45(2)	1.47(2)
N(1)–S(3)	1.49(2)	1.54(2)	C(4)–C(5)	1.46(2)	1.44(3)

Table 11

Bond Angles in the Two Independent Molecules of $Tc(NS)Cl_2(S_2CN(C_2H_5)_2)_2$ [17].

angle	molecule 1 in °	molecule 2 in °	angle	molecule 1 in °	molecule 2 in °
N(1)–Tc–Cl(1)	178.9(8)	176.8(9)	Tc–N(1)–S(3)	177(2)	174(2)
N(1)–Tc–Cl(2)	92.1(7)	87.3(9)	Tc–S(1)–C(1)	91.5(5)	91.3(5)
N(1)–Tc–S(1)	92.7(2)	92.1(2)	Tc–S(2)–C(1)	91.6(6)	90.3(6)
N(1)–Tc–S(2)	91.5(6)	93.7(7)	S(1)–C(1)–S(2)	108.8(9)	109.6(8)
Cl(1)–Tc–Cl(2)	89.0(2)	89.5(3)	S(1)–C(1)–N(2)	127(1)	126(1)
Cl(1)–Tc–S(1)	87.6(1)	87.1(1)	S(2)–C(1)–N(2)	124(1)	124(2)
Cl(1)–Tc–S(2)	87.6(2)	88.9(2)	C(1)–N(2)–C(2)	119(1)	121(2)
Cl(2)–Tc–S(1)	76.2(1)	75.5(1)	C(1)–N(2)–C(4)	121(1)	122(2)
Cl(2)–Tc–S(2)	144.2(1)	144.2(1)	C(2)–N(2)–C(4)	119(1)	117(1)
S(1)–Tc–S(2)	68.1(2)	68.7(2)	N(2)–C(2)–C(3)	111(2)	114(2)
S(2)–Tc–S(2)*)	71.2(2)	71.6(2)	N(2)–C(4)–C(5)	112(2)	117(2)

*) Coordinate transformation (x, 0.5−y, z).

The major differences between the independent molecules are in the Tc=N, N=S, and Tc–Cl bond distances. In the molecule 1 shown in Fig. 7, Tc=N and N=S bond lengths are 1.80(2) and 1.49(2) Å, whereas in the molecule 2 they are 1.70(2) and 1.54(2) Å, respectively. Of interest is the similarity of the overall Tc···S distance in both molecules (3.29 Å, 3.24 Å) and the near linearity of the Tc=N=S bonding (Tc=N=S angles of 177(2)° and 174(2)° in molecules 1 and 2, respectively). Atomic positional coordinates for crystal 2 and intramolecular nonbonded distances and intermolecular contact distances for the two independent molecules for crystal 2 are given in the paper [17]. For the more disordered crystal 1 structure the Tc=N and N=S bond lengths and Tc=N=S angles are 1.63(6) and 1.3(2) Å and 112(7)°, respectively in molecule 1 and 1.30(5), 2.1(1) Å and 154(5)°, respectively in molecule 2.

The IR spectrum (KBr) shows eleven absorptions in the range 1528 to 852 cm^{-1}; ν(NS) = 1248 cm^{-1}. The molar susceptibility of the complex, corrected for diamagnetism of the ligands (−3.02 mm^3/mol) was determined as 2.66 mm^3/mol at room temperature. This small paramagnetic susceptibility is consistent with Tc(III) [17].

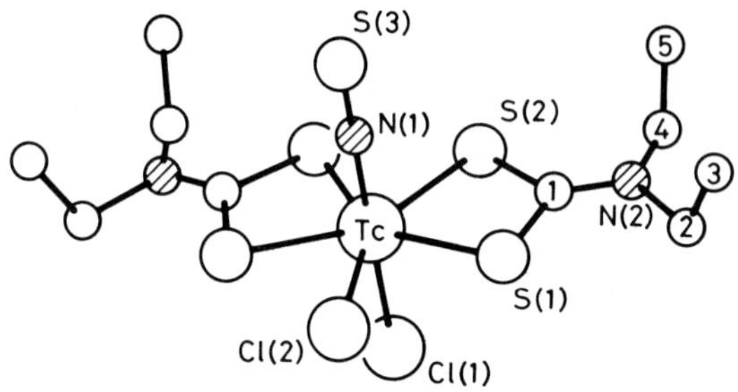

Fig. 7. Molecular structure of one of the independent $Tc(NS)Cl_2(S_2CN(C_2H_5)_2)_2$ molecules [17].

1.4.7 Thionitrosyl Complexes of Re

Thionitrosyl-Phosphine Complexes of Re

The preparation of the complexes is given in Table 12. Physical properties are compiled in Table 13, p. 64.

^1H NMR spectra ($CDCl_3$/TMS), δ in ppm, J in Hz:

$Re(NS)Cl_2(P(CH_3)_2C_6H_5)_3$ (No. **1**): 2.00 (t, 2 CH_3), | $^4J(P,H) + ^2J(P,H)$ | = 4.3; 1.84 (t, 2 CH_3), | $^4J(P,H) + ^2J(P,H)$ | = 4.3; 1.36 (d, 2 CH_3), | $^2J(P,H)$ | = 9.5; 7.0 to 7.5 (m, 3C_6H_5).

This spectrum indicates a meridional configuration for the phosphine ligands with the NS group trans to Cl.

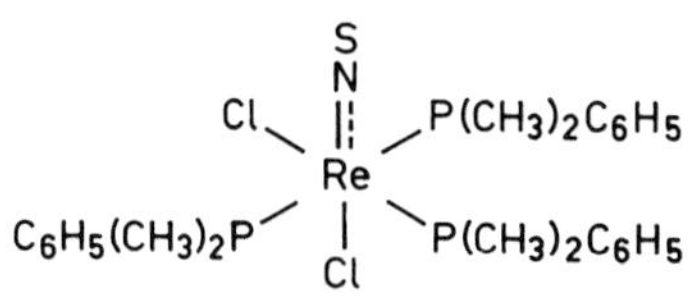

No. 1

$Re(NS)Cl_2(P(C_2H_5)_2C_6H_5)_3$ (No. **2**): 0.5 to 3.0 (m, 6C_2H_5); 6.95 to 7.5 (m, 3C_6H_5).

Table 12

Preparation of Thionitrosyl-Phosphine Complexes of Re.

No.	compound	yield in %	starting compounds	mole ratio (amounts)	solvent	temp., time	Ref.
1	$Re(NS)Cl_2(P(CH_3)_2C_6H_5)_3$	84	$Re(N)Cl_2(P(CH_3)_2C_6H_5)_3$,$S_2Cl_2$	2:1	CH_2Cl_2	r.t., 0.5 h	[10, 18]
2	$Re(NS)Cl_2(P(C_2H_5)_2C_6H_5)_3$	86	$Re(N)Cl_2(P(C_2H_5)_2C_6H_5)_3$,$S_2Cl_2$	2:1	CH_2Cl_2	r.t., 0.5 h	[10]
3	$Re(NS)Cl_2(PCH_3(C_6H_5)_2)_3$	76	$Re(N)Cl_2(PCH_3(C_6H_5)_2)_3$,$S_2Cl_2$	2:1	CH_2Cl_2	r.t., 0.5 h	[10, 18]
4	$Re(NS)Br(Cl)(P(C_2H_5)_2C_6H_5)_3$	87	$Re(N)Br_2(P(C_2H_5)_2C_6H_5)_3$,$S_2Cl_2$	2:1	CH_2Cl_2	r.t., 0.5 h	[10]
5	$Re(NS)Cl(SCN)(P(CH_3)_2C_6H_5)_3$	84	No. 1, NaNCS	(0.3 g, 0.3 g)	CH_3OH	reflux, 1 h	[10]
6	$Re(NS)Cl(S_2CN(CH_3)_2)(P(CH_3)_2C_6H_5)_2$	61	No. 1, $NaS_2CN(CH_3)_2 \cdot 3H_2O$	(0.4 g, 0.5 g)	$(CH_3)_2CO$	reflux, 1.5 h	[10]
7	$[Re(NS)Cl((C_6H_5)_2PCH_2CH_2P(C_6H_5)_2)_2]^+Cl^-$	80	$[Re(N)Cl((C_6H_5)_2PCH_2CH_2P(C_6H_5)_2)]^+Cl^-$,$S_2Cl_2$	(0.4 g, 0.1 mL)	CH_2Cl_2	r.t., 2 h	[10, 18]
8	$[Re(NS)Cl((C_6H_5)_2PCH_2CH_2P(C_6H_5)_2)_2]^+PF_6^-$	81	No. 7, $NH_4^+PF_6^-$	(0.35 g, 0.5 g)	CH_3OH-H_2O	r.t.	[10]
9	$[Re(NS)Cl((C_6H_5)_2PCH_2CH_2P(C_6H_5)_2)_2]^+FeCl_4^-$	95	No. 7, $FeCl_3$	(0.2 g, 0.2 g)	$C_2H_5OH-H_2O$	r.t.	[10]
10	$[Re(NS)Cl((C_6H_5)_2PCH_2CH_2P(C_6H_5)_2)_2]^+(C_2H_5)_2NCS_2^-$	72	No. 7, $NaS_2CN(C_2H_5)_2$	(0.3 g, 0.3 g)	$(CH_3)_2CO$	reflux, 1.5 h	[10]
11	$Re(NS)Cl_3(P(CH_3)_2C_6H_5)_2$	53	$Re(N)Cl_2(P(CH_3)_2C_6H_5)_3$, S_2Cl_2	?	CH_2Cl_2	r.t., 1 h	[10, 18]
12	$Re(NS)Cl_3(P(C_2H_5)_2C_6H_5)_2$	~60	$Re(N)Cl_2(P(C_2H_5)_2C_6H_5)_3$, S_2Cl_2	?	CH_2Cl_2	r.t., 1 h	[10]
13	$Re(NS)Cl_3(PCH_3(C_6H_5)_2)_2$	~60	$Re(N)Cl_2(PCH_3(C_6H_5)_2)_3$, S_2Cl_2	?	CH_2Cl_2	r.t., 1 h	[10, 18]
14	$Re(NS)Cl_3(P(n\text{-}C_3H_7)(C_6H_5)_2)_2$	89	$Re(N)Cl_2(P(n\text{-}C_3H_7)(C_6H_5)_2)_2$, S_2Cl_2	(~0.6 g, 0.033 mL)	CH_2Cl_2	r.t., 1.5 h	[10, 18]
15	$Re(NS)Cl_3(P(C_6H_5)_3)_2$	55	$Re(N)Cl_2(P(C_6H_5)_3)_2$, S_2Cl_2	(0.6 g, 0.033 mL)	CH_2Cl_2	r.t., 1.5 h	[10]

References on p. 86

Table 13
Physical Properties of Thionitrosyl-Phosphine Complexes of Re.

No.	compound	m.p. in °C	recrystallized from	color	magnetic moment[a] in B.M.	IR (KBr) ν(NS) in cm^{-1}	Ref.
1[b]	$Re(NS)Cl_2(P(CH_3)_2C_6H_5)_3$	166 to 167	$CH_2ClCH_2Cl–C_5H_{12}$	pink	D	1180	[10, 18]
2[b]	$Re(NS)Cl_2(P(C_2H_5)_2C_6H_5)_3$	—	$CH_2ClCH_2Cl–C_5H_{12}$	pink	D	1167	[10]
3[b]	$Re(NS)Cl_2(PCH_3(C_6H_5)_2)_3$	160 to 162	$CH_2ClCH_2Cl–C_5H_{12}$	pink	D	1172	[10, 18]
4	$Re(NS)Br(Cl)(P(C_2H_5)_2C_6H_5)_3$	—	$CH_2ClCH_2Cl–C_5H_{12}$	pink	D	1168	[10]
5	$Re(NS)Cl(SCN)(P(CH_3)_2C_6H_5)_3$	—	$CH_2ClCH_2Cl–(CH_3)_2CO$	pink-red	D	1177	[10]
6[b]	$Re(NS)Cl(S_2CN(CH_3)_2)(P(CH_3)_2C_6H_5)_2$	—	$CH_2ClCH_2Cl–C_6H_{14}$	yellow-brown	D	1150	[10]
7[c]	$[Re(NS)Cl((C_6H_5)_2PCH_2CH_2P(C_6H_5)_2)_2]^+Cl^-$	202 to 204	$CH_2ClCH_2Cl–(CH_3)_2CO$	yellow	D	1185	[10, 18]
8[b]	$[Re(NS)Cl((C_6H_5)_2PCH_2CH_2P(C_6H_5)_2)_2]^+PF_6^-$	—	—	yellow	D	1177	[10]
9	$[Re(NS)Cl((C_6H_5)_2PCH_2CH_2P(C_6H_5)_2)_2]^+FeCl_4^-$	—	—	yellow	D	1173	[10]
10[b]	$[Re(NS)Cl((C_6H_5)_2PCH_2CH_2P(C_6H_5)_2)_2]^+$-$(C_2H_5)_2NCS_2^-$	—	$CH_2ClCH_2Cl–(CH_3)_2CO$	yellow-orange	D	1183	[10]
11	$Re(NS)Cl_3(P(CH_3)_2C_6H_5)_2$	172 to 174	$CH_2ClCH_2Cl–(CH_3)_2CO$	purple	1.84	1228	[10, 18]
12	$Re(NS)Cl_3(P(C_2H_5)_2C_6H_5)_2$	—	$CH_2ClCH_2Cl–(CH_3)_2CO$	purple	1.86	1230	[10]
13	$Re(NS)Cl_3(PCH_3(C_6H_5)_2)_2$	176 to 178	$CH_2ClCH_2Cl–(CH_3)_2CO$	purple	2.01	1220	[10, 18]
14	$Re(NS)Cl_3(P(n\text{-}C_3H_7)(C_6H_5)_2)_2$	—	$(CH_3)_2CO$	green-purple	1.81	1226	[10]
15	$Re(NS)Cl_3(P(C_6H_5)_3)_2$	—	$(CH_3)_2CO$	purple-black	1.84	1214	[10]

[a] D = diamagnetic. – [b] ^1H NMR spectrum, see pp. 62, 65. – [c] ^{31}P NMR spectrum, see p. 65.

$Re(NS)Cl_2(PCH_3(C_6H_5)_2)_3$ (No. **3**): 2.16 (t, $2CH_3$), $|{}^4J(P,H)+{}^2J(P,H)|=4.2$; 1.81 (d, CH_3), $|{}^2J(P,H)|=9.0$; 7.0 to 7.6 (m, $6C_6H_5$).

$Re(NS)Cl(S_2CN(CH_3)_2)(P(CH_3)_2C_6H_5)_2$ (No. **6**): 1.82 (d, $P(CH_3)_2$), $|{}^2J(P,H)|=9$; 1.80 (d,$P(CH_3)_2$), $|{}^2J\ (P,H)|=9$; 7.3 to 7.7 (m, $2C_6H_5$); 3.28 (s,$N(CH_3)_2$). The equivalence of the dithiocarbamate methyl groups indicates the configuration given below.

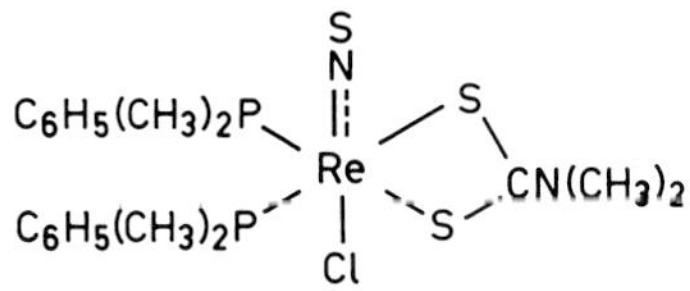

No. 6

$[Re(NS)Cl((C_6H_5)_2PCH_2CH_2P(C_6H_5)_2)_2]^+PF_6^-$ (No. **8**): 2.6 to 3.2 (m, 2 CH_2CH_2); 6.9 to 7.5 (m, $8C_6H_5$).

$[Re(NS)Cl((C_6H_5)_2PCH_2CH_2P(C_6H_5)_2)_2]^+(C_2H_5)_2NCS_2^-$ (No. **10**): 2.6 to 3.2 (m, $2CH_2CH_2$); 6.9 to 7.5 (m, $8C_6H_5$), 4.18 (q, 2 NCH_2); 1.17 (t, $2CH_3$).

^{31}P NMR ($CDCl_3$/external H_3PO_4). $[Re(NS)Cl((C_6H_5)_2PCH_2CH_2P(C_6H_5)_2)_2]^+Cl^-$ (No. **7**): $\delta\approx120$ ppm (s). This indicates that the four phosphorus atoms lie in the plane with the NS and Cl groups in trans position.

The complexes No. **1** to **4** and No. **11** to **15** in Table 12, p. 63, are stable in air being monomeric and nonelectrolytes in CH_2ClCH_2Cl. The salts No. **7** to **9** in Table 12 have conductivities in $C_6H_5NO_2$ typical of 1:1 electrolytes [10].

Treatment of $Re(NS)Cl_2(P(CH_3)_2C_6H_5)_3$ (No. **1**) with an excess of $(C_6H_5)_2PCH_2CH_2P(C_6H_5)_2$ under reflux in C_6H_6 results in loss of thionitrosyl sulfur and formation of $[Re(N)Cl((C_6H_5)_2$-$PCH_2CH_2P(C_6H_5)_2)_2]^+Cl^-$ in high yield [18]. The reaction of $Re(NS)Cl_2(P(CH_3)_2C_6H_5)_3$ with Na^+-NCS^- in refluxing CH_3OH produces $Re(NS)Cl(SCN)(P(CH_3)_2C_6H_5)_3$. Heating the complex and $Na^+S_2CN(CH_3)_2^-\cdot3H_2O$ in $(CH_3)_2CO$ under reflux gives $Re(NS)Cl(S_2CN(CH_3)_2)(P(CH_3)_2C_6H_5)_2$ (Nos. **5** and **6** in Table 12, p. 63).

$[Re(NS)Cl((C_6H_5)_2PCH_2CH_2P(C_6H_5)_2)_2]^+Cl^-$ (No. **7**) can be readily converted into PF_6^-, $FeCl_4^-$, and $(C_2H_5)_2NCS_2^-$ salts (Nos. **8**, **9**, **10** in Table 12, p. 63) [10].

$(Re(NS)Cl_4)_x$

A complex of composition $Re(NS)Cl_4$ was obtained by reaction of 9.5 mmol of Re_2Cl_{10} and 19 mmol of S_4N_4 in 150 mL CH_2Cl_2 at room temperature for 3 h. The brown precipitate was filtered off, washed with CH_2Cl_2, and dried in vacuum for 10 h. The yield is 6.3 g. The crude product decomposes at 210°C.

The IR spectrum exhibits 16 absorption bands in the range 1264 to 301 cm^{-1}. In the mass spectrum the following ions are observed: SN^+, SCl^+, $ClSN^+$, S_2Cl^+, SCl_2^+, $S_2Cl_2^+$, and $ReCl_5^+$.

Heating a mixture of the crude complex and $(C_6H_5)_4As^+Cl^-$ in CH_2Cl_2 under reflux yields $(C_6H_5)_4As^+[Re(NS)Cl_5]^-$ (see below) [19].

$(C_6H_5)_4As^+[Re(NS)Cl_5]^-$

The complex was prepared by refluxing a mixture of 6.3 g of crude $(Re(NS)Cl_4)_x$ and 4 g (9.5 mmol) of $(C_6H_5)_4As^+Cl^-$ for 8 h. After cooling and filtration the green-red solution was layered with 150 mL C_6H_{14}. Subsequent standing of the mixture at −5°C for 20 d caused separation of red-orange hexagonal platelets. The yield is 2.2 g.

References on p. 86

The IR spectrum shows absorption frequencies assigned to the anion at 1268 m, 1212 m, 1199 m, 361 m, and 298 cm^{-1} vs [19].

$Re(NS)(NSCl)Cl_2(NC_5H_5)_2$

The complex was prepared in 69% yield by adding 1.5 mL of C_5H_5N to a suspension of 2.53 mmol of $ReCl_3(NSCl)_2(POCl_3)$ in CH_2Cl_2 with stirring and cooling with ice. After warming to room temperature and stirring for further 12 h, the red-brown precipitate was filtered. It was also obtained in 72% yield by the analogous reaction of 0.95 g of $(ReCl_3(NSCl)_2)_2 \cdot (\mu\text{-}N_2S_2)$ with 1.5 mL of C_5H_5N in CH_2Cl_2. Black compact single crystals were produced by adding a few mL of CCl_4 to the deep red filtrate and cooling it to 5°C for several days.

The crystal structure was determined by X-ray diffraction. The crystals are monoclinic, space group $P2_1/c-C^5_{2h}$ (No. 14), with a = 1551.6(4), b = 694.9(2), c = 1513.2(2) pm, β = 96.50(2)°, $V = 1621 \cdot 10^{-30} m^3$. Z = 4. D_x = 2.22 g/cm³. R = 0.027 for 2105 observed reflexions. The molecular structure is shown in **Fig. 8**. Bond lengths and bond angles are given in Table 14.

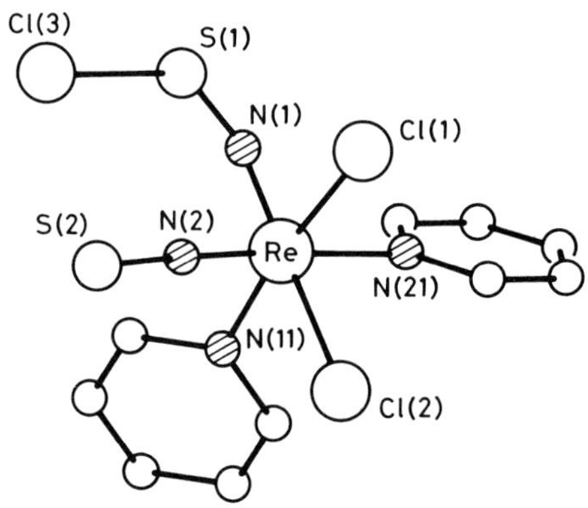

Fig. 8. Molecular structure of $Re(NS)(NSCl)Cl_2(NC_5H_5)_2$ [20].

Table 14
Bond Lengths and Bond Angles of $Re(NS)(NSCl)Cl_2(NC_5H_5)_2$ [20].

bond length	in pm	bond angle	in°	bond angle	in°
Re–Cl(1)	238.0(2)	Cl(1)–Re–Cl(2)	89.1(1)	N(11)–Re–N(21)	86.2(1)
Re–Cl(2)	240.4(2)	Cl(1)–Re–N(11)	172.0(1)	N(11)–Re–N(1)	89.7(2)
Re–N(11)	213.6(4)	Cl(1)–Re–N(21)	86.8(1)	N(11)–Re–N(2)	91.7(2)
Re–N(21)	223.4(4)	Cl(1)–Re–N(1)	93.8(2)	N(21)–Re–N(1)	86.0(2)
Re–N(1)	186.8(5)	Cl(1)–Re–N(2)	95.2(2)	N(21)–Re–N(2)	177.1(2)
Re–N(2)	177.2(5)	Cl(2)–Re–N(11)	86.5(1)	N(1)–Re–N(2)	96.1(2)
N(1)–S(1)	149.0(6)	Cl(2)–Re–N(21)	87.0(1)	Re–N(1)–S(1)	159.4(4)
S(1)–Cl(3)	220.6(3)	Cl(2)–Re–N(1)	172.3(2)	Re–N(2)–S(2)	176.3(4)
N(2)–S(2)	152.0(6)	Cl(2)–Re–N(2)	90.8(2)	N(1)–S(1)–Cl(3)	109.9(2)

In the monomer molecules, each Re atom has a distorted octahedral coordination of two cis-Cl atoms, two cis-N atoms of the pyridine ligands, and two cis-N atoms of the thionitrosyl and chloro thionitrene ligands. One of the pyridine groups is in trans position to the nearly linear thionitrosyl group (angle Re–N–S 176°) with distance Re–N of 177 pm and N–S of 152 pm. However, in the trans position of the chloro thionitrene group (angle Re–N–S 159°, distances Re–N 189 pm, N–S 149 pm) a chlorine atom is found. Atomic coordinates and parameters of the thermic vibration are given in the paper [20].

IR (solvent?): ν (in cm^{-1}) = 1625 w, 1610 s, 1485 m, 1451 s, 1445 s, all NC_5H_5; 1240 sh; 1230 vs, ν(NS) of ReNS; 1220 s, NC_5H_5; 1178 vs, ν(NS) of ReNSCl; 1165 w, 1068 vs, 1050 m, 1020 m, 980 w, 870 vw, 760 s, 690 s, 651 m, 641 m, all NC_5H_5; 510 sh; 500 m, ν(SCl); 460 vw, 445 vw, all $\nu(ReNC_5H_5)$; 420 vw; 325 w; 310 s, 295 vs, all ν(ReCl); 240 m. The moisture-sensitive complex is sparingly soluble in CH_2Cl_2 [20].

Reaction with excess $(CH_3)_3SiBr$ in boiling CH_2Br_2 gives the corresponding bromo compound (see p. 68) according to the equation $Re(NS)(NSCl)Cl_2(NC_5H_5)_2 + 3\,(CH_3)_3SiBr \rightarrow Re(NS)(NSBr)Br_2(NC_5H_5)_2 + 3\,(CH_3)_3SiCl$ [43].

$((C_6H_5)_4P^+)_2\ [Re(NS)(NSCl)Cl_4]^{2-} \cdot CH_2Cl_2$

The complex was prepared by adding equimolar amounts of $(C_6H_5)_4P^+Cl^-$ in CH_2Cl_2 to a stirred solution of $(C_6H_5)_4P^+[Re(S_2N_2)Cl_4]^-$ in CH_2Cl_2. Then the solution was layered with n-C_6H_{14} and allowed to stand at room temperature. Black moisture-sensitive crystals formed, which are recrystallized from CH_2Cl_2–CCl_4. The yield is 78%.

IR spectrum (Nujol) without the bands of the cation: ν (in cm^{-1}) = 1225 m and 1208 m, ν(NS) of ReNS; 1185 m, ν(NS) of ReNSCl; 500 vw, ν(SCl); 450 m, δ(ReNS); 318 w, 300 s, 263 m, all ν(ReCl) [43].

$((C_6H_5)_4As^+)_2\ [Re(NS)(NSCl)Cl_4]^{2-} \cdot CH_2Cl_2$

The complex salt was first prepared as a crude compound of composition $[Re(NS)(NSCl)Cl_2]_x$ by stirring a mixture of 2.56 g (7.48 mmol) of $Re(N)Cl_4$ and 1.38 g (7.48 mmol) of S_4N_4 in 100 mL CH_2Cl_2 at room temperature for 3 h. The brown product precipitated from the deep red solution. Then 3.13 g (7.48 mmol) of $(C_6H_5)_4\ As^+Cl^-$ was added and the mixture is stirred for a further 12 h. The red-brown filtered solution was layered with 70 mL C_6H_{14} and was allowed to stand at −5°C for 15 d. Dark brown crystals formed, which were recrystallized twice from CH_2Cl_2–C_6H_{14}. The yield is 2.6 g [19].

The compound was also prepared by reaction of $ReCl_4(NSCl)(POCl_3)$ with equimolar amounts of $(CH_3)_3SiN{=}S{=}NSi(CH_3)_3$ in CH_2Cl_2 at reflux temperature for 2 h which gave a brown precipitate. Extraction with hot CH_2ClCH_2Cl yielded a product of the composition $ReCl_3N_3S_2$. Subsequent treatment of the intermediate product with a solution of $(C_6H_5)_4As^+Cl^-$ in CH_2Cl_2 gave a dark red solution which was allowed to stand for 12 h. After evaporation of the solvent the residue was crystallized from CH_2Cl_2 – n-C_6H_{14} to give rod-shaped black crystals in 84% yield. The overall reaction is: $ReCl_4(NSCl)(POCl_3) + (CH_3)_3SiN{=}S{=}NSi(CH_3)_3 + 2\,(C_6H_5)_4As^+Cl^- \xrightarrow{CH_2Cl_2} ((C_6H_5)_4As^+)_2[Re(NS)(NSCl)Cl_4]^{2-} \cdot CH_2Cl_2 + 2\,(CH_3)_3SiCl + POCl_3 + 1/2\,N_2$ [43].

The crystal structure was determined by X-ray diffraction. The complex crystallizes in the triclinic space group $P\bar{1}$–C_i^1 (No. 2) with a = 10.434(2), b = 12.1454(6), c = 21.125(2) Å, α = 81.210(6)°, β = 86.70(1)°, γ = 76.624(8)°, V = 2573.1(7) $Å^3$, Z = 2. D_x = 1.687 g/cm^3. R = 0.028 for 4417 independent reflections. The molecular structure of the $[Re(NS)(NSCl)Cl_4]^{2-}$ ion is shown in **Fig. 9**, p. 68.

Bond lengths and bond angles are given in Table 15, p. 68.

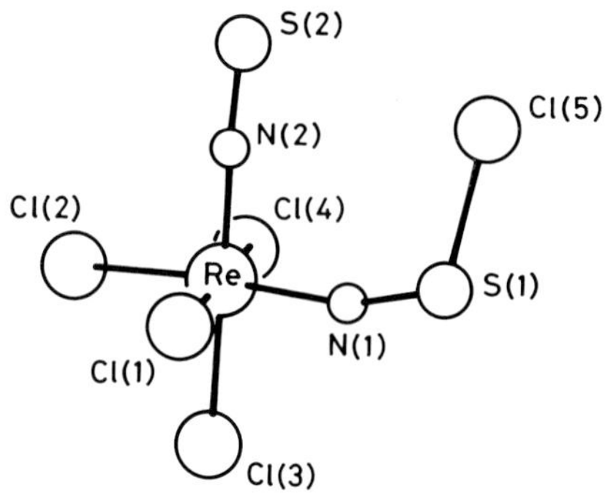

Fig. 9. Molecular structure of the anion in $((C_6H_5)_4As^+)_2$ $[Re(NS)(NSCl)Cl_4]^{2-} \cdot CH_2Cl_2$ [19].

Table 15
Bond Lengths and Bond Angles of the Anion in $((C_6H_5)_4As^+)_2$ $[Re(NS)(NSCl)Cl_4]^{2-}$ [19].

bond length	in Å	bond angle	in °	bond angle	in °
Re–Cl(1)	2.398(1)	Cl(1)–Re–Cl(2)	87.20(5)	Cl(3)–Re–Cl(4)	89.39(5)
Re–Cl(2)	2.433(1)	Cl(1)–Re–Cl(3)	86.38(5)	Cl(3)–Re–N(1)	87.21(13)
Re–Cl(3)	2.433(1)	Cl(1)–Re–Cl(4)	173.24(5)	Cl(3)–Re–N(2)	178.97(12)
Re–Cl(4)	2.391(1)	Cl(1)–Re–N(1)	92.66(12)	Cl(4)–Re–N(1)	92.40(12)
Re–N(1)	1.894(4)	Cl(1)–Re–N(2)	93.24(11)	Cl(4)–Re–N(2)	90.89(11)
		Cl(2)–Re–Cl(3)	87.96(5)	N(1)–Re–N(2)	93.77(17)
Re–N(2)	1.837(4)	Cl(2)–Re–Cl(4)	87.38(5)	N(1)–S(1)–Cl(5)	112.57(19)
S(1)–N(1)	1.443(4)	Cl(2)–Re–N(1)	175.17(13)	Re–N(1)–S(1)	163.46(29)
S(1)–Cl(5)	2.256(2)	Cl(2)–Re–N(2)	91.06(12)	Re–N(2)–S(2)	177.09(25)
S(2)–N(2)	1.456(4)				

The coordination geometry at the Re atom is a distorted octahedron. The longer Re–N distance of 1.894 Å in the case of the NSCl ligand suggests a lower bond order to the metal than in the case of the NS ligand which has an Re–N distance of 1.837 Å. The SN bond lengths in the NSCl and NS ligands are 1.443 Å and 1.456 Å, respectively, only little longer than in the SN^+ ion.

The dark brown crystals melt at 194°C. The IR spectrum shows the following absorption bands assigned to the anion at 1270 w, 1222 w, 768 w, 351 vs, and 310 cm^{-1} s [19].

$Re(NS)(NSBr)Br_2(NC_5H_5)_2 \cdot CH_2Br_2$

The compound was prepared by bromination of $Re(NS)(NSCl)Cl_2(NC_5H_5)_2 \cdot CH_2Cl_2$ with $(CH_3)_3SiBr$ in the mole ratio 1:10 in boiling CH_2Cl_2 for 2 h in 55% yield.

The dark red-violet crystals are sensitive to humidity.

IR (Nujol) without bands of NC_5H_5: ν (in cm^{-1}) = 1230 s, ν(NS) of ReNS and ReNSBr; 580 w, ν(ReN); 480 m, ν(SBr); 225 s, ν(ReBr) [43.]

$((C_6H_5)_4P^+)_2$ $[Re(NS)(NSBr)Br_4]^{2-} \cdot CH_2Br_2$

The title compound was obtained in 98% yield by addition of a solution of $(C_6H_5)_4P^+Br^-$ in CH_2Br_2 to a stirred solution of $(C_6H_5)_4P^+$ $[ReBr_4(N_2S_2)]^-$ in CH_2Br_2 (mole ratio 1:1) at room temperature. Red needle-shaped, moisture-sensitive crystals crystallize from the red solution at 5°C which were washed with n-C_6H_{14}.

The crystal structure was determined at 23°C by X-ray diffraction. The compound crystallizes in the triclinic space group $P\bar{1}$–C_i^1 (No. 2) with a = 1039.7(2), b = 1232.5(4), c = 2158.4(5) pm; α = 81.59(3)°, β = 87.05(2)°, γ = 77.06(3)°; V = 2665 × 10^{-30} m^3; Z = 2. D_x = 1.91 g/cm^3. R = 0.0587 for 6493 independent reflections. The molecular structure of the $[Re(NSBr)(NS)Br_4]^{2-}$ ion is shown in **Fig. 10**. Bond lengths and bond angles are compiled in Table 16. In the anion, the Re atom is coordinated in a distorted octahedron by four Br atoms and two N atoms of the NS and NSBr ligands in cis position. The structure is isotypic with $((C_6H_5)_4As^+)_2$ $[Re(NS)(NSCl)Cl_4]^{2-} \cdot CH_2Cl_2$, see p. 68. Whereas the thionitrosyl group [Re=N–S] is nearly linear, the bromothionitrene group [Re=N=SBr] has an ReNS angle of 165°.

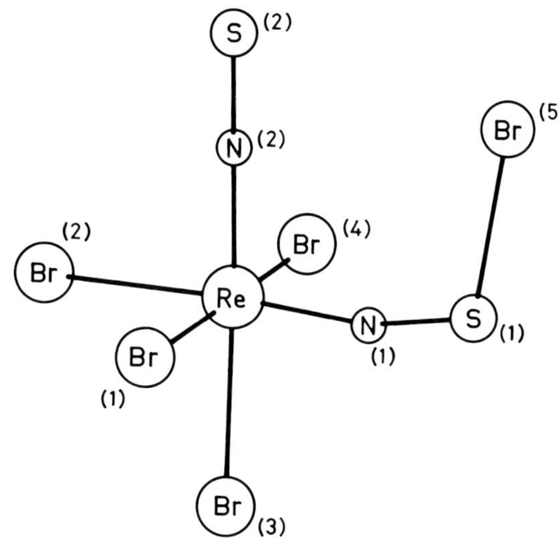

Fig. 10. Molecular structure of the anion in $((C_6H_5)_4P^+)_2[Re(NS)(NSBr)Br_4]^{2-}$ [43].

Table 16
Bond Lengths and Bond Angles of the Anion in $((C_6H_5)_4P^+)_2$ $[Re(NS)(NSBr)Br_4]^{2-}$ [43].

bond length	in pm	bond angle	in °	bond angle	in °
Re–Br(1)	254.3(2)	Br(1)–Re–Br(2)	87.4(1)	Br(3)–Re–Br(4)	89.8(1)
Re–Br(2)	258.7(2)	Br(1)–Re–Br(3)	87.0(1)	Br(3)–Re–N(1)	86.9(4)
Re–Br(3)	259.2(2)	Br(1)–Re–Br(4)	174.0(1)	Br(3)–Re–N(2)	177.8(3)
Re–Br(4)	253.9(2)	Br(1)–Re–N(1)	93.3(4)	Br(4)–Re–N(1)	91.5(4)
Re–N(1)	186(2)	Br(1)–Re–N(2)	91.3(3)	Br(4)–Re–N(2)	91.7(3)
Re–N(2)	186(2)	Br(2)–Re–Br(3)	88.0(1)	N(1)–Re–N(2)	94.6(5)
N(1)–S(1)	148(2)	Br(2)–Re–Br(4)	87.4(1)	N(1)–S(1)–Br(5)	110.6(6)
N(2)–S(2)	138(2)	Br(2)–Re–N(1)	174.8(4)	Re–N(1)–S(1)	164.6(9)
S(1)–Br(5)	238.6(6)	Br(2)–Re–N(2)	90.5(4)	Re–N(2)–S(2)	178.2(8)

IR spectrum (Nujol) without absorption frequencies of the cation: ν (in cm^{-1}) = 1225 m, ν(NS) of ReNS; 1182 s, ν(NS) of ReNSBr; 580 w, ν(ReN); 492 vw, ν(SBr); 470 w, δ(ReNS); 245 m, 220 s, 202 w, all ν(ReBr) [43].

References on p. 86

$[Re(NS)(CO)_5]^{2+}(AsF_6^-)_2$

The complex was prepared in almost quantitative yield by mixing SO_2 solutions of $[Re(CO)_5(SO_2)]^+AsF_6^-$ and $SN^+AsF_6^-$ in the mole ratio 1:1 at −60°C. After warming to room temperature the reaction solution was stirred for 3 d. The solid was obtained after evaporation of the solvent [15].

The compound was also obtained in quantitative yield by stirring 1.596 mmol of $[Re(CO)_5(N{\equiv}SF)]^+AsF_6^-$ and 1.666 mmol of AsF_5 in 10 mL SO_2 at −30°C for 1 h [21].

The complex salt formed in about 40% yield when $[Re(CO)_5(N{\equiv}SF)]^+AsF_6^-$ was reacted with SnI_4 in the mole ratio 4:1 in liquid SO_2 [22].

The light red solid decomposes at 136°C [15] or 130°C [21]. IR (Nujol): ν (in cm^{-1}) = 2219 m A_1, 2191 w B, 2157 s E, 2073 m A_1, all ν(CO); 1371 ν(NS); 546 m, 524 m, 491 w, 318 w, all δ(ReCO) and ν(ReC). The complex is insoluble in SO_2 [21].

The reactions with equimolar amounts of CsCl and CsBr in liquid SO_2 at −15°C give $[Re(CO)_5(N{\equiv}SCl)]^+AsF_6^-$ in almost quantitative yield and $[Re(CO)_5(N{\equiv}SBr)]^+AsF_6^-$ in 65% yield [22].

$[Re(NS)(CO)_5]^{2+}(SbF_6^-)_2$

The complex formed in low yield by reaction of $Re(CO)_5Br$ with equimolar amounts of $SN^+SbF_6^-$ in liquid SO_2 at −30°C.

IR (solvent?): ν (in cm^{-1}) = 2218 m A_1, 2191 w B, 2152 s E, 2067 m A_1, all ν(CO); 545 m, 524 m, 491 w, 320 w, all δ(ReCO) and ν(ReC); 1366 m ν(NS) [21].

1.4.8 Thionitrosyl Complexes of Fe, Co, and Ni

$[(\eta^5\text{-}C_5H_5)Fe(NS)(CO)_2]^{2+}(AsF_6^-)_2$

The complex formed in almost quantitative yield by mixing an SO_2 solution of $[(\eta^5\text{-}C_5H_5)Fe(CO)_2(SO_2)]^+AsF_6^-$ and $SN^+AsF_6^-$ in the mole ratio 1:1 at −60°C. The reaction solution was stirred for 12 h after warming to room temperature.

The rose-colored solid decomposes at 172°C.

IR (Nujol-Kel-F trituration): ν (in cm^{-1}) = 2178, ν(CO); 2165, ν(CO); 1388, ν(NS). The complex is difficultly soluble in SO_2 [15].

$Co(NS)Cl_2(P(OC_6H_5)_3)_2$

The formation of this complex as an intermediate by reaction of $CoH(P(OC_6H_5)_3)_4$ with $S_3N_3Cl_3$ in the mole ratio 1:1 in THF at 30°C under pure and dry Ar was observed by IR spectroscopy (ν(NS) = 1130 cm^{-1}). The reaction proceeds according to the equation $CoH(P(OC_6H_5)_3)_4 + S_3N_3Cl_3 \rightarrow Co(NS)Cl_2(P(OC_6H_5)_3)_2 + 2P(OC_6H_5)_3 + H^+ + Cl^- + 2SN$.

The air-sensitive intermediate in the THF solution is oxidized in the presence of air or O_2 to give $Co(NSO)Cl_2(P(OC_6H_5)_3)_2$ [23].

$Ni(NS)Cl(P(C_6H_5)_3)_2$

The complex formed by reaction of $NiCl_2(P(C_6H_5)_3)_2$ with $S_3N_3Cl_3$ in THF in the presence of $Na^+BH_4^-$.

IR (solvent?): 1186 cm^{-1}, ν(NS) [24].

References on p. 86

1.4.9 Thionitrosyl Complexes of Ru

$Ru(NS)Cl_3$

The compound was obtained by adding dropwise a solution of 11 mmol of $S_3N_3Cl_3$ in THF to a stirred solution of 10 mmol of $RuCl_3$ in THF under N_2 at room temperature. The reaction mixture was stirred for 5 h, filtered, and partially dried under reduced pressure. The brown residue was washed five times with n-C_6H_{14} and dried.

IR (Nujol): $\nu = 1290\ cm^{-1}$, ν(NS). Addition of $(C_6H_5)_4P^+Cl^-$ in H_2O to an aqueous solution of $Ru(NS)Cl_3$ gave a brown precipitate of $((C_6H_5)_4P^+)_2[Ru(NS)Cl_5]^{2-}$ and $(C_6H_5)_4P^+$-$[Ru(NS)Cl_4(H_2O)]^-$ (see below) [25].

$(C_6H_5)_4P^+[Ru(NS)Cl_4(H_2O)]^-$

The complex was produced by adding 3 g of $(C_6H_5)_4P^+Cl^-$ in 15 mL H_2O to a solution of 1 g of $Ru(NS)Cl_3$ in 50 mL H_2O. The brown precipitate which separated is a mixture of $((C_6H_5)_4P^+)_2[Ru(NS)Cl_5]^{2-}$ (90%) and $(C_6H_5)_4P^+[Ru(NS)Cl_4(H_2O)]^-$. It was filtered, dissolved in H_2O–CH_3OH (1:1), and allowed to stand at room temperature for 10 d. Brown crystals of $(C_6H_5)_4P^+[Ru(NS)Cl_4(H_2O)]^-$ deposited on the wall of the vessel.

The crystal and molecular structure was obtained by X-ray diffraction. The crystals are triclinic, space group $P\bar{1}$–C_i^1 (No. 2), with a = 10.173(5), b = 11.756(4), c = 12.793(10) Å, $\alpha = 66.24(6)°$, $\beta = 78.89(6)°$, $\gamma = 72.58(4)°$, V = 1332(1) Å^3, Z = 2, $D_x = 1.612\ g/cm^3$. R = 0.040 for 3243 independent reflections.

The geometry of the dimer anion formed by two monomers joined by hydrogen bonding is shown in **Fig. 11**; the bond lengths and bond angles are given in Table 17, p. 72.

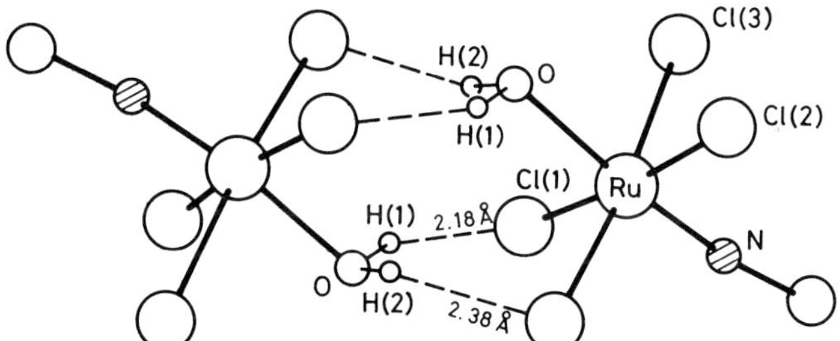

Fig. 11. Molecular structure of the hydrogen-bonded dimer anion in $(C_6H_5)_4P^+[Ru(NS)Cl_4(H_2O)]^-$ [25].

The Ru atom is octahedrally coordinated by four equatorial Cl atoms, an approximately linear NS group bonding via the N atom, and an H_2O molecule. Two $[Ru(NS)Cl_4(H_2O)]^-$ groups form centrosymmetric dimers via O–H ··· Cl hydrogen bonds. The dimers are interconnected by Cl ··· S van der Waals contacts (S ··· Cl(2) 3.240(2) Å and S ··· Cl(3) 3.300(2) Å). The N–S bond distance of 1.504(4) Å is similar to that in $Os(NS)Cl_3(P(C_6H_5)_3)_2$, 1.504(11) Å, see p. 80, but longer than that in $SN^+Sb_2F_{11}^-$, 1.42(1) Å, see p. 46. The longer NS distances in $[Ru(NS)Cl_4(H_2O)]^-$ in comparison to the SN^+ cation might be explained by considerable π-back bonding from Ru to NS. Atomic positional parameters are given in the paper [25].

References on p. 86

Table 17

Bond Lengths and Bond Angles of the Anion in $(C_6H_5)_4P^+[Ru(NS)Cl_4(H_2O)]^-$ [25].

bond length	in Å	bond angle	in °	bond angle	in °
Ru–Cl(1)	2.386(1)	Cl(1)–Ru–Cl(2)	172.76(5)	Cl(3)–Ru–Cl(4)	173.06(5)
Ru–Cl(2)	2.367(1)	Cl(1)–Ru–Cl(3)	90.38(5)	Cl(3)–Ru–O	85.25(9)
Ru–Cl(3)	2.362(1)	Cl(1)–Ru–Cl(4)	89.88(4)	Cl(3)–Ru–N	89.88(13)
Ru–Cl(4)	2.376(1)	Cl(1)–Ru–O	86.95(8)	Cl(4)–Ru–O	87.84(9)
Ru–N	1.729(4)	Cl(1)–Ru–N	94.24(12)	Cl(4)–Ru–N	97.02(13)
Ru–O	2.112(3)	Cl(2)–Ru–Cl(3)	89.84(4)	O–Ru–N	175.00(15)
N–S	1.504(4)	Cl(2)–Ru–Cl(4)	89.03(4)	Ru–N–S	170.9(3)
O–H(1)	0.99	Cl(2)–Ru–O	85.86(8)		
O–H(2)	0.86	Cl(2)–Ru–N	93.00(12)		

IR (KBr): ν (in cm^{-1}) = 1318, ν(NS); 305, ν(Ru–Cl) [25].

$((C_6H_5)_4P^+)_2[Ru(NS)Cl_4]_2^{2-}\cdot 4\,CH_2Cl_2$

$((C_6H_5)_4P^+)_2[Ru(NS)Cl_4]_2^{2-}$ was prepared in quantitative yield by vacuum pyrolysis of $(C_6H_5)_4P^+[RuCl_4(NO)(NSCl)]^-$ at 200 to 220°C/10^{-4} Torr over a period of about 8 h. Red crystals with four molecules of CH_2Cl_2 crystallized from a saturated solution in CH_2Cl_2 cooled to 5°C. Warming the crystals in the vacuum caused liberation of CH_2Cl_2.

The crystal and molecular structure of the title compound was determined by X-ray diffraction analysis. The complex crystallizes in the triclinic space group $P\bar{1}$–C_i^1 (No. 2), with a = 952.9(1), b = 1380.0(2), c = 1458.5(2) pm; α = 112.91(1)°, β = 106.67(1)°, γ = 92.61(1)°; V = 1665×10^{-30} m^3, Z = 1. D_x = 1.59 g/cm^3. R = 0.053 for 3760 observed reflections. The molecular structure of the $[Ru(NS)Cl_4]_2^{2-}$ ion is shown in **Fig. 12**. Bond lengths and bond angles are compiled in Table 18.

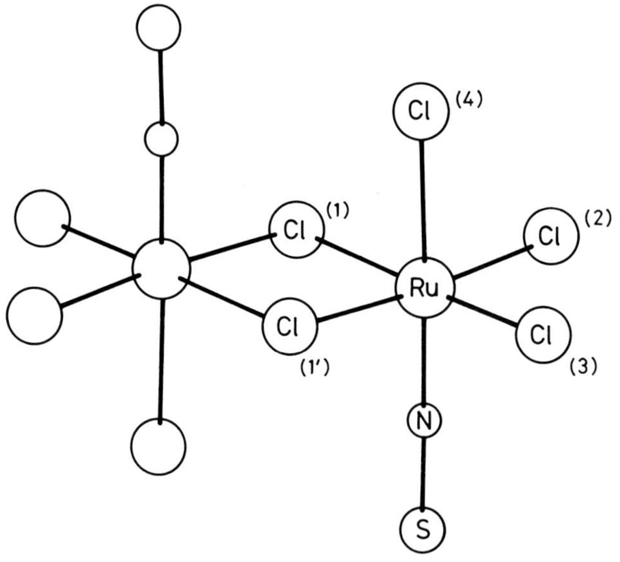

Fig. 12. Molecular structure of the anion in $((C_6H_5)_4P^+)_2[Ru(NS)Cl_4]_2^{2-}\cdot 4\,CH_2Cl_2$ [44].

Atomic coordinates and parameters of the thermic vibration are given in the paper [44]. The dimeric centrosymmetric $[Ru(NS)Cl_4]_2^{2-}$ ion has chloro bridges and NS ligands in axial positions. The RuNS bond angle of 177° and the bond lengths Ru–N of 175 pm and N–S of 147 pm indicate double bonds in the Ru=N=S group.

Table 18

Bond Lengths and Bond Angles of the Anion in $((C_6H_5)_4P^+)_2[Ru(NS)Cl_4]_2^{2-} \cdot 4CH_2Cl_2$ [44].

bond length	in pm	bond angle	in °	bond angle	in °
Ru–Cl(1)	241.5(2)	Cl(1)–Ru–Cl(1′)	85.1(1)	Cl(2)–Ru–Cl(4)	90.1(1)
Ru–Cl(1′)	241.8(2)	Cl(1)–Ru–Cl(2)	91.5(1)	Cl(3)–Ru–Cl(4)	89.2(1)
Ru–Cl(2)	235.2(2)	Cl(1)–Ru–Cl(3)	175.0(1)	N–Ru–Cl(1)	92.0(2)
Ru–Cl(3)	233.8(2)	Cl(1)–Ru–Cl(4)	87.2(1)	N–Ru–Cl(1′)	92.7(2)
Ru–Cl(4)	235.1(2)	Cl(1′)–Ru–Cl(2)	175.6(1)	N–Ru–Cl(2)	89.9(2)
Ru–N	175.2(6)	Cl(1′)–Ru–Cl(3)	91.3(1)	N–Ru–Cl(3)	91.6(2)
N–S	146.6(7)	Cl(1′)–Ru–Cl(4)	87.3(1)	N–Ru–Cl(4)	179.1(2)
Ru · · · Ru	356.1(1)	Cl(2)–Ru–Cl(3)	92.0(1)	Ru–N–S	177.3(5)
				Ru–Cl(1)–Ru	94.9(1)

IR spectrum (Nujol): ν (in cm^{-1}) = 1280 s, ν(NS); 335 sh, 321 vs, 304 s, $(Ru–Cl_{terminal})$; 270, $\nu(Ru–Cl_{bridge})$ [44].

$(NH_4^+)_2[Ru(NS)Cl_5]^{2-}$

The salt formed as a by-product in variable amount by the reaction of $S_3N_3Cl_3$ with $RuCl_3$ in THF at room temperature under N_2. The main product of this reaction is $Ru(NS)Cl_3$ [25].

$((C_6H_5)_4P^+)_2[Ru(NS)Cl_5]^{2-}$

The complex was obtained by adding $(C_6H_5)_4P^+Cl^-$ in H_2O to an aqueous solution of $Ru(NS)Cl_3$. It precipitated together with $(C_6H_5)_4P^+[Ru(NS)Cl_4(H_2O)]^-$ as a brown solid.

IR (KBr): 1300 cm^{-1}, ν(NS).

On hydrolysis, substitution occurs immediately at room temperature and in trans position with respect to the NS group to give $[Ru(NS)Cl_4(H_2O)]^-$ [25].

$Ru(NS)Cl_3(P(C_6H_5)_3)_2$

The compound was prepared in 46% yield by adding a THF solution of 0.17 mmol $S_3N_3Cl_3$ dropwise to a solution of 0.4 mmol of $RuCl_2(P(C_6H_5)_3)_3$ in THF at room temperature. The mixture was stirred for a further 5 h. After evaporation, the solid was recrystallized from CH_2Cl_2–CH_3OH (1:3) [26]. This preparation was also carried out by adding a THF solution of ca. 4 mmol of $S_3N_3Cl_3$ to a solution of 0.3 mmol of $RuCl_2(P(C_6H_5)_3)_3$ in CH_2Cl_2 followed by vigorous stirring for 0.5 h. Crystals were obtained by treatment of the resulting solution with CH_3OH [27, 28]. The complex was also obtained by reaction of an ethanolic solution containing $RuCl_3 \cdot H_2O$ and $P(C_6H_5)_3$ with a saturated solution of $S_3N_3Cl_3$ in THF at room temperature for several hours. The separated crystals were recrystallized from CH_2ClCH_2Cl–C_6H_{14} (1:2) [29].

The crystals are monoclinic, space group I2/a–C_{2h}^6 (No. 15), with a = 1613.5(15), b = 941.8(4), c = 2249.3(15) pm, β = 102.94(3)°. V = 3.331 nm^3, Z = 4. D_x = 1.55 g/cm^3. Conventional setting: C2/c, a = 2457.1(15), b = 941.8(4), c = 2249.3(15) pm, β = 140.21(3)°. The complex is isostructural with $Os(NS)Cl_3(P(C_6H_5)_3)_2$ [26].

References on p. 86

The reddish brown [27], brown [29], orange-brown [26], or shiny brown [28] crystals melt at 197 to 198 [27], 220 to 222 [29], and 192 to 194°C [26].

^{31}P NMR (solvent?/external 85% aqueous H_3PO_4): $\delta = 13.3$ ppm [28].

IR (KBr): ν (in cm^{-1}) = 1316, ν(NS); 334, 304, all ν(Ru–Cl) [28]. 1310, ν(NS) [27]; 1297, ν(NS) [29].

UV (CH_2Cl_2): $\lambda_{max} = 293$, 233 nm [27].

The diamagnetic complex is air-stable and soluble in CH_2Cl_2, CH_2ClCH_2Cl, and CH_3COCH_3 [27, 29]. UV photolysis in CH_2Cl_2 solution in the presence of O_2 proceeds according to the equation sequence $Ru(NS)Cl_3(P(C_6H_5)_3)_2 \rightleftharpoons Ru(NS)Cl_3(P(C_6H_5)_3) + P(C_6H_5)_3$, $P(C_6H_5)_3 + 1/2\ O_2 \rightarrow OP(C_6H_5)_3$.

When the complex is refluxed with $As(C_6H_5)_3$ (mole ratio 1:5) in $CHCl_3$ for 8 h substitution occurs producing $Ru(NS)Cl_3(P(C_6H_5)_3)(As(C_6H_5)_3)$. All efforts to replace $P(C_6H_5)_3$ by $Sb(C_6H_5)_3$ under these experimental conditions failed.

Passing N_2O_3 gas through a CH_2Cl_2 solution of the complex for 0.5 h, adding CH_3OH and an excess of $P(C_6H_5)_3$, and heating the solution on a water bath for about 10 min produces $Ru(NO)Cl_2(NO_3)(P(C_6H_5)_3)_2$.

Reactions with NOCl in CH_2Cl_2 in the presence of excess $P(C_6H_5)_3$, $As(C_6H_5)_3$, $Sb(C_6H_5)_3$, and pyridine, C_5H_5N, under refluxing conditions gives $Ru(NO)Cl_3(P(C_6H_5)_3)L$ with $L = P(C_6H_5)_3$, $As(C_6H_5)_3$, $Sb(C_6H_5)_3$, and NC_5H_5, respectively. Treatment with excess NOBr in CH_2Cl_2 in the presence of excess $P(C_6H_5)_3$ on heating on a water bath yields $Ru(NO)BrCl_2(P(C_6H_5)_3)_2$. The reaction with $NOBr_3$ under the same conditions yields $Ru(NO)Br_2Cl(P(C_6H_5)_3)_2$ [27].

$Ru(NS)Cl_3(As(C_6H_5)_3)_2$

The complex was prepared by adding a THF solution of ca. 4 mmol $S_3N_3Cl_3$ to a solution of 0.3 mmol of $RuCl_3(As(C_6H_5)_3)_2 \cdot CH_3OH$ in CH_2Cl_2 followed by vigorous stirring for 0.5 h. After addition of CH_3OH, CH_2Cl_2 was removed by heating on a water bath and $As(C_6H_5)_3$ was added, whereby the complex precipitates. It was recrystallized from boiling CH_2Cl_2 on addition of hot CH_3OH [27]. The compound was also obtained by addition of an ethanolic solution containing $RuCl_3 \cdot H_2O$ and $As(C_6H_5)_3$ to a saturated solution of $S_3N_3Cl_3$ in THF and stirring the mixture at room temperature. The crystals were washed with H_2O, C_2H_5OH, $(C_2H_5)_2O$ and dried in vacuum [29].

The reddish brown crystals melt at >260°C [27] and >280°C [29].

IR (KBr): ν (in cm^{-1}) = 1305, ν(NS) [27]; 1295, ν(NS) [29].

The diamagnetic complex is air-stable and soluble in CH_2Cl_2, CH_2ClCH_2Cl, and CH_3COCH_3 [27, 29].

Passing N_2O_3 gas through a CH_2Cl_2 solution of the complex for 0.5 h, adding CH_3OH and an excess of $As(C_6H_5)_3$, and heating the solution on a water bath for about 10 min gives $Ru(NO)Cl_2(NO_3)(As(C_6H_5)_3)_2$.

Irradiation of a CH_2Cl_2 solution containing $Ru(NS)Cl_3(As(C_6H_5)_3)_2$ and $Sb(C_6H_5)_3$ in the mole ratio 1:6 with a 376 nm UV source for 1 h and addition of CH_3OH to the resulting solution gives crystals of $Ru(NS)Cl_3(As(C_6H_5)_3)(Sb(C_6H_5)_3)$ and $(C_6H_5)_3As{=}O$. Reactions with NOCl in CH_2Cl_2 in the presence of excess $P(C_6H_5)_3$, $As(C_6H_5)_3$, $Sb(C_6H_5)_3$, and C_5H_5N under refluxing conditions gives $Ru(NO)Cl_3(As(C_6H_5)_3)L$ with $L = P(C_6H_5)_3$, $As(C_6H_5)_3$, $Sb(C_6H_5)_3$, and NC_5H_5, respectively. When the complex is treated with excess NOBr and $NOBr_3$ in CH_2Cl_2 in the presence of excess $As(C_6H_5)_3$ and the resulting solution is heated on a water bath, $Ru(NO)BrCl_2(As(C_6H_5)_3)_2$ and $Ru(NO)Br_2Cl(As(C_6H_5)_3)_2$, respectively, are formed [27].

References on p. 86

$Ru(NS)Cl_3(P(C_6H_5)_3)(As(C_6H_5)_3)$

The complex formed by refluxing a mixture of $Ru(NS)Cl_3(P(C_6H_5)_3)_2$ and $As(C_6H_5)_3$ (mole ratio 1:5) in $CHCl_3$ for 8 h. Concentration of the solution and addition of hot CH_3OH gave crystals which were recrystallized from CH_2Cl_2.

The reddish brown crystals melt at 247 to 248°C. IR (KBr): $\nu = 1300\ cm^{-1}$, $\nu(NS)$ [27].

$Ru(NS)Cl_3(As(C_6H_5)_3)(Sb(C_6H_5)_3)$

The compound was obtained by reaction of $RuCl_3(As(C_6H_5)_3)_2 \cdot CH_3OH$ with $S_3N_3Cl_3$ in the mole ratio ca. 1:10 in CH_2Cl_2 at ambient temperature and by addition of $Sb(C_6H_5)_3$ to the resulting solution. The precipitate was recrystallized from CH_2Cl_2 solution on addition of hot CH_3OH.

The complex formed also by irradiation of a CH_2Cl_2 solution of $Ru(NS)Cl_3(As(C_6H_5)_3)_2$ and $Sb(C_6H_5)_3$ (mole ratio 1:6) with a 376 nm UV source for 1 h. Crystals appeared on addition of CH_3OH.

The reddish brown crystals melt at 218 to 220°C. IR (KBr): $\nu = 1299\ cm^{-1}$, $\nu(NS)$ [27].

$Ru(NS)Br_2Cl(P(C_6H_5)_3)_2$

The complex was obtained by the reaction of $RuBr_2(P(C_6H_5)_3)_3$ with $S_3N_3Cl_3$ (mole ratio ca. 1:10) in CH_2Cl_2 at ambient temperature. The procedure is the same as those given for $Ru(NS)Cl_3(P(C_6H_5)_3)_2$.

The reddish brown crystals melt at 195 to 196°C. IR (KBr): $\nu = 1308\ cm^{-1}$, $\nu(NS)$.

Passing N_2O_3 gas through a CH_2Cl_2 solution of the complex and addition of excess $P(C_6H_5)_3$ gave $Ru(NO)BrCl(NO_3)(P(C_6H_5)_3)_2$. Reactions with NOCl in CH_2Cl_2 in the presence of $P(C_6H_5)_3$, $As(C_6H_5)_3$, $Sb(C_6H_5)_3$, and NC_5H_5 yielded $Ru(NO)Br_2Cl(P(C_6H_5)_3)L$ with $L = P(C_6H_5)_3$, $As(C_6H_5)_3$, $Sb(C_6H_5)_3$, and NC_5H_5, respectively. Treating with excess NOBr and $NOBr_3$ in CH_2Cl_2 in the presence of excess $P(C_6H_5)_3$ produced $Ru(NO)Br_3(P(C_6H_5)_3)_2$ [27].

$Ru(NS)Br_2Cl(As(C_6H_5)_3)_2$

The complex formed by reaction of $RuBr_3(As(C_6H_5)_3)_2 \cdot CH_3OH$ with $S_3N_3Cl_3$ (mole ratio ca. 1:10) in CH_2Cl_2 at ambient temperature. The procedure is analogous to those given for $Ru(NS)Cl_3(As(C_6H_5)_3)_2$.

The reddish brown solid melts at >260°C. IR (KBr): $\nu = 1300\ cm^{-1}$, $\nu(NS)$.

Reaction with N_2O_3 gas in CH_2Cl_2 in the presence of excess $As(C_6H_5)_3$ yielded $Ru(NO)BrCl(NO_3)(As(C_6H_5)_3)_2$. When the complex was reacted with NOCl in CH_2Cl_2 in the presence of $As(C_6H_5)_3$, $P(C_6H_5)_3$, $Sb(C_6H_5)_3$, and NC_5H_5 the following compounds formed: $Ru(NO)Br_2Cl(As(C_6H_5)_3)L$ with $L = As(C_6H_5)_3$, $P(C_6H_5)_3$, $Sb(C_6H_5)_3$, and NC_5H_5, respectively. Treatment with excess NOBr and $NOBr_3$ in CH_2Cl_2 in the presence of excess $As(C_6H_5)_3$ yielded $Ru(NO)Br_3(As(C_6H_5)_3)_2$ [27].

$Ru(NS)_2Cl_4$

The compound was prepared in 97% yield by heating to reflux a mixture of equimolar amounts of $RuCl_3$ which was obtained by dehydration of $RuCl_3 \cdot \sim 3H_2O$ with $SOCl_2$ and $S_3N_3Cl_3$ in $SOCl_2$ for 3 h. The reaction proceeds according to the equation: $RuCl_3 + 2/3\ S_3N_3Cl_3 \rightarrow Ru(NS)_2Cl_4 + 1/2\ Cl_2$.

References on p. 86

The black-brown powder is sensitive to humidity and highly soluble in $SOCl_2$, CH_2Cl_2, and CCl_4.

IR (Nujol): ν (in cm^{-1}) = 1325 vs, ν(NS); 1228 s, ν(NS); 472 w; 435 m; 385 w; 322 s, 290 sh, ν(RuCl); 280 sh, δ(RuNS).

The presence of two absorptions of the NS stretching vibration in the IR spectrum shows the cis position of both NS groups in the complex according to C_{2v} symmetry.

Reaction of a suspension of $Ru(NS)_2Cl_4$ in CH_2Cl_2 with equimolar amounts of $(C_6H_5)_3CH_3P^+Cl^-$ at room temperature gives $(C_6H_5)_3CH_3P^+[Ru((NS)_2Cl)Cl_4]^-$.

$Ru(NS)Cl_4$ reacts with equimolar amounts of $(C_6H_5)_4P^+Br^-$ in CH_2Cl_2 at room temperature for 24 h according to the equation 2 $Ru(NS)_2Cl_4$ + 2 $(C_6H_5)_4P^+Br^- \rightarrow ((C_6H_5)_4P^+)_2[(Ru(NS)Cl_4)_2$-$(\mu$-$S_2N_2)]^{2-}$ + Br_2. In the centrosymmetric anion the two Ru atoms are linked via the N atoms of a planar S_2N_2 ring [42].

$(C_6H_5)_3CH_3P^+[Ru((NS)_2Cl)Cl_4]^-$

The complex was produced in 90% yield by reaction of $Ru(NS)_2Cl_4$ with equimolar amounts of $(C_6H_5)_3CH_3P^+Cl^-$ in CH_2Cl_2 for about 0.5 h.

The red-brown microcrystalline powder is soluble in CH_2Cl_2 and insoluble in CCl_4.

IR spectrum (Nujol), not including the absorption frequencies of the cation: ν (in cm^{-1}) = 1320 s, ν(NS); 1255 s, ν(NS); 479 vw; 345 m, 332 vw, 320 m, 308 s, 292 m, all ν(RuCl); 277 m to s, δ(RuNS).

The IR spectrum is analogous to that of the corresponding Os complex $(C_6H_5)_4As^+$-$[Os((NS)_2Cl)Cl_4]^-$ whose structure was determined by X-ray diffraction; see p. 85.

It is supposed that relatively weak S · · · Cl bonds to both NS ligands are present as in the Os complex [42].

1.4.10 Thionitrosyl Complexes of Rh

Although a number of thionitrosyl complexes of Rh have been claimed [30 to 36], their existence is doubtful [28]. Only the microanalyses, melting points, and IR absorption frequencies at 1118 ± 3 cm^{-1} and 840 cm^{-1} erroneously assigned to ν(NS) terminal and ν(NS) bridge, respectively, are reported to support the existence of thionitrosyl complexes of Rh [30, 32 to 36]. The preparation procedures and physical properties of the claimed complexes are given in Table 19 and Table 20, p. 78.

Table 19
Preparation Procedures for Claimed Thionitrosyl Complexes of Rh.

compound	starting compounds	solvent	temp., time	Ref.
$Rh(NS)Cl_2(P(C_6H_5)_3)_2$	$RhCl_3 \cdot 3H_2O$, $S_3N_3Cl_3$, $P(C_6H_5)_3$	C_2H_5OH–THF	elevated	[35]
	$RhCl(P(C_6H_5)_3)_3$, $S_3N_3Cl_3$	THF	r.t.	[35]
	$(Rh(NS)Cl_2(P(C_6H_5)_3))_2$, $P(C_6H_5)_3$	CH_2Cl_2	reflux, 1 to 2 h	[30, 32, 34]

Table 19 (continued)

compound	starting compounds	solvent	temp., time	Ref.
$Rh(NS)Cl_2(P(C_6H_5)_3)$-$(As(C_6H_5)_3)$	$(Rh(NS)Cl_2(P(C_6H_5)_3))_2$, $As(C_6H_5)_3$	CH_2Cl_2	reflux, 3 to 4 h	[32, 33, 34]
	$(Rh(NS)Cl_2(As(C_6H_5)_3))_2$, $P(C_6H_5)_3$	CH_2Cl_2	reflux, 3 h	[19]
$Rh(NS)Cl_2(As(C_6H_5)_3)_2$	$RhCl_3 \cdot 3H_2O$, $S_3N_3Cl_3$, $As(C_6H_5)_3$	C_2H_5OH–THF	elevated	[26]
	$(Rh(NS)Cl_2(As(C_6H_5)_3))_2$, $As(C_6H_5)_3$	CH_2Cl_2	reflux, 3 h	[34]
$Rh(NS)ClBr(P(C_6H_5)_3)_2$	$RhBr(P(C_6H_5)_3)_3$, $S_3N_3Cl_3$	THF	r.t.	[35]
$(Rh(NS)Cl_2(P(C_6H_5)_3))_2$	$RhH(P(C_6H_5)_3)_4$, $S_3N_3Cl_3$	THF	5 or 25°C, 2 h	[30, 32]
	$Rh(NO)(P(C_6H_5)_3)_3$, $S_3N_3Cl_3$	CCl_4–$CHCl_3$	r.t., 1 h	[34]
	$Rh(NO)Cl_2(P(C_6H_5)_3)_2$, $S_3N_3Cl_3$	THF or CCl_4–$CHCl_3$	10°C, 1 h	[33]
	$Rh(NO)Br_2(P(C_6H_5)_3)_2$, $S_3N_3Cl_3$	THF	5°C	[30]
$(Rh(NS)Cl_2(As(C_6H_5)_3))_2$	$Rh(NO)Cl_2(As(C_6H_5)_3)_2$, $S_3N_3Cl_3$	CCl_4–$CHCl_3$	r.t.	[34]
$(Rh(NS)Br_2(P(C_6H_5)_3))_2$	$Rh(NO)Br_2(P(C_6H_5)_3)_2$, $S_3N_3Cl_3$	THF or CCl_4–$CHCl_3$	10°C, 1 h	[33]
$Rh(NS)(CO)Cl_2(P(C_6H_5)_3)_2$	$Rh(CO)Cl(P(C_6H_5)_3)_2$, $S_3N_3Cl_3$	THF	r.t., ~1 h	[35, 36]
	$(Rh(NS)(CO)Cl_2(P(C_6H_5)_3))_2$, $P(C_6H_5)_3$	CH_2Cl_2	r.t., 2 h	[36]
$Rh(NS)(CO)Cl_2$-$(As(C_6H_5)_3)_2$	$Rh(CO)Cl(As(C_6H_5)_3)_2$, $S_3N_3Cl_3$	THF	r.t., ~1.5 h	[35, 36]
	$(Rh(NS)(CO)Cl_2(As(C_6H_5)_3))_2$, $As(C_6H_5)_3$	CH_2Cl_2	reflux, 3 h	[36]
	$(Rh(NS)(CO)Cl_2(P(C_6H_5)_3))_2$, $As(C_6H_5)_3$	CH_2Cl_2	reflux, 3 h	[36]
$(Rh(NS)(CO)Cl_2$-$(P(C_6H_5)_3))_2$	$RhH(CO)(P(C_6H_5)_3)_3$, $S_3N_3Cl_3$	CCl_4–$CHCl_3$	r.t., 1 h	[36]
	$Rh(CO)Cl(P(C_6H_5)_3)_2$, $S_3N_3Cl_3$	THF	r.t., ~1 h	[36]
$(Rh(NS)(CO)Cl_2$-$(As(C_6H_5)_3))_2$	$Rh(CO)Cl(As(C_6H_5)_3)_2$, $S_3N_3Cl_3$	THF	r.t., ~1.5 h	[36]
$Rh(NS)(CO)(P(C_6H_5)_3)_2$	$(Rh(NS)(CO)Cl_2(P(C_6H_5)_3))_2$, $P(C_6H_5)_3$	CH_2Cl_2	reflux, ~1 h	[36]

Severe objections were made to the existence of thionitrosyl complexes of Rh on the basis of preparative work in which the various products were characterized by X-ray crystallography, ^{31}P NMR, and IR spectroscopy [28]. Reaction between $RhCl(P(C_6H_5)_3)_3$ and $S_3N_3Cl_3$ in THF was reported to form $Rh(NS)Cl_2(P(C_6H_5)_3)_2$ [35] but other workers [28] were unable to obtain any compounds that contain phosphorus bound to Rh, as judged by the absence of Rh–P couplings in the ^{31}P NMR spectra. Thus, when the reaction is carried out in a 1:1 mole ratio

References on p. 86

$(C_6H_5)_3PS$ (25%), $(C_6H_5)_3PO$ (10%), $[(C_6H_5)_3PNH_2]^+Cl^-$ (35%), and $(C_6H_5)_3PNH$ (25%) account for the majority of the phosphorus observed, a situation which is also evident when an excess of $S_3N_3Cl_3$ is used or when the reaction is carried out in the presence of Zn [28]. In addition, reaction of $Rh(NO)(P(C_6H_5)_3)_3$ with $S_3N_3Cl_3$ is reported as resulting in an NS bridged species $Rh(NS)Cl_2(P(C_6H_5)_3)_2$, assuming the IR band at 840 cm^{-1} as ν(NS) bridge [34]. But other workers obtained only $(C_6H_5)_3PO$ and $[(C_6H_5)_3PNH_2]^+Cl^-$ from this reaction [28].

Table 20

Physical Properties for Claimed Thionitrosyl Complexes of Rh.

compound	m.p. in °C	color	IR (KBr) ν in cm^{-1}	Ref.
$Rh(NS)Cl_2(P(C_6H_5)_3)_2$	160 to 165 [35], 265 [35]	brown	1120 ν(NS) [32, 35] 335 ν(RhCl) [32]	[32, 35]
$Rh(NS)Cl_2(P(C_6H_5)_3)(As(C_6H_5)_3)$	165 [32]	red-brown	1116 to 1120 ν(NS)	[32, 34]
$Rh(NS)Cl_2(As(C_6H_5)_3)_2$	168 [35]	red-brown	1116 to 1120 ν(NS)	[34, 35]
$Rh(NS)ClBr(P(C_6H_5)_3)_2$	160 to 162	brown	1120 ν(NS)	[35]
$(Rh(NS)Cl_2(P(C_6H_5)_3))_2$	215 [32], >210 [33]	brown	840 ν(NS) bridge	[30, 33, 34]
$(Rh(NS)Cl_2(As(C_6H_5)_3))_2$	—	red-brown	840 ν(NS) bridge	[34]
$(Rh(NS)Br_2(P(C_6H_5)_3))_2$	—	brown	840 ν(NS) bridge	[33]
$Rh(NS)(CO)Cl_2(P(C_6H_5)_3)_2$	128 to 130	red-brown	2090 ν(CO), 1118 ν(NS)	[35]
	165	red-brown	2105 ν(CO), 1120 ν(NS), 335 ν(RhCl)	[36]
$Rh(NS)(CO)Cl_2(As(C_6H_5)_3)_2$	159	red-brown	2070 ν(CO), 1115 ν(NS)	[35]
	170	red-brown	2100 ν(CO), 1120 ν(NS), 335 ν(RhCl)	[36]
$(Rh(NS)(CO)Cl_2(P(C_6H_5)_3))_2$	170	green	2110 ν(CO); 1118 ν(NS); 340, 260 ν(RhCl)	[36]
$(Rh(NS)(CO)Cl_2(As(C_6H_5)_3))_2$	176	yellow-green	2110 ν(CO); 1118 ν(NS); 340, 260 ν(RhCl)	[36]
$Rh(NS)(CO)(P(C_6H_5)_3)_2$	145	red-brown	1970 ν(CO), 1122 ν(NS)	[36]

Treatment of $Rh(CO)Cl(P(C_6H_5)_3)_2$ with $S_3N_3Cl_3$ in the mole ratio 3:1 in THF gave only $(C_6H_5)_3PS$ (20%), $[(C_6H_5)_3PNH_2]^+Cl^-$ (5%), $(C_6H_5)_3PNH$ (18%), $Rh(CO)Cl_3(P(C_6H_5)_3)_2$ (40%), and a trace of $(C_6H_5)_3PO$, together with some (10%) unreacted starting material [28]. In the same reaction, formation of the complexes $(Rh(NS)(CO)Cl_2(P(C_6H_5)_3))_2$ and $Rh(NS)(CO)Cl_2(P(C_6H_5)_3)_2$ was claimed and the IR bands at 2110 and 1118 cm^{-1} and 2105 and 1120 cm^{-1}, respectively were interpreted as ν(CO) of $Rh^{III}(CO)$ and ν(NS) [36]. In the view of other workers these IR bands are due to ν(CO) in $Rh(CO)Cl_3(P(C_6H_5)_3)_2$ and the phosphonium band ν(PN) of

References on p. 86

$[(C_6H_5)_3PNH_2]^+Cl^-$ [28]. $[(C_6H_5)_3PNH_2]^+Cl^-$ would coprecipitate from THF, in which it is only moderately soluble, and since no further purifications were carried out in many of the reactions, any final product would thus be contaminated with $[(C_6H_5)_3PNH_2]^+Cl^-$. Sometimes a quite large variation in melting point was found for the same compound, see Table 20, p. 78. These inconsistencies cast doubt on the existence of many of the claimed thionitrosyl complexes [28].

1.4.11 Thionitrosyl Complexes of Os

$Os(NS)Cl_3$

The compound of composition $Os(NS)Cl_3$ formed in 86% yield by adding $S_3N_3Cl_3$ in THF dropwise to a stirred solution of $OsCl_3$ in THF (mole ratio 1:3) at room temperature and stirring for a further 2 h.

The green solid, which is probably a chlorine-bridged dimer or oligomer, exhibits a typical NS stretching frequency at 1290 cm^{-1} in the IR (KBr).

Addition of $(C_6H_5)_4P^+Cl^-$ in H_2O to $Os(NS)Cl_3$ in H_2O (mole ratio 2:1) produced $((C_6H_5)_4P^+)_2$-$[Os(NS)Cl_5]^{2-}$ in 78% yield [37].

$Os(NS)_2Cl_4$

The compound formed in almost quantitative yield by dropping a solution of 2.40 mmol of $GaCl_3$ in 15 mL CH_2Cl_2 into a stirred solution of 2.20 mmol of $(C_6H_5)_4As^+[Os((NS)_2Cl)Cl_4]^-$ in 25 mL CH_2Cl_2. The brown crystalline powder is difficultly soluble and moisture-sensitive.

The IR spectrum (Nujol) is assigned according to symmetry C_{2v}: ν (in cm^{-1}) = 1388 m, ν_s(NS) A_1; 1326 s and 1308 s, ν_{as}(NS) B_1; 385 m, $\nu_s Os(NS)_2$ A_1; 369 m, ν_{as}Os(NS) B_1; 328 sh, 312 vs, 300 sh, and 280 sh, $\nu(OsCl_4)$ 2A_1, B_1, B_2.

Reaction with the equimolar amount of $(C_6H_5)As^+Cl^-$ in CH_2Cl_2 yields $(C_6H_5)_4As^+$-$[Os((NS)_2Cl)Cl_4]^-$ [38].

$Os(NS)Cl_4(POCl_3)$

The compound formed in about 5% yield by stirring a mixture of $OsCl_5$ and $S_3N_3Cl_3$ (mole ratio 3:2) in $POCl_3$–CCl_4 (1:3) for 3 d. After evaporating CCl_4 and excess $POCl_3$ in vacuum a black-brown powder remained. The CH_2Cl_2 soluble part was the $POCl_3$ solvate which was isolated after evaporation of the solvent as a red-violet powder. The complex was not obtained analytically pure. It reacts with almost equimolar amounts of $(C_6H_5)_4As^+Cl^-$ in CH_2Cl_2 according to the equation $Os(NS)Cl_4(POCl_3) + (C_6H_5)_4As^+Cl^- \rightarrow (C_6H_5)_4As^+[Os(NS)Cl_5]^- + POCl_3$ [39].

$Os(NS)Cl_3L^1_2$ and $Os(NS)Br_2ClL^2_2$, $L^1_2 = (P(C_6H_5)_3)_2$, $(P(CH_3)_2C_6H_5)_2$, $(NC_5H_5)_2$, $N_2C_{10}H_8$, and $(As(C_6H_5)_3)_2$; $L^2_2 = N_2C_{10}H_8$ and $(As(C_6H_5)_3)_2$

$Os(NS)Cl_3(P(C_6H_5)_3)_2$

The compound was prepared in 55% yield by dropping a solution of 0.7 mmol $S_3N_3Cl_3$ in 20 mL THF into a stirred solution of 2.0 mmol of $OsCl_2(P(C_6H_5)_3)_3$ in 100 mL THF at room temperature and stirring the mixture for a further 5 h. The resulting solution was evaporated under vacuum and extracted with CH_2Cl_2. On addition of hot CH_3OH to the CH_2Cl_2 solution, the

References on p. 86

complex separated out in the form of shiny green plates and yellowish orange needles. Recrystallization from CH_2Cl_2–CH_3OH (1:2) converted the product entirely to the green form [26]. The complex was also formed in low yield by adding, with stirring, a saturated solution of $S_3N_3Cl_3$ in THF to a mixture of $OsCl_3$ and $P(C_6H_5)_3$ in C_2H_5OH, followed by refluxing for 0.5 h then filtering. Brown crystals separated overnight which were recrystallized from CH_2ClCH_2Cl–C_6H_{14} (1:2) as green crystals [29].

The crystals are monoclinic, space group I2/a–C_{2h}^6 (No. 15), with a = 1615.9(6), b = 943.7(5), c = 2250.6(10) pm; β = 103.02(3)°. V = 3.344 nm^3, Z = 4, D_x = 1.722 g/cm^3. Conventional setting: C2/c, a = 2457.2(11), b = 943.7(5), c = 2250.6(10) pm; β = 140.15(3)°. R = 0.042 for 2076 unique reflections. The molecular structure is shown in **Fig. 13**. Bond lengths and bond angles are compiled in Table 21.

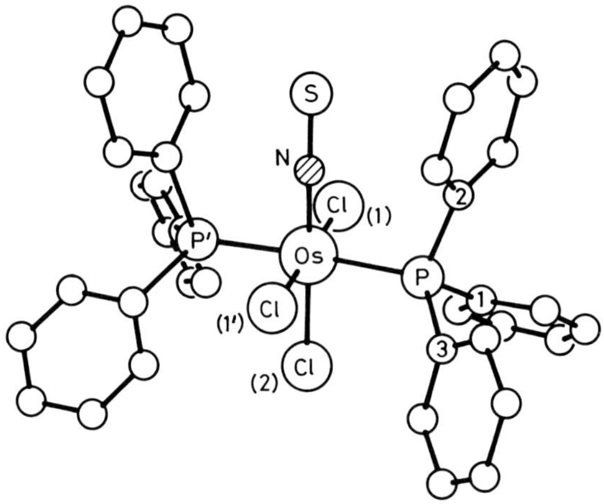

Fig. 13. Molecular structure of $Os(NS)Cl_3(P(C_6H_5)_3)_2$ [26].

Table 21
Selected Bond Lengths and Bond Angles of $Os(NS)Cl_3(P(C_6H_5)_3)_2$ [26].

bond length	in pm	bond angle	in °	bond angle	in °
Os–N	177.9(9)	N–Os–Cl(2)	180.0	N–Os–Cl(1)	89.7(1)
Os–Cl(2)	239.9(3)	N–Os–P	91.3(1)	Cl(1)–Os–Cl(2)	90.3(1)
Os–Cl(1)	238.7(3)	Cl(2)–Os–P	88.7(1)	Cl(1)–Os–P	93.4(1)
Os–P	245.9(2)	Cl(1)–Os–Cl(1′)	179.4(1)	P–Os–P′	177.4(1)
N–S	150.3(10)	P–Os–Cl(1′)	86.6(1)	Os–N–S	180.0(1)
P–C(2)	181.3(8)	Os–P–C(1)	119.0(2)	Os–P–C(2)	107.7(2)
P–C(1)	183.0(9)	C(1)–P–C(2)	105.9(4)	Os–P–C(3)	117.7(3)
P–C(3)	183.5(8)	C(1)–P–C(3)	100.9(4)	C(2)–P–C(3)	104.1(4)

A prime denotes atoms generated by the molecular twofold symmetry axis.

The Os atom is octahedrally coordinated with the phosphines cis to the NS group. A crystallographic twofold axis along the Cl–Os–N–S bonds imposes linearity at nitrogen. The atomic parameters are compiled in the paper [26].

The green crystals melt at >250°C [26] or 145 to 147°C (?) [29].

References on p. 86

IR (KBr): ν (in cm^{-1}) = 1310 and 1290, ν(NS) [26]; 1294, ν(NS) [29].

The complex is air-stable, diamagnetic, and soluble in CH_2ClCH_2Cl, CH_2Cl_2, and CH_3COCH_3 [29].

Reactions with NOCl and NOBr in CH_2Cl_2 gave the corresponding NO complexes $Os(NO)Cl_3(P(C_6H_5)_3)_2$ or $Os(NO)BrCl_2(P(C_6H_5)_3)_2$, respectively [26].

The preparation and physical properties of other complexes of the type $Os(NS)Cl_3L^1_2$ and $Os(NS)Br_2ClL^2_2$ are given in Table 22, p. 82 and Table 23, p. 83.

$(C_6H_5)_4P^+[Os(NS)Cl_4(H_2O)]^-$

The complex formed in 25% yield as green crystals when a solution of 1.0 g of $((C_6H_5)_4P^+)_2[Os(NS)Cl_5]^{2-}$ in 100 mL H_2O–CH_3OH (1:1) was allowed to stand at room temperature for 15 d.

The crystal structure is determined by X-ray diffraction. The crystals are monoclinic, space group $P2_1/c$–C^5_{2h} (No. 14), with a = 1103.3(2), b = 1959.8(3), c = 1254.3(2) pm; β = 96.51(3)°. V = 2.6946 nm^3, Z = 4, D_x = 1.813 g/cm^3. R = 0.029 for 3177 unique reflections. The molecular structure of the anion is shown in **Fig. 14**. Bond lengths and bond angles are given in Table 24, p. 84. The Os atom is octahedrally coordinated with H_2O trans to the NS group. The Os–N–S unit is almost linear. The anions form dimeric units in the crystal, each held together by four O–H···Cl hydrogen bonds. A packing diagram of the crystal, atomic coordinates, and isotropic thermal parameters are given in the paper [37].

IR (KBr): ν (in cm^{-1}) = 1285, ν(NS); 305, ν(OsCl) [37].

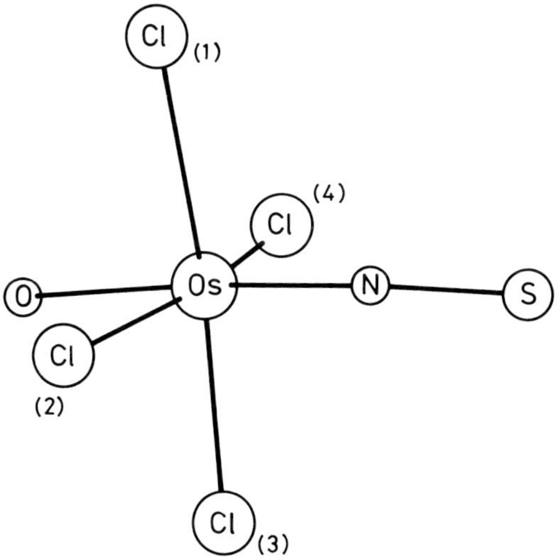

Fig. 14. Molecular structure of the anion in $(C_6H_5)_4P^+[Os(NS)Cl_4(H_2O)]^-$ [37].

$((C_6H_5)_4P^+)_2[Os(NS)Cl_5]^{2-}$

The complex was obtained in 78% yield as a green solid by addition of $(C_6H_5)_4P^+Cl^-$ to $Os(NS)Cl_3$ in the mole ratio 2:1 both in aqueous solution.

IR (KBr): ν (in cm^{-1}) = 1255, ν(NS); 300, ν(OsCl).

A solution of the complex in H_2O–CH_3OH standing at room temperature for 15 d gives $(C_6H_5)_4P^+[Os(NS)Cl_4(H_2O)]^-$ in 25% yield [37].

References on p. 86.

Table 22

Preparation of Other $Os(NS)Cl_3L^1_2$ and $Os(NS)Br_2ClL^2_2$ Complexes.
$L^1_2 = (P(CH_3)_2C_6H_5)_2$, $(NC_5H_5)_2$, $N_2C_{10}H_8$, $(As(C_6H_5)_3)_2$. $L^2_2 = N_2C_{10}H_8$, $(As(C_6H_5)_3)_2$.

compound	yield in %	starting compounds	mole ratio	solvent	temp., time	Ref.
$Os(NS)Cl_3(P(CH_3)_2C_6H_5)_2$	63	$Os(N)Cl_3(P(CH_3)_2C_6H_5)_2$, S_2Cl_2	2:1	CH_2Cl_2	r.t., 3 h	[10,18]
$Os(NS)Cl_3(NC_5H_5)_2$ [a)]	53	$(n\text{-}C_4H_9)_4N^+[Os(N)Cl_4]^-$, S_2Cl_2	2:1	CH_2Cl_2	r.t., 4 h	[10]
$Os(NS)Cl_3(N_2C_{10}H_8)$ [b)]	85	$Os(N)Cl_3(N_2C_{10}H_8)$ [b)], S_2Cl_2	2:1	CH_2Cl_2	r.t., 10 h	[10]
$Os(NS)Cl_3(As(C_6H_5)_3)_2$	71	$Os(N)Cl_3(As(C_6H_5)_3)_2$, S_2Cl_2	2:1	CH_2Cl_2	r.t., 3 h	[10,18]
	—	$OsCl_3$, $S_3N_3Cl_3$, $As(C_6H_5)_3$	?	C_2H_5OH–THF	r.t., 1.5 h	[29]
	—	$((C_6H_5)_4P^+)_2[Os(NS)(NCS)_5]^{2-}$, S_2Cl_2, $As(C_6H_5)_3$	—	CH_2Cl_2	r.t.	[40]
$Os(NS)Br_2Cl(N_2C_{10}H_8)$ [b)]	—	$Os(N)Br_3(N_2C_{10}H_8)$ [b)], S_2Cl_2	2:1	CH_2Cl_2	r.t., 3 h	[10]
$Os(NS)Br_2Cl(As(C_6H_5)_3)_2$	—	$Os(N)Br_3(As(C_6H_5)_3)_2$, S_2Cl_2	2:1	CH_2Cl_2	r.t., 3 h	[10]

[a)] NC_5H_5 = pyridine. — [b)] $N_2C_{10}H_8$ = 2,2′-bipyridine.

References on p. 86

Table 23

Physical Properties of Other $Os(NS)Cl_3L^1_2$ and $Os(NS)Br_2ClL^2_2$ Complexes.
$L^1_2 = (P(CH_3)_2C_6H_5)_2$, $(NC_5H_5)_2$, $N_2C_{10}H_8$, $(As(C_6H_5)_3)_2$. $L^2_2 = N_2C_{10}H_8$, $(As(C_6H_5)_3)_2$.

compound	recrystallized from	m.p. in °C	color	IR (Nujol) ν(NS) in cm^{-1}	Ref.
$Os(NS)Cl_3(P(CH_3)_2C_6H_5)_2$ [a]	$CH_2ClCH_2Cl–C_5H_{12}$	189 to 192	green	1285	[10, 18]
$Os(NS)Cl_3(NC_5H_5)_2$ [b]	$CH_2ClCH_2Cl–C_5H_{12}$	—	yellow-green	1284	[10]
$Os(NS)Cl_3(N_2C_{10}H_8)$ [c]	—	—	green	1282	[10]
$Os(NS)Cl_3(As(C_6H_5)_3)_2$	$CH_2ClCH_2Cl–C_6H_{14}$	138 to 141	green	1282	[10, 18]
	$CH_2ClCH_2Cl–C_6H_{14}$	150 to 152	light green	1290	[29]
$Os(NS)Br_2Cl(N_2C_{10}H_8)$ [c]	—	—	green	1280	[10]
$Os(NS)Br_2Cl(As(C_6H_5)_3)_2$	—	—	green	1270	[10]

[a] ^1H NMR ($CDCl_3$/TMS): δ (in ppm) = 2.08 (t, CH_3), 7.3 to 7.7 (m, C_6H_5); $|^4J(P,CH_3) + {}^2J(P,CH_3)| = 4.5$ Hz. Consistent with trans virtually-coupled phosphorus ligands. – [b] NC_5H_5 = pyridine. – [c] $N_2C_{10}H_8$ = 2,2′-bipyridine.

References on p. 86

Table 24
Bond Lengths and Bond Angles of the Anion in $(C_6H_5)_4P^+[Os(NS)Cl_4(H_2O)]^-$ [37].

bond length	in pm	bond angle	in °	bond angle	in °
Os–Cl(2)	235.8(2)	Cl(2)–Os–Cl(3)	89.9(1)	Cl(2)–Os–Cl(4)	168.8(1)
Os–Cl(4)	234.5(3)	Cl(3)–Os–Cl(4)	90.4(1)	Cl(2)–Os–N	96.9(2)
Os–O	216.5(4)	Cl(3)–Os–N	91.6(2)	Cl(4)–Os–N	94.3(2)
Os–Cl(3)	237.4(2)	Cl(2)–Os–O	84.0(1)	Cl(3)–Os–O	85.0(1)
Os–N	173.1(4)	Cl(4)–Os–O	84.9(1)	N–Os–O	176.5(2)
Os–Cl(1)	234.9(2)	Cl(2)–Os–Cl(1)	88.9(1)	Cl(3)–Os–Cl(1)	171.8(1)
N–S	151.4(5)	Cl(4)–Os–Cl(1)	89.2(1)	N–Os–Cl(1)	96.6(2)
		O–Os–Cl(1)	86.8(1)	Os–N–S	174.9(3)

$(C_6H_5)_4As^+[Os(NS)Cl_5]^-$

The compound formed in high yield by reaction of $Os(NS)Cl_4(POCl_3)$ with almost equimolar amounts of $(C_6H_5)_4As^+Cl^-$ in CH_2Cl_2. After addition of some CCl_4 to the solution and cooling to 5°C, black crystals and thin red platelets formed.

IR (solvent?): ν (in cm^{-1}) = 1340 s, ν(NS); 528 w, ν(OsN); 332 vs, ν($OsCl_4$); 313 sh, ν(OsCl′).

The complex anion probably contains Os of oxidation number +V according to a d^3 configuration.

The compound is moisture-sensitive. Alkaline hydrolysis gives NH_3 [39].

$((C_6H_5)_4P^+)_2[Os(NS)(NCS)_5]^{2-}$

The complex was prepared in 50% yield by boiling under reflux $(C_6H_5)_4P^+[Os(N)Cl_4]^-$ (1.5 g) and excess $(C_6H_5)_4P^+NCS^-$ (6 g) in a mixture of $(CH_3)_2CO$–H_2O (10:1) for 6 h. The green-brown platelets which deposited from solution were recrystallized from CH_3CN as hexagonal platelets. It was also obtained in 10% yield by boiling $Os(N)Cl_3(As(C_6H_5)_3)_2$ (0.25 g) with $(C_6H_5)_4P^+NCS^-$ (0.5 g) and K^+NCS^- (2 g) under reflux in a $(CH_3)_2CO$–H_2O mixture.

IR (paraffin): ν (in cm^{-1}) = 2125 sh, 2080 s, 2050 s, all ν(OsN); 1278 s, ν(NS); 850 w, ν(CS).

A suspension of $((C_6H_5)_4P^+)_2[Os(NS)(NCS)_5]^{2-}$ in CH_3OH (10 mL) containing $As(C_6H_5)_3$ (0.5 g) and HCl (1 mL) saturated with Cl_2 gas reacts, on standing, to give $Os(N)Cl_3(As(C_6H_5)_3)_2$. Stirring 0.2 g of the complex in 10 mL CH_2Cl_2 containing 0.5 g of $As(C_6H_5)_3$, followed by addition of 1 mL S_2Cl_2 causes, separation of crystalline $Os(NS)Cl_3(As(C_6H_5)_3)_2$ [40].

$(C_6H_5)_4As^+[Os((NS)_2Cl)Cl_4]^-$

The complex salt was prepared in almost quantitative yield by dropping a solution of $(C_6H_5)_4As^+Cl^-$ in CH_2Cl_2 into a stirred suspension of equimolar amounts of $Os(N{\equiv}SCl)_2Cl_4$ in CH_2Cl_2. After a few hours, a red-brown solution formed from which the complex precipitated by addition of CCl_4. It was also obtained by reaction of $Os(NS)_2Cl_4$ with equimolar amounts of $(C_6H_5)_4As^+Cl^-$ in CH_2Cl_2. Compact black monocrystals were obtained from CH_2Cl_2 solution by dropping CCl_4 into the solution until a slight turbidity occurred, followed by cooling to 5°C for 1 to 2 d.

References on p. 86

The crystal structure was determined by X-ray diffraction. The crystals are monoclinic, space group $P2_1/c-C_{2h}^5$ (No. 14), with a = 1735(1), b = 1058(1), c = 1578(1) pm; β = 95.46°; V = 2.884 nm³. Z = 4. R = 0.033 for 2161 observed reflections. The molecular structure is shown in **Fig. 15**. Bond lengths and bond angles are given in Table 25. In the $[Os((NS)_2Cl)Cl_4]^-$ ion the Os atom is octahedrally coordinated by four Cl atoms and two NS groups in a cis arrangement. The NS groups are essentially linear (with the bond lengths OsN 184 pm and NS 146 pm). A Cl atom is loosely attached to one of the S atoms. In the crystal the Cl atom belongs statistically to both S atoms with an occupation probability of 54 and 46%, respectively. Possibly, the Cl^- ion fluctuates dynamicly between the positions Cl(5) and Cl(6). The NS bond length of 146 pm is strikingly short. The binding can be described by the following resonance structures $\overset{\ominus}{Os}=\overset{\oplus}{N}=S\rangle \leftrightarrow \overset{2\ominus}{Os}-\overset{\oplus}{N}\equiv\overset{\oplus}{S}|$.

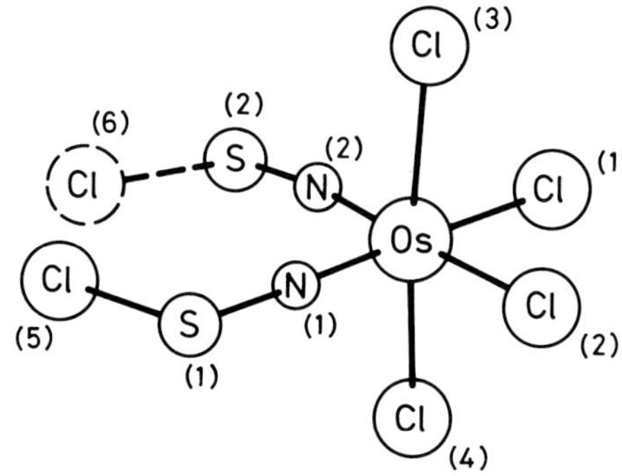

Fig. 15. Molecular structure of the anion in $(C_6H_5)_4As^+[Os((NS)_2Cl)Cl_4]^-$ [38].

Table 25

Bond Lengths and Bond Angles of the Anion in $(C_6H_5)_4As^+[Os((NS)_2Cl)Cl_4]^-$ [38].

bond length	in pm	bond angle	in °	bond angle	in °
Os–Cl(1)	236.3(2)	Cl(1)–Os–Cl(2)	87.9(1)	Os–N(2)–S(2)	169.1(5)
Os–Cl(2)	238.0(2)	Cl(1)–Os–Cl(3)	87.4(1)	N(1)–S(1)–Cl(5)	104.2(4)
Os–Cl(3)	236.0(2)	Cl(1)–Os–Cl(4)	88.1(1)	N(2)–S(2)–Cl(6)	102.9(4)
Os–Cl(4)	236.8(2)	Cl(2)–Os–Cl(3)	88.3(1)		
Os–N(1)	184.2(8)	Cl(2)–Os–Cl(4)	89.3(1)		
Os–N(2)	182.8(8)	Cl(3)–Os–Cl(4)	175.1(1)		
N(1)–S(1)	146(1)	Cl(1)–Os–N(2)	89.4(2)		
S(1)–Cl(5)	228(1)	Cl(2)–Os–N(1)	87.6(2)		
N(2)–S(2)	146(1)	N(1)–Os–N(2)	95.1(3)		
S(2)–Cl(6)	227(1)	Os–N(1)–S(1)	169.9(5)		

Dropping a solution of $GaCl_3$ in CH_2Cl_2 to an equimolar stirred solution of $(C_6H_5)_4As^+$-$[Os((NS)_2Cl)Cl_4]^-$ gives $Os(NS)_2Cl_4$ in almost quantitative yield [38].

References on p. 86

1.4.12 Thionitrosyl Complexes of Ir

The complexes were identified only by their IR spectra. Uncertainties about the assignment to ν(NS) and to the existence of thionitrosyl complexes, similar to those of the corresponding Rh complexes (see p. 76), may also be valid for the Ir complexes.

$Ir(NS)Cl_2(P(C_6H_5)_3)_2$

The complex was prepared by adding a solution of $S_3N_3Cl_3$ in THF dropwise to a stirred solution of $IrHCl_2(P(C_6H_5)_3)_3$ in THF. The resulting solution was stirred for 1 h more then evaporated. The residue was extracted with C_6H_6. On adding C_6H_{14} to the extract the yellow-brown complex precipitated.

IR (solvent?): ν (in cm^{-1}) = 1120, ν(NS); 325, ν(IrCl). The complex is stable in air, soluble in $CHCl_3$, CH_2Cl_2, and C_6H_6, and is diamagnetic and nonconducting in $C_6H_5NO_2$ [30].

$Ir(NS)(CO)Cl_2(P(C_6H_5)_3)_2$

The complex was prepared in 32% yield by slowly adding a solution of $S_3N_3Cl_3$ in THF to a stirred solution of $Ir(CO)Cl(P(C_6H_5)_3)_2$ in THF under Ar at room temperature. The mixture was stirred for 1 h more then evaporated. The residue was extracted with C_6H_6. When petroleum ether was added to the extract, a brown complex precipitated out which was recrystallized from C_6H_6-petroleum ether. The complex was also obtained by reaction of $S_3N_3Cl_3$ with $IrH(CO)(P(C_6H_5)_3)_3$ in THF.

IR (solvent?): ν (in cm^{-1}) = 2055, ν(CO); 1115, ν(NS); 320, ν(IrCl).

The complex is diamagnetic and nonconducting in CH_3NO_2 [41].

References:

[1] Greenhough, T. J.; Kolthammer, B. W. S.; Legzdins, P.; Trotter, J. (Inorg. Chem. **18** [1979] 3548/54).
[2] Kolthammer, B. W. S.; Legzdins, P. (J. Am. Chem. Soc. **100** [1978] 2247/8).
[3] Greenhough, T. J.; Kolthammer, B. W. S.; Legzdins, P.; Trotter, J. (J. Chem. Soc. Chem. Commun. **1978** 1036/7).
[4] Herberhold, M.; Smith, P. D.; Alt, H. G. (J. Organometal. Chem. **191** [1980] 79/83).
[5] Minelli, M.; Hubbard, J. L.; Lichtenberger, D. L.; Enemark, J. H. (Inorg. Chem. **23** [1984] 2721/2).
[6] Hubbard, J. L.; Lichtenberger, D. L. (Inorg. Chem. **19** [1980] 1388/90).
[7] Lichtenberger, D. L.; Hubbard, J. L. (Inorg. Chem. **24** [1985] 3835/41).
[8] Herrmann, W. A.; Rohrmann, J.; Nöth, H.; Nanila, C. K.; Bernal, I.; Draux, M. (J. Organometal. Chem. **284** [1985] 189/211, 192, 204/5).
[9] Herberhold, M.; Haumaier, L. (Z. Naturforsch. **35 b** [1980] 1277/80).
[10] Bishop, M. W.; Chatt, J.; Dilworth, J. R. (J. Chem. Soc. Dalton Trans. **1979** 1/5).

[11] Chatt, J.; Dilworth, J. R. (J. Chem. Soc. Chem. Commun. **1974** 508).
[12] Hursthouse, M. B.; Motevalli, M. (J. Chem. Soc. Dalton Trans. **1979** 1362/6).
[13] Lichtenberger, D. L.; Hubbard, J. L. (Inorg. Chem. **23** [1984] 2718/20).
[14] Bevan, P. C.; Chatt, J.; Dilworth, J. R.; Henderson, R. A.; Leigh, G. J. (J. Chem. Soc. Dalton Trans. **1982** 821/4).
[15] Hartmann, G.; Mews, R. (Angew. Chem. **97** [1985] 218/9; Angew. Chem. Intern. Ed. Engl. **24** [1985] 202).
[16] Kaden, L.; Lorenz, B.; Kirmse, R.; Stach, J. (Z. Chem. [Leipzig] **25** [1985] 29/30).

[17] Baldas, J.; Bonnyman, J.; Mackay, M. F.; Williams, G. A. (Australian J. Chem. **37** [1984] 751/9).
[18] Bishop, M. W.; Chatt, J.; Dilworth, J. R. (J. Chem. Soc. Chem. Commun. **1975** 780/1).
[19] Anhaus, J.; Siddiqi, Z. A.; Roesky, H. W.; Bats, J. W.; Elerman, Y. (Z. Naturforsch. **40b** [1985] 740/4).
[20] Hauck, H. G.; Willing, W.; Müller, U.; Dehnicke, K. (Z. Anorg. Allgem. Chem. **534** [1986] 77/84).

[21] Mews, R.; Liu, C. S. (Angew. Chem. **95** [1983] 156; Angew. Chem. Intern. Ed. Engl. **22** [1983] 158).
[22] Hartmann, G.; Mews, R. (Z. Naturforsch. **40b** [1985] 343/6).
[23] Tiwari, R. D.; Pandey, K. K.; Agarwala, U. C. (Inorg. Chem. **21** [1982] 845/6).
[24] Pandey, K. K. (unpublished observations in: Roesky, H. W.; Pandey, K. K., Advan. Inorg. Chem. Radiochem. **26** [1983] 337/56, 350).
[25] Bats, J. W.; Pandey, K. K.; Roesky, H. W. (J. Chem. Soc. Dalton Trans. **1984** 2081/3).
[26] Roesky, H. W.; Pandey, K. K.; Clegg, W.; Noltemeyer, M.; Sheldrick, G. M. (J. Chem. Soc. Dalton Trans. **1984** 719/21).
[27] Udupa, K. N.; Jain, K. C.; Khan, M. I.; Agarwala, U. C. (Inorg. Chem. Acta **74** [1983] 191/7).
[28] Hursthouse, M. B.; Walker, N. P. C.; Warrens, C. P.; Woollins, J. D. (J. Chem. Soc. Dalton Trans. **1985** 1043/7).
[29] Pandey, K. K.; Agarwala, U. C. (Z. Anorg. Allgem. Chem. **461** [1980] 231/3).
[30] Jain, K. C.; Agarwala, U. C. (Indian J. Chem. A **22** [1983] 336/7).

[31] Pandey, K. K. (unpublished work in: Roesky, H. W.; Pandey, K. K., Advan. Inorg. Chem. Radiochem. **26** [1983] 337/56, 347/8).
[32] Pandey, K. K.; Agarwala, U. C. (Indian J. Chem. A **21** [1982] 77/8).
[33] Pandey, K. K.; Agarwala, U. C. (Indian J. Chem. A **20** [1981] 74/5).
[34] Pandey, K. K.; Agarwala, U. C. (Inorg. Chem. **20** [1981] 1308/10).
[35] Pandey, K. K.; Datta, S.; Agarwala, U. C. (Z. Anorg. Allgem. Chem. **468** [1980] 228/30).
[36] Pandey, K. K.; Jain, K. C.; Agarwala, U. C. (Inorg. Chim. Acta **48** [1981] 23/7).
[37] Pandey, K. K.; Roesky, H. W.; Noltemeyer, M.; Sheldrick, G. M. (Z. Naturforsch. **39b** [1984] 590/3).
[38] Weber, R.; Müller, U.; Dehnicke, K. (Z. Anorg. Allgem. Chem. **504** [1983] 13/22).
[39] Weber, R.; Dehnicke, K. (Z. Naturforsch. **39b** [1984] 262/4).
[40] Wright, M. J.; Griffith, W. P. (Transition Metal Chem. [Weinheim] **7** [1982] 53/8).

[41] Pandey, K. K.; Agarwala, U. C. (Indian J. Chem. A **20** [1981] 906/7).
[42] Demant, U.; Willing, W.; Müller, U.; Dehnicke, K. (Z. Anorg. Allgem. Chem. **532** [1986] 175/83).
[43] Hauck, H.-G.; Willing, W.; Müller, U.; Dehnicke, K. (Z. Naturforsch. **41b** [1986] 825/30).
[44] Willing, W.; Müller, U.; Demant, U.; Dehnicke, K. (Z. Naturforsch. **41b** [1986] 560/6).

2 Sulfur-Nitrogen Compounds with 2-Coordinate Sulfur

2.1 Poly(sulfur nitride), $(SN)_x$

2.1.1 General Remarks

The inorganic compound, $(SN)_x$, formed from nonmetallic elements, was the first polymeric material which showed metallic and superconductivity properties [1, 2].

This exciting behavior was observed only recently, although the compound has been known since 1910 [3], because only "single crystals" exhibit the extraordinary properties. Such crystals are available now by a refinement of an old procedure [4]. Several groups studied the chemical and physical properties of $(SN)_x$ intensively, and a large number of publications resulted. The conclusion is that $(SN)_x$ is regarded as a highly anisotropic three-dimensional semimetal. Simultaneously much theoretical and experimental work on other new highly conducting materials was stimulated. These developments are still in progress and some technical applications are expected. The present knowledge of $(SN)_x$ has been dealt with in many reviews, see for example [5 to 26]. The literature on syntheses and properties of $(SN)_x$ through 1959 can be found in "Schwefel" B 3, 1963, pp. 1536/7.

References:

[1] Walatka, V. V., Jr.; Labes, M. M.; Perlstein, J. H. (Phys. Rev. Letters **31** [1973] 1139/42).
[2] Greene, R. L.; Street, G. B.; Suter, L. J. (Phys. Rev. Letters **34** [1975] 577/9).
[3] Burt, F. P. (J. Chem. Soc. **97** [1910] 1171/4).
[4] Douillard, A.; May, J. F.; Vallet, G. (Ann. Chim. [Paris] [14] **6** [1971] 257/61).
[5] Heal, H. G. (Advan. Inorg. Chem. Radiochem. **15** [1972] 375/412, 395/6).
[6] MacDiarmid, A. G.; Mikulski, C. M.; Saran, M. S.; Russo, P. J.; Cohen, M. J.; Bright, A. A.; Garito, A. F.; Heeger, A. J. (Advan. Chem. Ser. No. 150 [1976] 63/72).
[7] Kamimura, H. (Phys. Semicond. Proc. 13th Intern. Conf., Rome 1976, pp. 51/60).
[8] Geserich, H. P.; Pintschovius, L. (Festkörperprobleme [Advan. Solid State Phys.] **16** [1976] 65/94).
[9] Grant, P. M.; Greene, R. L.; Gill, W. D.; Rudge, W. E.; Street, G. B. (Mol. Cryst. Liquid Cryst. **32** [1976] 171/6).
[10] Street, G. B.; Greene, R. L. (IBM J. Res. Develop. **21** [1977] 99/110).

[11] Greene, R. L.; Street, G. B. (NATO Advan. Study. Inst. B **25** [1977] 167/95).
[12] Schechtman, B. H.; Gill, W. D.; Grant, P. M.; Greene, R. L.; Mengel, P.; Rudge, W. E.; Street, G. B. (Proc. Intern. Symp. Electrets Dielectr., San Carlos, Brazil, 1975 [1977], pp. 405/12).
[13] Akhtar, M.; Chiang, C. K.; Cohen, M. J.; Heeger, A. J.; Kleppinger, J.; MacDiarmid, A. G.; Milliken, J.; Moran, M. J.; Peebles, D. L. (Organometal. Polym. Symp., New Orleans 1977 [1978], pp. 301/12).
[14] Pintschovius, L. (Colloid. Polym. Sci. **256** [1978] 883/92).
[15] Bernard, C.; Robert, G. (Bull. Soc. Chim. France **1978** I 395/I 401).
[16] Labes, M. M.; Love, P.; Nichols, L. F. (Chem. Rev. **79** [1979] 1/15).
[17] Street, G. B.; Gill, W. D. (NATO Conf. Ser. VI **1** [1979] 301/26).
[18] Baughman, R. H.; Greene, R. L.; MacDiarmid, A. G.; Schmidt, M.; Soos, Z.; Street, G. B.; Vogel, F. L. (NATO Conf. Ser. VI **1** [1979] 505/10).
[19] Street, G. B.; Clarke, T. C. (Advan. Chem. Ser. No. 186 [1980] 177/94).
[20] Love, P. (Polym. News **7** [1981] 200/7).

[21] Peters, E. N. (in: Kirk-Othmer Encycl. Chem. Technol. 3rd Ed. **13** [1982] 406/7).
[22] Roch, K. M. (Macromol. Chem. [London] **2** [1982] 162/73, 170/3).
[23] Gill, W. D.; Clarke, T. C.; Street, G. B. (Appl. Phys. Commun. **2** [1982/83] 211/82).
[24] Nakada, I. (Kagaku Sosetsu No. 42 [1983] 113/9).
[25] Kaneto, K.; Yoshino, K.; Inuishi, Y. (in: Monceau, P., Electronic Properties of Inorganic Quasi-One-Dimensional Compounds, Pt. 2, Reidel, Dordrecht, Neth., 1985, pp. 69/109).
[26] Kagoshima, S.; Nagasawa, H.; Sambongi, T. (One-Dimensional Conductors, Springer, Berlin 1988, pp. 199/205).

2.1.2 Historical

The first formation of $(SN)_x$ was described in 1910 by Burt [1]. $(SN)_x$ in the form of blue films and crystals were obtained by passing S_4N_4 vapor over silver gauze or quartz wool at 100 to 300°C. A blue substance was also observed by Davis on heating $S_4N_4 \cdot SbCl_5$ as cited in [1]. Strutt reported that a deep indigo blue solid was generated by the action of active nitrogen on S_8 or CS_2 which he identified as $(SN)_x$ [2]. In 1925 Usher confirmed the results of Burt using a slightly modified experimental arrangement [3]. Several years later Goehring and Voigt [4 to 7] made the first fundamental studies of the structure and properties of $(SN)_x$. They recognized that the thermal decomposition of S_4N_4 leads to the intermediate species S_2N_2, a white volatile crystalline solid which polymerizes to $(SN)_x$. Electrical resistivity measurement of pressed powder showed it to be a semiconductor [4 to 7]. In 1962 and 1964 further physical studies dealing with the electronic properties of $(SN)_x$ were performed which confirmed the semiconducting behavior [8, 9]. A few years later $(SN)_x$ crystals were prepared which allowed structural studies by electron diffraction [10 to 12]. The physical properties were reinvestigated and metallic behavior of $(SN)_x$ down to liquid helium temperature was observed [13, 14]. Additional physical measurements of specific heat [15], optical reflectivity [16], and magnetic susceptibility [17] revealed metallic properties down to 1.2 K. Finally in 1975 superconductivity in $(SN)_x$ at a transition temperature of 0.26 ± 0.03 K was reported [18]. Since that time a flood of papers dealing with the extraordinary properties of $(SN)_x$ has been published.

References:

[1] Burt, F. P. (J. Chem. Soc. **97** [1910] 1171/4).
[2] Strutt, R. J. (Proc. Roy. Soc. [London] A **88** [1913] 539/49, 543).
[3] Usher, F. L. (J. Chem. Soc. **127** [1925] 730/5).
[4] Goehring, M.; Voigt, D. (Naturwissenschaften **40** [1953] 482).
[5] Goehring, M.; Voigt, D. (Z. Anorg. Allgem. Chem. **285** [1956] 181/90).
[6] Goehring, M. (Ergebnisse und Probleme der Chemie der Schwefelstickstoffverbindungen, Akad.-Verlag, Berlin 1957, pp. 1/173, 20/2).
[7] Goehring, M. (Quart. Rev. [London] **10** [1956] 437/50, 440).
[8] Kronick, P. L.; Kaye, H.; Chapman, E. F.; Mainthia, S. B.; Labes, M. M. (J. Chem. Phys. **36** [1962] 2235/7).
[9] Chapman, D.; Warn, R. J.; Fitzgerald, A. G.; Yoffe, A. D. (Trans. Faraday Soc. **60** [1964] 294/300).
[10] Douillard, A.; May, J. F.; Vallet, G. (Compt. Rend. C **269** [1969] 212/5).

[11] Douillard, A.; May, J. F.; Vallet, G. (Ann. Chim. [Paris] [14] **6** [1971] 259/61).
[12] Boudeulle, M.; Douillard, A.; Michel, P.; Vallet, G. (Compt. Rend. C **272** [1971] 2137/8).
[13] Walatka, V. V., Jr.; Labes, M. M.; Perlstein, J. H. (Phys. Rev. Letters **31** [1973] 1139/42).
[14] Hsu, C.; Labes, M. M. (J. Chem. Phys. **61** [1974] 4640/5).

[15] Greene, R. L.; Grant, P. M.; Street, G. B. (Phys. Rev. Letters **34** [1975] 89/92).
[16] Bright, A. A.; Cohen, M. J.; Garito, A. F.; Heeger, A. J.; Mikulski, C. M.; Russo, P. J.; MacDiarmid, A. G. (Phys. Rev. Letters **34** [1975] 206/9).
[17] Salvo, F. J.; Walatka, V. V., Jr.; Labes, M. M. (unpublished results, from Labes, M. M.; Love, P.; Nichols, L. F., Chem. Rev. **79** [1979] 1/15, 2).
[18] Greene, R. L.; Street, G. B.; Suter, L. J. (Phys. Rev. Letters **34** [1975] 577/9).

2.1.3 Preparation. Formation

$(SN)_x$ is produced in form of "single crystals" (crystalline microfibrils which are fully oriented in a common chain direction), unoriented and epitaxially oriented films and microcrystalline powders. In work published before ca. 1960 $(SN)_x$ was obtained predominantly as a polycrystalline material, see "Schwefel" B 3, 1963, pp. 1536/7.

2.1.3.1 Preparation of $(SN)_x$ Crystals by Solid State Polymerization of S_2N_2

The standard method for preparing $(SN)_x$ "single crystals" is the solid-state polymerization of vapor grown S_2N_2 crystals [1 to 20]: $x/2\,S_2N_2(s) \rightarrow (SN)_x(s)$. Photographs of $(SN)_x$ crystals are illustrated for example in [3, 4]. S_2N_2 is obtained by the thermal decomposition of S_4N_4 vapor passing through heated silver gauze; see "Sulfur-Nitrogen Compounds" Part 2, 1985, p. 1; but caution: S_4N_4 as well as S_2N_2 was reported to explode under certain conditions. See "Sulfur-Nitrogen Compounds" Part 2, 1985, pp. 12 and 127. The reader will also find some information about the dimerization and polymerization of S_2N_2 there on pp. 14/6. The technique used today is a variation of the procedure by Burt [6]. It was observed by Douillard that S_2N_2 crystals yield $(SN)_x$ crystals [8]. S_2N_2, carefully purified by distillation, was allowed to polymerize at ambient temperature. $(SN)_x$ was obtained in form of brillant little crystals [8, 20]. $(SN)_x$ crystals were also generated from S_2N_2 on cold fingers at 0 to 8°C [9, 10] or by allowing unpurified S_2N_2 to polymerize during a programed slow warm-up from −77°C to room temperature [10]. The crystals consist of highly oriented bundles of parallel, impure $(SN)_x$ fibres [9]. $(SN)_x$ crystals were also formed by polymerization of S_2N_2 crystals (grown at 0°C) at temperatures varying from −196 to 30°C [5]. Good quality $(SN)_x$ crystals were formed by warming crude S_2N_2 from −196 to 0°C followed by polymerization during 5 d. The polymerization process continued at room temperature for at least 14 d and preferably longer [3, 11]. Crystals in equidimensional form or as needles (5 to 10 mm long, 0.5 mm wide) were obtained [11]. Crystals of S_2N_2, grown in 48 h and collected at 0°C in a trap with rectangular walls, polymerized in the solid state at room temperature over a 60 h period to form lustrous, golden $(SN)_x$ crystals. Unpolymerized S_2N_2 was removed by pumping in a sequence consisting of 2 h at room temperature, 1 h as the temperature is raised to 75°C, and finally for 4 h at 75°C [1, 2]. Other workers purified the crystals by washing with benzene and acetone [9, 10]. Each method may leave voids in the $(SN)_x$ crystals [3]. Better results were obtained when S_2N_2 polymerized for 2 d at 75°C under 70 atm of N_2 [3]. In general, distortions in the $(SN)_x$ crystals are prevented if S_2N_2 polymerizes at room temperature rather than higher temperatures [3]. Brass-colored $(SN)_x$ crystals several millimeters thick and about 1 cm long ($V \sim 80$ mm^3) were reported to form from S_2N_2 crystals which were crystallized by a special localized cooling technique with a copper rod. The basic procedure is analogous to that used by others. The S_2N_2 crystals polymerized in the course of some days at 28°C. Residues of S_2N_2 were removed by pumping for 24 h [12]. A similar technique for preparation of sizeable $(SN)_x$ crystals is described in which a heat pipe was used for localized cooling. The $(SN)_x$ crystals grow during a week at room temperature. To remove the unpolymerized S_2N_2, the crystals were transferred to a glass tube which was then sealed with 0.5 atm argon. Annealing at 60°C for 10 d completed the polymerization [13]. $(SN)_x$ whiskers with diameters down to 400 nm and lengths up to 2 mm form by adding specific impurities (borneol or camphene) to a highly pure S_2N_2 material. The solid state polymerization to $(SN)_x$ was complete after three weeks at room temperature [14].

In some papers which deal with new procedures for the preparation of S_2N_2 (see "Sulfur-Nitrogen Compounds" Part 2, 1985, p. 2) the solid-state polymerization to $(SN)_x$ is also described. S_2N_2 which was obtained by the Ag catalyzed pyrolysis of $S_4N_3^+Cl^-$ vapor (130°C, ca. 10^{-6} Torr) transforms after a period of six weeks into brass-yellow $(SN)_x$ crystals ranging from submillimeter to millimeter sizes [15, 16]. The thermal decomposition of $S_3N_3N{=}As(C_6H_5)_3$ also yields S_2N_2 which converts to lustrous golden crystals (1 to 3 mm in length) [17]. $(SN)_x$ "single crystals" were also obtained from slow decomposition of S_4N_2 crystals at 263 K within 3 or 4 weeks. A mechanism of this reaction similar to that assumed for the S_2N_2 polymerization was discussed [18]. The formation of $(SN)_x$ as a by-product was observed in the reaction of S_4N_4 with $S_2N_2 \cdot 2AlCl_3$ at 80°C in a high vacuum (10^{-5} mbar) [19].

References:

[1] Mikulski, C. M.; Russo, P. J.; Saran, M. S.; MacDiarmid, A. G.; Garito, A. F.; Heeger, A. J. (J. Am. Chem. Soc. **97** [1975] 6358/63).
[2] MacDiarmid, A. G.; Mikulski, C. M.; Heeger, A. J.; Garito, A. F. (Inorg. Syn. **22** [1983] 143/9).
[3] Street, G. B.; Greene, R. L. (IBM J. Res. Develop. **21** [1977] 99/110).
[4] Nakada, I. (Kagaku Sosetsu No. 42 [1983] 113/9).
[5] Labes, M. M.; Love, P.; Nichols, L. F. (Chem. Rev. **79** [1979] 1/15, 3).
[6] Burt, F. P. (J. Chem. Soc. **97** [1910] 1171/4).
[7] Goehring, M.; Voigt, D. (Z. Anorg. Allgem. Chem. **285** [1956] 181/90).
[8] Douillard, A.; May, J. F.; Vallet, G. (Ann. Chem. [Paris] [14] **6** [1971] 257/61).
[9] Walatka, V. V., Jr.; Labes, M. M.; Perlstein, J. H. (Phys. Rev. Letters **31** [1973] 1139/42).
[10] Hsu, C.; Labes, M. M. (J. Chem. Phys. **61** [1974] 4640/5).

[11] Street, G. B.; Arnal, H.; Gill, W. D.; Grant, P. M.; Greene, R. L. (Mater. Res. Bull. **10** [1975] 877/81).
[12] Kahlert, H.; Kundu, B. (Mater. Res. Bull. **11** [1976] 967/72).
[13] Nakada, I. (J. Cryst. Growth **55** [1981] 447/52).
[14] Stejny, J.; Trinder, R. W.; Dlugosz, J. (J. Mater. Sci. **16** [1981] 3161/70).
[15] Banister, A. J.; Hauptman, Z. V. (J. Chem. Soc. Dalton Trans. **1980** 731/5).
[16] Banister, A. J.; Fiedler, A. J.; Hauptman, Z. V.; Smith, N. R. M. (U.S. 4268491 [1981]; C.A. **95** [1981] No. 83144).
[17] Chivers, T.; Cordes, A. W.; Oakley, R. T.; Swepston, P. N. (Inorg. Chem. **20** [1981] 2376/80).
[18] Small, R. W. H.; Banister, A. J.; Hauptman, Z. V. (J. Chem. Soc. Dalton Trans. **1981** 2188/91).
[19] Roesky, H. W.; Anhaus, J. (Chem. Ber. **115** [1982] 3682/4).
[20] Bernard, C.; Robert, G. (Bull. Soc. Chim. France **1978** I395/I401).

2.1.3.2 Preparation of Solution Grown $(SN)_x$ Crystals

$(SN)_x$ crystals were also prepared from S_2N_2 solution [1, 2]. Solvents such as benzene [1], ether, and tetrahydrofuran [2] were used. Slow evaporation of a tetrahydrofuran solution, saturated with benzophenone, in a purged glove box at room temperature yields shiny, bronze-colored, very well formed crystals [2].

References:

[1] Douillard, A.; May, J. F.; Vallet, G. (Ann. Chim. [Paris] [14] **6** [1971] 257/62).
[2] Hsu, C.; Labes, M. M. (J. Chem. Phys. **61** [1974] 4640/5).

2.1.3.3 Preparation of $(SN)_x$ Crystals by Photopolymerization of S_2N_2

Photopolymerization of partly dissolved S_2N_2 into $(SN)_x$ crystals was described in "Sulfur-Nitrogen Compounds" Part 2, 1985, p. 15. Photopolymerization of S_2N_2 (partly dissolved, grown from solution, or condensed in vacuum) is also claimed in a patent [1] and reported in [2].

References:

[1] Love, P.; Labes, M. M. (U.S. 4170477 [1978/79]; C.A. **92** [1980] No. 13663).
[2] Love, P.; Myer, G. H.; Kao, H. I.; Labes, M. M.; Junker, W. R.; Elbaum, C. (Org. Coat. Plast. Chem. **38** [1978] 621/2).

2.1.3.4 Formation of $(SN)_x$ Crystals from Pyrolysis Products of S_4N_4

An intensive investigation of the pyrolysis products of S_4N_4 showed that, besides S_2N_2, two other species formed which were supposed to be $SN^{\bullet}$ monomer and a dark brown $(SN)_2$ monomer. These materials polymerize quickly to a "red" or "brown $(SN)_x$", respectively, both forming lustrous, golden crystals [1 to 3]. Properties are described in the appropriate chapters, however, a better crystal quality in both cases is necessary for detailed studies. In a recent paper the authors generated the red volatile material by passing S_4N_4 vapor over silver selenide heated at ~160°C. However, an $SN^{\bullet}$ species could not be detected by an IR spectroscopic study of the trapped vapor in an inert matrix at 20 K. The existence of $SN^{\bullet}$ was, however, not ruled out because the species may be too reactive to be trapped under the used experimental conditions [4].

References:

[1] Love, P.; Myer, G.; Kao, H. I.; Labes, M. M.; Junker, W. R.; Elbaum, C. (Ann. N.Y. Acad. Sci. **313** [1978] 745/58).
[2] Labes, M. M.; Love, P.; Nichols, L. F. (Chem. Rev. **79** [1979] 1/15, 6).
[3] Love, P.; Myer, G. H.; Kao, H. I.; Labes, M. M.; Junker, W. R.; Elbaum, C. (Org. Coat. Plast. Chem. **38** [1978] 621/2).
[4] Almond, M. J.; Downs, A. J.; Jeffery, T. L. (Polyhedron **7** [1988] 629/34).

2.1.3.5 Preparation of $(SN)_x$ Films

Two general methods for the preparation of $(SN)_x$ films are described in literature. $(SN)_x$ films were produced by the thermal decomposition of S_4N_4 catalyzed by Ag, quartz or pyrex wool, for example see [1 to 3], and otherwise by the sublimation of $(SN)_x$ itself [4 to 6]. In addition several special synthetic routes to $(SN)_x$ films have been published [7 to 10].

The pyrolysis of S_4N_4 leads to $(SN)_x$ films of different thickness depending on the experimental conditions [2]. Films of 5000 Å were obtained by subliming S_4N_4 at 140°C through Ag wool of 200°C onto a substrate cooled to 8°C. Films 1000 Å thick form by subliming S_4N_4 at 100°C through Ag wool at 200°C onto glass cooled to 12°C. Both types of films had a shiny bronze-gold appearance and retained their luster for several days at room temperature even when exposed to air [2]. Golden $(SN)_x$ films form on glass held at 10 to 30°C when S_4N_4 is heated at 70°C, and the vapor is pumped over pyrex or quartz wool at ca. 275°C [3].

A blue $(SN)_x$ film was obtained by subliming $(SN)_x$ at 130°C in an inert atmosphere [4]. Thin $(SN)_x$ films, a few microns thick, form by the sublimation of crystalline $(SN)_x$ in a vacuum sublimator ($<10^{-4}$ Torr, 130 to 150°C) onto glass substrates at 27°C. The films consist of randomly oriented crystallites with a grain size of about 1 μm [5]. $(SN)_x$ films on graphite substrates ("grafoil" sheet) were obtained by subliming high purity crystalline $(SN)_x$ from a pyrex vessel at 350 K onto grafoil at ~130 K [6]. A blue-black $(SN)_x$ film was also prepared by sublimation of $(SN)_x$ powder at 100°C/10^{-5} Torr [7, 11].

$(SN)_x$ films were also formed by thermal decomposition of $S_5N_5^+FeCl_4^-$ or $S_4N_3^+Cl^-$ catalyzed by Ag wool [8, 9]. For example $S_5N_5^+FeCl_4^-$ was heated to 130°C and the vapor was passed over the silver catalyst at 220°C. A film, first showing a blue appearance and after a time developing a golden bronze luster, deposited on the cold finger [8]. $(SN)_x$ films were generated by a 10 MHz radiofrequency discharge of S_4N_4 sublimed at 70 to 80°C into a helium plasma (see also "Sulfur-Nitrogen Compounds" Part 2, 1985, p. 163) [10]. The polymerization of the red $SN^•$ monomer (see p. 92) at ~0°C leads to a golden, lustrous $(SN)_x$ film [12].

References:

[1] Burt, F. P. (J. Chem. Soc. **97** [1910] 1171/4).
[2] Hsu, C.; Labes, M. M. (J. Chem. Phys. **61** [1974] 4640/5).
[3] Louis, E. J.; MacDiarmid, A. G.; Garito, A. F.; Heeger, A. J. (J. Chem. Soc. Chem. Commun. **1976** 426/7).
[4] Douillard, A.; May, J.; Vallet, G. (Compt. Rend. C **269** [1969] 212/5).
[5] Bright, A. A.; Cohen, M. J.; Garito, A. F.; Heeger, A. J.; Mikulski, C. M.; Russo, P. J.; MacDiarmid, A. G. (Phys. Rev. Letters **34** [1975] 206/9).
[6] Kelemen, S. R.; Herschbach, D. R. (J. Phys. Chem. **86** [1982] 4388/92).
[7] Passmore, J.; Rao, M. N. S. (J. Chem. Soc. Chem. Commun. **1980** 1268/9).
[8] Banister, A. J.; Smith, N. R. M. (J. Chem. Educ. **59** [1982] 1058/62).
[9] Banister, A. J.; Fielder, A. J.; Hauptman, Z. V.; Hey, R. G.; Smith, N. R. M. (Ger. 2943160 [1978/80]; C.A. **93** [1980] No. 170471).
[10] Witt, M. W. R.; Bailey, W. I., Jr.; Lagow, R. J. (J. Am. Chem. Soc. **105** [1983] 1668/9).

[11] Banister, A. J.; Hauptman, Z. V.; Passmore, J.; Wong, C.-M.; White, P. S. (J. Chem. Soc. Dalton Trans. **1986** 2371/9).
[12] Labes, M. M.; Love, P.; Nichols, L. F. (Chem. Rev. **79** [1979] 1/15, 7).

2.1.3.6 Formation of Epitaxial $(SN)_x$ Films and Thin Single-Crystal Platelets

Epitaxial $(SN)_x$ films form by the following procedure [1, 2]. Crystalline $(SN)_x$ of high purity was heated at 150°C in a vacuum of 2×10^{-5} Torr or better. Substrates such as Teflon, Mylar, and polyethylene were attached to a water-cooled finger (10 to 20°C). Films 5000 to 10000 Å thick grow in 12 to 18 h on substrate surfaces with a temperature below 25°C. Powder X-ray diffraction studies show an agreement between the structure of $(SN)_x$ crystals and epitaxial films. The films show a gold appearance, and microscopic studies exhibit a high-quality reflective surface. Best results are reported for the use of Mylar substrates [1, 2]. A slightly different procedure was published for the preparation of epitaxial $(SN)_x$ films on Mylar substrates. The $(SN)_x$ crystals were heated in vacuum to 158°C. After two days the heat was removed and pumping was continued for 15 h [3].

Crystalline epitaxial $(SN)_x$ films were obtained also by the thermal decomposition of S_4N_4 vapor. S_4N_4 was heated at 70°C, the vapor was pumped over pyrex or quartz wool at ~275°C and golden $(SN)_x$ films deposited on polymer substrates at 10 to 30°C. The films are reported to be identical with those obtained by sublimation of $(SN)_x$, consisting of completely aligned, parallel $(SN)_x$ polymer chains [4].

The preparation of epitaxial $(SN)_x$ single-crystal platelets was successful on monovalent alkali halide substrates (NaF, NaCl, NaBr, NaI, KF, KCl, KBr, KI, RbI) and on AgCl. S_2N_2 was directly condensed onto the cooled salt plates which were attached to the bottom of cold fingers. The crystals were cleaved before use to ensure fresh (100) surfaces. The cold finger was then allowed to warm up to room temperature [5 to 9]. To complete the polymerization the authors annealed the crystals at 25°C for 18 to 24 h [5]. Epitaxial layers of $(SN)_x$ on (100) faces of KBr crystals were also prepared by high-vacuum sublimation of $(SN)_x$ powder [10]. In addition to the normal β phase of $(SN)_x$, three other crystalline phases of $(SN)_x$ were detected in epitaxial films (see Table 27, p. 106, and Section 2.1.4.3, p. 102) [7 to 9].

References:

[1] Bright, A. A.; Cohen, M. J.; Garito, A. F.; Heeger, A. J.; Mikulski, C. M.; MacDiarmid, A. G. (Appl. Phys. Letters **26** [1975] 612/5).

[2] Garito, A. F.; Heeger, A. J.; MacDiarmid, A. G.; Bright, A. A.; Cohen, M. J.; Mikulski, C. M. (U.S. 4016331 [1976]; C.A. **86** [1977] No. 172655).

[3] Rubinson, J. F.; Bethymer, T. D.; Mark, H. B., Jr.; Nowak, R. J. (J. Electrochem. Soc. **130** [1983] 121/4).

[4] Louis, E. J.; MacDiarmid, A. G.; Garito, A. F.; Heeger, A. J. (J. Chem. Soc. Chem. Commun. **1976** 426/7).

[5] Rickert, S. E.; Lando, J. B.; Hopfinger, A. J.; Baer, E. (Macromolecules **12** [1979] 1053/7).

[6] Ishida, H.; Rickert, S. E.; Hopfinger, A. J.; Lando, J. B.; Baer, E.; Koenig, J. L. (J. Appl. Phys. **51** [1980] 5188/93).

[7] Rickert, S. E.; Ishida, H.; Lando, J. B.; Koenig, J. L.; Baer, E. (J. Appl. Phys. **51** [1980] 5194/200).

[8] Lando, J. B.; Baer, E.; Rickert, S. E.; Nae, H.; Ching, S. (ACS Symp. Ser. No. 212 [1983] 89/100).

[9] Lando, J. B. (Physicochem. Aspects Polym. Surf. Proc. Intern. Symp., New York 1981 [1983], Vol. 1, pp. 559/67, 560/61).

[10] Banister, A. J.; Hauptman, Z. V.; Passmore, J.; Wong, C.-M.; White, P. S. (J. Chem. Soc. Dalton Trans. **1986** 2371/9, 2378/9).

2.1.3.7 Formation of $(SN)_x$ Powders

Alternative routes to $(SN)_x$ were developed for three reasons. (1) The thermal decomposition of S_4N_4 to S_2N_2 is dangerous [1]. (2) The solid state polymerization of S_2N_2 is tedious. (3) The polymerization leads to defects (for example, extensive twinning) [2]. In most cases these reactions produce $(SN)_x$ in form of an impure crystalline powder. A blue-black $(SN)_x$

powder was isolated as by-product, in addition to S_4N_4, in the reaction of SCl_2 with $(CH_3)_3SiN{=}S{=}NSi(CH_3)_3$ (mole ratio 1:2) in CH_2Cl_2 at −70°C [3]. Reacting the same components in the mole ratio 1:1 gave a better yield of $(SN)_x$ [4, 5]:

$$SCl_2 + (CH_3)_3SiN{=}S{=}NSi(CH_3)_3 \rightarrow 2/x\,(SN)_x + 2\,(CH_3)_3SiCl\ (+\,S_4N_4)$$

The reaction of $S_3N_3Cl_3$ with $(CH_3)_3SiN_3$ [6, 7, 8], or NaN_3 [6] in CH_3CN at −18°C [8] has been reported to lead to the formation of $(SN)_x$ according to [8]:

$$S_3N_3Cl_3 + 3\,(CH_3)_3SiN_3 \rightarrow 3/x\,(SN)_x + 3\,(CH_3)_3SiCl + 9/2\ N_2\ (+\,S_4N_4)$$

Gram quantities of the polymer were isolated using a mole ratio of 1:7.5. As in the reaction mentioned above, no dangerous intermediates form. As a by-product ~30 to 50% S_4N_4 are formed. The experimental conditions of the reaction and the technique for purification were intensively studied. Best results were achieved with rigorous exclusion of moisture. Impurities were exhaustively extracted with liquid SO_2. (Evaporation of the filtrate yielded S_4N_4 and the cyclic $(CH_3\overline{CNSSN})_2$. After purification the elemental analysis showed 1.2% Si, 0.6% Cl, and 0.9% C.) The black $(SN)_x$ powder (~14 to 22 m^2/g surface areas) was characterized by several analytical techniques (for example X-ray diffraction, mass spectral analysis, and thermal studies). It should be mentioned that the compression of the powder is hazardous under certain conditions. $(SN)_x$ films and occasionally crystals were obtained by sublimation of the powder [8]. The reaction of $S_3N_2Cl^+Cl^-$ and $(CH_3)_3SiN_3$ under similar conditions [7, 8] yields an inferior product (excess sulfur and up to 7% silicon) [8]. $(SN)_x$ powders with a higher grade of impurity were also obtained by reaction of S_3N_2Cl with $(CH_3)_3SiN_3$ in CH_3CN at −15°C [7], reaction of $S_2N^+AsF_6^-$ and CsN_3 in SO_2 at −20°C [7], reaction of S_2Cl_2 with NaN_3 in CH_3CN by subsequent treatment with excess $(CH_3)_3SiN_3$ [7], and by reaction of $(S_3N_2)_2N^+AsF_6^-$ with CsN_3 in SO_2 [9]. Good quality $(SN)_x$, along with S_4N_4, was obtained when $COCl_2$ gas was passed into a solution of $Cs^+S_3N_3^-$ in CH_3CN at −78°C (22% yield) [10]. Oxidation of $Cs^+S_3N_3^-$ with AsF_5 in CH_2Cl_2 at −78°C yielded $(SN)_x$ in 13% yield [10]. $(SN)_x$ formed on warming the matrix reaction products of NH_3 and SO_2 at 10 K; it was identified by Raman spectroscopy [11].

References:

[1] MacDiarmid, A. G.; Mikulski, C. M.; Heeger, A. J.; Garito, A. F. (Inorg. Syn. **22** [1983] 143/9, 144).

[2] Street, G. B.; Greene, R. L. (IBM J. Res. Develop. **21** [1977] 99/110).

[3] Lidy, W.; Sundermeyer, W.; Verbeek, W. (Z. Anorg. Allgem. Chem. **406** [1974] 228/34).

[4] Milliken, J. W. (Diss. Univ. Pennsylvania 1980; Diss. Abstr. Intern. B **41** [1980] 951).

[5] Wolmershäuser, G.; Fuhrmann, J.; Jotter, R.; Wilhelm, T.; Scherer, O. J. (Mol. Cryst. Liquid Cryst. **118** [1985] 435/8).

[6] Passmore, J.; Rao, M. N. S. (J. Chem. Soc. Chem. Commun. **1980** 1268/9).

[7] Kennett, F. A.; MacLean, G. K.; Passmore, J.; Rao, M. N. S. (J. Chem. Soc. Dalton Trans. **1982** 851/7).

[8] Banister, A. J.; Hauptman, Z. V.; Passmore, J.; Wong, C.-M.; White, P. S. (J. Chem. Soc. Dalton Trans. **1986** 2371/9).

[9] MacLean, G.; Passmore, J.; White, P. S. (J. Chem. Soc. Dalton Trans. **1984** 211/7).

[10] Chivers, T.; Rao, M. N. S. (Can. J. Chem. **61** [1983] 1957/62).

[11] Meyer, B.; Mulliken, B.; Weeks, H. (Phosphorus Sulfur **8** [1980] 291/9).

2.1.3.8 Electrochemical Preparation of $(SN)_x$ Deposits

The electrochemical preparation of $(SN)_x$ by cathodic reduction of $S_5N_5^+$ salts was described independently by two groups [1 to 4]. $S_5N_5^+Cl^-$ was reduced in liquid SO_2 [1 to 3], $S_5N_5^+BF_4^-$ in CH_3CN at 0°C [2, 3], and $S_5N_5^+FeCl_4^-$ or $S_5N_5^+AlCl_4^-$ in CH_3CN [1, 2, 4]. $(SN)_x$ was prepared electrochemically from $S_5N_5^+FeCl_4^-$ in 1M $(C_4H_9)_4NBF_4$–CH_2Cl_2 solution at −19°C in 62% yield (SN unit/electron) [4]. Major reduction peaks ($E_{p/2}^{red}$ vs. SCE) observed at 0.34 V (in SO_2) [3], 0.23 V (in CH_3CN) [3], and 0.46 V (in CH_2Cl_2) [4] were assigned to the reduction of the $S_5N_5^+$ ion. The yields were temperature-dependent for $S_5N_5^+FeCl_4^-$, varying from 43% at room temperature to 64% at −40°C. A table showing experimental conditions was presented in [4].

The $(SN)_x$ polymer formed was described as a black, microcrystalline solid (crystallites of 5 to 10 μm in size) [1] or as a bronze-colored compact deposit [3]. Layers deposited on $(SN)_x$-coated electrodes had a higher degree of crystallinity than those deposited on bright platinum [3]. (Scanning electron microscope photographs of $(SN)_x$ deposits were presented [3].) The electrochemically formed $(SN)_x$ was characterized by chemical analysis [1, 4], IR spectra [1, 4], mass spectroscopy [4], and X-ray powder spectroscopy [1, 4]. The reaction mechanism inferred from cyclic voltammetric data (presented in a table) and subsequent bulk electrolysis was summarized in [3]. An $S_5N_5^{\bullet}$ radical (which may be cleaved into a linear $S_2N_2^{\bullet}$ and $S_3N_3^{\bullet}$ radical) was assumed to be an intermediate species; no evidence for cyclic S_2N_2 was found [3].

References:

[1] Banister, A. J.; Hauptman, Z. V.; Kendrick, A. G. (J. Chem. Soc. Chem. Commun. **1983** 1016/8).

[2] Banister, A. J.; Hauptman, Z. V.; Kendrick, A. G. (Ger. Offen. 3430949 [1984/85]; C.A. **102** [1985] No. 187455).

[3] Banister, A. J.; Hauptman, Z. V.; Kendrick, A. G. (J. Chem. Soc. Dalton Trans. **1987** 915/24).

[4] Fritz, H. P.; Bruchhaus, R. (Z. Naturforsch. **38b** [1983] 1375/82).

2.1.3.9 Formation of $(SN)_x$ by Reactions of Active Nitrogen with Elemental Sulfur or Sulfur Compounds

Early investigations showed that $(SN)_x$ was produced as a blue material by reaction of active nitrogen produced by high-voltage discharge of N_2 with sulfur, H_2S, and CS_2 [1]. Later work confirmed this result. Atomic nitrogen produced by microwave electrodeless or condensed electrode discharge of N_2 reacts with elementary sulfur to yield $(SN)_x$ along with S_4N_4 and S_4N_2 [2, 3]. Evidence for $(SN)_x$ was deduced from the dark blue color, IR absorption, and the conversion of the material to S_4N_4 in hot solvents [3]. (In contrast to these results, an earlier work reported $(S_2N)_x$, S_4N_4, and S_4N_5 as reaction products based on elemental analysis [4].) $(SN)_x$ was also produced by reactions of atomic nitrogen with H_2S, CS_2 [3, 5] (similar to [1]) and COS [3]. The action of atomic nitrogen on S_2Cl_2 produced traces of $(SN)_x$ [3]. Amorphous $(SN)_x$ layers on various substrates were prepared recently by plasma or glow-discharge decomposition of H_2S in a partial vacuum containing NH_3 or N_2 [6].

References:

[1] Strutt, R. J. (Proc. Roy. Soc. [London] A **88** [1913] 539/49).

[2] Bett, J. A. S.; Winkler, C. A. (J. Phys. Chem. **68** [1964] 2501/8).

[3] Smith, J. J.; Jolly, W. J. (Inorg. Chem. **4** [1965] 1006/10).

[4] Moldenhauer, W.; Zimmerman, A. (Ber. Deut. Chem. Ges. **62** [1929] 2390/2).

[5] Westbury, R. A.; Winkler, C. A. (Can. J. Chem. **38** [1960] 334/42).

[6] Ovshinsky, S. R.; Madan, A. (Ger. Offen. 3004009 [1980]; C.A. **93** [1980] No. 160058).

2.1.3.10 Formation of "$(SN)_x$-Channel Complexes"

Attempts to prepare $(SN)_x$ chains constrained in dielectric matrices have not yet been successful [1 to 3]. Benzophenone, thiourea, perhydrotriphenylene (PHTP), and zeolite mordenite were used as clathrating hosts. The instability of S_2N_2 prevented the formation of the inclusion complexes before polymerization occurred with the exception of thiourea-S_2N_2 which only yielded an amorphous thiourea-$(SN)_x$ complex by irradiation [1, 2]. The formation and properties of "PHTP-$(SN)_x$ " was described in detail, however it is doubtful if $(SN)_x$ is truly clathrated [3].

References:

[1] Labes, M. M.; Walatka, V. V., Jr. (AD-772741-5GA [1973] 1/15; C.A. **81** [1974] No. 7442).

[2] Labes, M. M. (Mol. Cryst. Liquid Cryst. **32** [1976] 161/3).

[3] Love, P.; Myer, G.; Kao, H. I.; Labes, M. M.; Junker, W. R.; Elbaum, C. (Ann. N.Y. Acad. Sci. **313** [1978] 745/57, 754/6).

2.1.3.11 Process of Polymerization

Several papers about the S_2N_2 polymerization process were summarized in "Sulfur-Nitrogen Compounds" Part 2, 1985, p. 15.

The polymerization of an S_2N_2 crystalline film to $(SN)_x$ was monitored by ESR spectroscopy [1 to 8]. Both pure S_2N_2 and $(SN)_x$ give no ESR signal. The time dependence of the observed ESR signal during the polymerization is shown in **Fig. 16**.

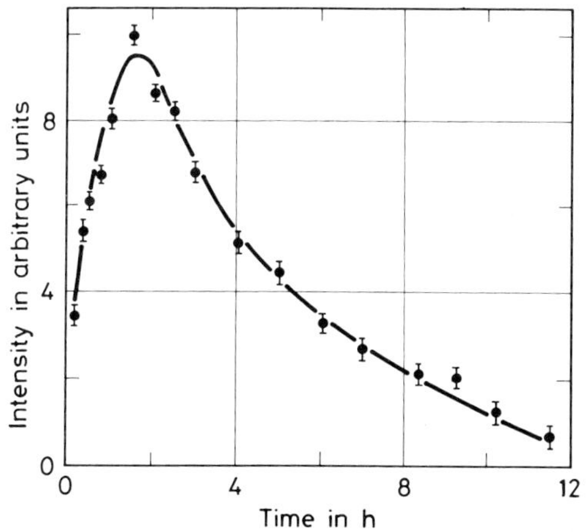

Fig. 16. The time dependence of the ESR signal of an S_2N_2 crystalline film (measured at $-140°C$) polymerized to $(SN)_x$ at room temperature [3].

The intensity of the signal first increased to a maximum value and then decreased. ESR absorption of paramagnetic species was still observed after a time of 12 h [3]. The observed behavior is consistent with a semiempirical INDO-type ASMO-SCF study. Polymerization proceeds via a deformed opened S_2N_2 molecule of $\pi\bar{\pi}^*$ (called $\pi\sigma^*$ in [19]) triplet biradical nature

(π and $\bar{\pi}$ represent π type MOs perpendicular to the molecular plane and in the molecular plane, respectively) [9]. Contrary to this, an ab initio SCF calculation (with double zeta basis augmented by polarization functions) yielded an $(SN)_2$ $\sigma\sigma^*$ triplet ($1^3A'$) as the most likely polymerization precursor [19] (for a previous paper see [20]). The single ESR line was explained by interactions between triplet molecules [5]. The average g value was estimated to be ~2.005 [3]. The ESR spectra (measured at −118°C) of partially polymerized $(SN)_x$ prepared by four different ways (from "normal", "red", "brown", and "sublimed" $(SN)_x$) were compared [6 to 8]. Several ESR parameters were determined and compiled in a table. The ESR traces are Lorentzian except that of "normal" $(SN)_x$. Differences in the ESR spectra are due to variations in the extent of order in the a and c unit cell directions as well as differences in defect densities [8].

Theoretical considerations for the lattice transformation during crystal growth of $(SN)_x$ from solid S_2N_2 were made. The least-motion principle shows that a nonunique reaction via one point ring cleavage leads to a cis-trans chain, in agreement with the observed structure of $(SN)_x$ [10 to 12]. The reaction was assumed to start at either of two equivalent, opposite S_2N_2 bonds, forming two distinguishable chains, see **Fig. 17**.

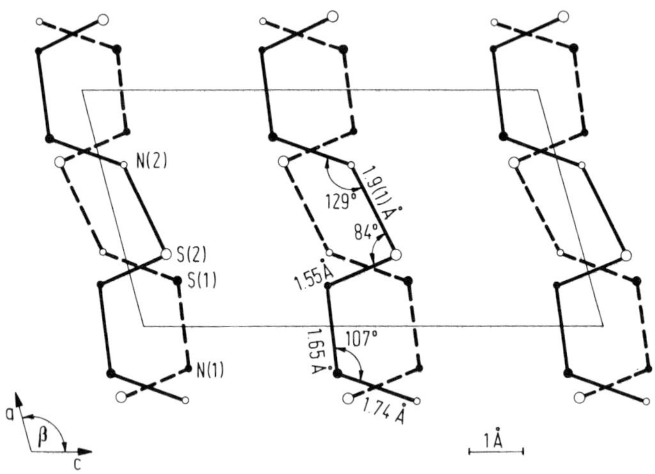

Fig. 17. The b axis projection of partially polymerized S_2N_2. The solid and dashed lines indicate the two symmetry related chains which are equally likely to result from the polymerization reaction. The major contribution to the S(1) and N(1) positions (filled circles) is due to unpolymerized S_2N_2 [11].

Chain coordinates, calculated on the basis of the least-motion principle, agree with observed atom coordinates from partially polymerized S_2N_2 and provide evidence for the proposed mechanism [10, 11].

Some information about the formation of solid $(SN)_x$ from the vapor phase was obtained from mass spectrometric studies of vaporized S_4N_4 and $(SN)_x$, see p. 137 [13, 14]. The gaseous "linear" $(SN)_4$ molecule in the vapor of $(SN)_x$, observed by mass spectroscopy [13, 14] and IR spectroscopy [15], polymerizes spontaneously on cool surfaces [13, 14]. It is assumed that above 200 K the polymerization proceeds largely via this precursor [15]. IR and Raman spectra recorded at 10 K of condensed vapor obtained by heating $(SN)_x$ to 140 and 160°C, and isolated in an Ar matrix indicate, besides noncyclic $(SN)_4$, other noncyclic species such as $(SN)_2$ or $(SN)_6$ [16]. On the basis of a mass spectroscopic study of the thermal decomposition of S_4N_4

(over and in the absence of silver wool) and of $(SN)_x$ vapor a mechanism of polymerization and repolymerization of $(SN)_x$, respectively has been proposed [17, 18]. The polymerization process involves the reaction of $(SN)_2$ (transformed from S_2N_2 in the presence of Ag wool) with SN originating from S_2N_2 [17]:

$$(SN)_2 \text{ or } (SN)_x + SN \rightarrow (SN)_{x+1}$$

$(SN)_x$ repolymerizes by combining acyclic $(SN)_x$ ($x \geqq 2$) units present in $(SN)_x$ vapor. Cyclic species decompose to SN and react with the polymer chain [18].

Special emphasis was placed on the discussion of the rate of polymerization and $(SN)_x$ crystal structure in relation to the different precursors proposed to be brown $(SN)_2$, red ($SN^{\bullet}$), and colorless S_2N_2. In the solid state the SN fragment forms only with difficulty [17, 18].

References:

[1] Douillard, A.; May, J.-F.; Vallet, G. (Ann. Chim. [Paris] [14] **6** [1971] 257/62).
[2] Mikulski, C. M.; Russo, P. J.; Saran, M. S.; MacDiarmid, A. G.; Garito, A. F.; Heeger, A. J. (J. Am. Chem. Soc. **97** [1975] 6358/63).
[3] Cohen, M. J.; Garito, A. F.; Heeger, A. J.; MacDiarmid, A. G.; Mikulski, C. M.; Saran, M. S.; Kleppinger, J. (J. Am. Chem. Soc. **98** [1976] 3844/8).
[4] Greene, R. L.; Grant, P. M.; Street, G. B. (Phys. Rev. Letters **34** [1975] 89/92).
[5] Moran, M. J. (Diss. Univ. Pennsylvania 1978; Diss. Abstr. Intern. B **39** [1979] 5374).
[6] Love, P.; Myer, G.; Kao, H. I.; Labes, M. M.; Junker, W. R.; Elbaum, C. (Ann. N.Y. Acad. Sci. **313** [1978] 745/57, 746/7).
[7] Love, P.; Myer, G. H.; Kao, H. I.; Labes, M. M.; Junker, W. R.; Elbaum, C. (Org. Coat. Plast. Chem. Preprints **38** [1978] 621/2).
[8] Love, P.; Labes, M. M. (J. Chem. Phys. **70** [1979] 5147/50).
[9] Yamabe, T.; Tanaka, K.; Fukai, K. (J. Phys. Chem. **81** [1977] 727/31).
[10] Baughman, R. H.; Chance, R. R. (J. Polym. Sci. Polym. Phys. Ed. **14** [1976] 2019/35).

[11] Baughman, R. H.; Chance, R. R.; Cohen, M. J. (J. Chem. Phys. **64** [1976] 1869/76).
[12] Baughman, R. H. (Contemp. Top. Polym. Sci. **2** [1977] 205/33).
[13] Smith, R. D.; Wyatt, J. R.; DeCorpo, J. J.; Saalfeld, F. E.; Moran, M. J.; MacDiarmid, A. G. (Chem. Phys. Letters **41** [1976] 362/4).
[14] Smith, R. D.; Wyatt, J. R.; DeCorpo, J. J.; Saalfeld, F. E.; Moran, M. J.; MacDiarmid, A. G. (J. Am. Chem. Soc. **99** [1977] 1726/30).
[15] Iqbal, Z.; Downs, D. S. (Chem. Phys. **44** [1979] 137/42).
[16] Teichman, R. A., III; Nixon, E. R. (Inorg. Chem. **15** [1976] 1993/4).
[17] Besenyei, E.; Eigendorf, G. K.; Frost, D. C. (Inorg. Chem. **25** [1986] 4404/8).
[18] Besenyei, E.; Eigendorf, G. K.; Frost, D. C. (J. Electron Spectrosc. Relat. Phenom. **41** [1986] 419/26).
[19] Palmer, M. H.; Guest, M. F. (Chem. Phys. **110** [1986] 187/94).
[20] Palmer, M. H.; Findlay, R. H. (J. Mol. Struct. **92** [1983] 373/83).

2.1.3.12 Impurities in $(SN)_x$

Mass spectral studies of the vapor phase of $(SN)_x$ show that hydride impurities in the order of 0.02 to 0.1 wt% are incorporated in "analytically pure" crystals [1 to 6]. In **Fig. 18**, p. 100, a mass spectrum of hydride ion species is shown ($(SN)_x^+$ ions were subtracted from the spectra of the $(SN)_x$ vapor).

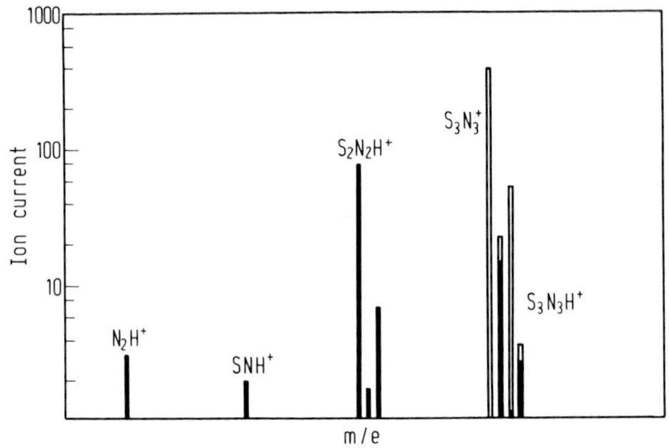

Fig. 18. Low-energy (14 eV) electron impact mass spectrum of the hydride impurity in $(SN)_x$ (solid bars). The light bars show the contribution at each mass peak due to $(SN)_x$ isotopic fragments of $S_3N_3^+$ [3, 8].

It was concluded from a phase-angle mass spectrometric study that all of the observed hydride ions originate from one single neutral species with a molecular weight of 215 ± 5 [3] or 220 ± 12 [4, 5]. The compound, HS_5N_4, is a reasonable precursor for the hydrogen species. This assumption was confirmed by a clustogram (measurement of ion current intensity vs. ionizing energy) [3, 5]. A linear geometry was ascribed to the HS_5N_4 molecule, based on the fact that it is not observable in FI or FD spectra [3]. NMR studies showed that hydrogen is chemically bonded [3] to nitrogen [6]. Hydrogen bonding between $(SN)_x$ chains is possible [6]. The hydride impurities probably originated from reaction with H_2O during the polymerization process [3], the water originating from outgassing of the glass surfaces, stopcock grease and small leaks in the vacuum system [4]. This is confirmed by identification of HNSO in the vapor of S_4N_4 after passing it over heated silver selenide. HNSO was isolated in the matrix [9]. Addition of H_2O to $(SN)_x$ crystals caused an increase in the hydride impurity [4] which did not, however, influence the electrical conductivity [5, 8].

Two other impurities, presumably $SNOH^+$ at m/e = 62 and 63 and $N_2S_3O^+$ at m/e = 140, were detected [1, 2]. These species are not inherent in $(SN)_x$ crystals, but instead may result from $(SN)_x$ surface contamination. The existence of $SNOH^+$ was not confirmed by an FT-IR study [7].

References:

[1] Smith, R. D.; Wyatt, J. R.; DeCorpo, J. J.; Saalfeld, F. E.; Moran, M. J.; MacDiarmid, A. G. (J. Am. Chem. Soc. **99** [1977] 1726/30).

[2] Smith, R. D.; Wyatt, J. R.; DeCorpo, J. J.; Saalfeld, F. E.; Moran, M. J.; MacDiarmid, A. G. (Chem. Phys. Letters **41** [1976] 362/4).

[3] Saalfeld, F. E.; DeCorpo, J. J.; Wyatt, J. R.; Mah, P. T.; Allen, W. N. (Ann. N.Y. Acad. Sci. **313** [1978] 559/66).

[4] Smith, R. D.; Wyatt, J. R.; Weber, D. C.; Saalfeld, F. E. (Inorg. Chem. **17** [1978] 1639/42).

[5] Saalfeld, F. E.; DeCorpo, J. J.; Allen, W. N.; Weber, D. C.; Wyatt, J. R. (Advan. Mass Spectrom. B **8** [1980] 1913/20).

[6] Street, G. B.; Greene, R. L. (IBM J. Res. Develop. **21** [1977] 99/110).

[7] Ishida, H.; Rickert, S. E.; Hopfinger, A. J.; Lando, J. B.; Baer, E.; Koenig, J. L. (J. Appl. Phys. **51** [1980] 5188/93).

[8] Saalfeld, F. E.; DeCorpo, J. J.; Wyatt, J. R. (in: Miller, J. S., Extended Linear Chain Compounds, Vol. I, Plenum, New York 1982, pp. 33/72, 50/58).
[9] Almond, M. J.; Downs, A. J.; Jeffery, T. L. (Polyhedron **7** [1988] 629/34).

2.1.3.13 Handling. Explosive Nature

Several papers mention the explosiveness of $(SN)_x$ under certain conditions. One group reported that $(SN)_x$ exploded, on grinding, compressing or heating it in the presence of air [1, 2]. Other workers observed that $(SN)_x$ decomposed violently when heated above 200°C [3]. Explosions were also observed when $(SN)_x$ was subjected to high pressures [3 to 5].

References:

[1] Street, G. B.; Greene, R. L. (IBM J. Res. Develop. **21** [1977] 99/110).
[2] Street, G. B.; Arnal, H.; Gill, W. D.; Grant, P. M.; Greene, R. L. (Mater. Res. Bull. **10** [1975] 877/82, 879).
[3] Mikulski, C. M.; Russo, P. J.; Saran, M. S.; MacDiarmid, A. G.; Garito, A. F.; Heeger, A. J. (J. Am. Chem. Soc. **97** [1975] 6358/63).
[4] Sargent, E. L. (Chem. Eng. News **54** No. 47 [1976] 5).
[5] Banister, A. J.; Hauptman, Z. V.; Passmore, J.; Wong, C.-M.; White, P. S. (J. Chem. Soc. Dalton Trans. **1986** 2371/9, 2374).

2.1.4 Structural Properties

2.1.4.1 "$(SN)_x$ Isomers"

Three possible isomers of $(SN)_x$ have been discussed [1, 2]:

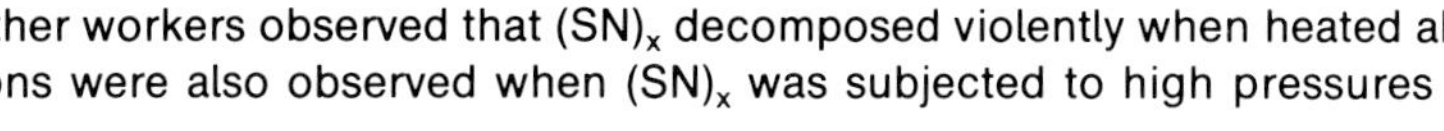

a (cis) b (cis) c (trans)

The energies of the three isomers (a to c) relative to the experimental geometry of a have been calculated (energies per S_2N_2 unit in kcal/mol) [2]:

a	b	c	Ref.
0	0.1	18.3	[1]
0	2.03	19.4	[2]

Only the cis isomer has been synthesized in agreement with a Woodward-Hoffmann orbital study which showed that the formation of the cis isomer is thermally allowed (and also under special conditions photochemically allowed) [2]. It was also shown by a Woodward-Hoffmann orbital correlation study that the formation of trans isomer would be unfavorable because of a high reaction barrier for the polymerization process and because *trans*-$(SN)_x$ is thermodynamically less stable than *cis*-$(SN)_x$ [3].

References:

[1] Whangbo, M.-H.; Hoffmann, R.; Woodward, R. B. (Proc. Roy. Soc. [London] A **366** [1979] 23/46, 41/44).

[2] Burdett, J. K. (J. Am. Chem. Soc. **102** [1980] 5458/62).
[3] Burdett, J. K.; Price, S. L. (Phys. Rev. B **25** [1982] 5778/89).

2.1.4.2 "Molecular Weight"

The conventional methods for the determination of the molecular weight of $(SN)_x$ cannot be employed because $(SN)_x$ is insoluble in common solvents, however, using a relationship between the average degree of polymerization and the vapor pressure, an average "molecular weight" of 2.9×10^{11} was calculated. It was concluded that considerable interchain bonding exists.

Reference:

Moran, M. J. (Diss. Univ. Pennsylvania 1978, pp. 1/443, 363/9).

2.1.4.3 $(SN)_x$ Crystals

Crystal Structure of Normal $(SN)_x$

Early crystallographic studies on $(SN)_x$ were performed using a polycrystalline material. A repeating pattern of 4.43 Å along, and 7.5 Å perpendicular to the chain axis was observed in rotating crystal X-ray photographs [1]. From electron diffraction measurements values different from those above were obtained (6.33 resp. 6.88 Å) [2]. However, the first data were confirmed by later studies. The first complete crystal structure determination of $(SN)_x$ was performed using electron diffraction [3 to 7]. The normal crystalline phase (described in this section) was called "β phase" [3 to 7]. (In some papers the term "α phase" is also used [17 to 20]. For other phases of $(SN)_x$ see p. 105.) The structure was also determined by X-ray diffraction [8, 9] and neutron diffraction [10]. All three studies agree that $(SN)_x$ has monoclinic structure with space group $P2_1/c-C^5_{2h}$ (No. 14). Other common results are: Four (SN) units per unit cell, an infinite alternating arrangement of S and N atoms, and a chain axis which coincides with the crystallographic b direction. Nearly the same lattice constants were found by all workers. The comparative Table 26 summarizes important fundamental structural parameters.

Table 26
Structural Parameters of $(SN)_x$ from Electron-, X-Ray-, and Neutron Diffraction Studies.

structural parameter	electron diffraction [3 to 7]	X-ray diffraction [8, 9]	neutron diffraction [10]
symmetry	monoclinic	monoclinic	monoclinic
space group	$P2_1/c$	$P2_1/c$	$P2_1/c$
a in Å	4.12	4.153(6)	4.11(1)
b in Å	4.43	4.439(5)	4.43(1)
c in Å	7.64	7.637(12)	7.63(1)

Table 26 (continued)

structural parameter	electron diffraction [3 to 7]	X-ray diffraction [8, 9]	neutron diffraction [10]
β in °	109.3	109.7	110.0
R factor	0.19	0.11	0.109
distance in Å intrachain			
S–N	1.575, 1.718	1.593, 1.628	1.585(10)
S···N	2.91	2.86	2.86(1)
N···N	2.72	2.58	2.55(1)
S···S	2.76	2.79	2.75(1)
interchain (parallel ($\bar{1}$02))			
S···N	3.59	3.24	3.26(1)
N···N	3.15	3.33	3.35(1)
S···S	3.10	3.48	3.47(1)
interchain (between ($\bar{1}$02) planes)			
S···N	3.60	3.37	3.38(1)
N···N	2.81	3.31	3.34(1)
S···S	3.55	3.72	3.70(1)
angle in ° (intrachain)			
S–N–S	113.5	119.9	120.3(5)
N–S–N	111.5	106.2	107.0(5)

X-ray powder diffraction data for $(SN)_x$ are given in [8]. Illustrations of the crystal structures of $(SN)_x$ and projections onto the ($\bar{1}$02) plane determined by X-ray diffraction are presented in **Fig. 19** and **20**, p. 104 [8, 9], by neutron diffraction in [10]. These structures, which are very similar, are believed today to be the better approximation to the true $(SN)_x$ structure than the electron diffraction structure [11].

The X-ray diffraction study was performed both at room temperature and at −145°C on a "single crystal" of $(SN)_x$ which showed only a very few reflections attributed to twinning. The structure was solved by the Patterson method. Atomic positions are given in the original paper [8]. Neutron diffraction results on an $(SN)_x$ "single crystal" (15 mm^3) performed at room temperature confirm the results of the X-ray study. The data are verified by a low R factor [10]. Additional fractionally occupied sites which imply a high degree of disorder on a molecular scale were observed [8]. Defect sites besides the N and S sites of ordinary strands were also detected [10]. The $(SN)_x$ chains were found from X-ray diffraction to be planar at −145°C and nearly planar at room temperature [8]. Neutron diffraction studies also show that each $(SN)_x$ chain is planar to the ($\bar{1}$02) plane and parallel to other chains [10]. A comparison between X-ray and electron-diffraction results shows discrepancies. The $(SN)_x$ chains from the two methods are shifted relative to one another along the b axis [12]. Alternating bond lengths corresponding to a single or double bond were reported using electron diffraction [3 to 7], but in contrast, nearly equal S–N bond lengths were determined using X-ray diffraction [8]. Neutron diffraction data show the intrachain SN distances to be equal (1.585(10) Å). The bond character (~1.4) lies between a single and a double bond [10]. In agreement with this, an "increased-valence" structure of $(SN)_x$ was presented, invoking double-bond character for all SN bonds in a chain [21]. X-ray and neutron diffraction show the SNS and the NSN angle to be similar [8, 9], but different from the electron diffraction results [3]. The electron-diffraction study reveals interchain distances at normal van der Waals distances (V.D.W. radii in Å: S–S: 3.70; N–N: 3.15; S–N: 3.35 [12]) except for lower values between sulfur atoms related by an inversion center, which indicates interchain coupling, but the X-ray data suggest fairly weak interchain coupling. Neutron diffraction data also show that the shortest interchain distances are in the range of the sum of the van der Waals radii, suggesting weak interchain bonding. Observed and calculated structure factors of $(SN)_x$ are presented in a table [10]. From IR and Raman data it was also concluded that $(SN)_x$ must be regarded as a system of interacting individual chains [10, 24]. Linear compressibility measurements confirm the anisotropic three-dimensional bonding character of $(SN)_x$ [23].

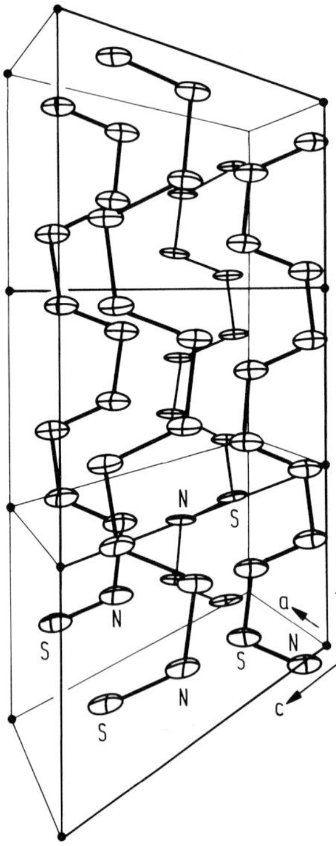

Fig. 19. X-ray crystal structure of $(SN)_x$ [9].

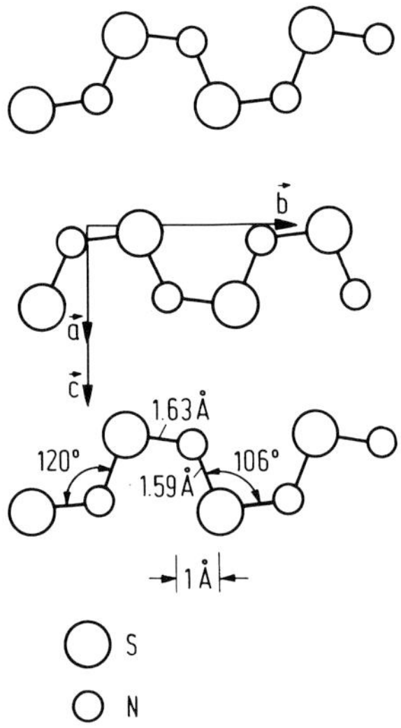

Fig. 20. Projection of $(SN)_x$ chains onto the ($\bar{1}$02) plane [9].

X-Ray Structure of "Red $(SN)_x$" and "Brown $(SN)_x$"

"Red $(SN)_x$" (see p. 92) was found by X-ray diffracion to have essentially the same unit cell parameters as observed for normal $(SN)_x$. The "b" and "c" parameters obtained for "brown $(SN)_x$" are in good agreement with those of normal $(SN)_x$ (see p. 102), but the "a" parameter is different (4.302 ± 0.002 Å), presumably indicating another crystal structure. A table in the original paper summarizes the unit cell data for $(SN)_x$ prepared from three different precursors [13].

X-Ray Structure of Partially Polymerized S_2N_2

The structure of an S_2N_2 crystal which was allowed to polymerize partially, was determined using X-ray diffraction [9]. The crystals were obtained by storing S_2N_2 crystals at 3°C for several days and subsequently warming to 23°C for 6 h. They exhibited the same golden luster and striated surface as normal $(SN)_x$. The X-ray analysis was performed at −130°C. Partially polymerized S_2N_2 crystals have monoclinic structure with space group $P2_1/c-C_{2h}^5$ (No. 14) and a = 4.460(4), b = 3.755(4) and c = 8.470(9) Å, β = 105.58(8)°. S–N bond distances of 1.650(4) and 1.652(3) Å, an SNS bond angle of 90.6(2)°, and an NSN angle of 89.4(2)° were determined. The crystal consisted of 25% crystalline S_2N_2. Like $(SN)_x$, partially polymerized S_2N_2 has a well-defined defect structure in the form of periodic interstitial atomic sites [9]. From a comparison of those data with the parameters of S_2N_2 and $(SN)_x$ crystals, it was concluded that the crystallographic a axis of S_2N_2 becomes the b axis in $(SN)_x$ [9].

Other Structural Phases of $(SN)_x$

In addition to the normal β phase of $(SN)_x$ the existence of a poorly crystalline α phase (monoclinic, a = 4.04 ± 0.05, b = 4.43 ± 0.05, and c = 7.84 ± 0.05 Å, β = 110° ± 1°) and of an α′ (also monoclinic) phase was discussed in the electron-diffraction study [3], however, these "phases" have never been confirmed by others until today [12].

A well-established new orthorhombic phase was produced by a martensitic-like phase transformation [14, 15]. The conversion was performed by simply grinding (or shearing) the normal phase of $(SN)_x$ in a mortar and pestle. This deformation, however, is incomplete. A part of the material is still monoclinic but with changed lattice parameters compared with unsheared $(SN)_x$: a = 4.040(2), b = 4.429(8), c = 7.849(3) Å, β = 109.3(1)° [15]. It should be noted that these data are very similar to those of the "β phase" detected in the electron-diffraction study (see above). The new orthorhombic phase belongs to the space group $P2_12_12_1-D_2^4$ (No. 19) having a = 6.251(3), b = 4.429(8), and c = 4.807(4) Å. The chain geometry was predicted to be quite similar to that of the normal phase. The principal difference between the normal and sheared $(SN)_x$ phase relates to the alternation of translationally inequivalent chains [15]. The transformation modes were calculated by a theory of shear-induced transformation modes in crystals [16]. The process is analogous to the stress-induced phase change in polyethylene. Exact properties of the orthorhombic phase were not evaluated, because the phase was impure. The new phase does not transform to the normal phase on heating even above 175°C [15].

Three other new phases of $(SN)_x$ were formed by epitaxial polymerization of S_2N_2 on freshly cleaved (100) faces of monovalent alkali halide single crystals [17 to 20]. $(SN)_x$ is highly oriented and shows no evidence of the used fibrillar morphology, being free of twinning. The structures are reported to differ markedly from the usual β phase (in the relevant papers named α phase). Polymerization on KCl yields the usual β form, free of twinning. On KI or NaF a phase referred to as γ phase has been obtained [17, 18] and on NaCl and KBr two further phases named δ and ε, respectively, were detected [17]. In Table 27, p. 106, the various crystal forms with lattice parameters which have been produced by epitaxial polymerization are compiled.

The intensities and structural factors of diffraction reflections for the γ, δ, and ε phases are presented in tables [17]. Lattice matching is a possible reason for the formation of three new $(SN)_x$ phases [20].

Differential thermal analysis of $(SN)_x$ crystals in a helium atmosphere showed no phase change between liquid nitrogen temperature and 200°C [22]. No phase change was visible in the Raman spectrum of $(SN)_x$ crystals when the temperature was lowered from 300 K to liquid helium temperature [25].

Table 27
Crystal Forms and Lattice Parameters of $(SN)_x$ Produced by Epitaxial Polymerization, from [19, 20].

substrate	form	crystal class	a in Å	b in Å	c in Å	β in °	space group
KCl	β[a]	monoclinic	4.15	7.64	4.44	110	$P2_1/c-C^5_{2h}$ (No. 14)
KI, NaF	γ	orthorhombic	9.20	0.72	4.93		$P2_12_12_1-D^4_2$ (No. 19)
NaCl	δ	—	6.80	—	5.24		$P2_1$[b]
KBr	ε	—	3.92	—	5.81		P2[b]

[a] In papers [17 to 20] called α. – [b] Full space group unknown.

References:

[1] Goehring, M.; Voigt, D. (Z. Anorg. Allgem. Chem. **285** [1956] 181/90, 184).
[2] Chapman, D.; Warn, R. J.; Fitzgerald, A. G.; Yoffe, A. D. (Trans. Faraday Soc. **60** [1964] 294/300).
[3] Boudeulle, M. (Diss. Claude-Bernard Univ. Lyon 1974).
[4] Boudeulle, M.; Douillard, A.; Michel, P.; Vallet, G. (Compt. Rend. C **272** [1971] 2137/8).
[5] Boudeulle, M.; Douillard, A. (J. Microsc. [Paris] **11** [1971] 3/4).
[6] Boudeulle, M.; Michel, P. (Acta Cryst. A **28** [1972] S199).
[7] Boudeulle, M. (Cryst. Struct. Commun. **4** [1975] 9/12).
[8] Mikulski, C. M.; Russo, P. J.; Saran, M. S.; MacDiarmid, A. G.; Garito, A. F.; Heeger, A. J. (J. Am. Chem. Soc. **97** [1975] 6358/63).
[9] Cohen, M. J.; Garito, A. F.; Heeger, A. J.; MacDiarmid, A. G.; Mikulski, C. M.; Saran, M. S.; Kleppinger, J. (J. Am. Chem. Soc. **98** [1976] 3844/8).
[10] Heger, G.; Klim, S.; Pintschovius, L.; Kahlert, H. (J. Solid State Chem. **23** [1970] 341/7).

[11] Greene, R. L.; Street, G. B. (NATO Advan. Study Intern. Ser. B **25** [1977] 167/95, 170).
[12] Street, G. B.; Greene, R. L. (IBM J. Res. Develop. **21** [1977] 99/110).
[13] Love, P.; Myer, G.; Kao, H. I.; Labes, M. M.; Junker, W. R.; Elbaum, C. (Ann. N.Y. Acad. Sci. **313** [1978] 745/57, 750).
[14] Baughman, R. H.; Apgar, P. A.; Chance, R. R.; MacDiarmid, A. G.; Garito, A. F. (J. Chem. Soc. Chem. Commun. **1977** 49/50).
[15] Baughman, R. H.; Apgar, P. A.; Chance, R. R.; MacDiarmid, A. G.; Garito, A. F. (J. Chem. Phys. **66** [1977] 401/9).
[16] Young, R. J.; Baughman, R. H. (J. Mater. Sci. **13** [1978] 55/61).
[17] Rickert, S. E.; Ishida, H.; Lando, J. B.; Koenig, J. L.; Baer, E. (J. Appl. Phys. **51** [1980] 5194/200).
[18] Rickert, S. E.; Lando, J. B.; Hopfinger, A. J.; Baer, E. (Macromolecules **12** [1979] 1053/7).
[19] Lando, J. B.; Baer, E.; Rickert, S. E.; Nae, H.; Ching, Y. (ACS Symp. Ser. No. 212 [1983] 89/100).
[20] Lando, J. B. (Physicochem. Aspects Polym. Surf. Proc. Intern. Symp., New York 1981 [1983], Vol. 1, pp. 559/67).

[21] Harcourt, R. D. (J. Inorg. Nucl. Chem. **39** [1977] 237/42).
[22] Street, G. B.; Arnal, H.; Gill, W. D.; Grant, P. M.; Greene, R. L. (Mater Res. Bull. **10** [1975] 877/82, 879).
[23] Clarke, R. (Solid State Commun. **25** [1978] 333/6).

[24] Stolz, H. J.; Otto, A.; Pintschovius, L. (Proc. 3rd. Intern. Conf. Light Scattering Solids, Campinas, Brazil, 1975 [1976], pp. 737/41).
[25] Temkin, H.; Fitchen, D. B. (Solid State Commun. **19** [1976] 1181/4).

2.1.4.4 Defects in $(SN)_x$ Crystals

Even the best $(SN)_x$ crystals prepared to date are highly imperfect [1, 2]. $(SN)_x$ crystals are fibrous, twinned, and disordered [2]. Only $(SN)_x$ formed by epitaxial polymerization on alkali halide single crystals is not fibrous and also not twinned [3], see p. 105. Using transmission electron microscopy (TEM), longitudinal striations of variable magnitude with a minimum of 240 to 250 nm were detected parallel to the fiber and polymer axes [7]. Electron microscopic studies and X-ray diffraction experiments reveal that $(SN)_x$ crystals prepared by the standard polymerization process via vapor grown S_2N_2, consist of a bundle of fibers. Different ranges were reported for the diameter of fibers: <200 Å using scanning electron microscopy (SEM) [4], 50 to 70 Å using transmission electron microscopy (TEM) [5], 250 to 1200 Å using SEM and TEM [6], 50 to 500 Å from linewidth of X-ray scans [2], 25 to 200 Å using electron-diffraction microscopy and TEM [7], 10 to 30 Å using SEM, and ~10 Å using TEM [8]. From fluctuation induced conductivity above T_c the fiber diameter has been estimated to be ~240 Å [16]. Also from a fluctuation conductivity study in combination with upper critical field results, the fiber radius was derived to lie in the range 74 to 200 Å (sample-dependent) [17]. A mosaic spread of 10° about both the fiber direction and the perpendicular to it was reported [8]. SEM images of $(SN)_x$ crystals are seen, for example, in [1, 2, 5 to 12], bright field and dark field electron micrographs are reproduced, for example, in [7, 10, 13], and high-resolution images of $(SN)_x$ are depicted in [14, 15]. Two different mechanisms for the fiber formation were proposed [6, 18]. In the first mechanism the fibrous texture was explained by assuming that polymer growth proceeds rapidly in the chain direction but stops in the lateral direction due to differences in the lattice constants of the monomer and polymer [6]. In the second mechanism, twinning is believed to be responsible for the fibrousity in $(SN)_x$ [18]. Other workers state, however, that neither mechanism alone can account for the periodic fibril arrangement [7]. High-resolution microscopy shows narrow microfibrils. The total end point dose (i.e., the electron irradiation dose necessary for complete disappearance of all crystalline reflections in the diffraction pattern) for $(SN)_x$ fibers is 0.5 to 0.8 Coulomb/cm² (electron energy 200 keV). The expected resolution is 0.5 to 0.4 nm [15].

Twinning is also a kind of structural defect in $(SN)_x$. From structure determination studies, a twinning mode was recogniced in which the twin was formed by a 180° rotation about the crystal c axis ("(100) twin mode") [19, 20]. The two twinning components were almost of the same size [20]. In addition to this generally accepted mode, others such as a $(\bar{1}02)^*$ and a $(\bar{1}01)$ mode were proposed [7]. A twin mode on $(\bar{1}02)^*$ was also postulated from rotating crystal X-ray diffraction data. Other modes are possible. Observed and theoretical periodicities (from X-ray and electron-diffraction study) were compiled into a comparative table. A figure and a table show relative X-ray intensities and additional data [8]. Additional electron-diffraction reflections were assigned to a new twinning mode on $(\bar{1}01)$ (besides the mode on (100)) and to unit cell doubling. A periodicity (2 μm = 200 Å) perpendicular to the chains was indicated from the streaked reflections which consist of regularly spaced dots [7]. Streaks perpendicular to the chain axis were observed using small-angle X-ray (SAXS) and electron scattering (SAES) technique. The streaks arise from regions of different density either as rods (50 to 100 nm in diameter) parallel to the chain axis or plates arrayed with cylindrical symmetry about the chain axis. They represent voids or incompletely polymerized zones. No scattering was observed from fine fibrils (<10 nm in diameter) [13]. Lattice fringes up to 0.22 nm, which varied from place to place, were observed (besides the fibrillar or mosaic texture) in $(SN)_x$ using high-resolution electron microscopy [14]. The lattice fringes were detected in domains extended along the molecular axis and were very narrow in the perpendicular direction. The authors concluded that an $(SN)_x$ crystal is composed of a large number of small crystallites varying in size but maintaining the b axis orientation [14].

Besides the ordinary S and N sites, another kind of disorder in $(SN)_x$ was detected with occupation half by S and half by N [2, 20]. Twinning ((100) mode), molecular scale disorder and also differing crystallite orientations are believed to be a result of a nonunique polymerization reaction of S_2N_2 [18, 21 to 23].

References:

[1] Greene, R. L.; Street, G. B. (NATO Advan. Study Inst. B **25** [1977] 167/95).
[2] Pintschovius, L. (Colloid Polym. Sci. **256** [1978] 883/92).
[3] Lando, J. B. (Physicochem. Aspects Polym. Surface Proc. Intern. Symp., New York 1981 [1983], Vol. 1, pp. 559/67, 662).
[4] Geiss, R. H.; Arnal, H. (Ind. Res. **18** [1976] 82/4).
[5] Gill, W. D.; Bludau, W.; Geiss, R. H.; Grant, P. M.; Greene, R. L.; Mayerle, J. J.; Street, G. B. (Phys. Rev. Letters **38** [1977] 1305/8).
[6] Street, G. B.; Greene, R. L. (IBM J. Res. Develop. **21** [1977] 99/110).
[7] Stejny, J.; Dlugosz, J.; Keller, A. (J. Mater. Sci. **14** [1979] 1291/300).
[8] Hoel, R. H.; Dingley, D. J. (J. Mater. Sci. **17** [1982] 2997/3008).
[9] Hoel, R. H.; Dingley, D. J. (J. Mater. Sci. **17** [1982] 2990/6).
[10] Boudeulle, M.; Douillard, A.; Michel, P.; Vallet, G. (Compt. Rend. C **272** [1971] 2137/9).

[11] Murray, C. (Chem. Eng. News **53** (21) [1975] 18/9 and cover).
[12] Kaneto, K.; Yamamoto, M.; Yoshino, K.; Inuishi, Y. (J. Phys. Soc. Japan **47** [1979] 167/75).
[13] Schultz, J. M.; Bentley, J.; Hendricks, R. W.; Lin, J. L.; Petermann, J. (J. Appl. Phys. **52** [1981] 5389/91).
[14] Katayama, K. (Colloid Polym. Sci. **262** [1984] 429/34).
[15] Katayama, K.; Isoda, S.; Tsuji, M.; Ohara, M.; Kawaguchi, A. (Bull. Inst. Chem. Res. Kyoto Univ. **62** [1984] 198/207, 205).
[16] Civiak, R. L.; Elbaum, C.; Nichols, L. F.; Kao, H. I.; Labes, M. M. (Phys. Rev. B **14** [1976] 5413/21).
[17] Barrett, P.; Herrenden-Harker, W. G.; Priestley, M. G.; da Silva, E. Z. (J. Phys. F **11** [1981] 873/82).
[18] Baughman, R. H.; Chance, R. R. (J. Polym. Sci. Polym. Phys. Ed. **14** [1976] 2019/35).
[19] Boudeulle, M. (Diss. Claude-Bernard Univ. Lyon 1974).
[20] Heger, G.; Klein, S.; Pintschovius, L.; Kahlert, H. (J. Solid State Chem. **23** [1978] 341/7).

[21] Cohen, M. J.; Garito, A. F.; Heeger, A. J.; MacDiarmid, A. G.; Mikulski, C. M.; Saran, M. S.; Kleppinger, J. (J. Am. Chem. Soc. **98** [1976] 3844/8).
[22] Baughman, R. H.; Chance, R. R.; Cohen, M. J. (J. Chem. Phys. **64** [1976] 1869/76).
[23] Baughman, R. H. (Contemp. Top. Polym. Sci. **2** [1977] 205/33).

2.1.4.5 Deformation of $(SN)_x$ Crystals

Crystals of $(SN)_x$, which were mechanically deformed either by pulling the crystals apart or by compression between glass slides, were examined using transmission electron microscopy [1]. The deformation process yields numerous fine kink bands. One of the typical features is an arbitrary angle of kinking. A model of kink-band formation under stress, whereby the chain lengths are maintained constant was discussed [1]. A further paper by the same authors deals with the regions of relatively high deformation [2]. Using transmission electron microscopy, fine striae (≧40 Å) running perpendicular to the chain axis and fine lineage (~60 Å) parallel to the chain axis were observed near the kink bands [2]. A fine fibrillation (~38 Å) parallel to the chain axis exists in areas in which rupture occurs. A model of repeatedly reversed twisting of chains was proposed as a possible explanation for the fine striae. The fibrillation at rupture sites and the fine lineage which have similar periodicity, are generated during the crystal growth or polymerization process [2]. Many subboundaries along the fiber axis, similar to the results above [2], were observed using dark field electron microscopy [3]. The mean separation between the boundaries is several hundred Ångström [3].

References:

[1] Petermann, J.; Schultz, J. M. (J. Mater. Sci. **14** [1979] 891/6).
[2] Schultz, J. M.; Petermann, J. (Phil. Mag. [8] A **40** [1979] 27/38).
[3] Oda, Y.; Takenaka, H.; Nagano, H.; Nakada, I. (Solid State Commun. **35** [1980] 887/90).

2.1.4.6 X-Ray Study of $(SN)_x$ Films

The structure of epitaxial films is described on pp. 105/6.

X-ray diffraction data of $(SN)_x$ films prepared by sublimation of high-purity crystalline material or by the thermal decomposition of S_4N_4 vapor agree with those of $(SN)_x$ crystals [1 to 4]. The films consist of randomly oriented crystallites with a grain size on the order of 1 μm. The ($\bar{1}02$) plane of the crystal which is parallel to the plane containing the $(SN)_x$ chain is also parallel to the plane of the polymer substrate. The conducting axis is assumed to lie in the plane of the film [1 to 4].

References:

[1] Bright, A. A.; Cohen, M. J.; Garito, A. F.; Heeger, A. J.; Mikulski, C. M.; Russo, P. J.; MacDiarmid, A. G. (Phys. Rev. Letters **34** [1975] 206/9).
[2] Bright, A. A.; Cohen, M. J.; Garito, A. F.; Heeger, A. J.; Mikulski, C. M.; Russo, P. J.; MacDiarmid, A. G. (Appl. Phys. Letters **26** [1975] 612/5).
[3] Louis, E. J.; MacDiarmid, A. G.; Garito, A. F.; Heeger, A. J. (J. Chem. Soc. Chem. Commun. **1976** 426).
[4] Hsu, C.; Labes, M. M. (J. Chem. Phys. **61** [1974] 4640/5).

2.1.5 Electronic Structure and Bonding

2.1.5.1 Survey

The electronic structure of $(SN)_x$ may be described by a threefold hierachy of interactions:

– first a strong metallic interaction in the direction of the chain axis (along b),

- second, perpendicular to it but within the planes of the chains, i.e., in the (100) plane, where strong interchain overlapping of π electron orbital occurs, a weaker interaction which is still just metallic, and
- finally between these planes, in the ($\bar{1}02$) plane, an interaction that is presumably no longer metallic but localized, see p. 123.

The bonding in $(SN)_x$ is characterized by a large charge transfer δ from sulfur to nitrogen, $(S^{+\delta}N^{-\delta})_x$ with $\delta \approx 0.5$, resulting in highly polar SN bonds. Such highly polar bonds were also found in other S–N compounds (such as $(SN)_2$). The polar character could be the (electrostatic) origin of the polymerization process via bond breaking in the energetically favored orientation of planar $(SN)_2$ rings, by thermal excitation to a low-lying high-spin state (possibly a triplet state, see p. 97), lengthening of the ring bonds and interaction of the excited species (cf. "Sulfur-Nitrogen Compounds" Part 2, 1985, pp. 14/7).

2.1.5.2 Core Electron Levels

XPS core level spectra of $(SN)_x$ were studied [1 to 6]. Core electron binding energies E_B (related to the Fermi level E_F) for sulfur and nitrogen in $(SN)_x$ are summarized in the following table [1, 2]:

level	$N(1s_{1/2})$	$S(2s_{1/2})$	$S(2p_{1/2})$	$S(2p_{3/2})$
E_B in eV	397.3	228.5	164.5	163.5

Other workers reported binding energy differences: $N(1s)-S(2p_{3/2})=233.4$ eV [4], $N(1s)-S(2s)=169.2$ eV [5].

The solid-state polymerization of S_2N_2 to $(SN)_x$ was also studied by core level spectroscopy (at −100, −40, 0, and 20°C). The N(1s), S(2s), and $S(2p_{1/2}, 2p_{3/2})$ core level spectra of $(SN)_x$ films (at 20°C) were presented [3] and there was substantial agreement with XPS spectral data of [4]. The N(1s) XPS spectrum of $(SN)_x$ crystals was recorded and discussed. The satellite structure (features at ~1.5, 3, 4.5, and 25 eV below the N(1s) line position) agrees well with EELS (electron-energy-loss spectroscopy) [4]. Wide-scan core level XPS spectra of $(SN)_x$ films (on glass), obtained by vaporization of pure $(SN)_x$ crystals or $(SN)_x$ powder (chemically-prepared) were compared. The spectra of the former films revealed a higher degree of purity and a greater stability against oxygen (E_B(N1s) = 399.2 eV) [6].

References:

[1] Mengel, P.; Grant, P. M.; Rudge, W. E.; Schechtman, B. H.; Price, D. W. (Phys. Rev. Letters **35** [1975] 1803/6).
[2] Mengel, P.; Ortenburger, J. B.; Rudge, W. E.; Grant, P. M. (Lect. Notes Phys. **65** [1977] 591/601).
[3] Sharma, J.; Downs, D. S.; Iqbal, Z.; Owens, F. J. (J. Chem. Phys. **67** [1977] 3045/9).
[4] Salaneck, W. R.; Lin, J. W.-P.; Epstein, A. F. (Phys. Rev. [3] B **13** [1976] 5574/7).
[5] Brant, P.; Weber, D. C.; Ewing, C. T.; Carter, F. L.; Hashmall, J. A. (Syn. Metals **1** [1980] 161/73).
[6] Banister, A. J.; Hauptman, Z. V.; Passmore, J.; Wong, C.-M.; White, P. S. (J. Chem. Soc. Dalton Trans. **1986** 2371/9).

2.1.5.3 Electron Energy Bands

2.1.5.3.1 General Remarks

The unusual electrical properties of $(SN)_x$ have stimulated a large number of band-structure calculations by different quantum-mechanical approximation methods. Several books and review articles have been published, for example [1 to 7]. The most important experimental results on $(SN)_x$ to be accounted for by a band model are [6]:

— the absence of a metal-insulator transition due to a Peierls distortion usually found in one-dimensional crystals;
— metallic conduction not only along but also perpendicular to the polymeric chains (the Fermi surfaces cannot be planes);
— the lowest band gap is 4 eV;
— compatibility of the calculated density-of-states with the measured photoelectron spectrum.

The band-structure calculations were based on the electron and X-ray diffraction structure: the $(SN)_x$ unit cell contains two SN chains with each two SN pairs related by a screw axis symmetry; the number of valence electrons is $4 \times 11 = 44$.

The band-structure calculations can be classified into two groups: one-dimensional (1D) types and three-dimensional (3D) types. The essential difference between these two types is based on the degree of s-p mixing. A small degree of s-p mixing (1D) leads to two overlapping conduction bands, which cross the Fermi level. On the other hand, with strong s-p mixing the two bands separáte, which results in a simple conduction band ("semimetal type" [7]). The absence of the Peierls distortion is accounted for by the band models in different ways. It appears to follow directly from the curvature of the Fermi surface [6].

References:

[1] Greene, R. L.; Street, G. B. (NATO Advan. Study Inst. Ser. B **25** [1977] 167/95, 172/6).
[2] Labes, M. M.; Love, P.; Nichols, L. F. (Chem. Rev. **79** [1979] 1/15, 5).
[3] Bredas, J. L. (Ann. Soc. Sci. Brux. Ser. I **94** [1980] 83/128).
[4] Kamimura, H.; Oshiyama, A. (Theoretical Aspects of Band Structures and Electronic Properties of Pseudo-One-Dimensional Solids, Reidel, Dordrecht 1985, pp. 123/62).
[5] Kaneto, K.; Yoshino, Y.; Inuishi, J. (in: Monceau, P., Electronic Properties of Inorganic Quasi-One-Dimensional Compounds, Pt. II, Reidel, Dordrecht 1985, pp. 69/109, 75/8).
[6] Geserich, H.-P.; Pintschovius, L. (Festkörperprobleme [Advan. Solid State Phys.] **16** [1976] 65/94, 89).
[7] Yoffe, A. D. (Chem. Soc. Rev. **5** [1976] 51/78, 62/3).

2.1.5.3.2 Band-Structure Calculations

2.1.5.3.2.1 One-Dimensional (1D)

Early band-structure calculations of $(SN)_x$ neglected interchain interactions. Only a one-dimensional $(SN)_x$ chain, with a geometry mostly based on the electron-diffraction structure was considered. These calculations, therefore, only explained metallic conduction parallel to the SN chains. The absence of the Peierls distortion was interpreted by the fact that there are two values for the Fermi wave vector k_F which are not a simple fraction of π/b although $k_{F_1} + k_{F_2} = \pi/b$ [9, 50].

Simple Hückel Method [1, 2]. The first band-structure consideration of $(SN)_x$ using the simple Hückel method was based on an $(SN)_x$ geometry assuming bond alternation [1]. The experimental semiconducting properties of the poorly crystalline $(SN)_x$ considered in that study were explained by a small energy gap. Later on the metallic properties of $(SN)_x$ were attributed to a half-filled band and the absence of a Peierls distortion was explained by weak interchain coupling [2].

Extended Hückel Method (EH) [3 to 8]. EH calculations were performed using a planar chain structure (differing from the electron-diffraction structure) [3], the electron-diffraction structure [4, 5], and the X-ray crystal structure [6, 7]. The results of all EH calculations are qualitatively similar. Degeneracy was found in pairs of all bands at the zone edge Z induced by a screw rotation axis. The metallic conductivity was ascribed to the degeneracy of the π valence and to conduction bands at the zone edge.

Tight-Binding Method (TB) [9 to 13]. The tight-binding method results in two overlapping bands crossing E_F (associated with metallic behavior). This is in contrast to EH calculations which result in a single conduction band (with E_F at $k=\pi/b \triangleq Z$) [9 to 13]. The two conduction bands have the character of p wave functions. Numerical errors [9 to 11] were corrected [12]. The absence of a Peierls distortion was explained by the incommensurate values (with respect to the reciprocal lattice vectors) of the two Fermi vectors. s-p mixing is important for the determination of the band structure [10].

LCAO Method [14 to 22, 29, 30]. The ab initio LCAO Hartree-Fock (HF) method was performed mainly on 1D $(SN)_x$ chain structures with equal bond lengths [17 to 19], alternating chains using the electron-diffraction structure [14 to 17] or the X-ray crystal structure [14, 17 to 21]. The geometry of the X-ray crystal structure was confirmed to be the most stable one among these chains [17, 19]. Ab initio LCAO Hartree-Fock crystal orbital (HF-CO) calculations using a double-zeta-type atomic-basis-set yielded values of the effective mass, electron state density and transferred charge which are in good agreement with experimental values (see Table 28, p. 116) [14, 20]. Other workers found that the main feature of the ab initio SCF LCAO-CO method is two overlapping (a σ and a π) partially filled bands [15]. This is in qualitative agreement with tight-binding results [13]. However, in the case of the LCAO calculation the width of the σ bond (6.9 eV) is larger [15]. Kertész et al. found good agreement between calculated and experimental values by the use of the ab initio LCAO HF-CO method [17]. An ab initio LCAO HF calculation was also performed with the use of a minimal Slater type basis set, each function expanded into three Gaussians (STO-3G) [16]. The energy bands obtained are shown in a figure. Metallic properties of $(SN)_x$ were associated with the twelfth half-filled band which has π character. The energy bands were compared with those obtained by an all-valence electron semiempirical calculation using a CNDO/2 type of parametrization [23]. A good agreement was found, with the difference that the half-filled band is of σ type in this method. Other workers performed an ab initio LCAO HF band-structure calculation of an infinite $(SN)_x$ chain using two basis sets which were employed in earlier works to study molecular precursors and molecular subchains of $(SN)_x$ (see Section 2.1.5.7, p. 125, and [24, 25]) [21]. The band structures from both basis sets closely resemble each other. A gap at the Fermi level (0.167, 0.149, 0.142, and 0.125 a.u. for four used basis sets, with inclusion and exclusion of d orbitals) was observed. The overall valence band width was calculated to be 33.3 and 32.5 eV, respectively (for comparison 23.8 [26] and 24.4 eV [27]). The d orbitals were found to stabilize the polymer reducing its ionicity [21].

Two further LCAO energy band structures for an $(SN)_x$ chain were described which correspond to the self-consistent (SCF) spin density wave (SDW) and to the self-consistent charge density wave (CDW) solution, calculated by the ab initio unrestricted (UHF) and restricted Hartree-Fock (RHF) CO method, respectively. A large energy gap of 0.535 a.u. and one of 0.120 a.u. were observed by SDW and CDW, respectively [18, 19]. The ab initio HF SCF LCAO method using a Christoffersen basis set and making use of floating spherical Gaussian orbitals (FSGO) was employed to find the Peierls ground state in five models of isolated $(SN)_x$ chains (two irregular, three regular). The Peierls ground state corresponds to a structure with a succession of two longer bonds and two shorter bonds; it is coherent with the CDW instabilities present in the regular models at the HF level (HF instabilities) [22].

Earlier the LCAO method using a basis set of usual atomic orbitals and a simple Gaussian, was performed to study the electronic structures of both an isolated $(SN)_x$ chain (X-ray crystal structure geometry) and of the three-dimensional crystal [29, 30]. The energy bands and density of states were compared. The band structure is in general similar to those based on EH methods [30].

CNDO Method (complete neglect of differential overlap) [8, 23, 31, 33]. A self-consistent LCAO tight-binding calculation using the CNDO/2 scheme was performed on the basis of two models, a planar zig-zag structure with large bond alternation (electron-diffraction structure) and on a nonalternant linear model [8, 31]. The general features of both band structures are similar. The metallic character was established by a partially occupied valence band. The alternant structure is more stable than the equal-bond structure. CNDO calculations reveal a low ionicity of the structure, which is reduced by considering 3d orbitals (EH calculations overestimate the ionicity). One paper expressed doubts, based on comparative studies of the electronic structure of S_2N_2, whether both EH or CNDO/2 methods are applicable to the $(SN)_x$ energy band problem [32]. A Peierls instability, which would produce a gap around the Fermi level and lower the total Hartree-Fock crystal energy, was not found in an investigation of a superlattice of model conformational distortions [8]. The DOS curve obtained from an SCF tight-binding calculation was in qualitative agreement with experimental data; DOS $(E_F) \neq 0$ indicated metallic character (see Table 28, p. 116) [33]. The all-valence electron CO method, applying the CNDO/2 approximation, was used with the electronic diffraction structure. A comparison made with the results of the ab initio calculation of [21] showed both methods to yield similar energy bands. This led to the conclusion that the CNDO/2 method should be applicable to corresponding 3D systems [23].

Pseudopotential Method. A valence effective Hamiltonian (VEH) pseudopotential band structure calculation of the $(SN)_x$ chain was performed using the geometry of the X-ray crystal structure [28]. The VEH results are similar to the ab initio self-consistent field Hartree-Fock LCAO band structure of [21]. A value of 4.13 eV was reported for the band width [28] (4.12 eV [24]). The VEH calculated density of states is in good agreement with experimental XPS results of [27].

Xα Method [26, 34, 36]. The intersecting spheres model was applied to an ab initio self-consistent calculation of the electronic structure of a $(SN)_x$ single chain based on both the X-ray and electron-diffraction structure yielding similar results for the two structures [26, 34]. A strict agreement with photoemission data was reported. The relative position and the general shape of the valence and of the conduction bands, apart from the doubling brought about by the interchain coupling, are also similar to those [30, 35] of 3D non-selfconsistent calculations [26, 34]. Molecular cluster calculations using the SCF-Xα-scattered wave method (in the overlapping sphere version) were performed on SN oligomers like S_2N_2, S_4N_4, and $(S_4N_4)_2$ (electron-diffraction geometry) [36]. The results for the S_4N_4 chain fragment are in good agreement with X-ray photoelectron results, indicating that the short chain already contains many of the essential features of $(SN)_x$. For a non-selfconsistent LCAO-Xα band structure of $(SN)_x$ see [30].

2.1.5.3.2.2 Three-Dimensional (3D)

Strictly 1D band structures are unstable with respect to a Peierls distortion [37]. Several 3D calculations were performed showing that interchain coupling in $(SN)_x$ stabilizes the metallic state, confirming the basic description of $(SN)_x$ as a highly anisotropic metallic conductor instead of a quasi-one-dimensional metal [4 to 7, 29, 30, 35, 39 to 42].

EH Method [4 to 7]. 3D-EH calculations of $(SN)_x$ were based on the electron-diffraction structure [5] and X-ray crystal structure [6, 7]. The only effect of the 3d orbitals is to narrow the bands near the Fermi level [5]. The single-chain band structure of $(SN)_x$ is unstable with respect to distortions which destroy the screw axis symmetry. This results in a band gap. Interchain coupling (S–S and N–N overlaps), however, splits the doubly degenerate energy bands. 3D-EH results were presented for different crystallographic directions. The results are in qualitative agreement with OPW results [5].

A similar 3D-EH band structure was obtained using the X-ray structure. The metallic properties were attributed to the half-filled conduction band [6]. The 3D-EH band structure based on the X-ray structure was compared with 1D and 2D calculations [7, 36]. Interchain coupling leads to band doubling with the consequence that the Fermi level intersects two bands; this gives a nonvanishing density of state at E_F, indicating metallic character [7]. The results of this calculation are qualitatively similar to others [5, 6], but the wider conduction band observed agrees with OPW results.

OPW Method [38, 39]. The OPW band structure results of $(SN)_x$ give a complete understanding of the unusual physical properties and are in accord with experimental results. 3D-OPW calculations were based on both the electron-diffraction and X-ray crystal structure [38]. **Fig. 21** shows the complete OPW energy bands using the X-ray crystal structure. The band structure based on the electron-diffraction structure is qualitatively the same [38]. Also shown in Fig. 21 is the resulting density of states (DOS) and, in the inset, the Brillouin zone. The important feature of this band structure is that overlapping bands, along symmetry lines corresponding to the polymer chain direction (Γ to Z) and perpendicular directions (Z to E), cross the Fermi level. In this way electron and hole pockets are formed at Z and E, respectively, see p. 121. Thus $(SN)_x$ should not be regarded as a quasi-one-dimensional, but as a highly anisotropic conductor or semimetal. The absence of a Peierls distortion can be explained by the closed Fermi surfaces introduced by interchain coupling (for two other possible explanations see [43]). The degree of splitting of the degeneracy of the bands at Z can be considered as a rough measure of the strength of interchain coupling. A comparison of the calculated DOS with specific heat data and optical measurements showed a good agreement. The OPW results coincide qualitative with other 3D calculations [4 to 7, 36] but not with results in [10]. A main common feature with EH calculations is that the conduction band is only due to π electrons [7, 36].

Pseudopotential Method [40, 41]. 3D semiempirical pseudopotential (PSE) calculations were performed based on the electron-diffraction structure [40]. The potential was taken from SnS_2 and GaN and AlN without change. The band structure, density of states, Fermi surface, and optical spectrum of $(SN)_x$ were studied. The main features of the band structure are qualitatively similar to tight-binding results and, after some corrections, also similar to OPW results. A figure of the band structure (together with the DOS) was given for the directions Γ-Z and Γ-B (parallel and perpendicular to the chain, respectively) [40, 41]. The band structure in the vicinity of the Fermi level E_F along various symmetry lines was also shown in a figure [41]. E_F is placed at a local minimum in the DOS. The bonding in $(SN)_x$ was discussed by means of charge density distribution illustrations and an angular momentum projection of the DOS. The states near the Fermi level are associated with predominantly nonbonding sulfur p-type character. The charge density maximum lies at the nitrogen atom. The DOS qualitatively agrees with XPS results. The calculated reflectivity data are in good accord with experimental data [41].

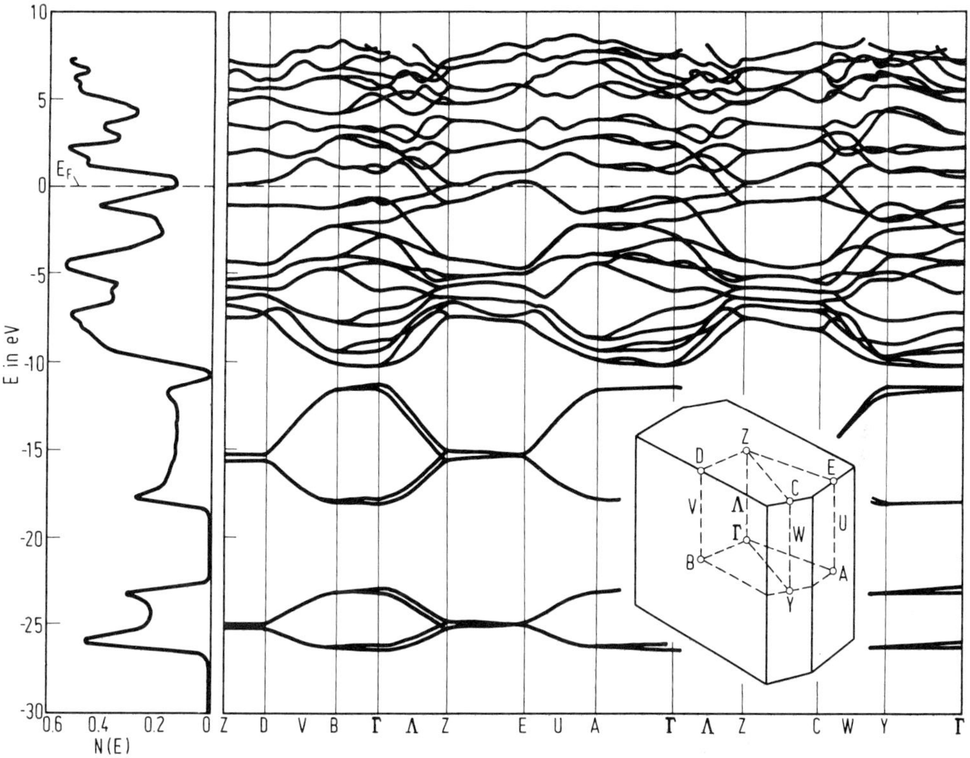

Fig. 21. Band structure of $(SN)_x$ from a 3D-OPW calculation with corresponding density of states N(E) in states·eV^{-1}·$spin^{-1}$·$molecule^{-1}$ on the left. The Brillouin zone is shown in the inset [38].

LCAO Method [29, 30, 35, 42]. A first principles calculation of the band structure of three-dimensional $(SN)_x$ was performed using the LCAO method based on the X-ray crystal structure [29, 30]. The basis functions used were the 1s, 2s, and 2p orbitals of N and the 1s, 2s, 2p, 3s, and 3p orbitals of S with additional single Gaussian orbitals. The energy bands along several symmetric lines were presented. The band structure is qualitatively similar to EH results. Some differences exist between these and OPW calculations. The DOS was calculated by means of the analytic tetrahedron linear energy method. A comparison with experimental XPS data shows agreement with respect to the positions of the major peaks. A very small finite value of the DOS was calculated for E_F. (0.01 states·eV^{-1}·$spin^{-1}$·$molecule^{-1}$), consistent with UPS results but much smaller than the value deduced from specific heat measurements (0.18 states·eV^{-1}·$spin^{-1}$·$molecule^{-1}$), see Table 28, p. 116. A steep DOS minimum near E_F (not observed in OPW and PSE studies) is believed to be a direct consequence of interchain coupling. The band structure of $(SN)_x$ was compared with that of S_2N_2 calculated by the same LCAO procedure [30]. The extended tight-binding method was also used for both structures [35]. S_2N_2 is an indirect-gap semiconductor in which the valence band maximum occurs near the Γ point and the conduction band minimum near the C point in the Brillouin zone (indirect gap: 1.7 eV;

direct gap at Γ: 2.7 eV). The transition from insulating S_2N_2 to metallic $(SN)_x$ and the absence of an energy gap in $(SN)_x$ was discussed in terms of bonding effects. S-3p and N-2p hybridization lead to delocalization of charge carriers [35]. The results of [35] and [30] gained by nearly the same technique agree very well. The contribution of the d orbitals to any energy state of $(SN)_x$ are only ≦5%. As in other 3D calculations, except [10], the absence of a Peierls distortion can be explained by strong three-dimensional interchain interaction. The calculated total DOS was compared with XPS results. The finite value of the DOS at E_F is consistent with the metallic character of $(SN)_x$ [35]. A first-principles self-consistent calculation using the numerical-basis-set LCAO method [42] is similar to non-self consistent 3D-LCAO band structure calculations [29, 30, 35]. The energy bands are highly dispersive along the chain direction while those along the interchain direction are quite flat; for the DOS see **Fig. 22**.

Table 28

Selected Results for $(SN)_x$ from Different Band-Structure Calculations and Some Experimental Data for Comparison. See also tables in [19] and [49].

method	dimensionality	conduction band width in eV	effective mass in m_0	DOS at E_F in states/(eV·spin·molecule)	charge transfer S→N	Ref.
EH	1, 3	1.5	1.5 (∥b) 4.5 (⊥b)	—	—	[6] from [36]
TB	1, 3	—	1.0 (∥b, E_1 band) 0.25 (∥b, E_2 band)	0.25	—	[12]
TB	1	1.5 to 4.5	—	[a]	—	[13]
LCAO	1, 3	1.78	—	0.01	—	[30]
LCAO	3	1.5 (∥b)[c]	—	0.14	0.4	[42]
LCAO	1	—	—	~0.09	—	[21]
LCAO	1	5.6	2.2	0.1	0.36	[17]
LCAO	1	4.12	1.7	0.14	0.4	[14]
PSE	3	—	—	0.12	0.6	[40, 41]
VEH	1	4.13	—	0.14	—	[28]
CNDO/2	1	~5	—	0.06	0.18	[8] from [19]
CNDO/2	1	—	—	0.04	0.231	[33]
Xα	1	—	—	~0.19[b]	—	[26]
Xα	1, 2, 3	~2.5	—	—	—	[36]
OPW	3	2 to 3	—	0.13	—	[38]
XPS	—	1.6±0.03	~1	0.12	—	[27]
XPS	—	—	—	0.04	0.3 to 0.42	[44, 48]
specific heat	—	≥0.9	—	0.18	—	[45]
specific heat	—	—	—	0.14	—	[46]
magn. suscept.	—	—	—	0.14±0.06	—	[47]

[a] DOS (E_F) values for σ and π electrons for three crystal directions are given. – [b] For electron-diffraction structure (X-ray: 0.23 states·eV^{-1}·$spin^{-1}$·$molecule^{-1}$). – [c] 0.36∥a, 0.04∥c.

For a comparison of calculated DOS with experimental XPS data see Fig. 22. A fairly good agreement in peak positions was reported. The important features of the DOS are: a finite (non-zero) number of states at the Fermi level E_F, and an energy gap between ~2.5 and 4 eV below E_F and between 4 and 6 eV above E_F, cf. DOS of Fig. 21, p. 115.

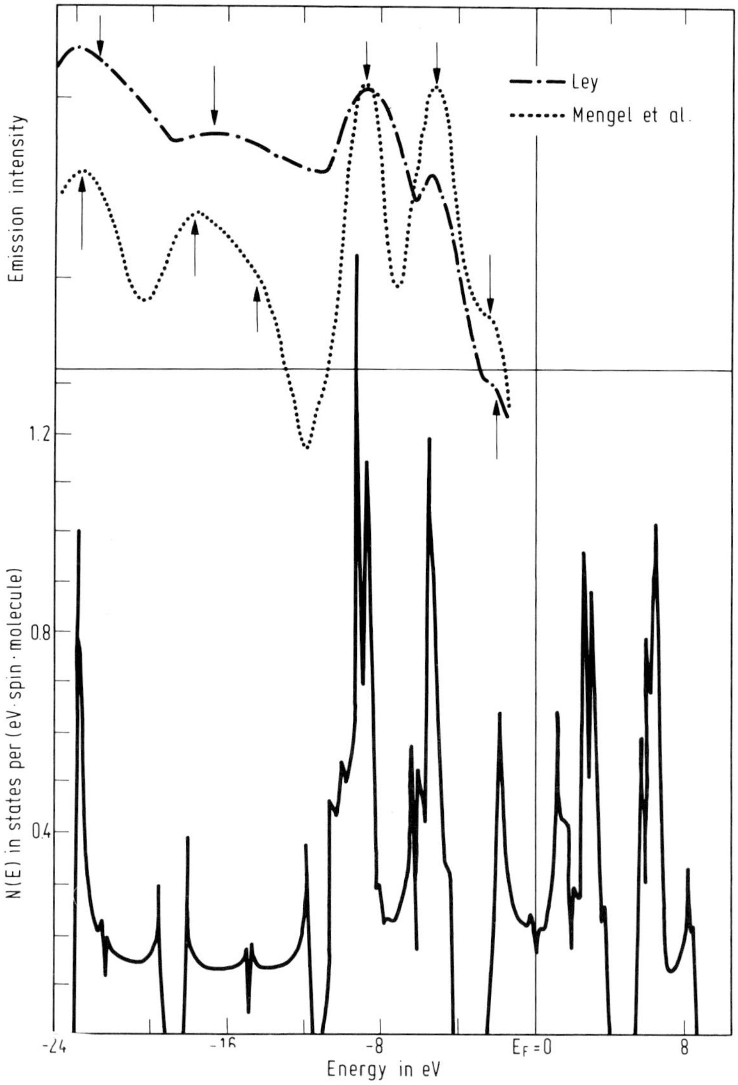

Fig. 22. Comparison of the density of states N(E) of $(SN)_x$ obtained by the LCAO (3D) method [42] with the XPS results from two experimental groups.

References:

[1] Chapman, D.; Warn, R. J.; Fitzgerald, A. G.; Yoffe, A. D. (Trans. Faraday Soc. **60** [1964] 294/300).
[2] Walatka, V. V., Jr.; Labes, M. M.; Perlstein, J. H. (Phys. Rev. Letters **31** [1973] 1139/42).
[3] Parry, D. E.; Thomas, J. M. (J. Phys. C **8** [1975] L45/L48).
[4] Friesen, W. I.; Berlinsky, A. J.; Bergersen, B.; Weiler, L.; Rice, T. M. (Bull. Am. Phys. Soc. **20** [1975] 360).
[5] Friesen, W. I.; Berlinsky, A. J.; Bergersen, B.; Weiler, L.; Rice, T. M. (J. Phys. C **8** [1975] 3549/57.
[6] Bright, A. A.; Soven, P. (Solid State Commun. **18** [1976] 317/20).
[7] Messmer, R. P.; Salahub, D. R. (Chem. Phys. Letters **41** [1976] 73/6).
[8] Zunger, A. (J. Chem. Phys. **63** [1975] 4854/60).
[9] Kamimura, H.; Grant, A. J.; Levy, F.; Yoffe, A. D.; Pitt, G. D. (Solid State Commun. **17** [1975] 49/53).
[10] Kamimura, H.; Glazer, A. M.; Grant, A. J.; Natsume, Y.; Schreiber, M.; Yoffe, A. D. (J. Phys. C **9** [1976] 291/303).

[11] Kamimura, H.; Grant, A. J.; Levy, F.; Yoffe, A. D. (Proc. 14th Intern. Conf. Low-Temp. Phys., Otaniemi, Finland 1975, Vol. 4, pp. 354/7).
[12] Kamimura, H. (Phys. Semicond. 13th Proc. Intern. Conf., Rome 1976, pp. 51/60).
[13] Rajan, V. T.; Falicov, L. M. (Phys. Rev. [3] B **12** [1975] 1240/6).
[14] Suhai, S.; Ladik, J. (Solid State Commun. **22** [1977] 227/9).
[15] Merkel, C.; Ladik, J. (Phys. Letters A **56** [1976] 395/6).
[16] Kertész, M.; Koller, J.; Ažman, A.; Suhai, S. (Phys. Letters A **55** [1975] 107/8).
[17] Kertész, M.; Koller, J.; Ažman, A. (Phys. Status Solidi B **77** [1976] K157/K160).
[18] Kertész, M.; Koller, J.; Ažman, A. (Intern. J. Quantum Chem. **14** [1978] 239/43).
[19] Kertész, M.; Ažman, A.; Kiss, A. I.; Koller, J. (Lect. Notes Phys. **65** [1977] 611/7).
[20] Ladik, J. (NATO Advan. Study Inst. Ser. C **39** [1978] 279/88).

[21] Dovesi, R.; Pisani, C.; Roetti, C.; Saunders, V. R. (J. Chem. Phys. **81** [1984] 2839/44).
[22] Bredas, J. L. (J. Phys. C **15** [1982] 3473/81).
[23] Suhai, S.; Kertész, M. (J. Phys. C **9** [1976] L347/L350).
[24] Palmer, M. H.; Findlay, R. H. (J. Mol. Struct. **92** [1983] 373/83).
[25] Haddon, R. C.; Wassermann, S. R.; Wudl F.; Williams, G. R. J. (J. Am. Chem. Soc. **102** [1980] 6687/93).
[26] Mihich, L. (Solid State Commun. **28** [1978] 521/4).
[27] Ley, L. (Phys. Rev. Letters **35** [1975] 1796/9).
[28] Bredas, J. L. (Chem. Phys. Letters **115** [1985] 119/23).
[29] Ching, W. Y.; Harrison, J. G.; Lin, C. C. (Bull. Am. Phys. Soc. [2] **21** [1976] 254/5).
[30] Ching, W. Y.; Harrison, J. G.; Lin, C. C. (Phys. Rev. [3] B **15** [1977] 5975/82).

[31] Zunger, A. (Ann. Soc. Sci. Bruxelles I **89** [1975] 231/51).
[32] Kertész, M.; Suhai, S.; Ažman, A.; Kocjan, D.; Kiss, A. I. (Chem. Phys. Letters **44** [1976] 53/7).
[33] Yamabe, T.; Tanaka, K.; Imamura, A.; Kato, H.; Fukui, K. (Bull. Chem. Soc. Japan **50** [1977] 798/801).
[34] Antoci, S.; Mihich, L. (Gazz. Chim. Ital. **108** [1978] 383/8).
[35] Batra, I. P.; Ciraci, S.; Rudge, W. E. (Phys. Rev. [3] B **15** [1977] 5858/68).
[36] Salahub, D. R.; Messmer, R. P. (Phys. Rev. [3] B **14** [1976] 2592/602).
[37] Berlinsky, A. J. (J. Phys. C **9** [1976] L283/L287).

[38] Rudge, W. E.; Grant, P. M. (Phys. Rev. Letters **35** [1975] 1799/1803).
[39] Rudge, W. E. (Bull. Am. Phys. Soc. [2] **20** [1975] 359/60).
[40] Schlüter, M.; Chelikowsky, J. R.; Cohen, M. L. (Phys. Rev. Letters **35** [1975] 869/72, **36** [1976] 452/3).

[41] Chelikowsky, J. R.; Schlüter, M.; Cohen, M. L. (Phys. Status Solidi B **82** [1977] 357/67).
[42] Oshiyama, A.; Kamimura, H. (J. Phys. C **14** [1981] 5091/107).
[43] Grant, P. M.; Rudge, W. E.; Ortenburger, I. B. (Lect. Notes Phys. **65** [1977] 575/88).
[44] Mengel, P.; Grant, P. M.; Rudge, W. E.; Schechtman, H. B.; Rice, D. W. (Phys. Rev. Letters **35** [1975] 1803/6).
[45] Greene, R. L.; Grant, P. M.; Street, G. B. (Phys. Rev. Letters **34** [1975] 89/92).
[46] Harper, J. M. E.; Greene, R. L.; Grant, P. M.; Street, G. B. (Phys. Rev. [3] B **15** [1977] 539/43).
[47] Kaneto, K.; Tanimura, K.; Yoshino, K.; Inuishi, Y. (Solid State Commun. **22** [1977] 383/5).
[48] Mengel, P.; Ortenburger, I. B.; Rudge, W. E.; Grant, P. M. (Lect. Notes Phys. **65** [1977] 591/601, 593).
[49] Bredas, J. L. (Ann. Soc. Sci. Bruxelles I **94** [1980] 83/128, 122).
[50] Yoffe, A. D. (Chem. Soc. Rev. **5** [1976] 51/78, 62/3).

2.1.5.3.3 Experimental Results

General Remarks

Energy bands as a function of $\vec{k}$ can be determined experimentally by angle-resolved photoemission (XPS, UPS), but not many such studies have yet been done. Other photoelectron studies permit only the determination of the density of states (DOS) from the measured energy distribution curves, and usually this only for the occupied part below E_F. Other methods give still fewer details. For instance, not much more than the DOS at E_F can be obtained from the electronic contribution of the specific heat at low temperature (constant γ), see Table 28, p. 116. The optical ε_2 spectra in the interband transition region (see p. 167) depend on the DOS in a complicated way. However, tunneling spectroscopy, see below, also gives some results for the unoccupied part of the band structure above E_F.

Photoelectron Spectra (XPS, UPS)

X-ray photoemission spectroscopic (XPS) measurements were performed for the valence band region on $(SN)_x$ films [1 to 4]. Five main peaks centered at 0.7, 3.6, 7.4, 15.2, and 21.0 eV below E_F [1] or at 1.0, 4.0, 7.5, 16.5, 21.5 eV (below E_F), and in addition, a feature at 13.2 eV [2, 3], were observed in agreement with [4]. Several calculations of the DOS by different quantum-mechanical procedures were compared with experimental XPS values, for example one-dimensional tight-binding computations [1 to 3], OPW [2, 3], TB (peaks at 0.9, 4.1, 7.1, 14.4, and 22.4 eV) [5, 6], and LCAO (see Fig. 22, p. 117) [7]. In general, a good agreement with respect to the locations of the peaks was observed but the intensities were different [1 to 3, 5 to 7], with the exception of the OPW results [2, 3]. A better agreement between theoretical and experimental values was obtained when the calculated DOS was modulated by photoionization-cross-section effects [8].

Ultraviolet photoemission spectra (UPS) were reported for $(SN)_x$ polycrystalline films [3, 9, 10] and "single crystals" using He I (21.2 eV) and He II (40.8 eV) resonance radiation [9]. Positions of structure (peaks or strong shoulders) observed in the photoelectron energy distribution curve obtained with He II from films and single crystals [9], compared with theoretical OPW density of states results [11], are given (in eV) in the following table [9]:

method, sample	peak and shoulder (sh) positions relative to E_F (in eV)									
UPS, single crystal, $k_{\parallel}$	−0.2	−1.2	−2.5 sh	—	−5.5	—	—	−10.0	−12.8	—
UPS, single crystal, $k_{\perp}$	−0.2	−1.2	−2.5 sh	−3.5	−4.9	−6.8	−8.4	−10.2	−14.3	−16.1
UPS, film	−0.2	−1.5	—	—	−4.8	−6.4	−8.2	−10.0	−13.0	−16.2
OPW	—	−1.2	−2.3	—	−4.6	−6.1	−7.4	− 9.0	−11.6	−17.2

$k_{\parallel}$ and $k_{\perp}$ refer to the two different orientations of the $(SN)_x$ fiber axes with respect to the axis of the electron energy analyzer.

The results of films and crystals are similar if one averages over the two directions of emission in the crystal case [9]. The UPS data were also compared with XPS values ([1, 2]), see figure in the original [9]. A small but finite DOS at the Fermi level (E_F) was observed in both XPS and UPS spectra [1 to 4, 7, 9, 10], see also Table 28, p. 116. Differences between these and theoretical values are not completely understood [3]. The small but finite value of DOS at E_F was also detected by the study of the S_2N_2 polymerization using XPS spectroscopy at 0 and 20°C [4].

Angular resolved photoemission experiments were also performed [9, 12]. The measured width (2.3 ± 0.3 eV) of the valence bands is slightly smaller than that calculated by the OPW method [12].

X-Ray Fluorescence Spectroscopy

The X-ray SK_{β} fluorescence spectrum of $(SN)_x$ was measured using AgL emission. The locations of the bands E (in eV) obtained from the EDC peaks (and shoulders) and the relative peak intensities, I_{rel}, are summarized in the following table [13]:

E − E(S 1s)	2471.2	2467.9	2464.8	2460.9	2458.5	2455.9	2452.9	2449.2
$E - E_F$	−0.4	−3.6	−6.7	−10.6	−13.0	−15.6	−18.6	−22.3
I_{rel}	0.2	0.6	0.8	0.2	0.1	<0.1	0.1	<0.1

Tunneling Spectroscopy

Tunneling spectroscopy is one of the effective methods for revealing the electron band structure near the Fermi level E_F. Several peaks or shoulders were observed in the voltage-current curves and differential conductance of $(SN)_x$–Al_2O_3–Al junctions for i parallel to the b axis, which indicates the existence of band edges or singular points at ~0.8, 0.4, and 0.03 eV above E_F and 0.4 and 0.8 eV below E_F [14].

References:

[1] Ley, L. (Phys. Rev. Letters **35** [1975] 1796/9).
[2] Mengel, P.; Grant, P. M.; Rudge, W. E.; Schechtman, B. H.; Price, D. W. (Phys. Rev. Letters **35** [1975] 1803/6).
[3] Mengel, P.; Ortenburger, I. B.; Rudge. W. E.; Grant, P. M. (Lect. Notes Phys. **65** [1977] 591/601).
[4] Sharma, J.; Downs, D. S.; Iqbal, Z.; Owens, F. J. (J. Chem. Phys. **67** [1977] 3045/9).
[5] Batra, I. P.; Ciraci, S.; Rudge, W. E. (Phys. Rev. [3] B **15** [1977] 5858/68, 64).
[6] Batra, I. P.; Ciraci, S. (Bull. Am. Phys. Soc. [2] **22** [1977] 372).
[7] Ching, W. Y.; Harrison, J. G.; Lin, C. C. (Phys. Rev. [3] B **15** [1977] 5975/82).

[8] Batra, I. P. (Phys. Rev. [3] B **17** [1978] 4114/6).
[9] Koch, E. E.; Grobman, W. D. (Solid State Commun. **23** [1977] 49/52).
[10] Mengel, P.; Grobman, W. D.; Ortenburger, I. B.; Grant, P. M.; Schechtman, B. H. (Bull. Am. Phys. Soc. [2] **21** [1976] 254).

[11] Rudge, W. E.; Grant, P. M. (Phys. Rev. Letters **35** [1975] 1799/1803).
[12] Ley, L.; Stolz, H. J.; Grandke, T.; Azoulay, J. (Extend. Abstr. 5th Intern. Conf. Vac. Ultraviolet Radiat. Phys., Montpellier 1977, Vol. 2, pp. 189/91).
[13] Zibarev, A. V.; Dolenko, G. N.; Pykhov, V. Z. (Dokl. Akad. Nauk SSR **255** [1980] 1162/5; Dokl. Phys. Chem. Proc. Acad. Sci. USSR **250/255** [1980] 1032/4).
[14] Tanimura, K.; Kaneto, K.; Yoshino, K.; Inushi, Y. (Technol. Rept. Osaka Univ. **29** [1979] 435/7).

2.1.5.4 Fermi Surface

The Fermi surface of $(SN)_x$ was studied by several groups [1 to 11]. Different sketches of the shapes of the Fermi surface were presented [1 to 3, 6, 8 to 11]. The Fermi surface, calculated by means of a pseudopotential interpolation of the OPW band-structure results, consists of electron and hole pockets (roughly ellipsoidal tubes) which give $(SN)_x$ a semimetallic character [1 to 3]. The closed Fermi surface as a result of interchain coupling easily explains the absence of a Peierls distortion. Other groups agree with a semimetallic Fermi surface [4, 5].

A slightly different result was obtained by the use of the LCAO method [6]; the Fermi surface shows a closed electron pocket near the ΓZ line and a hole pocket near the CY line of the Brillouin zone. In contrast to these results, the Fermi surfaces determined by a tight-binding method, where interchain interaction is treated by the perturbation theory, have the shape of plane-like parallel sheets [8 to 10].

The Fermi surfaces determined by the SC-numerical basis-set LCAO method confirm the main features of the non-selfconsistent OPW results, see **Fig. 23**, p. 123 [11]; the Fermi surfaces consist of two electron and two hole pockets which are located around the top surface of the Brillouin zone (plane ZDCE).

The results of Doppler broadening studies of positron annihilation using H parameter were compared with known calculated Fermi surfaces. Based on their experimental results, the authors conclude that the Fermi surface is closed in the k_b (chain direction) and k_a direction, and opened in the k_c direction. They classified $(SN)_x$ as quasi-one-dimensional [12, 13]. Similar results were obtained by the pseudopotential method [7]. The Fermi surface was reported to consist of four separate, nearly planar sheets running perpendicular to the chain axis. Sketches of electron and hole Fermi surfaces corresponding to the electron spectrum of $(SN)_x$ obtained via theoretical group analysis were presented. The result resembles that of a semimetal [14].

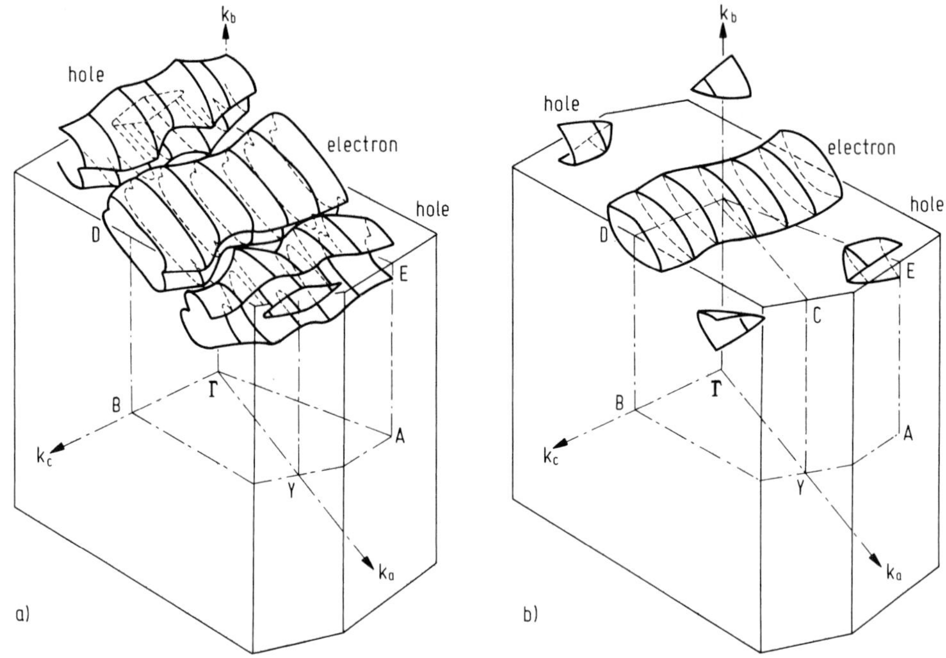

Fig. 23. Fermi surfaces of $(SN)_x$ obtained by the LCAO method. The outermost surfaces are shown in a) and the inner most ones in b). The two electron pockets are connected with each other on the plane ZDCE and on the line BD. The two hole pockets are also connected with each other on the same plane and on the line AE [11].

References:

[1] Rudge, W. E.; Ortenburger, I. B. (Bull. Am. Phys. Soc. [2] **21** [1976] 255).

[2] Grant, P. M.; Rudge, W. E.; Ortenburger, I. B. (Lect. Notes Phys. **65** [1977] 575/88).

[3] Greene, R. L.; Street, G. B. (NATO Advan. Inst. Ser. B **25** [1977] 167/95, 174/6).

[4] Friesen, W. I.; Berlinsky, A. J.; Bergersen, B.; Weiler, L.; Rice, T. M. (J. Phys. C **8** [1975] 3549/57).

[5] Bright, A. A.; Soven, P. (Solid State Commun. **18** [1976] 317/20).

[6] Ching, W. Y.; Harrison, J. G.; Lin, C. C. (Phys. Rev. [3] B **15** [1977] 5975/82).

[7] Schlüter, M.; Chelikowsky, J. R.; Cohen, M. L. (Phys. Rev. Letters **35** [1975] 869/72, **36** [1976] 452/3).

[8] Kamimura, H.; Oshiyama, A. (Theoretical Aspects of Band Structures and Electronic Properties of Pseudo-One-Dimensional Solids, Reidel, Dordrecht 1985, pp. 123/62).

[9] Kamimura, H.; Glazer, A. M.; Grant, A. J.; Natsume, Y.; Schreiber, M.; Yoffe, A. D. (J. Phys. C **9** [1976] 291/303).

[10] Kamimura, H. (Phys. Semicond. Proc. 13th Intern. Conf., Rome 1976, pp. 51/60, 55/6).

[11] Oshiyama, A.; Kamimura, H. (J. Phys. C **14** [1981] 5091/107).

[12] Suzuki, Y.; Suzuki, J.; Matsusaki, S.; Doyama, M.; Nakata, I. (Positron Annihilation Proc. 6th Intern. Conf., Arlington, Tex., 1982, pp. 248/50).

[13] Suzuki, Y.; Ishibashi, S.; Matsusaki, S.; Doyama, M.; Nakata, I. (Syn. Metals **18** [1987] 601/4).

[14] Gordyunin, S. A.; Dorokhov, O. N. (J. Phys. C **12** [1979] 2015/23).

2.1.5.5 Charge Transfer

Both band-structure calculations as well as experimental data from the XPS core level spectra and the X-ray fluorescence spectra yield a charge transfer δ from the S atom to the N atom in $(SN)_x$, $(S^{+\delta}N^{-\delta})$, showing highly polar SN bonds. Some theoretical values for δ: 0.402 [1], 0.5 [2], 0.9 [3]; the last value was criticized in [4], and improved to 0.6 in [5]; values from XPS core spectra: 0.30 to 0.42 [6], ~0.4 [7], 0.5 [4, 8], 0.42 (+δ), 0.40_5 (−δ) [4, 9]; $\delta = 0.25 \pm 0.02$ from the X-ray fluorescence spectrum [10]. See also Table 28, p. 116. Hence, in general, theoretical and experimental values agree.

References:

[1] Suhai, S.; Ladik, J. (Solid State Commun. **22** [1977] 227/9).
[2] Salahub, D. R.; Messmer, R. P. (Phys. Rev. [3] B **14** [1976] 2592/602, 2599).
[3] Schlüter. M.; Chelikowsky, J. R.; Cohen, M. L. (Phys. Rev. Letters **35** [1975] 861/72).
[4] Salaneck, W. R.; Lin, J. W.-P.; Epstein, A. F. (Phys. Rev. [3] B **13** [1976] 5574/7).
[5] Chelikowsky, J. R.; Schlüter, M.; Cohen, M. L. (Phys. Status Solidi B **82** [1977] 357/67).
[6] Mengel, P.; Grant, P. M.; Rudge, W. E.; Schechtman, B. H.; Price, D. W. (Phys. Rev. Letters **35** [1975] 1803/6).
[7] Mengel, P.; Ortenburger, I. B.; Rudge, W. E.; Grant, P. M. (Lect. Notes Phys. **65** [1977] 591/601).
[8] Epstein, A. J.; Salaneck, W. R.; Lipari, N. O.; Lin, J. W.-P. (Bull. Am. Phys. Soc. [2] **21** [1976] 254).
[9] Larsson, R. (J. Electron Spectrosc. Relat. Phenom. **24** [1981] 37/42).
[10] Zibarev, A. V.; Dolenko, G. N.; Pykhov, V. Z. (Dokl. Akad. Nauk SSR **255** [1980] 1162/5; Dokl. Phys. Chem. Proc. Acad. Sci. USSR **250/255** [1980] 1032/4).

2.1.5.6 Interchain Coupling

Interchain interactions in $(SN)_x$ become evident in conductivity behavior, inhibiting a metal-semiconductor transition. Several papers were published dealing with this special question [1 to 8]. From a comparison of IR and Raman spectra of $(SN)_x$ it was concluded that $(SN)_x$ cannot be considered as a system of weakly interacting individual chains, but rather as an anisotropic 3D material [1]. A study of the interaction of two linear S_4N_4 units in a geometry appropriate to the $(SN)_x$ crystal structure was performed using the self-consistent-field-Xα-scattered wave method [2 to 4]. This study served as a model for $(SN)_x$ chain interaction.

The most important interchain interactions are those in the (100) plane which come preliminarily from π electrons yielding a splitting of the π bands of ~1 eV [2] or 0.4 eV [3]. The interactions in the ($\bar{1}$02) plane are less important, at least for the conduction bands (for electronic transport theory mainly the p orbitals of S are responsible), and lead to a band splitting for the upper bands of ~1 eV [2] or 0.3 eV [3] for σ type and 0.5 eV for π type [2]. The dominant σ interchain interactions are not due to S–S interactions [3]. The dominance of the interchain interaction in the (100) plane was confirmed by a one-dimensional tight-binding SCF MO calculation of two $(SN)_x$ chains (geometry of the X-ray crystal structure) in the ($\bar{1}$02) and (100) planes, see **Fig. 24**, p. 124. However, S–S interchain coupling was found to be dominant in the (100) and (102) plane [5].

The CNDO/2 approximation (including 3 d orbitals of sulfur) was employed to study chain interaction [8]. The band structures and densities of states of two interacting chains in the ($\bar{1}$02) and the (100) plane are presented in a figure together with those of the infinitely separated chains. A split between highest occupied and lowest unoccupied band of 0.04 eV and 2.027 eV was reported for the ($\bar{1}$02) and (100) case, respectively. The AO densities, π-electron densities, atomic net charges, and the total energies per unit cell (in the ($\bar{1}$02) plane: −2427.028 eV, in the (100) plane: −2457.010 eV, two infinitely separated monomers (in most stable square form) of S_2N_2: −2456.812 eV) were tabulated, cf. [8]. Another table presents the contributions of decomposed terms of interatomic interaction energies (core resonance, exchange, and electrostatic). Intracell and intercell interchain energies are also presented in a table [5].

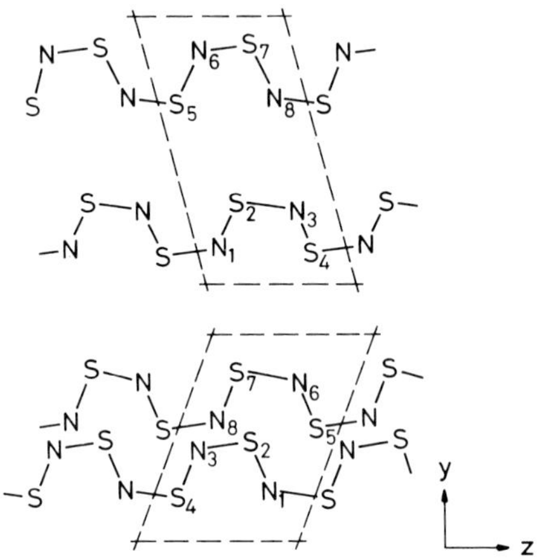

Fig. 24. The geometries of two chains of $(SN)_x$: a) in the ($\bar{1}$02) plane; b) in the (100) plane. Note that mean planes of the two chains are separated by 3.25 Å in b). The unit cell employed in the calculation is indicated by the broken line in each case [5].

A polarized reflection microspectroscopy study on both the natural lateral faces parallel to the b axis and the polished faces normal to b gives further experimental evidence for interchain interaction due to orientation of the π-electron orbitals in $(SN)_x$. A value of 0.3 eV was reported (using the simple tight-binding method) for the band splitting t⊥b caused by π-electron interaction, high enough to explain the absence of a Peierls distortion [6]. Electrostatic interactions of $(SN)_x$ (S_2N_2 and S_4N_4) were calculated for the X-ray and electron-diffraction structure. The net electrostatic interaction between polymer chains is small but still significant [7].

References:

[1] Stolz, H. J.; Otto, A.; Pintschovius, L. (Proc. 3rd Intern. Conf. Light Scattering Solids, Campinas, Brazil, 1976, pp. 737/41).
[2] Messmer, R. P.; Salahub, D. R. (Chem. Phys. Letters **41** [1976] 73/6).
[3] Salahub, D. R.; Messmer, R. P. (Phys. Rev. [3] B **14** [1976] 2592/602).
[4] Salahub, D. R.; Messmer, R. P. (Bull. Am. Phys. Soc. [2] **21** [1976] 254).
[5] Tanaka, K.; Yamabe, T.; Fukui, K.; Imamura, A.; Kato, H. (Chem. Phys. Letters **53** [1978] 452/6).
[6] Gutman, A. I.; Agroskin, L. S.; Papayan, G. V.; Rautian, L. P.; Schachnina, O. S. (Solid State Commun. **49** [1984] 187/9).
[7] Epstein, A. J.; Lipari, N. O. (Phys. Letters A **56** [1976] 235/7).
[8] Yamabe, T.; Tanaka, K.; Imamura, A.; Kato, H.; Fukui, K. (Bull. Chem. Soc. Japan **50** [1977] 798/801).

2.1.5.7 $(SN)_x$ Oligomers ($x=1$ to 10). Changes on Polymerization

Several papers deal with theoretical studies on $(SN)_x$ oligomers [1 to 8]. Minimal-basis-set LCAO-SCF calculations were performed on $(SN)_2$ and $(SN)_3$ [1]. The chain-structure form $(SN)_2$ has a higher energy than the planar ring form S_2N_2. Using an ab initio MO method, bond alternation was found in the species $H(SN)_2H$ and $H(SN)_4H$ in comparison to $(SN)_x$ [2]. The results of an SCF–Xα–SW calculation on an $(SN)_4$ chain showed good agreement with the X-ray photoelectron spectrum of $(SN)_x$ polymer, indicating that the small $(SN)_4$ fragment contains many of the essential features of $(SN)_x$ polymer [3]. Semiempirical INDO (Intermediate Neglect of Differential Overlap) type ASMO-SCF calculations were performed on $(SN)_4$ and $(SN)_6$ species to study the mechanism of $(SN)_x$ polymerization. No triplet ground states were detected, which indicates that the system becomes diamagnetic as the polymerization proceeds (see p. 97) [4]. The total energies and atomic populations for various electronic states of some $(SN)_x$ units with $x=1$ to 6, 10, were calculated by an ab initio MO method ((N(7s3p) S(10s6p1d) basis). Steady electron distribution is apparent in $(SN)_x$ for $x=10$ for the closed-shell state but is less apparent for several triplet states [5]. An MNDO (Modified Neglect of Differential Overlap) study of some $(SN)_x$ fragments were performed and a low-lying BS (Broken Symmetry) structure with repeating $(SN)_2$ units was detected in $(SN)_{10}S$ [6].

References:

[1] Deutsch, P. W.; Curtiss, L. A. (Chem. Phys. Letters **51** [1977] 125/31).
[2] Haddon, R. C.; Wasserman, S. R.; Wudl, F.; Williams, G. R. J. (J. Am. Chem. Soc. **102** [1980] 6687/93).
[3] Messmer, R. P.; Salahub, D. R. (Chem. Phys. Letters **41** [1976] 73/6).
[4] Yamabe, T.; Tanaka, K.; Fukui, K.; Kato, H. (J. Phys. Chem. **81** [1977] 727/31).
[5] Palmer, M. L.; Findlay, R. H. (J. Mol. Struct. **92** [1983] 373/83).
[6] Aquino, A. J. A.; Soares, L. A., II; da Silva, A. B. F.; Trsic, M. (J. Mol. Struct. **139** [1986] 327/32).
[7] Palmer, M. H.; Guest, M. F. (Chem. Phys. **110** [1986] 187/94).
[8] Tachibana, A.; Inoue, T.; Yamabe, T. (Intern. J. Quantum Chem. **30** [1986] 575/9).

2.1.5.8 Influence of Impurities

Hydrogen Impurities

Hydrogen was found (4 to 8 mol%) in $(SN)_x$ [1], which leads to a change in the density of states and a shift of the Fermi level [2]. The coherent potential approximation (CPA) was applied to calculate the density of states (DOS) curves for the valence band of the mixed system consisting of $(SN)_x$ and $(SNH)_x$ linear chains containing hydrogen impurities in the range 3 to 50 mol% ($f=0.03$ to 0.50) [2 to 6]. DOS curves of pure $(SN)_x$, $(SNH)_x$, and of mixed systems with $f=0.03$, 0.05, 0.07, 0.10, 0.20, 0.30, 0.40, and 0.50 were given [2, 3]. The main feature of the DOS curves is the complicated structure of the impurity band, which exhibits (even at $f=0.03$) spikes and dips in the region -4.4 to -8.0 eV below E_F where the density of states goes to zero. The Fermi level of $(SN)_x$ lies at -1.90 eV and the upper limit of the completely filled valence band of $(SNH)_x$ lies at -4.38 eV ($E_F=-1.6$ eV for $0.03 \leq f \leq 0.1$; $E_F=-1.9$ eV for $f=0.2$; $E_F=-2.6$ eV at $f=0.50$). The DOS at the Fermi level increases monotonically with increase of f [2].

Substitutional Impurities

The effects of substitutional impurities in $(SN)_x$ chains on the electronic properties were studied by the extended Hückel method for an idealized polymer [7]. A figure, based on the X-ray diffraction data, presents the band structures of $(SN)_x$, $[(S_2N_2)_4(SCH_2)_2]_n$ (fivefold superlattice system, detailed band structure is illustrated only near the Fermi level), and $[S_2N_2 \cdot (SCH_2)_2]_n$. The introduction of random substitutional $(SCH_2)_2$ segments results in a retention of metallic behavior with a lowering of conductivity to an extent depending on the impurity concentration (sulfur positions are retained and some of nitrogen atoms are replaced by CH_2 groups). p-type semiconductivity was predicted for a system at a high level of substitutional impurities [7]. The band structure of the hypothetical "$(SCH)_x$", studied by an SCF tight-binding method was compared with that of $(SN)_x$. "$(SCH)_x$", was predicted to be metallic if a Peierls transition was excluded [8]. In an MO study of sulfur-nitrogen and sulfur-carbon conjugation it was shown that a –HC=S=CH– unit is unstable with respect to a dimerization reaction. This is not the case for the corresponding NSN unit if d orbitals are involved [9].

References:

[1] Smith, R. D.; Wyatt, J. R.; Weber, D.; DeCorpo, J. J.; Saalfeld, F. E. (Inorg. Chem. **17** [1978] 1639/42).
[2] Seel, M.; Collins, T. C.; Martino, F.; Rai, D. K.; Ladik, J. (Phys. Rev. [3] B **18** [1978] 6460/4).
[3] Seel, M. (Lect. Notes Phys. **113** [1980] 271/90, 277/281).
[4] Ladik, J.; Suhai, S.; Seel, M. (ACS Symp. Ser. No. 162 [1981] 73/80).
[5] Andre, J. M.; Ladik, J. (Lect. Notes Phys. **113** [1980] 299/306, 303).
[6] Ladik, J. (NATO Advan. Study Inst. Ser. C **39** [1978] 279/88).
[7] Parry, D. E.; Thomas, J. M. (J. Polym. Sci. Polym. Phys. Ed. **16** [1978] 1023/9).
[8] Yamabe, T.; Tanaka, K.; Imamura, A.; Kato, H.; Fukui, K. (Bull. Chem. Soc. Japan **50** [1977] 798/801).
[9] Haddon, R. C.; Wasserman, S. R.; Wudl, F.; Williams, G. R. J. (J. Am. Chem. Soc. **102** [1980] 6687/93).

2.1.6 Lattice Dynamics

2.1.6.1 Survey

Important information about the strength of the intra- and interchain binding forces of $(SN)_x$ can be obtained from the study of its lattice dynamics. Provided that there is a clear distinction in the strength of intra- and interchain forces, the factor group analysis for long-wavelength phonons leads to three acoustic phonons, five external optical phonons (the chains may be assumed as quasi-rigid for these modes) and eight internal modes of the single chain. Each internal mode is split into two because there are two chains in the unit cell [1].

Dates for the acoustic phonons were obtained from inelastic neutron scattering. IR and Raman spectra gave information about the optical external and internal modes. The g-u splittings of the internal modes should be a measure of the interchain coupling, see p. 123. A_g modes are polarized, B_g modes are depolarized. From the dates, using a valence force model, force constants were evaluated. Debye temperature, see on p. 135, electron-phonon-interaction on p. 131.

2.1.6.2 Acoustic Phonons. Inelastic Neutron Scattering

Studies of the acoustic phonons of $(SN)_x$ by inelastic neutron scattering were accomplished with great difficulty owing to the small size and poor quality of the crystals [2 to 7].

The dispersion of the acoustic phonons was determined for small wave vectors parallel, and in two directions perpendicular to the chain axis b [2, 4, 7]. **Fig. 25** shows dispersion curves for the three acoustic branches and a low-lying optical branch for phonons traveling in the chain direction. Important results are the large difference in the slopes of the longitudinal and transverse branches and also the unusual upward curvature of the transverse branches, indicating a clear anisotropy in the strength of intra- and interchain forces [2 to 4, 7] (see Section 2.1.6.4, p. 131). Two small dips observable at $\zeta = q \cdot b/2\pi = 0.14$ and at $\zeta = 0.25$ on the longitudinal acoustic branch in the first half of the Brillouin zone are associated with Kohn anomalies (see Fig. 25) [6]. Characteristics of these anomalies (position, strength, temperature behavior, and the behavior when going away from the chain direction) are not consistent with those expected for a quasi-1 D conductor. However, the experimental results are in agreement with several 3 D band-structure calculations which showed that the Fermi surface of $(SN)_x$ is not that of a quasi-1 D metal [6].

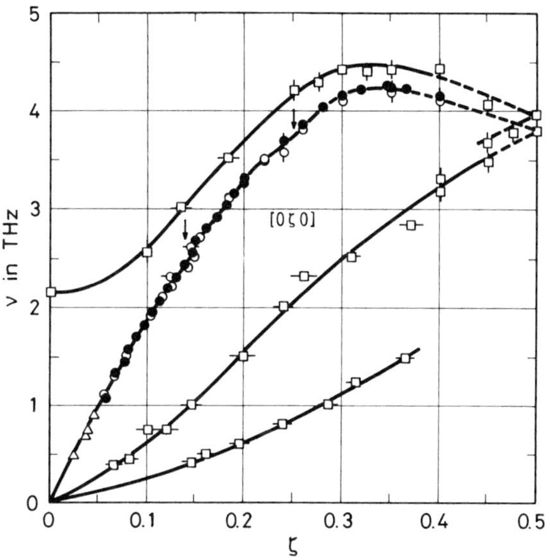

Fig. 25. Dispersion curves of low-frequency phonons in $(SN)_x$ which propagate in the chain direction at room temperature (ζ = reduced wave vector coordinate). The arrows indicate Kohn anomalies. The upper two branches are (at $\zeta \rightarrow 0$) longitudinal modes (LO, LA), the lower two are transverse modes (TA, TA) [7].

References:

[1] Geserich, H. P.; Pintschovius, L. (Festkörperprobleme [Advan. Solid State Phys.] **16** [1976] 65/93, 86/9).

[2] Stolz, H. J.; Wendel, H.; Otto, A.; Pintschovius, L.; Kahlert, H. (Phys. Status Solidi B **78** [1976] 277/86, 280/82).

[3] Pintschovius, L.; Wendel, H.; Kahlert, H. (Lect. Notes Phys. **65** [1977] 589/90).

[4] Pintschovius, L.; Currat, R.; Wendel, H. (Proc. Intern. Conf. Lattice Dyn., Paris 1977 [1978], pp. 579/81).

[5] Pintschovius, L. (Europhys. Conf. Abstr. **2 F C** [1978] Paper C 17 from C.A. **89** [1978] No. 204377).
[6] Pintschovius, L.; Pynn, R. (Lect. Notes Phys. **96** [1979] 421/6).
[7] Pintschovius, L. (Colloid Polym. Sci. **256** [1978] 883/92, 889/90).

2.1.6.3 Optical Phonons. IR and Raman Spectra

The first IR spectroscopic data (range 1400 to 600 cm^{-1}) of $(SN)_x$ were published in 1964 [1]. In the last few years several papers dealing with IR and Raman spectra of $(SN)_x$ crystals and films appeared [2 to 14].

Good qualitative IR spectra down to 600 cm^{-1} were recorded by transmission measurements through thin $(SN)_x$ films [2]. The IR spectra of $(SN)_x$ films on CsI above 400 cm^{-1} and below 400 cm^{-1} are shown in **Fig. 26** and **27**.

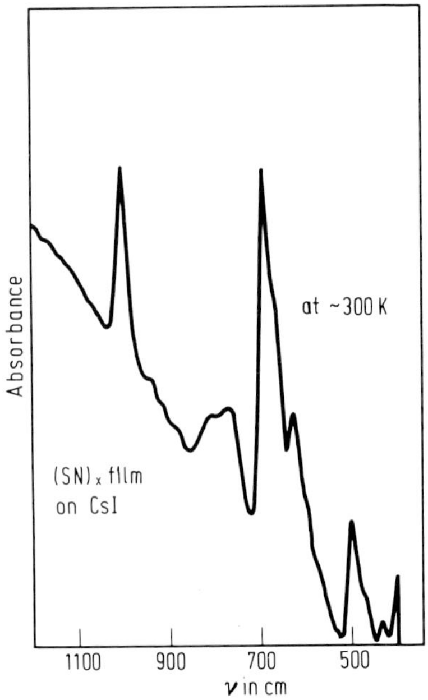

Fig. 26. IR spectrum of $(SN)_x$ above 400 cm^{-1}. The spectrum spans 0 to 0.25 absorbance units [2].

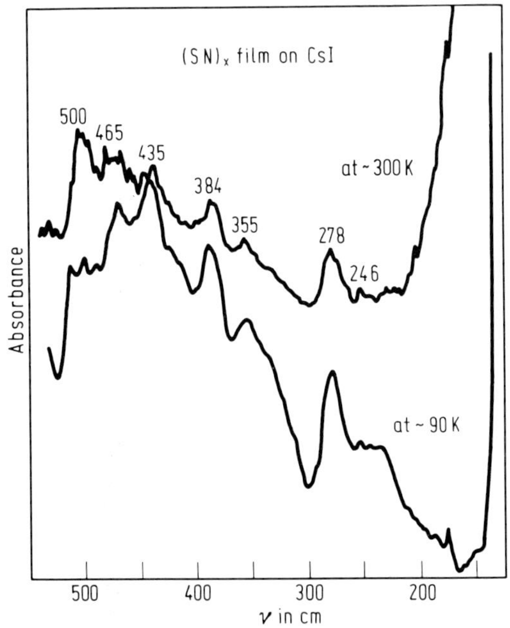

Fig. 27. IR spectra of $(SN)_x$ below 400 cm^{-1}. Both spectra span 0 to 0.25 absorbance units [2].

The Fourier transform IR data (range 246 to 1000 cm^{-1}) are compiled in Table 29 together with Raman data (from [3 to 7]). They are assigned by comparison to spectra of brominated $(SN)_x$ [2].

The strong absorptions at 695 and 995 cm^{-1} (corresponding to 689 and 1001 cm^{-1} in Table 29) were also assigned to SN stretching modes by other authors. Observed lines above 1000 cm^{-1} in the spectrum of air-exposed $(SN)_x$ are due to hydrolysis [8].

The number of fundamental vibrations of crystalline $(SN)_x$ was determined by a factor group analysis [11]. The unit cell was chosen to include two adjacent translationally inequivalent S_2N_2 units. Of the 24 representations for the unit cell vibrations, 3 belong to translations (acoustic) and 5 to external optical phonons. This leaves 16 assignable to internal optical frequencies of $(SN)_x$ [2], see Table 30.

Table 29
IR and Raman Frequencies of $(SN)_x$ (symmetry C_2) Observed at 300 K [2].
(Frequencies in parentheses vary in relative intensity in spectra of different $(SN)_x$ samples.)

frequencies in cm^{-1} IR	Raman	suggested assignments by class in symmetry group C_2 and approximate description
1001	1004	ν_1A sym. str.
918 vw, br	~930	$2\times\nu_3$ and $\nu_4+\nu_2$
(812)	—	384 + 435
768	786	ν_5B antisym. str.
689	~700	ν_2A sym. str.
(662 sh)	658 vs	ν_2A sym. str.
629	627	ν_6B in-plane deform.
590 sh	586	245 + 355
560 sh	—	$2\times\nu_4$
500	—	ν_3A in-plane deform.
(465)	456 vs	ν_3A in-plane deform.
(435)	—	?
(384)	—	ν_7B in-plane deform.
(355)	—	ν_8B out-of-plane deform.
(330)	—	ν_8B out-of-plane deform.
278	292	ν_4A out-of-plane deform.
(246)	248	?

Table 30
Long-Wavelength Phonons of $(SN)_x$ [11].

single chain	double chain		
internal modes:	internal modes:		
$4\times A$ (IR, Ra)	$4\times A_g$ (Ra)		
	$4\times A_u$ (IR ∥ b)		
$4\times B$ (IR, Ra)	$4\times B_g$ (Ra)		
	$4\times B_u$ (IR ⊥ b)		
	external modes:		
	translations	trans.-libr.	librations
	$2\times B_g$ (Ra)	$2\times A_g$ (Ra)	$1\times A_u$ (IR)

Polarized Raman and IR reflectance spectra allow the assignment to Raman active A_g, B_g modes and IR active A_u, B_u modes, respectively [11]. The results are listed in Table 31, p. 130.

The underlined frequencies in Table 31, p. 130, are the most probable choise of phonons of the different symmetry types at Γ (center of the Brillouin zone) for $(SN)_x$. Other polarized Raman spectra of $(SN)_x$ crystals were measured in the range 1200 to 200 cm^{-1} at laser wavelengths of 4579 [3, 4, 6], 5145 [3], and 6005 Å [3] and in the range 1100 to 100 cm^{-1} with the 6116 Å laser line [7]. The Raman active peaks at 454, 621, 658, and 782 cm^{-1} [3] are strong, very anisotropic and better resolved than in thin films ([6, 10]). The Raman lines have only been observed using α_{yy} polarization (incident and scattered light are both polarized parallel to the polymer chain axis) [3].

Table 31
Optical Phonons[a)] in $(SN)_x$ (in cm^{-1}) [11].

A_g	B_g[b)]	A_u	B_u
1004		1000	
930		818	
791[c)]		700	756
660		675	652
627		635	612
586		500	
464[c)]			
292			
248			

[a)] Taken with the 4579 Å line of Ar^+ laser. – [b)] Phonons of B_g symmetry are omitted because they were regarded as experimentally not sufficiently verified. – [c)] The lines 791 and 464 cm^{-1} (A_g) are said by other authors to be not of the first-order Raman [11, 13].

The IR spectra of $(SN)_x$ deposited on various supports at room- and liquid nitrogen temperature show only differences in the relative intensities of some of the bands [2].

Raman data of $(SN)_x$ films on fused silica substrates were reported and compared with IR spectroscopic results (4579 and 5245 Å laser Ar line; range 1400 to 200 cm^{-1}). Raman and IR spectra showed no changes when cooled to 77 or 4.2 K [10]. The line at 658 cm^{-1} for $(SN)_x$ crystals narrows with decreasing temperature (from 15 cm^{-1} at 300 K to 11 cm^{-1} at 4.2 K) and with increasing pressure (to 7 cm^{-1} at 8.2 kbar) but no shift of energy was observed [3, 4, 6]. For an IR spectrum of $(SN)_x$ in KBr matrix at 295 K (range 1243 to 435 cm^{-1}) see [9].

IR reflection and transmission spectra were recorded for epitaxial $(SN)_x$ films (on fused silica substrates, thin silicon wafers, NaCl and KBr substrates) [10] and IR reflection spectroscopic data were obtained for $(SN)_x$ single crystals [11]. Transmission IR absorption spectra of thin $(SN)_x$ films on CsI between 1600 and 300 cm^{-1} are shown in [12]. An IR spectrum in the range 1600 to 300 cm^{-1} from a gold-colored $(SN)_x$ film and one from crushed $(SN)_x$ crystals (exposed to the atmosphere for ca. 1 h) at 295 K have been published [8]. Calculated non-zero optic frequencies based on a four parameter valence force model were reported together with proper symmetry assignments [11, 13]. The assignments above 600 cm^{-1} agree (except for the band at 629 cm^{-1}) with those obtained by this calculation [2]. The differences in strength of experimental Raman modes at 1004 and 660 cm^{-1} [11] are explained by a smaller change of the polarizability for the former vibration [11, 13]. Additional information about the vibrational modes was obtained using a "pseudo-lattice method" [14].

References:

[1] Chapman, D.; Warn, R. J.; Fitzgerald, A. G.; Yoffe, A. D. (Trans. Faraday Soc. **60** [1964] 294/300).
[2] Macklin, J. W.; Street, G. B.; Gill, W. D. (J. Chem. Phys. **70** [1979] 2425/30).
[3] Temkin, H.; Fitchen, D. B. (Solid State Commun. **19** [1976] 1181/4).
[4] Temkin, H.; Fitchen, D. B. (Bull. Am. Phys. Soc. [2] **21** [1976] 666).
[5] Temkin, H.; Street, G. B. (Solid State Commun. **26** [1978] 455/60).
[6] Temkin, H.; Fitchen, D. B. (Proc. 5th Intern. Conf. Raman Spectrosc., Freiburg 1976, pp. 644/5).
[7] Temkin, H.; Fitchen, D. B.; Street, G. B.; Gill, W. D. (Ann. N.Y. Acad. Sci. **313** [1978] 771/4).
[8] Iqbal, Z.; Downs, D. S. (Chem. Phys. **44** [1979] 137/42).
[9] Iqbal, Z.; Christoe, C. W.; Owens, F. J. (Ferroelectrics **16** [1977] 219/222).
[10] Stolz, H. J.; Otto, A.; Pintschovius, L. (Proc. 3rd Intern. Conf. Light Scattering Solids, Campinas, Braz., 1975 [1976], pp. 737/41).

[11] Stolz, H. J.; Wendel, H.; Otto, A.; Pintschovius, L.; Kahlert, H. (Phys. Status Solidi B **78** [1976] 277/86).
[12] Iqbal, Z.; Downs, D. S. (Proc. Intern. Conf. Lattice Dynamics, Paris 1977 [1978], pp. 582/4).
[13] Wendel, H. (J. Phys. C **10** [1977] L1/L5).
[14] Wieting, T. J. (Proc. Intern. Conf. Lattice Dynamics, Paris 1977 [1978], pp. 621/4).

2.1.6.4 Force Constants

Force constants of $(SN)_x$ were determined from the analysis of valence force models based on simplified $(SN)_x$ structures using optical and acoustic phonon data. Two four-parameter models using a two-dimensional structure ($10\bar{2}$ plane) [1] and a simplified three-dimensional $(SN)_x$ structure [2] were studied. Calculated force constants were presented [1, 2]. It was concluded that interchain interactions between S and N atoms are important and that internal strains play an essential role [1]. Another calculation [3] of force constants was also based on phonon data. A valence force model took into account central interactions for atoms within the same chain up to the 4th nearest neighbors and 10 force constants representing central interactions between the chains. The largest intrachain force constant (3.85×10^5 dyn·cm^{-1} for S–N at r(S–N) = 1.59 Å) is higher by a factor of ~50 than the largest interchain force constant (0.08×10^5 dyn·cm^{-1} for S–S at r = 3.47 Å) [3].

References:

[1] Stolz, H. J.; Wendel, H.; Otto, A.; Pintschovius, L.; Kahlert, H. (Phys. Status Solidi B **78** [1976] 277/86).
[2] Wendel, H. (J. Phys. C **10** [1977] L1/L5).
[3] Pintschovius, L.; Currat, R.; Wendel, H. (Proc. Intern. Conf. Lattice Dynamics, Paris 1977 [1978], pp. 579/81).

2.1.6.5 Electron-Phonon Coupling Constant

The electron-phonon coupling constant $\lambda = 0.31$ was estimated by the McMillan formula for the superconducting transition temperature, using the Debye temperature 170 K obtained from specific heat data [1]. The similar value 0.35 was determined based on calculated deformation potentials (for $T_c = 0.26$ K) [2].

References:

[1] Greene, R. L.; Street, G. B.; Suter, L. J. (Phys. Rev. Letters **34** [1975] 577/9).
[2] Chelikowsky, J. R.; Schlüter, M. S.; Cohen, M. L. (Phys. Status Solidi B **82** [1977] 357/67).

2.1.7 Mechanical Properties

2.1.7.1 Density

The density of $(SN)_x$ has been determined from X-ray crystallographic lattice parameters to be 2.34 [1] and 2.30 g/cm^3 [2] and by experimentally methods to be 2.19 (at 25°C; pycnometrically) [3], 2.20 (pycnometrically) [4], 2.3106 ± 0.0007 (at 23 ± 1°C; density gradient column) [5], and 2.30 ± 0.02 g/cm^3 (at room temperature; heavy-liquid method) [6].

The density of the sheared phase I of $(SN)_x$ (i.e., orthorhombic phase II of $(SN)_x$) is 2.307 ± 0.003 g/cm^3 [5]. From X-ray data a value of 2.2985(72) g/cm^3 was estimated [5]. A density of 2.30 g/cm^3 was determined for the powdered polymer (β phase) compressed to a tablet under 40 kbar [7].

References:

[1] Boudeulle, M. (Diss. Claude-Bernard Univ. Lyon 1974).
[2] Mikulski, C. M.; Russo, P. J.; Saran, M. S.; MacDiarmid, A. G.; Garito, A. F.; Heeger, A. J. (J. Am. Chem. Soc. **97** [1975] 6358/63).
[3] Goehring, M.; Voigt, D. (Z. Anorg. Allgem. Chem. **285** [1956] 181/90, 188).
[4] Boudeulle, M.; Douillard, A.; Michel, P.; Vallet, G. (Compt. Rend. C **272** [1971] 2137/8).
[5] Baughman, R. H.; Apgar, P. A.; Chance, R. R.; MacDiarmid, A. G.; Garito, A. F. (J. Chem. Phys. **66** [1977] 401/9, 404).
[6] Oda, Y.; Takenaka, H.; Nagano, H.; Nakada, I. (Solid State Commun. **35** [1980] 887/90).
[7] Banister, A. J.; Hauptman, Z. V.; Passmore, J.; Wong, C.-M.; White, P. S. (J. Chem. Soc. Dalton Trans. **1986** 2371/9, 2374).

2.1.7.2 Elastic Properties

Mechanical properties, Young's modulus [1 to 3], yield strength [1, 3], yield strain [1, 3], breaking stress [1 to 4], and linear compressibilities [5] were measured by various groups. The main published data are compiled in Table 32.

Table 32
Mechanical Properties of $(SN)_x$.

Young's modulus (in 10^{10} N/m^2)				
∥ b axis	1.13 to 3.09	2.1 (±50%)	2.98	1.15 to 1.49
⊥ b axis	0.026 to 0.094	0.14 (±50%)	0.48	
yield strength (in 10^7 N/m^2)	4.68 to 14.50			10 to 16
yield strain (in %)	0.39 to 0.54			0.7 to 1.1
breaking stress (in 10^7 N/m^2)				
∥ b axis	5.5 to 36.6	1.5 to 7.4		18.4 to 23.0[c)]
⊥ b axis	0.011 to 0.22	1.4 to 2.44		
Ref.	[1]	[2]	[2][a)]	[3][b)]
Linear compressibility (in 10^{-12} cm^2/dyn)				
1 bar	5.5 (k_a)	2.2±0.2 (k_b)	5.0±0.5 (k_c)	[5]
20 bar	0.7 (k_a)	0.31 (k_b)	0.7 (k_c)	

a) From velocity of sounds measurements of [6]. – b) Mean values from two crystals. – c) A still higher value of 30.0×10^7 N/m^2 was found [4].

Stress-strain measurements were performed on different $(SN)_x$ crystals [1 to 3] and results were compared with those of polyethylene terephthalate and trigonal selenium [1]. The influence of the method of polymerization and crystal perfection on the results is large. Young's modulus, yield strength, and breaking stress increase with a slower rate and longer time of polymerization. $(SN)_x$ is a moderately stiff, weak material comparable with drawn polyethylene and viscose rayon. $(SN)_x$ is stiffer than trigonal Se and less brittle owing to its fibrous nature [1].

Other workers studied the difference in the stress-strain behavior, comparing application of stress parallel or perpendicular to the chain axis [2]. For stress applied parallel to the b axis (after flattening the sample), the unloading-reloading behavior is elastic, as is typical for metals. For stress perpendicular to the b axis, the observed stress-strain behavior depended on the crystal orientation. There is probably only one easy glide system, (100) [001], in $(SN)_x$ [2].

The most accurate values for the mechanical properties of $(SN)_x$ were obtained by using a miniature tensometer in a scanning electron microscope. Load-strain curves for two $(SN)_x$ needles are presented, see **Fig. 28** [3].

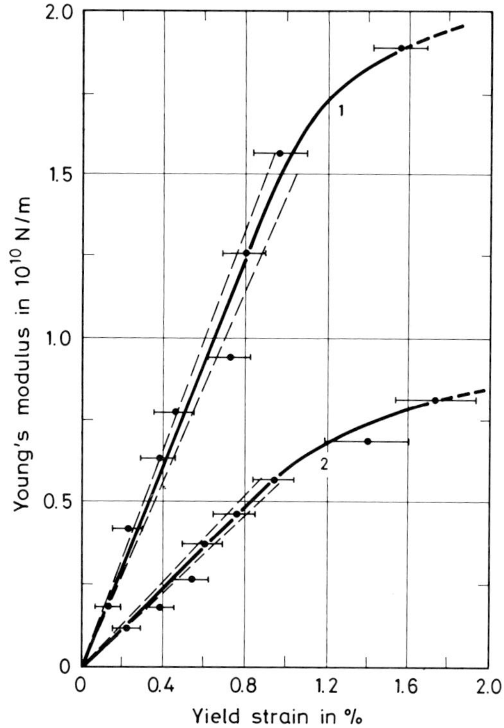

Fig. 28. Load-strain curves of two $(SN)_x$ needles [3].

The results were compared with other published data [1, 2] (see Table 32) and there is a general agreement in the values for Young's modulus. It was concluded that $(SN)_x$ crystals are 50 times less stiff than metals, but have approximately the same breaking strength of a metal such as aluminium [3]. Compressive strain curves as a function of hydrostatic pressure (for the axial lattice spacings and the spacing d_{102}) up to 22 kbar were presented. The X-ray diffraction patterns showed no evidence of a change in structure [5].

Compressed $(SN)_x$ crystals (perpendicular to the fiber axis) are soft and malleable and may be flattened into thin golden sheets. Long fibrous strands of $(SN)_x$, mechanically stripped from the crystal, are strong and easily bent and twisted [7].

References:

[1] Davidson, A. T.; Yoffe, A. D. (Phil. Mag. [8] **36** [1977] 1083/97, 1086).
[2] Brown, N.; Chiang, C. K. (J. Mater. Sci. **14** [1979] 49/58).
[3] Hoel, R. H.; Dingley, D. J. (J. Mater. Sci. **17** [1982] 2990/6).

[4] Chiang, C. K.; Heeger, A. J.; MacDiarmid, A. G. (Phys. Letters A **60** [1977] 375/7).
[5] Clarke, R. (Solid State Commun. **25** [1978] 333/6).
[6] Stolz, H. J.; Wendel, H.; Otto, A.; Pintschovius, L.; Kahlert, H. (Phys. Status Solidi B **78** [1976] 277/86).
[7] Mikulski, C. M.; Russo, P. J.; Saran, M. S.; MacDiarmid, A. G.; Garito, A. F.; Heeger, A. J. (J. Am. Chem. Soc. **97** [1975] 6358/63, 6361/2).

2.1.7.3 Sound Velocity

Longitudinal and transverse velocities of sound for different crystallographic directions were experimentally measured and calculated, see Table 33 [1 to 3].

Table 33
Experimental and Calculated Velocities of Sound (in m/s) in $(SN)_x$ for Different Crystallographic Directions in Longitudinal (V_L) and Transverse (V_T) Configuration.

direction	V_L (exp)	V_L (calc)	V_T (exp)	V_T (calc)	Ref.
020	8300±800	7885	2000±100	1934	[1, 3]
			1200±100		[1]
102	4000±500	4212	2400±300	1980	[1, 3]
			1350±150		[1]
$10\bar{2}$	3200±200	2480			[1, 2]
	3200±200	2990		2699	[3]
$\bar{2}02$	2600±300		1300±200		[1]
	2600±300	2410	1300±200	1050	[2]
010	8700±300	8490	2200±200	2270	[2]
			1200±100	720	

References:

[1] Stolz, H. J.; Wendel, W.; Otto, A.; Pintschovius, L.; Kahlert, H. (Phys. Status Solidi B **78** [1976] 277/86, 280/2).
[2] Pintschovius, L.; Currat, R.; Wendel, H. (Proc. Intern. Conf. Lattice Dyn., Paris 1977 [1978], pp. 579/81).
[3] Wendel, H. (J. Phys. C **10** [1977] L1/L5).

2.1.8 Thermal Properties

2.1.8.1 Heat Capacity

The heat capacity of $(SN)_x$ was determined in the temperature ranges 1.5 to 10 K [1], 1.5 to 55 K [2], 1.5 to 80 K [3], and 0.1 to 1.5 K [4]. The heat capacity data below 1 K (see **Fig. 29**) confirmed that superconductivity in $(SN)_x$ is a bulk effect [4].

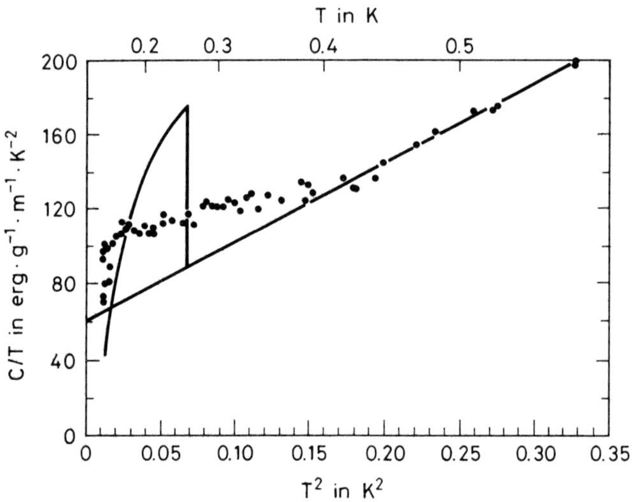

Fig. 29. Specific heat of $(SN)_x$ from 0.1 to 0.6 K [4].

A broad hump in heat capacity is observed, starting below 0.4 K. The entropy, which is associated with this hump, was compared with that of a simulated BCS (Bardeen-Cooper-Schrieffer) anomaly with $T_c = 0.26$ K (solid curve in Fig. 29) [4]. In the range 1.5 to 3.2 K a Debye behavior ($C = \gamma T + \beta T^3$ with $\gamma = (18 \pm 2.0) \times 10^{-6}$ J·g^{-1}·K^{-2} and $\beta = (8.8 \pm 0.4) \times 10^{-6}$ J·g^{-1}·K^{-4} [1]) of the lattice contribution to the heat capacity, which is typical for metals was reported [1 to 3]. Above 3.2 K deviation from T^3 (Debye behavior) to about $T^{2.7}$ (range 4 to 20 K) and then to a T dependence (range 40 to 80 K) was observed [2]. The unusual result is due to the anisotropy in the lattice dynamics of $(SN)_x$. Several models for the lattice specific heat of polymers were discussed [2]. The estimated value of the Debye temperature is 103 [4] or 148 K [2], leading to the conclusion that lattice interchain interaction exists [2].

References:

[1] Greene, R. L.; Grant, P. M.; Street, G. B. (Phys. Rev. Letters **34** [1975] 89/92).
[2] Harper, J. M. E.; Greene, R. L.; Grant, P. M.; Street, G. B. (Phys. Rev. B **15** [1977] 539/43).
[3] Harper, J. M. E.; Grant, P. M.; Street, G. B.; Greene, R. L. (Bull. Am. Phys. Soc. **20** [1975] 360).
[4] Lou, L. F.; Garito, A. F. (Preprint from Greene, R. L.; Street, G. B., NATO Advan. Study Inst. B **25** [1977] 167/95, 187/8, and unpublished results from [2]).

2.1.8.2 Thermal Conductivity

A thermal conductivity value of 6.4×10^{-10} w·cm^{-1}·deg^{-1} was reported for a compressed sample at 1750 atm, using the Ioffe dynamic method in a temperature range of 3 to 43°C [1]. The thermal conductivity was also measured in magnetic fields up to 18 kOe in the temperature range 0.08 to 4 K [2, 3] and 0.88 to 4 K [4]. An "anomalously large value" was reported but no further detailed data were published [3]. Phonons are the main contribution to the thermal conductivity over the entire temperature range [4]. In the normal conducting state the electronic contribution to the thermal conductivity is the same as that calculated from the Wiedemann-Franz law using the measured electrical resistivity [2].

References:

[1] Kronick, P. L.; Kaye, H.; Chapman, E. F.; Mainthia, S. B.; Labes, M. M. (J. Chem. Phys. **36** [1962] 2235/6).
[2] Azevedo, L. J.; Clark, W. G. (Bull. Am. Phys. Soc. **23** [1978] 304).
[3] Azevedo, L. J.; Chaikin, P. M.; Clark, W. G.; Fuller, W. W.; Hammann, J. (J. Phys. Colloq. [Paris] **39** [1978] C6-446/C6-447).
[4] Azevedo, L. J.; Clark, W. G. (Lect. Notes Phys. **96** [1978] 431).

2.1.8.3 Vaporization

2.1.8.3.1 Vapor Pressure. Thermodynamics of Vaporization

$(SN)_x$ sublimes at 140 to 150°C yielding lustrous epitaxial films [1]. The vapor pressure of $(SN)_x$ together with the vaporization coefficient were calculated in the range 100 to 150°C by a modification of the direct Knudsen method [2]. **Fig. 30** shows the experimental pressures above $(SN)_x$ as a function of temperature for four different orifice sizes along with the corresponding degrees of undersaturation. (The Knudsen equation was employed using the molecular weight of an $(SN)_4$ species representing the vapor phase.) The vapor pressure is expressed by $p = 37.64 - 16351/T$. A small vaporization coefficient of 0.0035 ± 0.0004 (from undersaturation data) or 0.0024 (from free-evaporation data) [2], ca. 2×10^{-3} [3] was determined. The small value indicates a difference between the structures in the gas phase and the bulk phase [2].

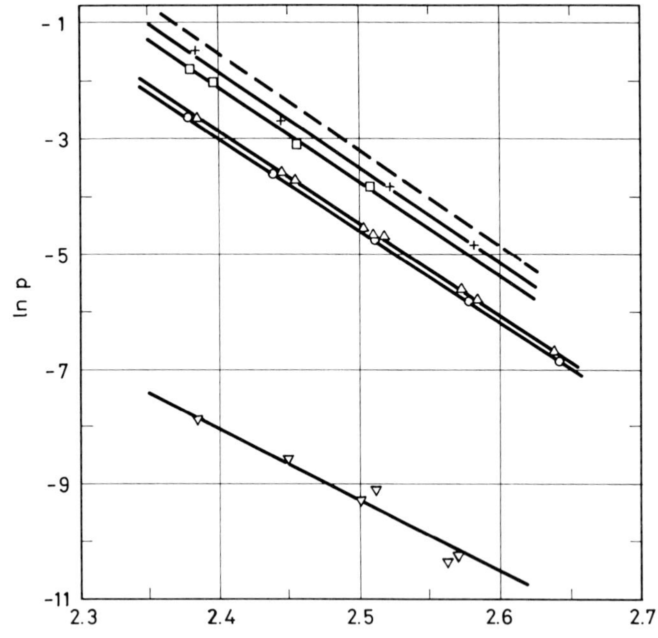

Fig. 30. Experimental pressures of $(SN)_x$ vs. reciprocal of absolute temperatures: saturation values, p_c (- - -); undersaturation values, p_o, at 0.23 (×), 0.36 (□), 0.55 (△), and 0.63 Torr (○); Langmuir free-evaporation pressures (▽) [2].

The enthalpy of vaporization is 32.49±0.99 kcal/mol [2] (vapor pressure studies), 29.0±0.5 kcal/mol [4] (field desorption mass spectrometry), and 29.0±3 kcal/mol [5] (TGA (thermogravimetric analysis)). The enthalpy of vaporization of "red $(SN)_x$" is 52±3 kcal/mol. The entropy of vaporization is 80.98±2.48 $cal \cdot mol^{-1} \cdot K^{-1}$ [2].

References:

[1] Mikulski, C. M.; Russo, P. J.; Saran, M. S.; MacDiarmid, A. G.; Garito, A. F.; Heeger, A. J. (J. Am. Chem. Soc. **97** [1975] 6358/63).

[2] Weber, D. C.; Ewing, C. T. (Inorg. Chem. **16** [1977] 3025/7).

[3] Cavanagh, R. R.; Altman, R. S.; Herschbach, D. R.; Klemperer, W. (J. Am. Chem. Soc. **101** [1979] 4734/5).

[4] Smith, R. D.; Wyatt, J. R.; DeCorpo, J. J.; Saalfeld, F. E.; Moran, M. J.; MacDiarmid, A. G. (J. Am. Chem. Soc. **99** [1977] 1726/30).

[5] Saalfeld, F. E.; DeCorpo, J. J.; Wyatt, J. R.; Mah, P. T.; Allen, W. N. (Ann. N.Y. Acad. Sci. **313** [1978] 759/66).

2.1.8.3.2 Constitution of Vaporized $(SN)_x$ Studied by Mass Spectroscopy

The exact nature of the vapor-phase species of $(SN)_x$ is still a subject of investigation. An open-chain $(SN)_4$ unit was discussed by several authors in a series of mass spectroscopic studies [1 to 6]. Later, in contrast, a cyclic S_3N_3 radical was reported to be the major species in $(SN)_x$ vapor under particular conditions [7, 8].

Electron impact (EI) at 14 eV, field ionization (FI), and field desorption (FD) mass spectra of $(SN)_x$ in comparison with those of S_4N_4 are given in **Fig. 31** [5, 6].

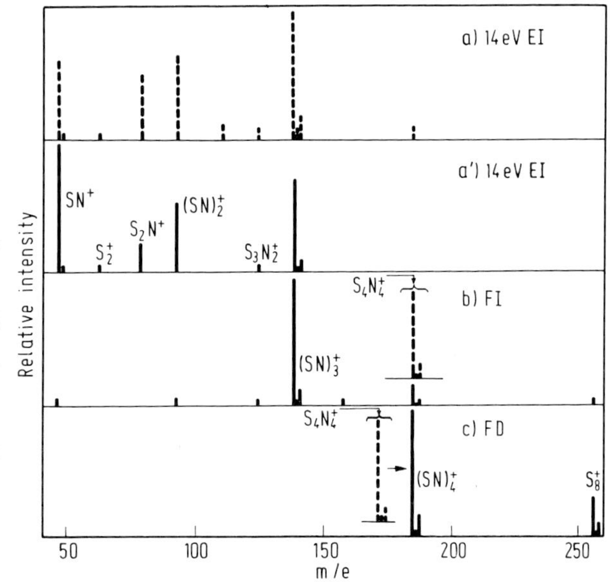

Fig. 31. Comparison of low-energy 14 eV electron impact (EI), field ionization (FI), and field desorption (FD) mass spectra of the vapors of $(SN)_x$ (–) and S_4N_4 (- - -). EI spectra of S_4N_4 and $(SN)_x$ vapors are given in a) and a'), respectively. The FI and FD spectra of S_4N_4 and $(SN)_x$ are compared in b) and c) [6].

Temperature studies on the mass spectra of $(SN)_x$ showed the relative intensities of fragment ions to be temperature-independent between 100 and 160°C [3] and in the range 150 to 180°C [9]. At higher temperatures, decomposition of $(SN)_x$ was indicated by increased contributions of ions due to molecular sulfur and nitrogen [3]. Phase angle data for $(SN)_x$ vapor ions together with those of hydride impurity ions and $(SNBr_{0.35})_x$ vapor ions are illustrated in a figure [4, 5]. From this study it was shown that more than 85% of observed ions originate from a neutral species having a molecular weight of 190±10 [1, 3 to 6]. Summarizing the mass spectroscopic data, the workers concluded that $(SN)_x$ vapor consists of one single species with the composition $(SN)_4$ (mol. wt. 184) [1 to 6]. Some authors find $(SN)_4$ to be the only species in FD spectrum, but it could not be observed in the EI, and appeared only weakly in the FI spectra [4]. The identity of the FD spectra of S_4N_4 and $(SN)_x$, both showing $S_4N_4^+$ as the main product, and the fact that linear molecules fragment under field ionization whereas cyclic compounds do not, lead the authors to the conclusion that $(SN)_4$ must have a "linear" structure [4, 5]. A "linear" structure was also proposed for the highly colored, chemically unstable species (possibly free radicals), which form from $(SN)_x$ vapors condensed at liquid nitrogen temperature [3]. It was noted that the red-purple, diamagnetic "linear" $(SN)_4$ is not identical to the red paramagnetic monomer obtained by the pyrolysis of S_4N_4 and S_4N_2 [10]. The gaseous "linear" molecule is very reactive; it spontaneously polymerizes on cool surfaces to produce $(SN)_x$ films [1, 3]. Other S_xN_y species such as S_3N_2, S_2N_2, S_2N, S_4N_2, S_3N_3, S_4N_4, and SN are believed to form from $(SN)_4$ on collisions with walls, or possibly with other molecules [1, 3]. The open-chain $(SN)_4$ species as the main vapor phase molecule is confirmed by other EI [9] and chemical-ionization mass-spectrometric studies [11]. Some workers assumed that $(SN)_x$ decomposes by vaporization to SN compounds like $(SN)_3$, $(SN)_2$ as well as $(SN)_4$ [9]. Others postulated an unstable acidic S_4N_4 which should fragment to cyclic $S_3N_3^+$ and SN^+ ions in the mass spectrometer [11]. The existence of a linear $(SN)_4$ oligomer in $(SN)_x$ vapor was confirmed in an IR spectroscopic study of an $(SN)_x$ film produced by trapping $(SN)_x$ vapor on CsI, KBr, sapphire or glass substrates at 12 K [12]. In a matrix isolation study of the $(SN)_x$ vapor phase, UV and IR analysis showed that mostly cyclic S_4N_4 and a small amount of an unidentified species were produced at 140 to 160°C. This unknown, intensively colored species converted completely to S_2N_2 by silver wool treatment. An intense red color and broad bands in both the IR and Raman spectrum of $(SN)_x$ vapor condensed in argon matrices at 10 K may be due to nonrigid, noncyclic, biradical isomers of S_2N_2, S_4N_4, and possibly S_6N_6 [13]. A study of the electrical polarity of the $(SN)_x$ vapor using the molecular beam deflection method agrees with mass spectral studies ([1 to 4]) in the fact that the $(SN)_x$ vapor consists predominantly of an $(SN)_4$ species. However, the $(SN)_4$ molecule should be nonpolar. (The linear isomer would have an electric dipole moment.) The open-chain form is not ruled out (since the Stark effect could be reduced below the level of detectability) but a cyclic isomer of the known S_4N_4 seemed to be more probable [14].

In contrast to the findings of prior studies [1 to 4] a photoelectron (PE)-photoionization mass spectrometric (PIMS) study showed that an S_3N_3 radical is the dominant species produced in the gas-phase vaporization of $(SN)_x$ under certain conditions [7, 8]. **Fig. 32** shows the PIMS spectra at different light-source energies. The highest observed mass peak at 138 amu (atomic mass units) was assigned to $(SN)_3^+$. The PE spectrum is produced mainly by only one compound. $(SN)_3^+$ is assumed to arise by direct ionization and not by fragmentation. A possible explanation for the discrepancy, in comparison to other MS studies [1 to 4], is that vaporization and condensation can lead to the reformation of $(SN)_x$ polymer and other colored materials. The revaporization yielded mass peaks in small amounts above 138 amu including 156 (S_4N_2) and 184 (S_4N_4). Several arguments which prove the formation of S_3N_3 were reported [8]. On the basis of ab initio calculations, a planar ring geometry was suggested for the structure. It was speculated that the compound is identical to the "red monomer" precursor of "red $(SN)_x$" (and not $SN^{\bullet}$ as proposed previously [10]). The same $S_3N_3^{\bullet}$ species was identified in the pyrolysis of S_4N_4, see "Sulfur-Nitrogen-Compounds" Part 2, 1985, pp. 160/2, and [15]. The study of vapor pressure also gives evidence for a nonlinear structure in the gas phase. A comparison of the heat of vaporization (~30 kcal/mol) and the SN bond strength (~50 to 60 kcal/mol) suggests that an exothermic molecular rearrangement of ca. 30 kcal/mol takes place, which may involve the formation of a cyclic species [16]. Relative abundances and appearance potentials of ions from S_4N_4 and $(SN)_x$ obtained by mass spectrometric analysis are shown in Table 34 and Table 35, p. 140.

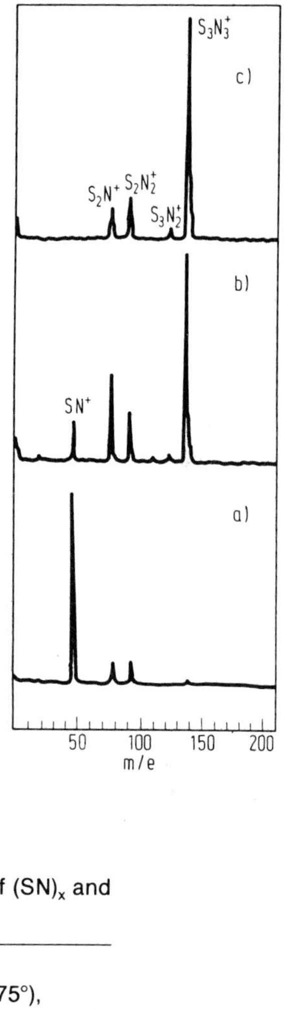

Fig. 32. Photoionization mass spectra (PIMS) of the vaporization products of $(SN)_x$ at 140°C at different light-source energies: a) HeI, 21.2 eV; b) $HL_{\alpha,\beta,\gamma}$, 10.2 to 13.6 eV; c) HL_α (filtered with LiF), 10.2 eV [8].

Table 34

Relative Abundances*) of Principal Ions in Mass Spectra (70 eV) of $(SN)_x$ and S_4N_4 [17].

ion	m/e	relative abundances		
		$(SN)_x$ (160°C)	S_4N_4 (75°C)	S_4N_4 (75°), silver wool (80°C)
SN^+	46	100	100	100
$HSNO^+$	63	10.2	1	2
SO_2^+, S_2^+	64	25	13	17
S_2N^+	78	17	18	12
$S_2N_2^+$	92	68	48	65
S_3N^+	110	1.2	2	—
$S_3N_2^+$	124	3.7	1.3	0.6
$S_3N_3^+$	138	49	68	42

Table 34 (continued)

ion	m/e	relative abundances		
		$(SN)_x$ (160°C)	S_4N_4 (75°C)	S_4N_4 (75°), silver wool (80°C)
$S_4N_2^+$	156	1.3	—	0.3
$S_4N_4^+$	184	2.7	12	6

*) All abundances represent the average of five or more measurements. Reproducibility is ±20% of the value.

Table 35

Appearance Potentials of Mass Fragments from $(SN)_x$ and S_4N_4 [4, 17].

ion	m/e	appearance potential in eV			
		$(SN)_x$ (130°C) [17]	S_4N_4 (60°C) [17]	$(SN)_x$ [4]	S_4N_4 [4]
SN^+	46	8.9	9.1	13.4	14.1
S_2^+	64	9.4	9.7	12.9	—
S_2N^+	78	9.7	9.7	10.1	—
$S_2N_3^+$	92	10.3	10.2	11.2	13.7
S_3N^+	110	9.8	9.9	—	—
$S_3N_2^+$	124	8.4	8.7	10.5	—
$S_3N_3^+$	138	8.4	8.8	9.5	12.3
$S_4N_3^+$	156	9.5	9.2	—	—
$S_4N_4^+$	184	9.4	9.5	9.8	10.4

Mass spectral data of $(SN)_x$ powder were also measured, see Table 36 [18].

Table 36

Mass Spectral Data (70 eV, ~80°C) of $(SN)_x$ Powder ($(SN)_x$ sample with S:N ~1.0) [18].

ion	m/e	rel. intensity	ion	m/e	rel. intensity
SN^+	46	100	$S_3N_3^+$	138	12
S_2^+	64	12	$S_4N_2^+$	156	1
S_2N^+	78	2.5	$S_4N_4^+$	184	1
$S_2N_2^+$	92	49			

References:

[1] Smith, R. D.; Wyatt, J. R.; DeCorpo, J. J.; Saalfeld, F. E.; Moran, M. J.; MacDiarmid, A. G. (Chem. Phys. Letters **41** [1976] 362/4).

[2] Smith, R. D.; DeCorpo, J. J.; Wyatt, J. R.; Saalfeld, F. E. (Intern. J. Mass. Spectrom. Ion Phys. **21** [1976] 411/4).

[3] Smith, R. D.; Wyatt, J. R.; DeCorpo, J. J.; Saalfeld, F. E.; Moran, M. J.; MacDiarmid, A. G. (J. Am. Chem. Soc. **99** [1977] 1726/30).

[4] Saalfeld, F. E.; DeCorpo, J. J.; Wyatt, J. R.; Mah, P. T.; Allen, W. N. (Ann. N.Y. Acad. Sci. **313** [1978] 759/66).
[5] Saalfeld, F. E.; DeCorpo, J. J.; Wyatt, J. R. (Extend. Linear Chain Compounds **1** [1982] 33/72, 50/8).
[6] Wyatt, J. R.; DeCorpo, J. J.; Smith, R. D.; Saalfeld, F. E. (Advan. Mass. Spectrom. A **7** [1978] 597/600).
[7] Lau, W. M.; Westwood, N. P. C.; Palmer, M. H. (J. Chem. Soc. Chem. Commun. **1985** 752/3).
[8] Lau, W. M.; Westwood, N. P. C.; Palmer, M. H. (J. Am. Chem. Soc. **108** [1986] 3229/37).
[9] Ueno, N.; Sugita, K.; Koga, O.; Suzuki, S.; Yoshino, K.; Kaneto, K.; Inuishi, Y. (Japan. J. Appl. Phys. **18** [1979] 1597/8).
[10] Labes, M. M.; Love, P.; Nichols, L. F. (Chem. Rev. **79** [1979] 1/15, 7).

[11] Siedle, A. R.; Hertz, H. S. (Mater. Res. Bull. **11** [1976] 1185/9).
[12] Iqbal, Z.; Downs, D. S. (Chem. Phys. **44** [1979] 137/42).
[13] Teichmann, R. A., III; Nixon, E. R. (Inorg. Chem. **15** [1976] 1993/4).
[14] Cavanagh, R. R.; Altman, R. S.; Herschbach, D. R.; Klemperer, W. (J. Am. Chem. Soc. **101** [1979] 4734/5).
[15] Besenyei, E.; Eigendorf, G. K.; Frost, D. C. (Inorg. Chem. **25** [1986] 4404/8).
[16] Weber, D. C.; Ewing, C. T. (Inorg. Chem. **16** [1977] 3025/7).
[17] Besenyei, E.; Eigendorf, G. K.; Frost, D. C. (J. Electron. Spectrosc. Relat. Phenom. **41** [1986] 419/26).
[18] Banister, A. J.; Hauptman, Z. V.; Passmore, J.; Wong, C.-M.; White, P. S. (J. Chem. Soc. Dalton Trans. **1986** 2371/9).

2.1.9 Magnetic and Electrical Properties

2.1.9.1 Magnetic Susceptibility

Magnetic properties of $(SN)_x$ were intensively studied for the transition into the superconducting state, see Meissner Effect, p. 153. In the normal state $(SN)_x$ crystals of high quality exhibit Pauli paramagnetism. The static magnetic susceptibility $\varkappa$ of $(SN)_x$ crystals is ca. 9×10^{-6} cm^3/mol ($(0.2\pm0.1)\times10^{-6}$ cm^3/g) [1] or $(5.5\pm1.0)\times10^{-6}$ cm^3/mol [2]. The values are temperature-independent [1, 2]. After γ irradiation (10^7 Röntgen) the susceptibility increases with decreasing temperature closely following the Curie-Weiss law. The density of magnetic dipoles in the irradiated samples was estimated to be 2.5×10^{20} cm^{-3} from the Curie constant [1, 3].

References:

[1] Kaneto, K.; Tanimura, K.; Yoshino, K.; Inuishi, Y. (Solid State Commun. **22** [1977] 383/5).
[2] Scott, J. C.; Kulick, J. D.; Street, G. B. (Bull. Am. Phys. Soc. [2] **23** [1978] 304).
[3] Yoshino, K.; Tachibana, S.; Makio, Y.; Yamamoto, M.; Kaneto, K.; Inuishi, Y. (Technol. Rept. Osaka Univ. **27** [1977] 437/43).

2.1.9.2 Electrical Conductivity

2.1.9.2.1 Survey on Conduction Mechanism

$(SN)_x$ exhibits metallic conductivity along the polymeric chains ($\parallel$b) ($\sigma_{\parallel}$ decreases with increasing T). Perpendicular to the chains (i $\perp$ b) σ is much lower ($\sigma_{\parallel}/\sigma_{\perp}$ =100 to 1000) but is still just metallic, as revealed by optical measurements. The value of $\sigma_\perp$ is strongly influenced by extrinsic effects due to imperfections of the chains and fibrous morphology.

Measurements of the thermoelectric power S confirm the metallic behavior along the chains. The low absolute value of S and its decrease with decreasing temperature correspond to a metal or a highly degenerate semiconductor. They show, as well, that the metallic state is retained down to low temperatures (no Peierls transition to an insulator). Moreover, both thermopower (negative sign of S) and galvanomagnetic properties indicate that the majority current carriers are electrons (and not holes) which also have the higher mobility, see p. 161. The metallic or semimetallic behavior is also revealed by band structure considerations (besides by other properties). There is a nonvanishing density of states at the Fermi level for the directions ΓZ ($\vec{k}_b$) and ΓY ($\vec{k}_a$), and the Fermi surface consists of electron and hole pockets (semimetal), closed in $\vec{k}_b$ and $\vec{k}_a$ direction, opened in $\vec{k}_c$ direction.

The observed T^2 dependence of the resistivity at low temperatures point to electron-electron Umklapp scattering as the predominant conduction mechanism (instead of the usual electron-phonon scattering, T^3 dependence). Conductivity and superconductivity (see p. 150) properties of $(SN)_x$ were described in context with other chemical and physical properties in several review papers [1 to 9].

The electronic **mean free paths,** $\bar{l}_{\parallel}$ and $\bar{l}_{\perp}$ at low temperature were determined to lie in the range 180 to ~2000 Å [10, 11] and 6 to 28 Å [10], respectively, using the relation $\bar{l} = 3D/v_F$ (D = electronic diffusion constant, see p. 154, v_F = Fermi velocity, taken as 1×10^8 cm/s). Other workers reported for l a value of ~700 Å at 4.2 K along the b axis using galvanomagnetic data [12].

References:

[1] Rothenberg, M. S. (Phys. Today **28** [1975] 17/20).

[2] Geserich, H. P.; Pintschovius, L. (Festkörperprobleme [Advan. Solid State Phys.] **16** [1976] 73/84).

[3] Greene, R. L.; Street, G. B. (NATO Advan. Study Inst. Ser. B **25** [1977] 167/95, 176/82).

[4] Street, G. B.; Greene, R. L. (IBM J. Res. Develop. **21** [1977] 99/110).

[5] Greene, R. L.; Gill, W. D.; Azevedo, L. J.; Clark, W. G.; Deutscher, G. (Lect. Notes Phys. **65** [1977] 603/10).

[6] Pintschovius, L. (Colloid Polym. Sci. **256** [1978] 883/92).

[7] Labes, M. M.; Love, P.; Nichols, L. F. (Chem. Rev. **79** [1979] 1/15, 8/9).

[8] Gill, W. D.; Clarke, T. C.; Street, G. B. (Appl. Phys. Commun. **2** [1982/83] 211/82, 217/21).

[9] Kaneto, K.; Yoshino, K.; Inuishi, Y. (in: Monceau, P., Electronic Properties of Inorganic Quasi-One-Dimensional Compounds, Reidel, Dordrecht 1985, pp. 69/109, 79/82).

[10] Barrett, P.; Herrenden-Harker, W. G.; Priestley, M. G.; da Silva, E. Z. (J. Phys. F **11** [1981] 873/82).

[11] Barrett, P.; Chambers, R. G.; Feenan, P.; Herrenden-Harker, W. G.; Priestley, M. G.; Trinder, R. W.; Read, S. F. J. (J. Phys. F **10** [1980] L89/L94).

[12] Beyer, W.; Gill, W. D.; Street, G. B. (Solid State Commun. **23** [1977] 577/81).

2.1.9.2.2 Single Crystals

Pressed pellets, which do not show intrinsic properties, provided the first conductivity measurements on $(SN)_x$ [1 to 4]. In these early works $(SN)_x$ was claimed to be a semiconductor, though indications for metallic conductivity were observed [2]. Reinvestigation of the electrical conductivity in $(SN)_x$ crystals, parallel to the $(SN)_x$ chain direction, established the metal-like nature in the temperature range 4.2 to 300 K [5]. The temperature dependence of the conductivity $\sigma_{\parallel}$ was studied from various $(SN)_x$ probes [6]; $\sigma_{\parallel}$ depends strongly on crystal quality [7 to 9] and is assumed to be influenced by chain breaks and defects, which may be minimized by annealing processes [6]. An observed resistivity minimum due to extrinsic effects associated with the preparation method [6], lying in the range 20 to 30 K [9, 10] or at 33 K [5], moves to lower temperatures [7 to 9] or is absent [11] as crystal perfection increases. No metal-insulator transition was observed down to 0.07 K [12]. Room temperature conductivity values along the fiber axis are in the range ca. 1000 [9] to 4000 $\Omega^{-1}\cdot cm^{-1}$ [4], depending on crystal perfection [4, 8]. Longer polymerization times increase the conductivity up to values of 1200 to 3700 $\Omega^{-1}\cdot cm^{-1}$ [13] and the resistance ratios ($\varrho_{RT}/\varrho_{4K}$) up to values of 50 to 205 [13] or 250 [6]. Twisted needles of $(SN)_x$ exhibit very high conductivities, $\sigma_{\parallel}\approx 10^6$ $\Omega^{-1}\cdot cm^{-1}$ at 4 K indicating a low concentration of defects. An ac conductivity value of $\sigma_{\parallel}=1000$ $\Omega^{-1}\cdot cm^{-1}$ for 300 K was reported [14].

Fewer conductivity data perpendicular to the chain axis were reported. The perpendicular conductivity $\sigma_{\perp}$ is smaller than the parallel conductivity [7, 15]. A $\sigma_{\perp}$ maximum was observed between 45 and 70 K [9] or at ca. 40 K [16]. The conductivity ratio $\sigma_{\parallel}/\sigma_{\perp}$ is ca. 50 [9] or 100 [15] at room temperature. In addition $\sigma_{\parallel}/\sigma_{\perp}$ ratios of 500 at 20 K [9] ~1000 between 4 K and room temperature [6] and values beween 1000 and 10000 at low temperatures [15] were published. **Fig. 33** shows measured curves for $\varrho_{\parallel}$ and $\varrho_{\perp}$ in the range ~4 to ~400 K [7].

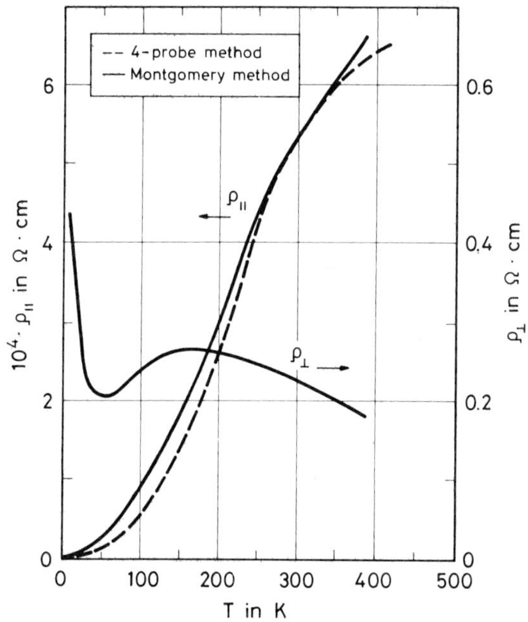

Fig. 33. Anisotropy in the dc conductivity of $(SN)_x$ crystals measured along the fiber or b axis ($\varrho_{\parallel}$) and in the perpendicular direction ($\varrho_{\perp}$) [7].

The temperature dependence of the b axis resistivity $\varrho_{\parallel}$ measured in the range 15 to 250 K [13, 17] and 20 to 250 K [6] can be expressed by a relationship $\varrho=\varrho_0+\varrho_i(T)=\varrho_0+AT^2$ (ϱ_0 is the sample-dependent residual resistivity, ϱ_i is intrinsic resistivity, A is a material constant with an average value $5.6\times10^{-9}\,\Omega\,cm/K^2$). The T^2 dependence is not typical for an ordinary metal [13], but several groups agree in this expression [6, 7, 14, 18 to 20]. An electron-electron Umklapp scattering mechanism instead of an electron-phonon mechanism was assumed responsible for the unusual T^2 behavior [13, 18 to 20]. The experimental results (T^2 dependence and magnitute of resistivity) were verified by theoretical work [21, 22]. The resistivity is proportional to T^2 for electron energy states $E-E_F \leq 0.001$ eV [22]. The observed T^2 behavior could also be explained by electron-hole scattering [21]. Conductivity measurements in the range 0.1 to 5 K (with applied magnetic field to avoid transition to superconductivity and fluctuations) show a $T^{1/2}$ dependence for $\sigma_{\parallel}$ and $\sigma_{\perp}$ [23].

The measured perpendicular conductivity $\sigma_{\perp}$ in $(SN)_x$ is predominantly extrinsic [7]. Its temperature dependence is neither that of a semiconductor nor that expected for a metal [15]. A simple chemical explanation of the electrical conductivity in $(SN)_x$ was given [7]; it arises from delocalization along the polymer chain of the unpaired electron in the $\pi^*_{p_z}$ antibonding level of each SN unit.

References:

[1] Goehring, M. (Z. Anorg. Allgem. Chem. **285** [1956] 181/90).
[2] Kronick, P. L.; Kaye, H.; Chapman, E. F.; Mainthia, S. B.; Labes, M. M. (J. Chem. Phys. **36** [1962] 2235/7).
[3] Chapman, D.; Warn, R. J.; Fitzgerald, A. G.; Yoffe, A. D. (Trans. Faraday Soc. **60** [1964] 294/300).
[4] Douillard, A.; May, J. F.; Vallet, G. (Compt. Rend. C **269** [1969] 212/5).
[5] Walatka, V. V., Jr.; Labes, M. M.; Perlstein, J. H. (Phys. Rev. Letters **31** [1973] 1139/42).
[6] Greene, R. L.; Street, G. B. (NATO Advan. Study. Inst. Ser. B **25** [1977] 167/95).
[7] Street, G. B.; Greene, R. L. (IBM J. Res. Develop. **21** [1977] 99/110).
[8] Labes, M. M.; Love, P.; Nichols, L. F. (Chem. Rev. **79** [1979] 1/15).
[9] Hsu, C.; Labes, M. M. (J. Chem. Phys. **61** [1974] 4640/5).
[10] Greene, R. L.; Street, G. B.; Suter, L. J. (Phys. Rev. Letters **34** [1975] 577/9).

[11] Street, G. B.; Arnal, H.; Gill, W. D.; Grant, P. M.; Greene, R. L. (Mater. Res. Bull. **10** [1975] 877/82).
[12] Greene, R. L.; Street, G. B.; Suter, L. J. (Bull. Am. Phys. Soc. [2] **20** [1975] 360).
[13] Chiang, C. K.; Cohen, M. J.; Garito, A. F.; Heeger, A. J.; Mikulski, C. M.; MacDiarmid, A. G. (Solid State Commun. **18** [1976] 1451/5).
[14] Kahlert, H.; Seeger, K. (Phys. Semicond. Proc. 13th Intern. Conf., Rome 1976, pp. 353/6).
[15] Pintschovius, L. (Colloid Poly. Sci. **256** [1978] 883/92).
[16] Grant, P. M.; Greene, R. L.; Gill, W. D.; Rudge, W. E.; Street, G. B. (Mol. Cryst. Liquid Cryst. **32** [1976] 171/6).
[17] Chiang, C. K.; Garito, A. F.; Heeger, A. J.; MacDiarmid, A. G. (Bull. Am. Phys. Soc. [2] **21** [1976] 255).
[18] Geserich, H. P.; Pintschovius, L. (Festkörperprobleme [Advan. Solid State Phys.] **16** [1976] 73/84).
[19] Friend, R. H.; Jerome, D.; Rehmatullah, S.; Yoffe, A. D. (J. Phys. C **10** [1977] 1001/12).
[20] Oshiyama, A.; Nakao, K.; Kamimura, H. (J. Phys. Soc. Japan **45** [1978] 1136/46).

[21] Oshiyama, A. (J. Phys. C **14** [1981] 5109/25).
[22] Love, P. (J. Phys. C **12** [1979] L627/L631).
[23] Brinslow, G. M.; Priestley, M. G. (Solid State Commun. **45** [1983] 1063/7).

2.1.9.2.3 Films

The electrical conductivity of $(SN)_x$ films was studied by several groups [1 to 12]. Its temperature dependence was determined on epitaxially crystalline films in the range 0.07 to 300 K [3], 0.03 to 8.9 K [4, 5], and 0.1 to 100 K [6]. In contrast to the resistivity $\varrho_\parallel$ of $(SN)_x$ crystals, that of epitaxially grown films decreases as the temperature increases [4, 5, 7], thus $(SN)_x$ films show a thermally activated conductivity with very low-activation energies [2 to 5, 8]. Activation energies in the range 0.004 to 0.014 eV were reported by [2], and a value of 0.1 eV by [7, 12]. The room temperature conductivities range from 10^{-2} [1, 2] to 10^2 $\Omega^{-1}\cdot cm^{-1}$ [2, 9]. At 4.2 K conductivity values of 2.5 to 2.7×10^{-3} $\Omega^{-1}\cdot cm^{-1}$ were measured [8]. The resisitivity is mainly dominated by interparticle barrier effects [8, 10]. The temperature dependence of the resistivity $\varrho_\perp$ perpendicular to the chain axis is essentially the same as that for $(SN)_x$ "single crystals" [3]. The resistivity ratio $\varrho_\perp/\varrho_\parallel$ is ca. 5 and is nearly temperature independent [3, 11]. Other reported values for this ratio on epitaxial films are 16 at room temperature and 10 at 4.2 K [1].

The deposition temperature T_s of $(SN)_x$ films strongly affects the conductivity properties [9, 11, 12], see **Fig. 34**. For low T_s low conductivities ($\sigma_{RT}\approx10^{-2}$ $\Omega^{-1}\cdot cm^{-1}$) with semiconductor-like temperature dependence were measured. Increasing T_s lead to an increase of the conductivity (for example $\sigma_{RT}\approx10^2$ $\Omega^{-1}\cdot cm^{-1}$ at $T_s=50°C$) and a slightly metallic temperature dependence [9], even at 77 K [11].

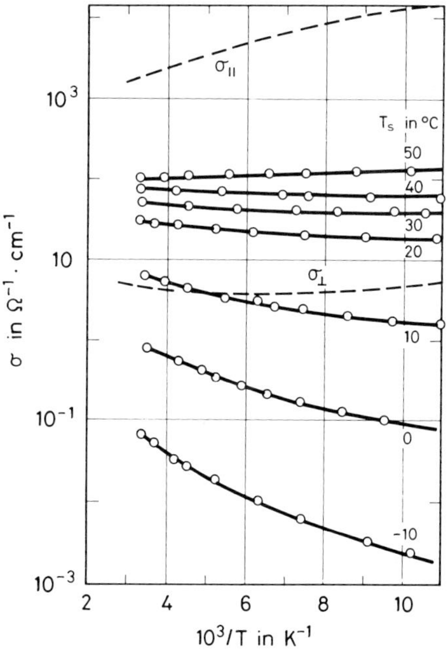

Fig. 34. Temperature dependence of the conductivity σ of $(SN)_x$ films prepared at different substrate temperatures T_s. Dashed lines refer to an $(SN)_x$ crystal [11].

The nature of the conductivity mechanism in $(SN)_x$ films is still not clarified. A possible intrinsic conduction mechanism which produces a finite conductivity $\sigma_\perp$ even at 0 K was suggested and an equation for the temperature dependence of σ was given: $\sigma(T)=\sigma(0)+\alpha e^{-\Delta/kT}$; $\sigma(0)$ represents the process that determines the conductivity at T = 0 K and the other term, the thermally activated conductivity with activity energy Δ. This equation fits the data between 0.07 K and 2 K [3], but not low-temperature data (from 0.030 K) of [4]. Down to 20 mK a superconducting transition was not found [3, 4]. Above 0.25 K σ is represented by a relationship ($R = R_0 \cdot \exp(AT^{-b})$) involving transport of electrons by tunneling between metallic particles [4]. A similar expression ($\lg R = A + (C/KT)^{1/2}$) was reported for the temperature range from 0.4 to 20 K corresponding to a tunnel mechanism from grain to grain [6]. Other workers reported an activated temperature dependence $\log \sigma \sim -T^{-1/4}$ due to different gaps in the density of states (heterogeneous model) for depositing temperatures ≤30°C, for the whole temperature range investigated [11] cf. [8]. One paper describes the dependence of electrical conduction during the preparation of $(SN)_x$ films and presents the time dependence of electrical conductance of evaporated material [7]. An increase of the frequency of an applied electrical field leads to an increase of the conductivity and to a decrease of the activation energy [12].

References:

[1] Hsu, C.; Labes, M. M. (Chem. Phys. **61** [1974] 4640/5).
[2] Bright, A. A.; Cohen, M. J.; Garito, A. F.; Heeger, A. J.; Mikulski, C. M.; MacDiarmid, A. G. (Appl. Phys. Letters **26** [1975] 612/5).
[3] de la Cruz, F.; Stolz, H. J. (Solid State Commun. **20** [1976] 241/3).
[4] Soulen, R. J.; Utton, D. B. (Solid State Commun. **21** [1977] 105/8).
[5] Soulen, R. J.; Utton, D. B. (Bull. Am. Phys. Soc. [2] **22** [1977] 372).
[6] Young, M. W.; Thomas, J. M. D.; Adkins, C. J.; Tate, J. W. (J. Phys. Colloq. [Paris] **39** [1978] C6-448/C6-450).
[7] Yoshino, K.; Makio, Y.; Kaneto, K.; Inuishi, Y. (Japan J. Appl. Phys. **16** [1977] 1047/8).
[8] Beyer, W.; Mell, H.; Gill, W. D. (Solid State Commun. **27** [1978] 185/9).
[9] Gill, W. D.; Beyer, W.; Street, G. B. (Bull. Am. Phys. Soc. [2] **22** [1977] 372).
[10] Gill, W. D.; Clarke, T. C.; Street, G. B. (Appl. Phys. Commun. **2** [1982/83] 211/82, 223/24).

[11] Beyer, W.; Gill, W. D.; Street, G. B. (Solid State Commun. **27** [1978] 343/8).
[12] Yoshino, K.; Tachibana, S.; Makio, Y.; Yamamoto, M.; Kaneto, K.; Inuishi, Y. (Technol. Rept. Osaka Univ. **27** [1977] 437/43).

2.1.9.2.4 Powders

$(SN)_x$ powders prepared via a chemical synthesis show drastically smaller conductivities than crystals or films. Values of $0.94 \times 10^{-2} \Omega^{-1} \cdot cm^{-1}$ [1] and $5 \times 10^{-2} \Omega^{-1} \cdot cm^{-1}$ [2] were reported for hand compacted powders. Gold-bronze layers obtained by subliming powdered $(SN)_x$ exhibit conductivities in the range 2 to 7 $\Omega^{-1} \cdot cm^{-1}$ [3].

References:

[1] Kennett, F. A.; MacLean, G. K.; Passmore, J.; Rao, M. N. S. (J. Chem. Soc. Dalton Trans. **1982** 851/7).
[2] Wolmershäuser, G.; Fuhrmann, J.; Jotter, R.; Wilhelm, T.; Scherer, O. J. (Mol. Cryst. Liquid Cryst. **118** [1985] 435/8).
[3] Banister, A. J.; Hauptman, Z. V. (J. Chem. Soc. Dalton Trans. **1986** 2371/9).

2.1.9.2.5 "Red $(SN)_x$"

The room temperature conductivity of "red $(SN)_x$" crystals is ~1200 $\Omega^{-1} \cdot cm^{-1}$. A weak conductivity maximum of magnitude ca. 2 times the room temperature value was observed at ~50 K.

Reference:

Love, P.; Myer, G.; Kao, H. I.; Labes, M. M.; Junker, W. R.; Elbaum, C. (Ann. N.Y. Acad. Sci. **313** [1978] 745/58, 752/3).

2.1.9.2.6 Pressure Dependence. Stress- and Strain Dependence

The pressure dependence of the normal state conductivity σ was the subject of several investigations [1 to 10]. **Fig. 35** shows the pressure dependence up to ca. 12 kbar for both parallel and perpendicular to the fiber axis (at T = 295 K). The reproducible results indicate that, in contrast to the temperature behavior, the pressure dependence of both $\sigma_{\parallel}$ and $\sigma_{\perp}$ is very similar, showing an experimental increase of the conductivities below 4 kbar and a tendence to saturation at about 12 kbar [1]. Similar measurements up to 15 kbar are in accord with this study [2]. Investigations of the pressure dependence of normal state conductivity were also performed at higher pressures up to 50 [3], 80 [4], and 400 kbar [8].

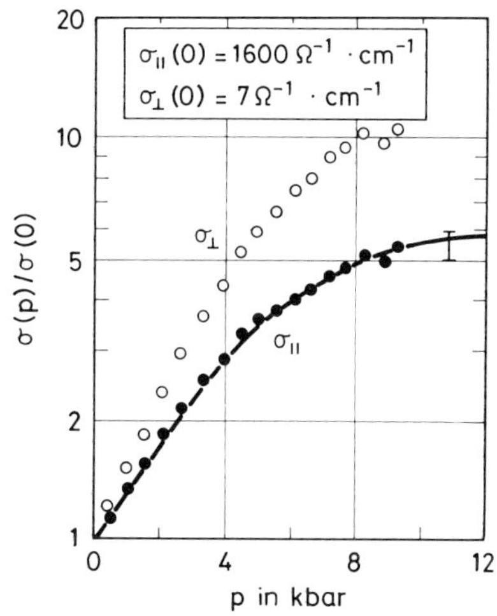

Fig. 35. Pressure dependence of the conductivity of $(SN)_x$ at T = 295 K. Data points are for a Montgomery-geometry crystal. The solid line represents $\sigma_{\parallel}$ from a linear 4-probe measurement [1].

The conductivity ratio $(\sigma(p)/\sigma(0))_{\parallel}/(\sigma(p)/\sigma(0))_{\perp}$ reached a constant value of 10 at pressures greater than 30 kbar [4]. Around 50 to 70 kbar a saturation of σ(p) occured [4, 8], followed by a monotonic decrease in conductivity on further application of pressure up to 400 kbar, see **Fig. 36**, p. 148. In the decreasing conductivity region the dependence on pressure is greater for $\sigma_{\parallel}$ than for $\sigma_{\perp}$. At 380 kbar $\sigma_{\parallel} \approx 1930$, $\sigma_{\perp} \approx 67\ \Omega^{-1}\cdot cm^{-1}$ [8].

Possible reasons for the initial increase of conductivity with increasing pressure were discussed in terms of a band-structure change [1, 8], a decrease of the effect of crystal imperfection [3], and a phase transition (monoclinic $P_1 \rightarrow$ orthorhombic P_2) [8], see p. 105. The unusual increase of σ with pressure was also interpreted as a confirmation that the electron-electron scattering is the predominant conduction mechanism [5]. The subsequent decrease of σ above 50 to 70 kbar may be due to a change in the electronic band structure of the new phase (σ decrease) and an increase of broken fibers ($\sigma_{\parallel}$ decrease is stronger than $\sigma_{\perp}$ decrease) [8].

Additional support for these conclusions comes from the temperature dependence of ϱ (p). Two regimes of low-temperature T^2 dependence (different coefficients of ϱ(T)) between 3 and

8 kbar were observed which both become saturated at room temperature [4]. Measurements of the resistance as a function of temperature from 2.3 to 300 K at various fixed pressures from 90 to 400 kbar [8] and up to ~560 K at 23±2 kbar [9] were also discussed. At 23±2 kbar normal metallic behavior of ϱ(T) was found up to about 503 K. The subsequent thermal decomposition resulted in a sharp decrease in resistance [9]. The resistivity at higher pressures is dominated by electron-phonon scattering [4]. The conductivity along the fiber axis b is directly proportional to the inverse compressibility ($\varkappa_b^{-1}$). In contrast, $\sigma_\perp$ depends on interfiber interactions and is not an intrinsic property [10].

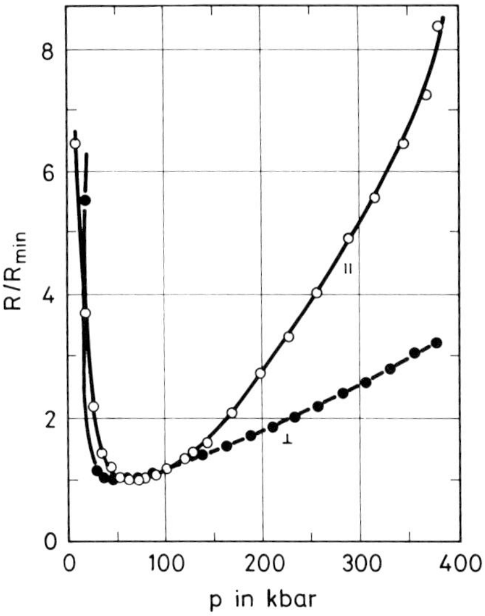

Fig. 36. Reduced electrical resistance R/R_{min} of $(SN)_x$ both perpendicular (⊥) and parallel (∥) to the b (fiber) axis vs. pressure at room temperature [8].

Application of **uniaxial tensional stress** led to a decrease in electrical conductivity at room temperature [11 to 13]. The change in $\sigma_\perp$ (with p⊥b) was greater than that in $\sigma_\parallel$ (with p∥b), a behavior which relates to the fibrous nature of $(SN)_x$ crystals [11] (ratio d (lnσ)/dp for ∥ and ⊥ case differ by a factor of 160). The observed dependence of $\sigma_\parallel$ (p) is typical of metals. The behavior was explained by a stress-induced change in the semimetal band structure [11]. For a tension of 2×10^8 N·m^{-2} a change in conductivity of about 15% was observed [13].

A filamentary sample was sharply bent at an angle of about 90° and returned to its original form. The low-temperature conductivity of the bent sample was two orders of magnitude lower than that of the undeformed material [14].

References:

[1] Gill, W. D.; Greene, R. L.; Street, G. B.; Little, W. A. (Phys. Rev. Letters **35** [1975] 1732/5).

[2] Kamimura, H.; Grant, A. J.; Levy, F.; Yoffe, A. D.; Pitt, G. D. (Solid State Commun. **17** [1975] 49/53).

[3] Müller, W. H. G.; Baumann, F.; Dammer, G.; Pintschovius, L. (Solid State Commun. **25** [1978] 119/22).

[4] Friend, R. H.; Jerome, D.; Rehmatullah, S.; Yoffe, A. D. (J. Phys. C **10** [1977] 1001/12).

[5] Gill, W. D.; Clarke, T. C.; Street, G. B. (Appl. Phys. Commun. **2** [1982/83] 211/82, 220).

[6] Labes, M. M.; Love, P.; Nichols, L. F. (Chem. Rev. **79** [1979] 1/15, 9).

[7] Gill, W. D.; Street, G. B. (Bull. Am. Phys. Soc. [2] **20** [1975] 360).

[8] Dunn, K. J.; Bundy, F. P.; Interrante, L. V. (Phys. Rev. [3] B **23** [1981] 106/11).

[9] Vezzoli, G. C.; Abel, J. E. (Mater. Res. Bull. **13** [1978] 151/8).
[10] Clarke, R. (Solid State Commun. **25** [1978] 333/6).

[11] Chiang, C. K.; Heeger, A. J.; MacDiarmid, A. G. (Phys. Letters A **60** [1977] 375/7).
[12] Chiang, C. K.; Garito, A. F.; Heeger, A. J.; MacDiarmid, A. G. (Bull. Am. Phys. Soc. [2] **21** [1976] 255).
[13] Davidson, A. T.; Yoffe, A. D. (Phil. Mag. [8] **36** [1977] 1083/97).
[14] Kahlert, H.; Seeger, K. (Phys. Semicond. Proc. 13th Intern. Conf., Rome 1976, pp. 353/6).

2.1.9.2.7 Effect of Irradiation

The effect of irradiation on the electrical conductivity of $(SN)_x$ was studied using γ-rays (from Co^{60} source) [1, 2], neutrons and heavy ions (at 20 K) [3], and electron beam [4, 5]. The temperature dependence of the resistivity of the original and γ-ray (~10^7 Röntgen) irradiated samples was compared [1, 2]. A resistivity minimum comparable to that found in low-quality crystals was observed only in γ-irradiated samples at 30 to 40 K [1] and in samples irradiated by electron beams [5]. The minimum was explained in terms of a localized magnetic moment (Kondo effect) [1]. The resistivity increased linearly with dosage on irradiation with neutrons (~1 MeV) [3]. The damage could be completely annealed on warming to room temperature, however the damage caused by heavy-ion bombardement (much stronger resistivity increase) could not be annealed [3]. Irradiation with an electron beam also led to damage that increased with dose [4]. For higher doses, saturation was found. The temperature dependence of the resistivity showed two distinct regions: T^2 dependence above 30 K, $T^{1/2}$ behavior below 30 K [5].

References:

[1] Kaneto, K.; Tanimura, K.; Yoshino, K.; Inuishi, Y. (Solid State Commun. **22** [1977] 383/5).
[2] Yoshino, K.; Tachibana, S.; Makio, Y.; Yamamoto, M.; Kaneto, K.; Inuishi, Y. (Technol. Rept. Osaka Univ. **27** [1977] 437/43).
[3] Zuppiroli, Z.; Friend, R. H. (Phil. Mag. [8] B **37** [1978] 321/8).
[4] Bandyopadhyay, A. K.; Bernard, C. (J. Mater. Sci. Letters **3** [1984] 901/3).
[5] Bandyopadhyay, A. K.; Bernard, C. (J. Mater. Sci. Letters **4** [1985] 71/4).

2.1.9.2.8 Microwave Conductivity

The microwave conductivity σ_{mw} of $(SN)_x$, measured at f = 9.3 and 17.8 GHz in the temperature range 30 to 300 K, agrees within 20% with the dc conductivity [2]. Such values were determined on crystals [1] and on $(SN)_x$ needles [2]. Between 30 and 300 K the measured conductivity σ_{mw} fits an approximately T^{-2} dependence. Below 30 K the microwave conductivity is higher than the dc conductivity. The difference was explained by the damage which was produced by contacting $(SN)_x$ crystals for dc measurements [2]. Microwave conductivity $\sigma_\perp$ perpendicular to the b axis at room temperature lies in the range 30 to 50 $\Omega^{-1}\cdot cm^{-1}$, which is about one order of magnitude larger than the dc value. In contrast to $\sigma_{dc\parallel}$, σ_{mw} increased monotonically with decreasing temperature and the intrinsic conductivity $\sigma_\parallel/\sigma_\perp$ is smaller than that estimated by dc technique [3]. Using a contactless microwave coaxial resonator technique (16 GHz), a conductivity value of 1400 $\Omega^{-1}\cdot cm^{-1}$ (at room temperature?) was reported [4].

References:

[1] Kahlert, H.; Seeger, K. (Phys. Semicond. Proc. 13th Intern. Conf., Rome 1976, pp. 353/6).
[2] Gill, J. C.; Herrenden-Harker, W. G.; Wills, H. H. (Lect. Notes Phys. **96** [1978] 427/30).
[3] Kaneto, K.; Yamamoto, M.; Yoshino, K.; Inuishi, Y. (J. Phys. Soc. Japan **47** [1979] 167/75).
[4] Bhattacharyya, T. J.; Kudaba, P. K. (J. Mater. Sci. Letters **2** [1983] 727/8).

2.1.9.3 Superconductivity

2.1.9.3.1 Survey

Below $T_c(H=0) \approx 0.3$ K $(SN)_x$ is a weakly coupled filamentary type II superconductor with an upper critical field H_{c2} $(T \rightarrow 0) \approx 8100$ G along the chains ($\parallel$b) and ≈ 900 G normal to them; the lower critical field H_{c1} (transition to the perfect, diamagnetic Meissner state) is extremely low: ~0.9 to 1.1 Oe$\perp$b, 0.3 to 0.4 Oe$\parallel$b, see p. 154. Thus, the mixed state in which flux trapping fluctuations appear (see also fluctuation conductivity, pp. 157/8) is extended. For type II superconductors in fields greater than H_{c1} the magnetization M decreases in absolute value (M is negative) for increasing fields up to H_{c2}, where it effectively vanishes. Hence, for $H_{c1} < H < H_{c2}$ in the absence of any hysteresis effects, $\chi = dM/dH$ should be positive. From the observed broad resistivity transition in combination with the critical field results, it was concluded that $(SN)_x$ is neither a 3D, nor a simple 1D or 2D system. Temperature-dependent dimensional cross-over greatly affects the physical behavior of $(SN)_x$.

2.1.9.3.2 Resistive Transition Temperature T_c

In 1975 Greene, Street, and Suter reported that crystalline $(SN)_x$ became superconducting along the fiber axis at a transition temperature of 0.26 ± 0.03 K [1]. $(SN)_x$ was the first polymer material, containing no metal atoms, to exhibit superconductivity. For short reviews of the superconducting properties of $(SN)_x$ see [2 to 5]. **Fig. 37** shows the superconducting resistive transition curve which were measured on various $(SN)_x$ "single crystals" by different groups [1, 6 to 8].

The transition temperature lies in the range 0.18 to 0.33 K for all measured samples. The variation of the transition temperature showed only weak dependence of the transition temperature on crystal quality [5, 8]. Other papers report an influence of extrinsic effects on the superconduction transition [9]. A better crystal growing technique yielded crystals with an increased superconducting transition of 0.35 K [10]. The midpoint of the superconducting resistive transition of "red $(SN)_x$" lies at ~0.220 K [23]. The generally broad transition in $(SN)_x$ is almost completely due to low dimensional fluctuations and not to sample disorder or inhomogenities [24].

Many papers reported that $(SN)_x$ films do not show a superconducting transition at ca. 0.3 K or at any temperature down to 0.025 K [11 to 13]. However weak evidence exists that some of the crystallites in discontinuous $(SN)_x$ films go superconducting [14].

The **pressure dependence** of the transition temperature T_c was studied at pressures p up to 9 [6], 15 [17, 18], 17 [7], and 400 kbar [16]. **Fig. 38** shows for example the T_c (p) data from [7]. The transition temperature and width increase with pressure up to 8 [7] or 9 kbar [6]. The dependence of T_c on pressure (from atmospheric pressure to ca. 9 kbar) is quadratic [6] on linear [7, 17]. This discrepancy was attributed to differences in measurement techniques [7]. At ca. 9 kbar a steep drop of T_c was observed (see Fig. 38) possibly due to a phase transition [7, 17].

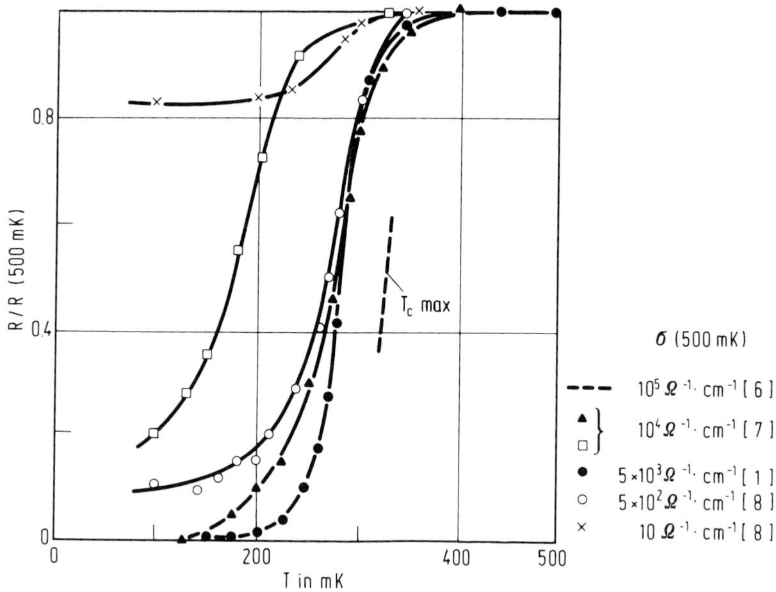

Fig. 37. Superconducting resistive transition in various $(SN)_x$ single crystals, normalized to R (500 mK). The corresponding absolute values of the normal conductivity at 500 mK are given in the figure [5].

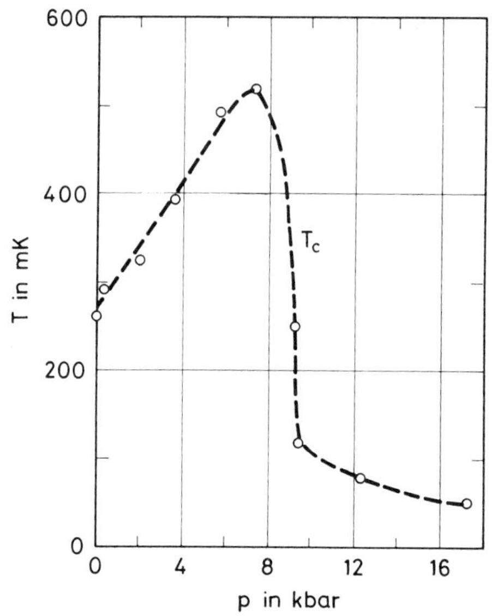

Fig. 38. Pressure dependence of the superconducting transition temperature T_c [7].

At higher pressure, 37 kbar down to 2 K [15], 50 kbar down to 1.1 K [7], and 100 kbar down to 1.2 K [19] no evidence for superconductivity was detected, however, at pressures in the range 130 to 400 kbar and at temperatures between 2 and 3 K, the onset of a superconducting transition occurred. T_c, though already much higher than at atmospheric pressure, increased further with increasing pressure [16].

The unusual increase of T_c up to ca. 8 kbar was attributed to a pressure-induced band structure change [5, 20] or to a stronger coupling between small superconducting fibers [2, 17]. A band-structure change was thought unlikely; a calculated pressure simulation yielded no significant change in the density of states at the Fermi level [21]. An increase in T_c due to a phonon spectrum change [21] also seems to be doubtful because the thermal expansion is normal and no modes were observed in Raman spectra that change under pressure or with a temperature increase [22]. The most plausible explanation so far seems to be a stronger coupling in superconducting fibers under pressure. This was supported by fluctuation and critical field measurements [2]. The renewed onset of superconductivity at pressures in the range 130 to 400 kbar was explained by an increase in electron-phonon coupling and a phase transition in the range 70 to 100 kbar which restores superconductivity [16].

References:

[1] Greene, R. L.; Street, G. B.; Suter, L. J. (Phys. Rev. Letters **34** [1975] 577/9).
[2] Greene, R. L.; Gill, W. D.; Azevedo, L. J.; Clark, W. G.; Deutscher, G. (Lect. Notes Phys. **65** [1977] 603/10).
[3] Labes, M. M.; Love, P.; Nichols, L. F. (Chem. Rev. **79** [1979] 1/15, 8/9).
[4] Rothenberg, M. S. (Phys. Today **28** [1975] 17/20).
[5] Geserich, H. P.; Pintschovius, L. (Festkörperprobleme [Advan. Solid State Phys.] **16** [1976] 65/94, 81/4).
[6] Gill, W. D.; Greene, R. L.; Street, G. B.; Little, W. A. (Phys. Rev. Letters **35** [1975] 1732/5).
[7] Müller, W. H. G.; Baumann, F.; Dammer, G.; Pintschovius, L. (Solid State Commun. **25** [1978] 119/22).
[8] Civiak, R.; Junker, W.; Elbaum, C.; Kao, H. I.; Labes, M. M. (Solid State Commun. **17** [1975] 1573/5).
[9] Street, G. B.; Greene, R. L. (IBM J. Res. Develop. **21** [1977] 99/110).
[10] Street, G. B.; Arnal, H.; Gill, W. D.; Grant, P. M.; Greene, R. L. (Mater. Res. Bull. **10** [1975] 877/82).

[11] de la Cruz, F.; Stolz, H. J. (Solid State Commun. **20** [1976] 241/3).
[12] Soulen R. J., Jr.; Utton, D. B. (Solid State Commun. **21** [1977] 105/8).
[13] Beyer, W.; Gill, W. D.; Street, G. B. (Solid State Commun. **27** [1978] 343/8).
[14] Young, M. W.; Thomas, J. M. D.; Adkins, C. J.; Tate, J. W. (J. Phys. Colloq. [Paris] **39** [1978] C6-448/C6-450).
[15] Friend, R. H.; Jerome, D.; Rehmatullah, S.; Yoffe, A. D. (J. Phys. C **10** [1977] 1001/12).
[16] Dunn, K. J.; Bundy, F. P.; Interrante, L. V. (Phys. Rev. [3] B **23** [1981] 106/11).
[17] Bickford, L. R.; Greene, R. L.; Gill, W. D. (Phys. Rev. [3] B **17** [1978] 3525/33).
[18] Bickford, L. R.; Gill, W. D.; Greene, R. L.; Street, G. B. (Bull. Am. Phys. Soc. [2] **22** [1977] 371).
[19] Chu, C. W. (private communication in [14]).
[20] Greene, R. L.; Gill, W. D.; Azevedo, L. J.; Clark, W. G. (Ferroelectrics **16** [1977] 243/5).

[21] Chelikowsky, J. R.; Schlüter, M.; Cohen, M. L. (Phys. Status Solidi B **82** [1977] 357/67).
[22] Greene, R. L.; Street, G. B. (NATO Advan. Study Inst. Ser. B **25** [1976/77] 167/95, 187; C.A. **90** [1979] No. 131230).
[23] Love, P.; Myer, G.; Kao, H. I.; Labes, M. M.; Junker, W. R.; Elbaum, C. (Ann. N.Y. Acad. Sci. **313** [1978] 745/58, 752/3).
[24] Kwak, J. F.; Greene, R. L.; Fuller, W. W. (Phys. Rev. [3] B **20** [1979] 2658/64).

2.1.9.3.3 Meissner Effect

The superconductivity of $(SN)_x$ is definitively established by the observation of a Meissner effect with the magnetic field applied perpendicular and parallel to the $(SN)_x$ fiber axis b. However, considerable flux trapping was observed both in dc [3, 8, 9] and in ac susceptibility measurements [1, 2, 4 to 6]. The Meissner effect is only partial [3 to 5] ($93 \pm 7\%$ of $\chi = -1/4\pi$ for $\perp$b, $65 \pm 10\%$ for $\parallel$b, in Gaussian units at 30 mK [5]). The magnetic transition at T′ occurs over a broad temperature range and the magnetization curves are similar to those expected for coupled filamentary type II superconductors [3 to 6].

The complex ac magnetic susceptibility ($\chi = \chi' + i\chi''$) of $(SN)_x$ was measured below 1 K for applied magnetic fields perpendicular and parallel to the fiber axis [1, 2, 4 to 6]. For instance, see **Fig. 39** in which the resistive transition is also shown [4]. In earlier ac studies [1] the Meissner effect was not detected because of flux trapping effects, stray fields, and the relatively high field applied to the specimen [3]. At temperatures near 0.2 K both the real (χ') and imaginary (χ'') component showed a marked change which was attributed to the onset of superconductivity [1]. Resistance measurements show the value of T′ to be somewhat lower than T_c, 230 or 280 K, depending upon the sample [1, 3, 10]. T′ was lowered if a dc magnetic field applied to the sample was increased [1, 2]. Values of χ' and χ'', studied by others, were sensitive to the magnitude of the ac magnetic field but not to the dc field [5, 6]. The authors discuss the dependence of susceptibility on temperature and magnetic field in terms of weakly coupled filamentary superconductors at which the shielding current passes Josephson junctions and the current loop expands as the temperature decreases [6]. The susceptibility data are in good agreement with another proposed model for $(SN)_x$ composed of finely divided normal conductive and superconductive elements in which superconducting fibers (diameter <10 nm, fiber radius $r < \xi$ (coherence length), $r \ll \lambda$ (penetration depth)) are separated by normal conducting sheets [4, 7].

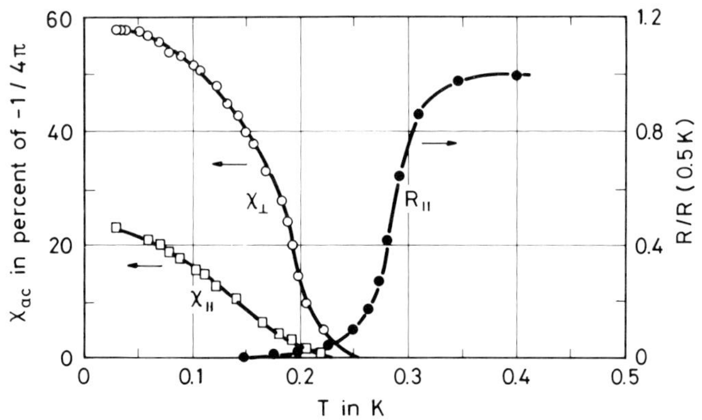

Fig. 39. The low-field ac susceptibility of an $(SN)_x$ sample measured with the fiber axis both parallel and perpendicular to the applied field. The measurements were made with $H_{ac} = 0.025$ Oe peak to peak, at a frequency of 0.07 Hz. Also shown is the resistive transition [4].

A study of the temperature dependence of the dc magnetic susceptibility of $(SN)_x$ (temperatures below 25 mK, magnetic fields from 0.003 to 500 Oe) shows a gradual diamagnetic transition in agreement with ac measurements [8]. The magnetization of $(SN)_x$ was studied as a function of applied field and temperature, and the results were compared with that of $(SNBr_{0.4})_x$.

Both materials appear to be filamentary superconductors with some coupling between fibers, but exhibiting significant quantitative differences [9, 10].

References:

[1] Dee, R. H.; Berlinsky, A. J.; Carolan, J. F.; Klein, E.; Store, N. J.; Turrell, B. G.; Street, G. B. (Solid State Commun. **22** [1977] 303/6).
[2] Dee, R. H.; Berlinsky, A. J.; Carolan, J. F.; Klein, E.; Stone, N. J.; Turrell, B. G. (Bull. Am. Phys. Soc. [2] **22** [1977] 371/2).
[3] Dee, R. H.; Dollard, D. H.; Turrell, B. G.; Carolan, J. F. (Solid State Commun. **24** [1977] 469/72).
[4] Bastuscheck, C. M.; Buhrman, R. A.; Scott, J. C. (Phys. Rev. [3] B **24** [1981] 6707/10).
[5] Oda, Y.; Takenaka, H.; Nagano, H.; Nakada, I. (Solid State Commun. **32** [1979] 659/63).
[6] Oda, Y.; Takenaka, H.; Nagano, H.; Nakada, I. (Solid State Commun. **35** [1980] 887/90).
[7] Bastuscheck, C. M.; Buhrman, R. A.; Kulick, J. D.; Scott, J. C. (Mol. Cryst. Liquid Cryst. **81** [1982] 155/61).
[8] Bastuscheck, C. M.; Buhrman, R. A.; Temkin, H.; Scott, J. C. (Bull. Am. Phys. Soc. [2] **23** [1978] 305).
[9] Dee, R. H.; Carolan, J. F.; Turrell, B. G.; Greene, R. L.; Street, G. B. (J. Phys. Colloq. [Paris] **39** [1978] C6-444/C6-445).
[10] Dee, R. H.; Carolan, J. F.; Turrell, B. G.; Greene, R. L. (Phys. Rev. [3] B **22** [1980] 174/9).

2.1.9.3.4 Upper Critical Magnetic Field H_{c2}

Values for the lower critical field $H_{c1}(\parallel b) = 0.2$ to 0.3 Oe and $H_{c1}(\perp b) = 0.9$ to 1.1 Oe were obtained [16].

The results of the first studies of the upper critical magnetic field H_{c2} of $(SN)_x$ were summarized and discussed by [1, 2]. Several groups investigated the temperature dependence [3 to 9], the angular dependence [3, 8], and the pressure dependence [10, 11] of H_{c2}. Some papers dealt with theoretical models to explain the unusual results of the H_{c2} measurements [12 to 15]. The results of a study of the temperature dependence of H_{c2} over the range 40 mK to 1 K for magnetic fields parallel and perpendicular to the fiber axis are illustrated in **Fig. 40** [8]. Earlier critical field $H_{c2}(T)$ measurements are pictured in [3].

Critical field data together with other superconduction parameters (coherence length ξ, see p. 156, diffusion constant D, see p. 158, r_o = fiber radius) of various $(SN)_x$ samples are compiled in the following table:

$H_{c2\parallel}(0)$ in G	$H_{c2\perp}(0)$ in G	$D_\parallel$ in $cm^2 \cdot s^{-1}$	$D_\perp$ in $cm^2 \cdot s^{-1}$	r_o in Å	$\xi_\perp(0)$ in Å	$\xi_\parallel(0)$ in Å	Ref.
8100 ± 200	870 ± 40	—	—	—	135 ± 15	3500 ± 1500	[3]
8200 ± 100	900 ± 30	910	22	62	1540	1.0×10^4	[7]
7390 ± 80	795 ± 20	369	8.6	109	970	6.4×10^3	[7]

The large anisotropy of H_{c2}, observed in measurements of the temperature dependence, was explained in terms of the polymeric crystal structure and fibrous morphology of $(SN)_x$. The material is a type II superconductor [3 to 5]. $H_{c2\parallel}(0)$ exceeds the Pauli limiting field $H_p = 18.4\,T_c = 5.4$ kOe [3]. Other workers reported similar H_{c2} data and observed that $H_{c2\parallel}/H_{c2\perp}$ increased markedly as $t = T_c(H)/T_c(0)$ approaches 1, reaching values of more than 50 [8], see

Fig. 40. A critical field anisotropy in the basal plane ac perpendicular to the b axis was observed by [7, 8, 10] and a maximal value $H_{c2}(\parallel c)/H_{c2}(\parallel a) \approx 1.3$ at 0 K was found [7]. Curves and a table visualize the results [8]. $H_{c2}(a) > H_{c2}(c)$ was found by [10]. The $H_{c2\perp}$-basal plane anisotropy must arise from the intrinsic band structure.

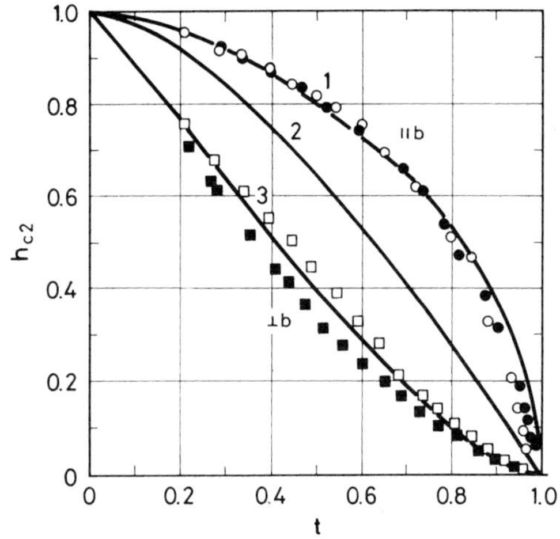

Fig. 40. Temperature dependence of the parallel and perpendicular upper superconducting critical fields in $(SN)_x$, plotted in reduced form with $h_{c2} = H_{c2}(T)/H_{c2}(0)$ and $t = T_c(H)/T_c(0)$. The theoretical curves are 1) SG (da Sylva, Gyorffy [14]) anisotropic fiber model for both $h_{c2\parallel}$ and $h_{c2\perp}$; 2) SG anisotropic bulk model; 3) an empirical curve with $h_{c2\perp}$ equal to $(1-t)$ times curve 1. Experimental points $h_{c2\parallel}$: ● sample with $H_{c2\parallel}(0) = 8200$ G, ○ sample with $H_{c2\parallel}(0) = 7390$ G; $h_{c2\perp}$: ■ sample with $H_{c2\perp}(0) = 900$ G; □ sample with $H_{c2\perp}(0) = 795$ G. The theoretical curves are uncorrected for sample bending [8].

The **pressure dependence** of the upper critical field in various crystallographic directions was investigated up to 9.5 and 15 kbar by [10] and [11], respectively; dates for pressures p of 0 (1 bar), 3.5, and 9.5 kbar, see in a table. Both $H_{c2\parallel}$ and $H_{c2\perp}$ decrease with p but ⊥ c much more so than ∥c. The results support the conception of $(SN)_x$ as a weakly coupled filamentary superconductor at these pressures which are not high enough to allow intrinsic properties to dominate. $H_{\parallel}(T)$ is paramagnetically limited at all of the temperatures and pressures investigated. The slope of the zero-pressure curves for $H_{\perp}$ was attributed to a progressive decoupling of the fibers as T is lowered, and that of 9.5 kbar curve to a system of fibers which are coupled at all temperatures [10].

Theoretical models were developed in order to explain the results of H_{c2} measurements [12 to 15]. Turkevich and Klemm treated $(SN)_x$ as a coupled infinitesimal filamentary system on the basis of the Ginzburg-Landau theory [12, 13]. If $T_c = 0.23$ K was chosen they obtained data which disagreed with experimental values [3], but calculation with $T_c = 0.285$ K gave agreement between theoretical and experimental data. Da Silva and Gyorffy [14] studied an anisotropic bulk and an isolated fiber model of $(SN)_x$ [14]. Their calculation was based on the Maki-de Gennes generalization of the Ginzburg-Landau theory. Comparison of their theoretical data for $H_{c2\parallel}$ with the experimental values [3] clearly favor the fiber model, in spite of the fact that coupling between fibers was neglected. In the case of $H_{c2\perp}$ neither of the two models gave a good fit. Also more recent experimental work [6 to 9] showed that the temperature dependence of $H_{c2\parallel}(T)$ agreed with the predicted behavior of the independent fiber model [14], see Fig. 40, however, $H_{c2\perp}(T)$ was still not in accord with this model [8].

The **angular dependence** of $H_{c2}(\theta)$ (θ measured from the fiber axis) was determined and compared with theoretical models [3, 8]. A comparison with the angular variation predicted by da Silva, Gyorffy [14] showed a good agreement for the range $\theta > 20°$ but not for θ near zero, however, an expression of Tinkham provides a very good fit [8], in agreement with other workers [3], see Fig. 40. Both the calculation of the temperature and angular dependence were made self-consistent by postulating a new form of the Cooper pair breaker which appears in the microscopic study [8]. The interfiber coupling was taken into account for H ⊥ b. Hida [15] also studied the upper critical field based on a theoretical model which incorporates the effect of interfiber coupling with the Ginzburg-Landau theory of the isolated fiber model. The model describes a cross-over behavior between the coupled fiber system to the isolated fiber. Theoretical and experimental results are in fairly good agreement [15].

References:

[1] Geserich, H. P.; Pintschovius, L. (Festkörperprobleme [Advan. Solid. State Phys.] **16** [1976] 65/94, 83/4).
[2] Greene, R. L.; Street, G. B. (NATO Advan. Study Inst. Ser. B **25** [1977] 167/95, 186).
[3] Azevedo, L. J.; Clark, W. G.; Deutscher, G.; Greene, R. L.; Street, G. B.; Suter, L. J. (Solid State Commun. **19** [1976] 197/201).
[4] Greene, R. L.; Gill, W. D.; Azevedo, L. J.; Clark, W. G. (Ferroelectrics **16** [1977] 243/5).
[5] Greene, R. L.; Gill, W. D.; Azevedo, L. J.; Clark, W. G.; Deutscher, G. (Lect. Notes Phys. **65** [1977] 603/10, 605).
[6] Barrett, P.; Chambers, R. G.; Feeman, P. J.; Herrenden-Harker, W. G.; Priestley, M. G.; Trinder, R. W. (Lect. Notes Phys. **96** [1979] 456).
[7] Barrett, P.; Chambers, R. G.; Feeman, P. J.; Herrenden-Harker, W. G.; Priestley, M. G.; Trinder, R.; Read, S. F. J. (J. Phys. F **10** [1980] L89/L94).
[8] Barrett, P.; Brinslow, G. M.; Herrenden-Harker, W. G.; Priestley, M. G. (J. Phys. F **11** [1981] 859/71).
[9] Barrett, P.; da Silva, E. Z.; Gyorffy, B. L.; Herrenden-Harker, W. G.; Priestley, M. G.; Wills, H. H. (AIP [Am. Inst. Phys.] Conf. Proc. No. 58 [1980] 282/7).
[10] Bickford, L. R.; Greene, R. L.; Gill, W. D. (Phys. Rev. [3] B **17** [1978] 3525/33).

[11] Bickford, L. R.; Gill, W. D.; Greene, R. L.; Street, G. B. (Bull. Am. Phys. Soc. **22** [1977] 371).
[12] Turkevich, L. A. (Bull. Am. Phys. Soc. [2] **22** [1977] 371).
[13] Turkevich, L. A.; Klemm, R. A. (Phys. Rev. [3] B **19** [1979] 2520/39).
[14] da Silva, E. Z.; Gyorffy, B. L. (Phys. Rev. [3] B **20** [1979] 147/58).
[15] Hida, K. (Physica B + C **107** [1981] 127/8).
[16] Kaneto, K.; Yoshino, K.; Inushi, Y. (in: Monceau, P., Electronic Properties of Inorganic Quasi-One-Dimensional Compounds, Pt. 2, Reidel, Dordrecht 1985, pp. 69/109, 90).

2.1.9.3.5 Superconducting Coherence Length. Penetration Depth

The superconducting coherence length parallel and perpendicular to the chain axis was determined from experimental upper critical field data using the Ginzburg-Landau (GL) theory. For the parallel GL coherence length the following values were reported: $\xi_{\parallel}(0) \approx 3500 \pm 1500$ Å [1], 4530 Å [2], and 3000 Å [3]. For the perpendicular GL coherence length: $\xi_{\perp}(0) \approx 135 \pm 15$ Å [1], 94 Å [2], and ~140 Å [3]. Other methods were also employed. From the relation $\xi_i(0) = 10^{-6} \cdot (\varrho_i \gamma T_c)^{-1/2}$ (ϱ_i = normal resistivity, γ = electronic contribution to heat capacity, T_c = transition temperature) $\xi_{\parallel}(0)$ values of ~3000 Å ($T_c = 0.29$ K, $\varrho_{\parallel} \approx 10^{-5}\,\Omega \cdot$cm at 4 K, $\gamma \approx 400$ erg·cm^{-3}·K^{-2}) [1], 3300 Å [2], and ~2000 Å [3], and $\xi_{\perp}(0)$ values of ~20 Å (atmospheric pressure) [3, 4] and 150 Å (9.5 kbar) [4] were estimated. Other determinations were performed using fluctuation conductivity and critical field results on the basis of a theory of da Silva and Gyorffy (for independent anisotropic metal fibers). Values of $\xi_{\parallel}(0)$ (in the range 2500 to 10000 Å) and values of $\xi_{\perp}(0)$ (in the range 570 to 1540 Å) were estimated for different samples using the relation $\xi_i(0) = [\pi\hbar D/(8\,kT)]^{1/2}$ (D = diffusion coefficient) [5, 6]. In general the coherence length is larger than the fiber diameter (see p. 158) indicating that $(SN)_x$ is a system of weakly coupled fibers [1].

The London penetration depth λ_L of $(SN)_x$ crystals (using optical data) is ~2700 Å. The effective London penetration depth λ_e (using magnetization data) is ~0.05 mm for $\lambda_e(H_\perp)$ and ~0.1 mm for $\lambda_e(H_\parallel)$ (for H = 0.025 Oe, T = 59 mK) [7].

References:

[1] Azevedo, L. J.; Clark, W. G.; Deutscher, G.; Greene, R. L.; Street, G. B.; Suter, L. J. (Solid State Commun. **19** [1976] 197/201).
[2] Turkevich, L. A.; Klemm, R. A. (Phys. Rev. [3] B **19** [1979] 2520/39).
[3] Kwak, J. F.; Greene, R. L.; Fuller, W. W. (Phys. Rev. [3] B **20** [1979] 2658/64).
[4] Bickford, L. R.; Greene, R. L.; Gill, W. D. (Phys. Rev. [3] B **17** [1978] 3525/33).
[5] Barrett, P.; Chambers, R. G.; Feenan, P.; Herrenden-Harker, W. G.; Priestley, M. G.; Trinder, R. W.; Read, S. F. J. (J. Phys. F **10** [1980] L89/L94).
[6] Barrett, P.; Herrenden-Harker, W. G.; Priestley, M. G.; da Silva, E. Z. (J. Phys. F **11** [1981] 873/82).
[7] Dee, R. H.; Carolan, J. F.; Turrell, B. G.; Greene, R. L. (Phys. Rev. [3] B **22** [1980] 174/9).

2.1.9.3.6 Superconductivity Energy Gap

Tunneling contacts were prepared between $(SN)_x$ and bulk In as well as between different $(SN)_x$ crystals. The differential tunneling resistance was measured for temperatures down to 50 mK and information on the energy gap of superconducting $(SN)_x$ was derived. A strong anisotropy of the gap with a value $\Delta_\parallel = 80 \pm 10$ μeV, $2\Delta_\parallel/k_BT_c = 6.51$ for tunneling parallel to the polymer chains was observed. The lower transverse gap $\Delta_\perp$ remained unobservable. The measured gap turned out to depend critically on the way the tunneling contact was prepared and reasons for an intrinsic or extrinsic origin of the gap anisotropy were discussed [1, 2]. Other studies of tunneling junctions consisting of $(SN)_x$ crystals and evaporated Pb, Sn, or Hg–In amalgam layers down to 60 mK failed to detect a superconducting energy gap in $(SN)_x$, obviously because the surface of the $(SN)_x$ electrode consisted of small nonsuperconducting particles [3, 4].

References:

[1] Binnig, G.; Hoenig, H. E. (J. Phys. Colloq. [Paris] **39** [1978] C6-441/C6-449).
[2] Binnig, G.; Hoenig, H. E. (Z. Physik B **32** [1978] 23/6).
[3] Chaikin, P. M.; Hansma, P. K.; Greene, R. L.; Engler, E. M. (Bull. Am. Phys. Soc. [2] **22** [1977] 256).
[4] Chaikin, P. M.; Hansma, P. K.; Greene, R. L. (Phys. Rev. [3] B **17** [1978] 179/89).

2.1.9.3.7 Fluctuation Conductivity

Just above the superconducting transition, $(SN)_x$ exhibits electrical conductivity (σ_f) in excess to the normal state conductivity (σ_n) [1 to 3]. Measurable fluctuation-induced conductivity (σ_f) was observed at temperatures as high as 700 mK [3]. The effect was attributed to thermodynamic fluctuations in the superconducting transition state ("Aslamazov-Larkin type") [1, 2]. The temperature dependence of the fluctuation conductivity is that of a 0 D (samples with lower normal state conductivity) or 1 D (samples with higher normal state conductivity) superconductor with respect to the temperature dependent coherence length.

The average diameter of "wires" is ca. 250 Å, which is smaller than the superconductivity coherence length (see p. 156). The observed dimensionality is not a result of weak coupling between individual chains, but instead is due to interaction among bundles of many individual fibers. This view of $(SN)_x$ is confirmed by band-structure calculations which show that $(SN)_x$ should be regarded as an anisotropic 3 D metallic polymer [1]. A theoretical calculation of the fluctuation conductivity and other superconducting properties of an anisotropic material like $(SN)_x$ was derived for two models [4, 5]. $(SN)_x$ was treated in the first model as an anisotropic bulk solid, in the second as a set of parallel noninteracting fibers [4, 5]. Experimental measurements of the fluctuation conductivity above T_c for transitions in magnetic fields, both parallel and perpendicular to the fiber axis clearly support the fiber model [6 to 8]. The fluctuation contribution to the conductivity just above T_c at constant magnetic field $H \parallel b$ agrees with $\sigma_f = C[(T-T_c)/T_c]^{-3/2}$ (C = constant) as predicted by [4, 5]. A dependence in the form, $\sigma_f \sim [(H-H_{c2})/H_{c2}]^{-3/2}$, for the fluctuation conductivity was found at constant temperature. The values of the fiber radius r_0 and the electronic diffusion coefficients $D_{\parallel}$ and $D_{\perp}$ were found by fitting the coefficients of the fluctuation relationship and combining the results with analysis of $H_{c2\parallel}(0)$ and optical data. $D_{\parallel}/D_{\perp} \approx 20$, and using the relation $D = \frac{1}{3} v_F^2 \tau$ (v_F = Fermi velocity, τ = relaxation time), $D_{\parallel} = 60 \pm 10$ and 280 ± 40, $D_{\perp} = 3 \pm 0.5$ and $14 \pm 2\ cm^2 \cdot s^{-1}$ for different samples [8], see also [4, 7]. A comparison of calculated with experimental values of electrical conductivity $\sigma(T)$ also shows clear evidence that superconducting fluctuations occur at very low temperatures prior to the occurrence of stable superconductivity. The theoretical data $\sigma(T)$ were obtained using the Fermi function [9].

References:

[1] Civiak, R. L.; Elbaum, C.; Junker, U.; Gough, C.; Kao, H. I.; Nichols, L. F.; Labes, M. M. (Solid State Commun. **18** [1976] 1205/8).

[2] Civiak, R. L.; Elbaum, C.; Nichols, L. F.; Kao, H. I.; Labes, M. M. (Phys. Rev. [3] B **14** [1976] 5413/21).

[3] Civiak, R. L.; Elbaum, C.; Kao, H. I.; Nichols, L. F.; Labes, M. M. (Bull. Am. Phys. Soc. [2] **21** [1976] 255).

[4] da Silva, E. Z.; Gyorffy, B. L. (Lect. Notes Phys. **96** [1979] 439/45).

[5] da Silva, E. Z.; Gyorffy, B. L. (Phys. Rev. [3] B **20** [1979] 147/58).

[6] Barrett, P.; da Silva, E. Z.; Gyorffy, B. L.; Herrenden-Harker, W. G.; Priestley, M. G. (AIP [Am. Inst. Phys.] Conf. Proc. No. 58 [1980] 282/7).

[7] Barrett, P.; Chambers, R. G.; Feeman, P.; Herrenden-Harker, W. G.; Priestley, M. G.; Trinder, R.; Read, S. F. J. (J. Phys. F **10** [1980] L89/L94).

[8] Barrett, P.; Herrenden-Harker, W. G.; Priestley, M. G.; da Silva, E. Z. (J. Phys. F **11** [1981] 873/82).

[9] Love, P. (Polym. Prepr. Am. Chem. Soc. Div. Polym. Chem. **25** [1984] 274/5).

2.1.9.4 Thermoelectric Power

The thermoelectric power (Seebeck coefficient S) of $(SN)_x$ crystals as a function of temperature was determined in the ranges 0.1 to 4.2 K [1 to 4], 4.2 to 300 K [5 to 7], 8 to ~300 K [8], and at room temperature [9]. S for $(SN)_x$ films, deposited at various temperatures (T_s = −10 to 50°C) was measured from 100 to 300 K [10]. Small negative thermopower values which are typical for metals were observed in the wide range from 4.2 to ~300 K, both parallel and perpendicular to the b axis of the crystals [5 to 9]. An observed maximum value of $|S|$ at ca. 20 K was associated with a phonon drag peak [5, 6, 8]. Room temperature thermopower values were estimated to be $-1.5\ \mu V \cdot K^{-1}$ [9] and ca. $-6\ \mu V \cdot K^{-1}$ [8] parallel and $-3.5\ \mu V \cdot K^{-1}$ [9]

perpendicular to the b axis. A value of $-6.8 \pm 0.6\ \mu V \cdot K^{-1}$ on pressed pellets was presented in an earlier report [11]. The negative sign of S throughout the investigated temperature range implies a larger contribution of electrons than holes to the electrical conduction. Results of the measurement below 4 K are shown in **Fig. 41** [2].

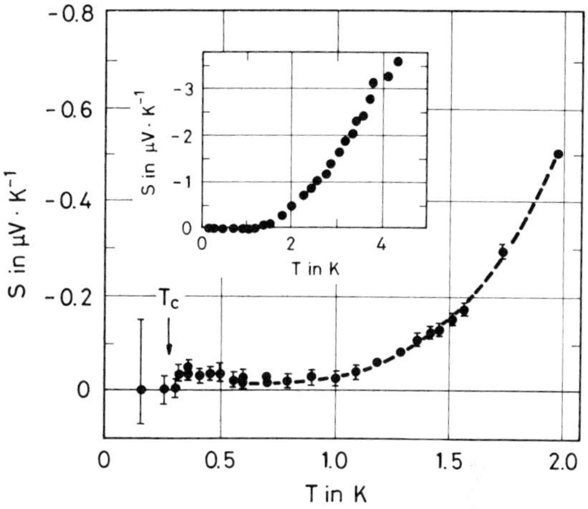

Fig. 41. Thermoelectric power S of $(SN)_x$ in zero magnetic field as a function of temperature. The phonon-drag tail is notable above 1 K. Near 0.3 K the superconducting transition is seen. The increase of |S| between 1 and 0.3 K is attributed to the Kondo effect [2].

Between 1 and 4.2 K |S| (in $\mu V \cdot K^{-1}$) varies with $0.043 \pm 0.003\ T^3$ (a linear term is negligible) [2, 4]. This rapid increase to negative thermopower values above 1 K (until ~20 K) was interpreted as phonon drag, indicating strong electron-phonon scattering. The increase of |S| with decreasing temperature below 1 K, which was attributed to the Kondo effect, is terminated at the superconducting transition temperature at 0.3 K where the thermoelectric power drops to zero [1 to 4].

The behavior of S in a magnetic field (up to 15 kOe) at constant temperature (up to 4.2 K) was investigated [1 to 4]. Curves of thermopower data, for fields both parallel and perpendicular to the crystal axis at 250 and 500 mK, were published [2, 4]. The complicated behavior was attributed to a combination of the quenching of superconductivity, superconducting fluctuations, magnetic impurities, and the Kondo effect.

The temperature dependence of the thermoelectric power of $(SN)_x$ films as a function of the deposition temperature (T_s) was presented. At high substrate temperatures the thermopower values of films are similar to those of $(SN)_x$ crystals, however, with decreasing T_s S changes its sign from negative to positive. An increase of S with rising temperature was observed for low substrate temperatures. Annealing of the films led to an S decrease for high and an S increase for low substrate temperature films, respectively. The results were explained in the same way as the conductivity results, in terms of a heterogeneous model with areas having different electronic gaps in the density of states, correlated with the $(SN)_x$ chain length [10].

References:

[1] Azevedo, L. J.; Chaikin, P. M.; Clark, W. G.; Fuller, W. W.; Hammann, J. (J. Phys. Colloq. [Paris] **39** [1976] C6-446/C6-447).

[2] Azevedo, L. J.; Chaikin, P. M.; Clark, W. G.; Fuller, W. W.; Hammann, J. (Phys. Rev. [3] B **20** [1979] 4450/6).

[3] Fuller, W. W.; Chaikin, P. M.; Azevedo, L. J.; Clark, W. G.; Hammann, J. (Bull. Am. Phys. Soc. [2] **23** [1978] 304).

[4] Azevedo, L. J.; Chaikin, P. M.; Clark, W. G.; Fuller, W. W.; Hammann, J. (AD-A 062709 [1978] **1** 25; C.A. **91** [1979] No. 100783).
[5] Walatka, V. V., Jr.; Labes, M. M.; Perlstein, J. H. (Phys. Rev. Letters **31** [1973] 1139/42).
[6] Kahlert, H.; Seeger, K. (Phys. Semicond. Proc. 13th Intern. Conf., Rome 1976, pp. 353/6).
[7] Gill, W. D. (NRL-MR-3906 [1979] 1/13, 8).
[8] Greene, R. L.; Grant, P. M.; Street, G. B. (Phys. Rev. Letters **34** [1975] 89/92).
[9] Cohen, M. J.; Chiang, C. K.; Garito, A. F.; Heeger, A. J.; MacDiarmid, A. G.; Mikulski, C. M. (Bull. Am. Phys. Soc. [2] **20** [1975] 360).
[10] Beyer, W.; Gill, W. D.; Street, G. B. (Solid State Commun. **27** [1978] 343/8).

[11] Kronick, P. L.; Kaye, H.; Chapman, E. F.; Mainthia, S. B.; Labes, M. M. (J. Chem. Phys. **36** [1962] 2235/7).

2.1.9.5 Galvanomagnetic Effects. Current Carrier Mobilities

The magnetoresistance behavior of $(SN)_x$ crystals [1 to 14] and films [15] ([11] is a review article) was studied. Transverse magnetoresistance (J⊥B) data of $(SN)_x$ crystals were presented for temperatures at 4.2 K (before and after mechanical deformation) [1] and at 77 and 300 K [2]. Measurements of the transverse and longitudinal magnetoresistance were performed for current flow in the b direction (J∥b) and in the a direction (J⊥b) at 4.2 K [3 to 5]. Transverse magnetoresistance results for current flow parallel to the b axis ([010] direction) and B∥[201] were reported for crystalline $(SN)_x$ in the temperature range 1.8 to 290 K [6, 7], 1.7 to ~290 K [9] and for various samples between 0.1 and 4.2 K [10, 13] and at 0.2, 1.0, and 4.2 K [14]. The angular dependence of the transverse magnetoresistance was determined at 4.2 K [4 to 7], 1.4 K [4, 5], 1.6 K [9], and in the range 0.1 to 4.2 K [10]. On films at 4.2 K (at higher temperatures the effect was below the sensitivity limit) the transverse magnetoresistance was also measured [15]. Results for the transverse magnetoresistance in an $(SN)_x$ crystal at temperatures of 1.7 to 290 K [11] are shown in **Fig. 42**.

The magnetoresistance of $(SN)_x$ crystals is composed of a normal positive component and a negative component [4 to 11, 13]. Only a negative region occurs at 500 mK for parallel fields (0 to 15 kOe), but a positive region was also observed for perpendicular fields [12]. $(SN)_x$ films do not exhibit a magnetic field region of negative magnetoresistance [15]. The positive magnetoresistance of $(SN)_x$ crystals increases in proportion to the square of the magnetic field (B^2) [4 to 11]. At higher magnetic fields, however, deviation from the B^2 dependence was observed when $\Delta\varrho/\varrho_0$ exceeds a value of $\sim 8\times10^{-3}$ [9]. The effect was pronounced below 1 K [10, 11, 14]. The magnetoresistance became proportional to B indicating open orbit magnetoresistance [10]. The negative magnetoresistance component was observed at temperatures below 4.2 K and at lower magnetic fields (<30 kG), but it was not observed above 77 K [6]. An increase of the ratio $\sigma_{4.2}/\sigma_{RT}$ diminishes the negative region [7].

Negative magnetoresistance data were measured below 40 kG after the bending of crystals [1]. Taking this effect into account the negative magnetoresistance was interpreted in terms of defects in the fibrous structure [4, 5]. In contrast, it was attributed by other workers to spin-dependent scattering of carriers at impurities or defects with localized spins [6, 7]. Confirmation of this theory was presented by comparing experimental values with theoretical predictions of Toyozawa [9].

Results of a determination of the angular dependence of magnetoresistance in $(SN)_x$ crystals for various samples (at 4.2 and ~1.4 K, J⊥b axis) were given by [4, 5]. The anisotropy ratio of $\Delta\varrho/\varrho_0$ for B⊥b to B∥b, determined in this way, is ~3.1, in reasonable agreement with plasma tensor calculation [4, 5]. This result has led to the conclusion that the low conductivity of $(SN)_x$ perpendicular to b is an extrinsic effect due to interfiber barriers [5]. On rotation of H in the ac plane (010) a maximum and a minimum of $\Delta\varrho/\varrho_0$ are observed periodically at the magnetic field applied perpendicular and parallel to the [201] direction (J∥b, B≈50 kG, T=1.6 and 4.2 K), respectively [7, 9].

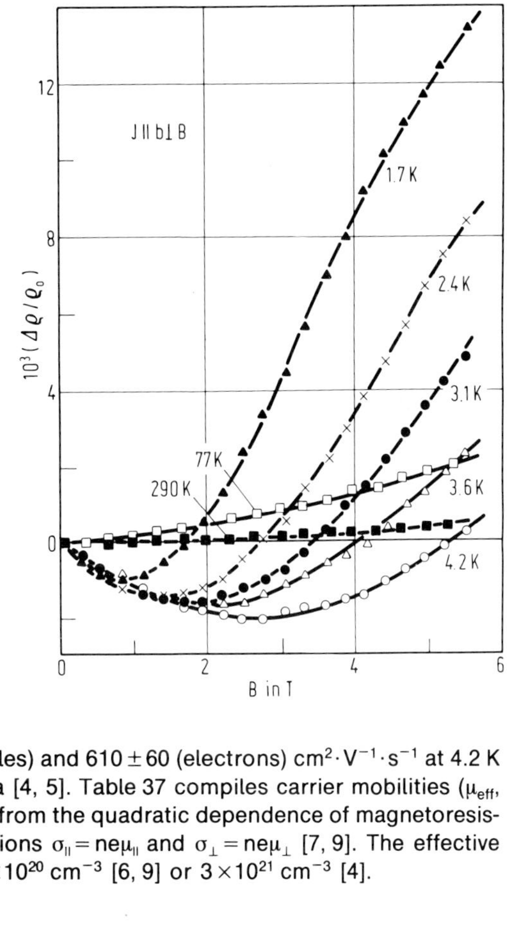

Fig. 42. Transverse magnetoresistance in an $(SN)_x$ crystal at various temperatures [10, 11].

b axis carrier **mobilities** $\mu_{\parallel}$ of 430±40 (holes) and 610±60 (electrons) $cm^2 \cdot V^{-1} \cdot s^{-1}$ at 4.2 K were obtained from magnetoresistance data [4, 5]. Table 37 compiles carrier mobilities (μ_{eff}, $\mu \parallel b$, $\mu \perp b$) at various temperatures obtained from the quadratic dependence of magnetoresistance ($\Delta\varrho/\varrho_0 = \mu_{\parallel}\mu_{\perp}B^2 = \mu_{eff}^2 B^2$) and the relations $\sigma_{\parallel} = ne\mu_{\parallel}$ and $\sigma_{\perp} = ne\mu_{\perp}$ [7, 9]. The effective carrier density n is estimated to be 1 to 3×10^{20} cm^{-3} [6, 9] or 3×10^{21} cm^{-3} [4].

Table 37

Temperature Dependence of Mobilities [$cm^2 \cdot V^{-1} \cdot s^{-1}$]. (dc and mw mean that the conductivities used to estimate the mobilities are obtained by dc and microwave method.)

T in K	μ_{eff}	$\mu_{\parallel}$ (dc)	$\mu_{\perp}$ (dc)	$\mu_{\parallel}$ (mw)	$\mu_{\perp}$ (mw)	Ref.
290	40	690	2.3	250	6.3	[9]
77	60	4100	0.88	—	—	[9]
4.2	120	23000	0.64	—	—	[9]

Earlier estimates of the dc mobilities show much less anisotropy: $\mu_{\parallel}=120$, $\mu_{\perp}=10$ at 290 K, $\mu_{\parallel}=4200$, $\mu_{\perp}=3.3$ at 4.2 K [6, 7]. In the ac plane (010) the carrier mobility along the [201] direction is larger than that perpendicular to [201] by about 40% at 4.2 K [7], 20 to 30% at 1.6 and 4.2 K [9]. Carrier mobilities in $(SN)_x$ films at 4.2 K are 390 ($T_s=30°C$), 500 ($T_s=15°C$), and

520 ($T_s = 10°C$) $cm^2 \cdot V^{-1} \cdot s^{-1}$ for B = 30 kG [15]. These mobility values and their temperature dependence together with the microscopic hole concentration of ca. 10^{21} cm^{-3} (in highly conductive areas) are close to data for $(SN)_x$ crystals [10].

A value of 1.5×10^{-13} s for the scattering time τ at $T \leqq 4.5$ K was evaluated, corresponding to a mean free path of ~700 Å [4, 5].

The **Hall effect** of $(SN)_x$ crystals was studied at 4.2 K [7, 9]. A positive value of the Hall coefficient R_H in the range 10^{-3} to ~10^{-2} $cm^3 \cdot A^{-1} \cdot s^{-1}$ [7] or 10^{-3} to 10^{-1} $cm^3 \cdot A^{-1} \cdot s^{-1}$ [9], depending on the sample, was reported. Based on these data, carrier densities in the range 10^{20} to 10^{21} cm^{-3} [7] or 10^{20} to 10^{22} cm^{-3} [9], and Hall mobilities in the range 10^2 to 10^3 [7] or 10^2 to 10^4 $cm^2 \cdot V^{-1} \cdot s^{-1}$ [9] were determined. The magnetic field dependence of the Hall voltage is shown in a figure [7].

The Hall effect of $(SN)_x$ films was also studied [15]. The films were deposited at substrate (sapphire, glass, and NaCl crystals) temperatures (T_s) between 77 and 300 K. The Hall coefficients are positive for all films (in agreement only with thermopower results for films with low T_s), indicating that the conduction in $(SN)_x$ films is dominated by holes. The temperature dependences of the Hall mobility (μ_H) and current carrier concentration ($\sigma/e\mu_H$) were measured for different substrate temperatures. μ_H decreases with T_s (approximately in proportion to σ), at room temperature from values near 1 at $T_s = 50°C$ to about 10^{-3} $cm^2 \cdot V^{-1} \cdot s^{-1}$ at $T_s = -10°C$. Similar values were determined for the Hall carrier concentration for all samples. The effective carrier density is ca. 10^{21} cm^{-3} in good agreement with values reported for $(SN)_x$ crystals (3.2×10^{21} cm^{-3} [16]). The small Hall mobility at 300 K ($\leqq 1$ $cm^2 \cdot V^{-1} \cdot s^{-1}$) and its activated temperature dependence were explained in terms of a heterogeneous model for $(SN)_x$ films with a network of thin carrier depletion regions separating highly conductive islands [15].

References:

[1] Kahlert, H.; Seeger, K. (Phys. Semicond. Proc. 13th Intern. Conf., Rome 1976, pp. 353/6).
[2] Möller, W.; Plante, R.; Ranvand, R.; Roth, S.; Roedhammer, P. (Solid State Commun. **19** [1976] 943/5).
[3] Beyer, W.; Gill, W. D. (Bull. Am. Phys. Soc. [2] **21** [1976] 255).
[4] Beyer, W.; Gill, W. D.; Street, G. B. (Solid State Commun. **23** [1977] 577/81).
[5] Gill, W. D.; Beyer, W.; Street, G. B. (Lect. Notes Phys. **65** [1977] 573/622, 619/22).
[6] Kaneto, K.; Yamamoto, M.; Yoshino, K.; Inuishi, Y. (Solid State Commun. **26** [1978] 311/4).
[7] Kaneto, K.; Yamamoto, M.; Yoshino, K.; Inuishi, Y. (Technol. Rept. Osaka Univ. **28** [1978] 127/34).
[8] Yamamoto, M.; Kaneto, K.; Yoshino, K.; Inuishi, Y. (Solid State Commun. **29** [1979] 541/3).
[9] Kaneto, K.; Yamamoto, M.; Yoshino, K.; Inuishi, Y. (J. Phys. Soc. Japan **47** [1979] 167/75).
[10] Sasa, S.; Kaneto, K.; Yoshino, K.; Inuishi, Y. (Technol. Rept. Osaka Univ. **32** [1982] 75/80).

[11] Kaneto, K.; Yoshino, K.; Inuishi, Y. (in: Monceau, P., Electronic Properties of Inorganic Quasi-One-Dimensional Compounds, Pt. II, Reidel, Dordrecht 1985, pp. 69/110, 94/100).
[12] Azevedo, L. J.; Chaikin, P. M.; Clark, W. G.; Fuller, W. W.; Hammann, J. (Phys. Rev. [2] B **20** [1979] 4450/6).
[13] Yoshino, K.; Tachibana, S.; Makio, Y.; Yamamoto, M.; Kaneto, K.; Inuishi, Y. (Technol. Rept. Osaka Univ. **27** [1977] 437/43).
[14] Kaneto, K.; Sasa, S.; Yoshino, K.; Inuishi, Y. (Mol. Cryst. Liquid Cryst. **86** [1982] 189/93).
[15] Beyer, W.; Mell, H.; Gill, W. D. (Solid State Commun. **27** [1978] 185/9).
[16] Kahlert, H. (NATO Advan. Study Inst. Ser. B **25** [1977] 167/95, 179).

2.1.9.6 Schottky Barriers. Solar Cells

$(SN)_x$ films were used to form Schottky barriers [1 to 4]. A possible technical use of $(SN)_x$ films in transistors or solar cells, however, is difficult because of the instability of $(SN)_x$. The processing and measurement of the Schottky barriers on various semiconductors were described in detail [3]. The barrier heights were determined by photoresponse, current voltage, and by the capacitance voltage method [1 to 4]. The photoresponse data were considered the most reliable. A summary of the results of barrier height measurements of $(SN)_x$ and Au contacts for nine air-cleaned semiconductors (n-GaAs, p-GaAs, n-InP, p-GaP, p-ZnTe, n-CdSe, n-ZnSe, n-CdS, and n-ZnS) was given [3]. The main result is that $(SN)_x$ produces higher barriers to n-type semiconductors than Au. The barrier height difference, $\Phi(SN)_x - \Phi(Au)$, for n-type ZnS is 0.9 to 1.0 eV [2 to 4]. For MS and MIS (Schottky and metal-insulator semiconductor) diodes, the barrier height differences, $\Phi(SN)_x - \Phi(Au)$, were estimated to be 0.75 eV and 0.85 eV [5]. For p-type semiconductors, $(SN)_x$ produces lower barriers than Au. $(SN)_x$ follows the "ionic-covalent transition" (the barrier height improvement is smaller on covalent semiconductors than on ionic semiconductors, showing an abrupt change) but not the "common anion rule". (The barrier height of Au depends only on the anion of the semiconductors.) $(SN)_x$ films were also used as the metal electrode in the fabrication of Schottky (MS) and metal-insulator semiconductor (MIS) diodes on ZnS [5]. In forward bias these devices showed a blue electroluminescence. The higher barrier height (about 0.75 eV) with $(SN)_x$ instead of Au leads to a blue shift of the total emission (two separate bands: emission at 2.558 eV (selfactivated) and at 2.815 eV (high-energy emission comming from the vicinity of the barrier)) to higher photon energies. In addition, the quantum efficiency of the high-energy emission increased by a factor of 100 [5].

The favorable properties of $(SN)_x$ Schottky barriers were used in **solar cells** which were the first with a polymer-semiconductor interface [6 to 11]. The cells were fabricated with thin $(SN)_x$ films deposited on GaAs [6 to 9] or amorphous Si [7, 11, 12]. An $(SNBr_{0.4})_x$–GaAs solar cell was also constructed [10]. The technique of vapor deposition for $(SN)_x$–GaAs solar cells was described in detail [6]. Open circuit voltages in the range 0.68 to 0.71 V on various $(SN)_x$–GaAs solar cells are more than 40% over the voltages commonly measured with metal-GaAs cells (0.492 V for Au–GaAs) [6 to 10]. Similar enhancements in voltages were obtained in $(SN)_x$–Si solar cells [7]. I-V and C-V characteristics of three $(SN)_x$–GaAs solar cells were presented and the short-circuit current density of the cells was determined to be 10.3 mA/cm^2 (Au–GaAs: 10.8 mA/cm^2). A value of about 0.2 eV higher than that of the Au–GaAs diode was obtained for the relative Schottky barrier height. The increased blue response expected from Schottky barrier cells was observed in measurements of the external photoresponse (i.e., uncorrected for reflection and absorption losses) of the Au–$(SN)_x$–GaAs and Au–GaAs cells [6], cf. [10]. The resistance at the interface of the metal-$(SN)_x$ junction with metals like Ag, Cu, Zn, and Al increases remarkably at lower temperatures. This may be due to the formation of semiconductive metal sulfides such as Ag_2S, Cu_2S, ZnS, and Al_2O_3 (Al_2S is unstable) at the interface. Confirmation comes from Auger analysis, but still more work is necessary [13].

References:

[1] Scranton, R. A.; Mooney, J. B.; McCaldin, J. O.; McGill, T. C.; Mead, C. A. (Appl. Phys. Letters **29** [1976] 47/8).

[2] Scranton, R. A.; Best, J. S.; McCaldin, J. O. (J. Vac. Sci. Technol. **14** [1977] 930/4).

[3] Scranton, R. A. (J. Appl. Phys. **48** [1977] 3838/42).

[4] Scranton, R. A.; McGill, T. C. (U.S. 4084172 [1977/78]; C.A. **89** [1978] No. 35387).

[5] Thomas, A. E.; Woods, J.; Hauptman, Z. V. (J. Phys. D **16** [1983] 1123/36).

[6] Cohen, M. J.; Harris, J. S., Jr. (Appl. Phys. Letters **33** [1978] 812/4).

[7] Cohen, M. J.; Harris, J. S., Jr. (Tech. Dig. Intern Electron Devices Meeting **1978** 247/9).
[8] Cohen, M. J.; Harris, J. S., Jr. (Lect. Notes Phys. **96** [1979] 417).
[9] Harris, J. S.; Cohen, M. J.; Manasevit, H. M.; Zehr, S. W. (SERI-TR-8041-1-T1 [1983]; Energy Res. Abstr. **9** [1984] Abstr. No. 4534; C.A. **100** [1984] No. 203814).
[10] Cohen, M. J.; Harris, J. S., Jr. (U.S. 4227943 [1979/80]; C.A. **94** [1981] No. 33758).

[11] Sharp Corp. (Japan Kokai Tokkyo Koho 80-160475 [1979/80] from C.A. **94** [1981] No. 106599).
[12] Fritz, H. P.; Bruchhaus, R. (Z. Naturforsch. **38b** [1983] 1375/82).
[13] Yoshino, K.; Ura, S.; Sasa, S.; Kaneto, K.; Inuishi, Y. (Japan J. Appl. Phys. **21** [1982] L507/L508).

2.1.9.7 Electron-Energy-Loss Spectroscopy

Inelastic electron scattering energy-loss spectroscopy with a 75 or 30 keV electron beam was used to study the plasmon excitation in $(SN)_x$ crystals [1] and in epitaxial $(SN)_x$ films [2] up to 30 eV. A sharp peak at ~2.5 eV and a broad peak at 1.5 eV were found for plasmon wave vector $\vec{q}$ parallel and perpendicular to the b axis, respectively; both are consistent with the plasma edges for the same polarizations from optical measurements [1]. Two weaker peaks at 6.6 and 8.7 eV attributed to interband transitions, and a broad peak at 23 eV (plasmon involving all valence electrons per unit cell, i.e., 11×4) were also observed for $\vec{q} \parallel b$ [1]. The observed angular dependence of the 2.5 eV plasmon peak is not in agreement with a cosine dependence ($\hbar\omega_{pl}(\theta) = \hbar\omega_{pl}(0)\cos(\theta)$) for a quasi-one-dimensional spectrum [1].

Energy-loss spectrum peaks occur at 1.7, 2.4, 4.4, 6.2, 8.2, 12.0, and 22.3 eV for the epitaxial films for incident electron beam wave vector k parallel to the b axis. Assignments of the peaks are: peak at 1.7 eV has contributions from surface and bulk plasmons; 2.4 eV peak is a plasmon peak caused by the four S-3p electrons per unit cell; 4.4, 6.2, 8.2, and 12 eV peaks are due to interband transitions; 22.3 is a bulk plasmon peak and a broad structure around 17 eV is the surface plasmon corresponding to the bulk plasmon at 22.3 eV [2]. For $k \perp b$ a much less pronounced and broader plasmon was obtained.

Experimental plasmon dispersion data were presented for $(SN)_x$ epitaxial films ((10$\bar{2}$) parallel film plane) up to wave vector transfers k of 1 $Å^{-1}$, using a linear and quadratic k-scale. The observed positive quadratic plasmon dispersion is typical of free electron behavior. At $k = 0.4$ $Å^{-1}$ a slight discontinuity in the dispersion of the bulk plasmon of the conduction electrons was found, and attributed to the interference with free-electron-like single particle excitations [2]. The dispersion of plasmons as a function of momentum, $\omega_{pl}(q)$, was also calculated by an RPA (random-phase approximation) dielectric function using a model of an electron gas with an anisotropic effective mass ($m_\perp/m_\parallel = 1.9$) [3, 4]. The theoretical plasmon dispersion data [4] are in a remarkable agreement with experimental values [1].

References:

[1] Chen, C. H.; Silcox, J.; Garito, A. F.; Heeger, A. J.; MacDiarmid, A. G. (Phys. Rev. Letters **36** [1976] 525/8).
[2] Stolz, H. J.; Petri, E.; Otto, A. (Phys. Status Solidi B **80** [1977] 289/94).
[3] Brosens, F.; Devreese, J. T.; Lemmens, L. F.; Ruvalds, J. (Bull. Am. Phys. Soc. [2] **22** [1977] 372).
[4] Ruvalds, J.; Brosens, F.; Lemmens, L. F.; Devreese, J. T. (Solid State Commun. **23** [1977] 243/6).

2.1.10 Optical Properties

2.1.10.1 Survey

Raman spectra and IR absorption spectra between $\tilde{\nu} \approx 200$ to 1000 cm^{-1} are based on optical external and internal phonons, see pp. 128/30. The optical properties of $(SN)_x$ are uniquely caused by transport properties up to $h\nu \approx 4$ eV: range of intraband transitions (Drude range). Interband transitions become important at higher photon energies which causes strong (intrinsic) absorption (and reflection) maxima between ~6 and ~10 eV. A strong conduction-electron plasmon loss (plasma edge of reflectivity) occurs for light polarized along the chain axis b at $h\nu \approx 2.7$ eV. This is also seen for $\vec{E} \perp b$, but is damped and lower lying (at ~1.7 eV). Another (also bulk) plasmon, caused by all valence electrons, seems to appear at $h\nu \approx 23$ eV. These features in the frequency dependences of the optical constants (reflectance, transmittance, absorption, dielectric functions) were also found in the electron-loss spectra, see p. 164.

Conduction-electron plasmon and all-valence electron plasmon were also detected in the fine and satellite structure of the N(1s) XPS core electron spectrum, see p. 110. Metallic conduction also in the direction $\perp b$ can be inferred from the value of the optical conductivity (in contrary to electrical measurements) [1 to 3].

2.1.10.2 Optical Constants

2.1.10.2.1 Compilation of Measurements

The optical reflectance spectra of $(SN)_x$ films [1 to 14, 29] and crystals [1 to 3, 15 to 22] ([1 to 3] are review articels) were studied intensively by several groups. The reflectivity data were very helpful in the understanding of the intrinsic properties of $(SN)_x$, clearly showing that $(SN)_x$ is not a "one-dimensional" metal, for example see [1, 3, 18]. The reflectance spectra of $(SN)_x$ films (at room temperature) were studied in the energy ranges 1.5 to 4.5 eV (for electrochemically prepared $(SN)_x$) [29], 1 to 3.5 eV (unpolarized spectrum) [4], 500 to 30000 cm^{-1} (epitaxial film) [7, 8], 3000 to 30000 cm^{-1} (epitaxial film) [9, 10], 0.3 to 5.5 eV [1, 13], 0.5 to 27 eV (unpolarized spectrum by the use of synchrotron radiation technique (SR)) [6], and 5 to 25 eV (SR) [14]. Reflectance spectra of $(SN)_x$ crystals were measured in the energy region 0.25 to 6 eV [15], 11000 to 27000 cm^{-1} [16], 3000 to 30000 cm^{-1} [9, 10], 0.25 to 4 eV [17], 0.1 to 5 eV [1, 18], 1 to 25 eV (SR) [21], and 1.7 to 4 eV (unpolarized reflection microspectroscopy) [22]. Reflectance spectra of $(SN)_x$ crystals were also recorded in the range ~1 to 3 eV at temperatures of 4.2, 77, and 300 K [20].

Optical transmittance spectra of $(SN)_x$ films were measured in the energy range 0.5 to 3.5 eV [29], 0.2 to 4.0 eV [4], 0.3 to 5.5 eV (reciprocal transmittance $1/T_{\parallel}$ and $1/T_{\perp}$) [1, 13], 0.5 to 4 eV (oriented films) [6], and 5 to 25 eV (also absorption coefficients) [14]. The optical transmission spectra presented in papers [4] and [14] are in good agreement with absorption coefficient data [6]. Polarized absorption spectra of oriented thin $(SN)_x$ films were presented for the visible region 250 to 800 nm [21] and for the range 0.5 to 5.0 eV [28]. Earlier absorption data [30] were attributed to an unidentified blue nonmetallic SN compound [4].

2.1.10.2.2 Range of Intraband Transitions

A polarized reflectance spectrum (range 0.1 to 5 eV) of $(SN)_x$ crystals at room temperature for normally incident light with E vector polarized parallel and perpendicular to the b axis is shown in **Fig. 43**, p. 166. The energy range below ~4 eV was attributed to intraband transitions (transport properties) [1]. For light polarized parallel to the chain axis b a steep plasma edge (see Fig. 43), typical of metallic reflectance, was observed in crystals at 2.7 eV [16, 20], near 22000 cm^{-1} [9, 10], at ~3 eV [15], and in films at ~22000 cm^{-1} [8 to 10], at ~2.6 eV [13], and at 3 eV [14]. The reflectivity data were analyzed by several groups in terms of a Drude dielectric function [4, 8, 15, 16, 20]. In Table 38 electronic properties of $(SN)_x$ are compiled. Additional data by other authors are tabulated in [3]; optical conductivity data are also included, see p. 169.

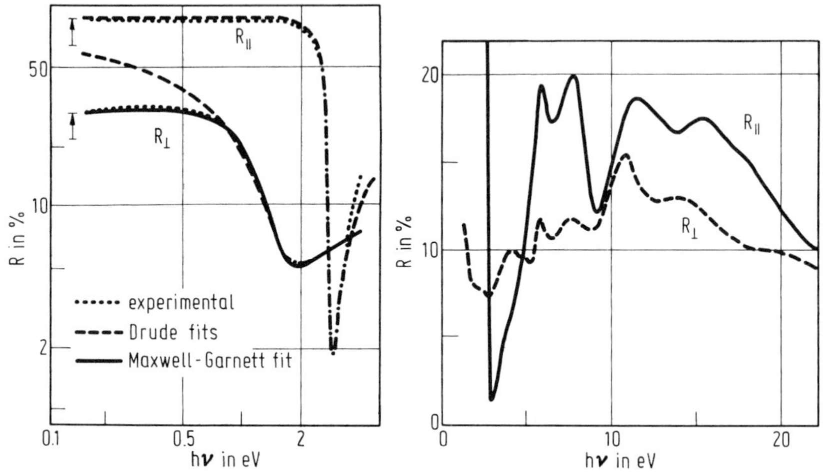

Fig. 43. Polarized reflectance spectra of $(SN)_x$ single crystals (at room temperature) for $\vec{E}\parallel b$ ($R_\parallel$) and $\vec{E}\perp b$ ($R_\perp$) in the intraband (left) [2, 17, 18] and interband (right) [21] region.

Table 38

Drude Parameters from Reflectance Spectra for E∥b and E⊥b in $(SN)_x$ Crystals and Film (ε_{core}: dielectric constant of the core polarizability ($\triangleq \varepsilon_\infty$); ω_p: plasma frequency; τ: relaxation time of free carrier; m*: effective mass; RT: room temperature).

	T in K	ε_{core}	ω_p in $10^{15}\ s^{-1}$	$\hbar\omega_p/\varepsilon_{core}^{1/2}$ in eV	τ in 10^{-15} s	$n/(m^*/m_e)$ in $10^{22}\ cm^{-3}$	Ref.
E∥b	300	4.4	8.3	2.16	1.3	2.2	[20]
	77	4.4	8.5	2.67	1.4	2.3	[20]
	4.2	4.4	8.6	2.70	1.9	2.3	[20]
	RT	6.5	10.4	2.7	2.6	3.4	[13]*)
E⊥b	300	4.2	4.5	1.5	0.48	0.68	[20]
	77	4.2	4.9	1.8	0.50	0.75	[20]
	4.2	4.2	5.2	1.9	0.52	0.85	[20]
	RT	4.0	4.8	1.6	0.26	0.74	[13]*)

*) Oriented thin film.

The reflectivity of light polarized perpendicularly to the chain axis is much lower and shows a broad drop-off rather than a sharp plasma edge (see Fig. 43) [1, 9, 13, 17, 18, 20]. The absorption spectra show a strong absorption for E∥b between ~1 and ~2 eV including a sharp maximum at 1.2 eV and for E⊥b a shallow maximum at 1 eV [6], cf. [13]. It follows from reflectivity results parallel and perpendicular to the chain axis that $(SN)_x$ has a low intrinsic anisotropy, in contrast to dc conductivity measurements [1, 3, 18]. It was concluded that the transverse dc conductivity is an extrinsic effect [18]. Lateral crystal face reflection spectra for $R_{\parallel}$ and $R_{\perp}$ indicate the presence of strong anisotropy of the interchain interaction due to the orientation of the π-electron orbitals, and suggest two-dimensional metallic behavior of the π-electron system in $(SN)_x$. R(ω) for E∥y (($\bar{1}$02) in the yz plane) shows no plasma edge [22].

The parallel reflectivity can be described by a Drude model whereas the perpendicular reflectivity data were fitted to a modified Maxwell-Garnett model (system consisting of metallic fibers embedded in an insulating matrix) taking the fibrous structure of $(SN)_x$ into account (see Fig. 43) [1, 13, 17, 18]. The experimental transmittance data in the range 0.3 to 3 eV are in agreement with a Maxwell-Garnett fit. Deviations were attributed to misoriented crystallites [1, 13]. A relatively good agreement was found [24] between the peak positions of the theoretical reflectivity curve, calculated by a Kramers-Kronig analysis from energy-loss spectrum (the dielectric functions ε_1 and ε_2 were also given), and the experimental data for crystals below 6 eV [15] and also with the synchrotron-radiation results of [6]. Further, calculated reflectivity data, based on the OPW method using plasma and dielectric tensor and the Fresnel reflectance equation, were compared with experimental reflectivity ($E_{\parallel}, E_{\perp}$) data for $(SN)_x$ crystals (range 0.1 to ~8.0 eV) and unoriented $(SN)_x$ films (range 0 to 12 eV); again a relatively good agreement was found [25, 26]. Another theoretical calculation of the reflectivity of $(SN)_x$ was performed by [27], but no comparison with experimental data was made.

2.1.10.2.3 Range of Interband Transitions

The onset of strong optically allowed interband transitions occurs at ca. 5 eV. Intra- and interband transitions are well separated by a deep minimum in ε_2. The onset of absorption is much steeper for the direction of polarization parallel to the chain axis [1]. Preliminary measurements on films in the vacuum UV show absorption structures at 6 and 7.2 eV, a minimum at 9 eV, a broad absorption band around 10 eV, a broad shoulder at 15 eV, and a very weak broad band at 23 eV [6]. The peaks at 6 and 7.4 eV, the band at 10 eV, and the shoulder at 15 eV were assigned to interband transitions near Γ by use of the band-structure calculation of Kamimura et al. [31]. The polarized reflectivity spectrum of $(SN)_x$ single crystals in the range ~2 to ~25 eV at room temperature is given in Fig. 43 [21]. The Kramers-Kronig relation was used to transform the experimental reflectivity data (also from films) to the imaginary part of the dielectric function ε_2 [6, 14, 21, 23] and the curves obtained were compared with theoretical ε_2 curves calculated by the OPW method [26] and by the self-consistent LCAO band model [32]. In the latter case a fairly good agreement for both $\vec{E}\parallel b$ and $\vec{E}\perp b$ was found. The observed peak energies were assigned to possible interband transitions across E_F. Most of the strong peaks of ε_2 above 5 eV (namely that at 5.8, 7.5, 11, and 15 eV) were ascribed to the transitions between the same types of band states, i.e., either σ–σ* or π–π*. From the assignments of the structures it was concluded that the higher conduction bands of $(SN)_x$ are predominantly of one-dimensional character above the Fermi level. Broad absorptions above 10 eV may be composed of several interband transitions [21].

2.1.10.2.4 Temperature and Pressure Dependence

The temperature dependence of the optical reflectivity around the Drude plasma edge of $(SN)_x$ films and crystals was studied from 15 to 300 K [12] and also with $(SN)_x$ crystals at 4.2, 77, and 300 K [20]. The results of these studies disagree. A red shift of the Drude plasma edge with decreasing temperature for $\vec{E}\parallel b$ was attributed to a movement of the Fermi level due to volume effects [12]. A blue shift of the plasma edge with decreasing temperature for $\vec{E}\parallel b$ and $\perp b$ axis was associated mainly with an increase of the plasma frequency ω_p [20]. The relaxation time τ_{opt} for $\vec{E}\parallel b$ (see Table 38, p. 166) increases with decreasing temperature (for τ_{dc} a much larger increase was calculated) [20].

The pressure dependence of the Drude edge was also measured. A weak red shift was observed for $\vec{E}\parallel b$ with increasing pressure. This observation was also interpreted as a volume effect [19].

References:

[1] Geserich, H. P.; Pintschovius, L. (Festkörperprobleme [Advan. Solid State Phys.] **16** [1976] 65/94, 77).
[2] Geserich, H. P. (in: Monceau, P., Electronic Properties of Inorganic Quasi-One-Dimensional Compounds, Pt. 2, Reidel, Dordrecht 1985, pp. 111/39, 128/31).
[3] Kaneto, K.; Yoshino, Y.; Inuishi, Y. (in: Monceau, P., Electronic Properties of Inorganic Quasi-One-Dimensional Compounds, Pt. 2, Reidel, Dordrecht 1985, pp. 69/109, 101/4).
[4] Kamimura, H.; Grant, A. J.; Levy, F.; Yoffe, A. D.; Pitt, G. D. (Solid State Comun. **17** [1975] 49/53).
[5] Kamimura, H.; Grant, A. J.; Levy, F.; Yoffe, A. D. (Proc. 14th Intern. Conf. Low Temp. Phys., Otaniemi, Finland, 1975, Vol. 4, pp. 354/7).
[6] Bordas, J.; Grant, A. J.; Hughes, H. P.; Jakobsson, A.; Kamimura, H.; Levy, F. A.; Nakao, K.; Natsume, Y.; Yoffe, A. D. (J. Phys. C **9** [1976] L277/L282).
[7] Bright, A. A.; Cohen, M. J.; Garito, A. F.; Heeger, A. J.; MacDiarmid, A. G.; Mikulski, C. M. (Bull. Am. Phys. Soc. [2] **20** [1975] 360).
[8] Bright, A. A.; Cohen, M. J.; Garito, A. F.; Heeger, A. J.; Mikulski, C. M.; Russo, P. J.; MacDiarmid, A. G. (Phys. Rev. Letters **34** [1975] 206/9).
[9] Bright, A. A.; Cohen, M. J.; Garito, A. F.; Heeger, C. M.; MacDiarmid, A. G. (Appl. Phys. Letters **26** [1975] 612/5).
[10] Garito, A. F.; Heeger, A. J.; MacDiarmid, A. G.; Bright, A. J.; Cohen, M. J.; Mikulski, C. M. (U.S. 4016331 [1976/77]; C.A. **86** [1977] No. 172655).

[11] Street, G. B.; Grant, P. M.; Greene, R. L. (Bull. Am. Phys. Soc. [2] **20** [1975] 360).
[12] Grant, P. M.; Mengel, P.; Engler, E. M.; Castro, G.; Street, G. B. (Bull. Am. Phys. Soc. [2] **21** [1976] 254).
[13] Möller, W.; Geserich, H. P.; Pintschovius, L. (Solid State Commun. **18** [1976] 791/4).
[14] Mitani, T.; Mori, H.; Suga, S.; Koda, T.; Shin, S.; Inoue, K.; Nakada, I.; Kanzaki, H. (J. Phys. Soc. Japan **47** [1979] 679/80).
[15] Grant, P. M.; Greene, R. L.; Street, G. B. (Phys. Rev. Letters **35** [1975] 1743/6).
[16] Pintschovius, L.; Geserich, H. P.; Möller, W. (Solid State Commun. **17** [1975] 477/9).
[17] Geserich, H. P.; Möller, W.; Scheiber, G.; Pintschovius, L. (Phys. Status Solidi B **80** [1977] 119/23).
[18] Geserich, H. P.; Scheiber, G.; Dürrler, M.; Potel, M.; Sergent, M.; Monceau, P. (Physica B **143** [1986] 234/6).
[19] Grant, P. M.; Welber, B. (Bull. Am. Phys. Soc. [2] **22** [1977] 372).
[20] Kaneto, K.; Yoshino, K.; Inuishi, Y. (J. Phys. Soc. Japan **43** [1977] 1013/5).

[21] Mitani, T.; Koyama, K.; Mori, H.; Koda, T.; Nakada, I. (J. Phys. Soc. Japan **51** [1982] 3197/205).
[22] Gutman, A. I.; Agroskin, L. S.; Papayan, G. V.; Rautian, L. P.; Schachnina, O. S. (Solid State Commun. **49** [1984] 187/9).

[23] Mitani, T.; Suga, S.; Tokura, Y.; Koyama, K.; Nakada, I.; Koda, T. (Intern. J. Quantum Chem. **18** [1980] 655/9).
[24] Stolz, H. J.; Petri, E.; Otto, A. (Phys. Status Solidi B **80** [1977] 289/94).
[25] Ortenburger, I. B.; Rudge, W. E.; Grant, P. M. (Bull. Am. Phys. Soc. [2] **21** [1976] 254).
[26] Grant, P. M.; Rudge, W. E.; Ortenburger, I. B. (Lect. Notes Phys. **65** [1977] 575/88).
[27] Numata, H. (Phys. Letters A **72** [1979] 453/5).
[28] Iqbal, Z.; Downs, D. S. (Solid State Commun. **20** [1976] 1147/50).
[29] Banister, A. J.; Hauptman, Z. V.; Kendrick, A. G.; Small, R. W. H. (J. Chem. Soc. Dalton Trans. **1987** 915/24).
[30] Chapman, D.; Warn, R. J.; Fitzgerald, A. G.; Yoffe, A. D. (Trans. Faraday Soc. **60** [1964] 294/300).

[31] Kamimura, H.; Glazer, A. M.; Grant, A. J.; Natsume, Y.; Schreiber, M.; Yoffe, A. D. (J. Phys. C **9** [1976] 291/303).
[32] Oshiyama, A.; Kamimura, H. (J. Phys. C **14** [1981] 5091/107).

2.1.10.3 Optical Conductivity

Optical conductivity values of $(SN)_x$ [1 to 10] were obtained from optical reflectivity data on polycrystalline films [5 to 7] and "single crystals" [8 to 10, 13]. Some data for $(SN)_x$ crystals and $(SN)_x$ thin oriented films are compared with dc values in Table 39. Additional values of optical conductivity were published in [1, 5, 6, 8, 9], [11, summarizing table], [12, 13].

Table 39
Optical and dc Conductivity of $(SN)_x$ Single Crystals (SC) and Oriented Films (F).

sample	T in K	$\sigma_{opt\parallel}$ in $\Omega^{-1}\cdot cm^{-1}$	$\sigma_{dc\parallel}$ in $\Omega^{-1}\cdot cm^{-1}$	$\sigma_{opt\perp}$ in $\Omega^{-1}\cdot cm^{-1}$	$\sigma_{dc\perp}$ in $\Omega^{-1}\cdot cm^{-1}$	Ref.
SC	300	7900	1500	850	140	[10]
SC	77	8900	65000	1100	160	[10]
SC	4.2	16000	120000	1200	90	[10]
F	RT	25000	—	5400	—	[7]

The optical conductivity of $(SN)_x$ crystals along the chain axis b was observed to be ca. one decade higher than the highest value from dc measurement at 300 K. The difference was ascribed to imperfections in the chains which decrease the dc conductivity. In contrast to dc conductivity determinations, the optical measurements indicate that metallic conduction also occurs perpendicularly to the chain axis. For this reason $(SN)_x$ should not be regarded as a one-dimensional metal [1, 8, 9]. The small values of $\sigma_\perp$ from dc measurement are related to the influence of potential barriers between neighboring chain bundles [10]. The anisotropy ratio $\sigma_\parallel/\sigma_\perp$ is reduced from the value 400 for dc conductivities to 50 for optical conductivities [1]. Other reported values for $(\sigma_\parallel/\sigma_\perp)_{opt}$ are 65 [4], 20 to 43 [9], and 10 [10]. The optical conductivities increase slightly with decreasing temperature, see Table 39 [10].

References:

[1] Geserich, H.-P.; Pintschovius, L. (Festkörperprobleme [Advan. Solid State Phys.] **16** [1976] 65/93, 75/8).
[2] Pintschovius, L. (Colloid Polym. Sci. **256** [1978] 883/92, 887/8).

[3] Labes, M. M.; Love, P.; Nichols, L. F. (Chem. Rev. **79** [1979] 1/15, 11).
[4] Greene, R. L.; Street, G. B. (NATO Advan. Study Inst. Ser. B **25** [1977] 167/95, 179).
[5] Kamimura, H.; Grant, A. J.; Levy, F.; Yoffe, A. D.; Pitt, G. D. (Solid State Commun. **17** [1975] 49/53).
[6] Bright, A. A.; Cohen, M. J.; Garito, A. F.; Heeger, A. J.; Mikulski, C. M.; Russo, P. J.; MacDiarmid, A. G. (Phys. Rev. Letters **34** [1975] 206/9).
[7] Möller, W.; Geserich, H.-P.; Pintschovius, L. (Solid State Commun. **18** [1976] 791/4).
[8] Pintschovius, L.; Geserich, H.-P.; Möller, W. (Solid State Commun. **17** [1975] 477/9).
[9] Geserich, H.-P.; Möller, W.; Scheiber, G.; Pintschovius, L. (Phys. Status Solidi B **80** [1977] 119/23).
[10] Kaneto, K.; Yoshino, K.; Inuishi, Y. (J. Phys. Soc. Japan **43** [1977] 10/3, 5).

[11] Kaneto, K.; Yoshino, K.; Inuishi, Y. (in: Monceau, P., Electronic Properties of Inorganic Quasi-One-Dimensional Compounds, Pt. 2, Reidel, Dordrecht 1985, pp. 69/109, 104).
[12] Grant, P. M.; Greene, R. L.; Street, G. B. (Phys. Rev. Letters **35** [1975] 1743/6).
[13] Gutman, A. I.; Agroskin, L. S.; Papayan, G. V.; Rautian, L. P.; Schachnina, O. S. (Solid State Commun. **49** [1984] 187/9).

2.1.11 Electrochemical Behavior

2.1.11.1 $(SN)_x$ as an Electrode Material

$(SN)_x$ was intensively studied as an electrode material [1]. $(SN)_x$ electrodes were fabricated using crystals [2 to 12], films [13 to 15], composites [8, 16], or pastes [17 to 19]. Studies were performed in aqueous [2 to 16, 18, 19] and in nonaqueous solvents [17, 20, 21]. Of particular interest in view of possible applications, is the observed catalytic effect of the modified $(SN)_x$ electrode surfaces [6, 11, 14, 22].

"Single-Crystal" Electrodes

The fabrication of $(SN)_x$ "single-crystal" electrodes was described in detail [3, 4]. $(SN)_x$ crystals were cemented to the end of a silver wire (0.040 cm radius, sealed in glass) with conducting silver paint and all but the crystal face of interest was masked off with Torr Seal® (surface area: 1 to 50×10^{-3} cm^2). The procedure allowed the construction of "parallel" and "perpendicular" electrodes which had the $(SN)_x$ fiber bundles parallel and perpendicular to the electrode/solution interface, respectively [3, 4].

$(SN)_x$ electrodes are stable at 70°C over long periods of time. No degradation or gas evolution at any temperature was observed [7]. Cyclic voltammetry studies on parallel electrodes show that background currents and breakdown potentials depend on pH, ionic strength, scan rate, scan number, and on the nature of the supporting electrolyte [3 to 5, 7]. For perpendicular electrodes the values of the potentials of anodic and cathodic current breakdown and the magnitude of the faradaic currents are independent of the supporting electrolyte [4]. Anodic breakdown potentials of perpendicular electrodes in aqueous solutions are 0.5 [3, 4] or 0.7 V vs. SCE [5] and cathodic breakdown potentials are −1.2 [3, 4] or −1.25 V vs. SCE [5]. The cathodic breakdown potential of parallel electrodes depends largely on the cation in aqueous solution: −1.54 for LiCl, −1.25 for NaCl, −1.15 for KCl, and −1.20 V vs. SCE for CsCl [3]. The general tendency is that a higher charge density on the alkali metal cation (crystallographic radii) causes a lower background current and a more negative cathodic decomposition potential on parallel electrodes [3, 4]. The same effects were observed for perpendicular electrodes on addition of Cr^{3+} ions to a solution of supporting electrolyte [3, 4].

The $Fe(CN)_6^{3-}/Fe(CN)_6^{4-}$ couple was studied by cyclic voltammetry and chronoamperometry at an $(SN)_x$ electrode and compared with its behavior at other solid electrode materials [3, 5, 7]. The diffusion-controlled nature of the current is indicated by the results shown in Table 40.

Table 40
Electrochemical Characterization of $(SN)_x$ Electrodes with Respect to $Fe(CN)_6^{3-}$ Solutions [3].

				electrode surface area $\times 10^3$ cm^2				
				cyclic chronoamperometry [c]				
electrode	c(KCl) in mol/L	$c(K_3Fe(CN)_6)$ in mol/L	$\frac{Ep_a+Ep_c}{2}$ [a] in V	cathodic decay (10 s)	anodic decay (10 s)	projected-calibrated microscope	$\frac{ip_c}{ip_a}$ [b]	$\frac{\partial \log i_p}{\partial \log v}$ [e]
$(SN)_x\parallel$	3.0	0.03	+0.260	4.57	4.47	2.31	0.96	0.44
$(SN)_x\parallel$	3.0	0.03	+0.265	8.71	8.10	—	0.99	0.54
$(SN)_x\perp$	0.5	0.1	+0.215	5.70	5.76	3.08	1.03	0.50
$(SN)_x\perp$	0.5	0.1	+0.215	8.22	8.02	7.01	1.09	0.50
Pt disk [d]	0.5	0.1	+0.215	5.01	4.87	4.95	1.05	0.47
Pt disk	3.0	0.03	+0.270	5.59	—	4.95	1.00	0.50

[a] E vs. SCE; Ep = peak potential, anodic (a) and cathodic (c), determined by cyclic voltammetry. – [b] ip = peak current. – [c] For the determination of the electrochemically active surface area by cyclic chronoamperometry a 10 s current decay period was chosen so that the diffusion layer thickness in chronoamperometry experiments was comparable to that in voltammetry experiments under the scan rate employed. – [d] Torr Seal mounted. – [e] v = scan rate in V/s.

Nearly the same behavior was observed for $(SN)_x$ electrodes as for Pt electrodes [3]. Other redox couples, e.g., $Ru(NH_3)_6^{3+}/Ru(NH_3)_6^{2+}$ or quinone/hydroquinone exhibited a similar behavior, either on $(SN)_x$ electrodes or on Pt disk electrodes [3]. The following ions are electrochemically inactive: $Co(CN)_6^{3-}$, $Cr(SCN)_6^{3-}$, $S_4O_6^{2-}$, $S_2O_8^{2-}$, Fe^{3+}, Eu^{3+}, and BrO_4^- [3]. The irreversible Pb^{2+} deposition on $(SN)_x$ electrodes was investigated by cyclic voltammetry [2, 3, 7], see **Fig. 44**, p. 172. The results measured for parallel and perpendicular $(SN)_x$ electrodes differ markedly. A more negative cathodic maximum for parallel $(SN)_x$ electrodes than for Ag, Cu, Au, or Pt electrodes was observed [2, 3].

The electrochemical behavior of $(SN)_x$ in H_2SO_4 solution was investigated by using an $(SN)_x$ graphite powder electrode [8, 16]. The voltammogram (linear sweep) showed a single oxidation peak (0.9 V vs. NHE at pH 0.3). For pH values lower than 5 the following oxidation process was assumed [8, 16]:

$$(SN)_x + x\ H_2O \rightarrow (SNO)_x + 2x\ H^+ + 2x\ e^-$$

Background current studies in 1 N H_2SO_4 show that oxygen chemisorbs on electrode surfaces [9, 10]. The principal product of anodic breakdown on fresh surfaces is oxygen but on prolonged cycling is "$(SN)_x$-hydroxides" [10]. The results of another investigation of background currents indicate chemical anisotropy of $(SN)_x$ perpendicular electrode surfaces in aqueous phosphate buffered solution; presumably with either a predominantly "sulfur" or a "nitrogen surface". The phenomenon may be a function of the $(SN)_x$ crystal preparation method [12].

The electrochemical behavior of $(SN)_x$ crystal electrodes has also been studied in several nonaqueous solutions such as Ag- and alkali metal salts in CH_3CN solutions [20] or $MClO_4$, KPF_6, and MBr (M = Li, Na, K, $(C_2H_5)_4N$) in propylene carbonate (4-methyl-1,3-dioxalane-2-one) solutions [21]. S_7N^- ions were probably formed by superficial degradation during cathodic polarization in alkali metal salt solutions [20]. In $AgClO_4$–CH_3CN solution, an Ag deposit formed which could be dissolved anodically. Oxidation peaks were observed during the first anodic polarization [20]. Alkali metal insertion compounds of $(SN)_x$ are postulated products of cathodic reactions in propylene carbonate solutions of KPF_6–$MClO_4$ and MBr (M = Li, Na, K). Bromine insertion in MBr solutions was stated also. The $(SN)_x$ electrode material in $MClO_4$-propylene carbonate solution decomposes at a cathodic voltage of about −2 V (vs. Ag/Ag^+) to produce a turquoise-blue product that can be oxidized to yield a red solution. The behavior of an $(SN)_x$ electrode treated chemically or electrochemically with C_4H_9Li and containing presumably an insertion compound $(SN(C_4H_9Li)_{0.2})_x$ was studied also [21].

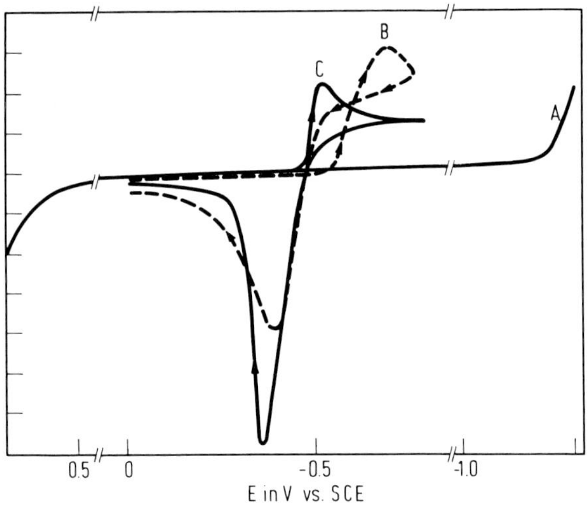

Fig. 44. Cyclic voltammograms of $Pb(NO_3)_2$ in H_2O at $(SN)_x$ electrodes in dependence on crystal orientation. (A) 0.1 M KNO_3, pH 5, v 70 mV/s, i 100 μA/division; (B) parallel electrode: 0.1M KNO_3, 20 mM $Pb(NO_3)_2$, pH 4, v 15 mV/s, i 10 μA/division; (C) perpendicular electrode: same conditions as (B) [2].

The electrochemical behavior of $(SN)_x$ films in aqueous solutions is essentially the same as that of $(SN)_x$ crystals [13, 15] but still further studies are necessary. Variation of factors such as pH, nature of cation in solution, ionic strength, on the $Fe(CN)_6^{3-}/Fe(CN)_6^{4-}$ redox couple were tested. The larger surface of $(SN)_x$ films would be of advantage for the use of $(SN)_x$ electrodes in catalytic reactions. An unresolved problem is the durability of the films themselves, as they tend to separate from the Mylar substrate after extended periods of immersion under electrolysis [15].

Paste Electrodes

Whereas the procedures of fabricating $(SN)_x$ electrodes by using $(SN)_x$ crystals, films or composites are often tedious and unreliable [17, 18], more successful results were obtained using $(SN)_x$ paste electrodes. These are easy to make in analogy to the carbon paste electrode using Apiezon® M grease, with a ratio $(SN)_x$: grease 1.5 : 1 [17] or 2 : 1 [18]. The electrochemical behavior of $(SN)_x$ paste electrodes was investigated in aqueous [18] and nonaqueous solvents [17, 19] with different supporting electrolytes. Before use in aqueous solution the $(SN)_x$ paste electrodes must be dipped in pure DMF (or CH_3CN) solution to activate the $(SN)_x$ surface [18]. In general the $(SN)_x$ paste electrodes are superior to "single crystal" and film $(SN)_x$ electrodes with respect to background currents, reproducibility on preparation and response, surface renewal and anisotropic behavior [18]. Cyclic voltammograms of the $Fe(CN)_6^{3-}/Fe(CN)_6^{4-}$ couple (at a DMF treated and untreated paste electrode) and of the Pb^{2+}/Pb^0 couple (both in aqueous solution) were presented [18]. The electrochemical behavior of $(SN)_x$ paste electrodes in nonaqeous solution is similar to that of metal electrodes [17]. C_2H_5OH, propylene carbonate, CH_3CN, and DMF were used as solvents [17, 18]. Breakdown potentials obtained in different nonaqueous solvent-supporting electrolyte solutions are compiled in Table 41 [17].

Table 41
Breakdown Potentials for $(SN)_x$ Paste Electrode [17].

solution [a]	cathodic breakdown [b]	anodic breakdown [b]
C_2H_5OH–LiCl	−0.57	0.82
C_2H_5OH–$(CH_3)_4NCl$	−0.51	0.80
propylene carbonate-$LiClO_4$	−0.81	0.80
propylene carbonate-$(C_2H_5)_4NClO_4$	−0.48	0.87
CH_3CN–$LiClO_4$	−0.74	0.96
CH_3CN–$(C_2H_5)_4NClO_4$	−0.40	0.95

[a] 0.1 M electrolyte. – [b] Potential, in V vs. saturated sodium chloride calomel electrode (SSCE), at which $|i| = 20\ \mu A/cm^2$. Scan rate 0.050 V/s.

The advantages compared to $(SN)_x$ crystal electrodes are the same as for aqueous solution. Cyclic voltammograms of TCNQ (tetracyanoquinonedimethane) and of ferrocene at $(SN)_x$ paste electrodes in $(C_2H_5)_4NClO_4$–CH_3CN solution were presented [17]. The electrochemical behavior of several types of organic compounds was studied at the $(SN)_x$ paste electrode in $(C_2H_5)_4NClO_4$–CH_3CN solution (stable potential range 0.95 to −0.4 V vs. SCE). The reduction of benzoquinone in the presence of protons shows a low current density in comparison to other electrodes (Pt, vitreous carbon), possibly due to slow protonation kinetics at the electrode surface. A high overpotential was measured for the proton reduction, both at $(SN)_x$ paste electrodes and at $(SN)_x$ "single-crystal" electrodes. The electrochemical characteristics of aromatic amines and hydrocarbons (phenothiazine, N,N-dimethyl-o-phenylenediamine, N,N-dimethylaniline, 2,3-di-phenylindole, 2,3-benzanthracene, and perylene) are similar on $(SN)_x$ paste electrodes to those observed at platinum electrodes. The oxidation of pyrrole at the anodic background limit of the $(SN)_x$ paste electrode yields black films with voltammetric properties comparable to those formed on other electrodes. Other heterocyclic compounds such as pyridine, quinoline, and acridine are not oxidizable in the present potential range. Addition of these compounds or several alkylpyridines to the electrolyte solution leads to a negative shift of the $(SN)_x$ oxidation itself. Thiazole did not show this effect [19].

Modified Electrodes

A partly chlorinated $(SN)_x$ powder was electrochemically characterized. Cathodic and anodic breakdown potentials on Pt-electrodes coated with poorly crystallized $(SN)_x$ in 0.1 M $(C_4H_9)_4N^+BF_4^-$–CH_2Cl_2 solution were determined (cyclovoltammogram peak potential E_p = −0.46 V vs. SCE and 1.4 V vs. SCE, respectively) and compared with other published data. Possible decomposition products were also discussed [23]. An important observation was that $(SN)_x$ electrodes, which had been modified by the interaction with certain cations (and IO_4^-), exhibit significant electrocatalysis with respect to redox processes. Surface modifications of $(SN)_x$ electrodes were performed by dipping them into solutions of Ag^+, Pd^{2+}, Cr^{3+}, and IO_4^-

[4, 5, 7, 11]. Before use, the electrodes were rinsed with distilled water. After pretreatment, catalysis of the irreversible iodate reduction was studied. The choice of the electrochemical iodate reduction (IO_3^-/I^-) [4, 5, 7, 9, 11] and the iodide oxidation [5 to 7, 9, 11] allowed the study (by a kinetic probe technique [5]) of two different potential regions, before and after chemical or electrochemical modification. Voltammograms of the IO_3^-/I^- couple were presented [4, 5, 11]. Pretreatment results in a positive shift in the value of the iodate reduction peak potential (~0.3 V with Pd^{2+}) [5, 7, 11]. Electrocatalysis with Cr^{3+} solutions was observed on perpendicular $(SN)_x$ electrodes but not on parallel electrodes [4]. Pd^{2+} and Ag^+ were chemically and/or electrochemically reduced at the $(SN)_x$ surface forming a metallic surface phase (via reduction by $(SN)_x$ surface functional group). Cr^{3+}, instead, was strongly bonded to parallel $(SN)_x$ electrodes in an ionic form, and IO_4^- caused an oxidation of the $(SN)_x$ surface [5, 11]. $(SN)_x$ electrodes for the study of the I_2/I^- system were modified with Pd^{2+}, Ag^+, and Hg^{2+} solutions [5, 7, 11]. The observed voltammetric behavior of the modified $(SN)_x$ electrode lay between the metallic state and unmodified $(SN)_x$. A permanent effect, however, was only achieved for Pd^{2+} pretreated electrodes [5, 11]. Pretreatment with $[Ru(NH_3)_5Cl]Cl_2$ solutions enhanced the peak currents of the I^-/I_2 couple for perpendicular $(SN)_x$ electrodes from 100 to 200% [6]. No catalytic behavior with respect to the I^-/I_2 couple was observed with Cr^{3+}, Eu^{3+}, Fe^{3+}, and Fe^{2+} [5, 11]. An $(SN)_x$ "single-crystal" perpendicular electrode, pretreated with a solution of $[Ru(bipy)_2(CH_3OCH_2CH_2OCH_3)](PF_6)_2$ in N_2-de-aerated $CH_3OCH_2CH_2OCH_3$ for 10 min, continously generated H_2 gas for a sustained period of time in acidic aqueous supporting electrolyte solution at −0.10 V vs. SCE when irradiated with visible light [22]. "Parallel" electrodes, treated or untreated with the Ru complex, did not exhibit H_2 evolution or current increase on illumination. Attempts with other Ru complexes were unsuccessful. The rate of hydrogen evolution was observed to decrease with increasing pH at fixed potentials. At fixed pH the rate of H_2 evolution increased with increasing negative potential up to the limit of 100 mV beyond the onset potential. Different explanations for the results were discussed; further work is necessary [22]. Molybdenum-modified $(SN)_x$ electrodes (parallel or perpendicular), produced by dipping crystals or films [13, 15] of $(SN)_x$ in aqueous solutions of sodium molybdate or $Mo_2O_2S_2^+$ "core", catalyze the transformation of acetylene to ethylene. A negative electrode potential of ca. −1.46 V vs. SCE and pH values in the range 5.4 to 14 were used in the experiments [14]. Molybdenum-iron treated electrodes produced the same rate of C_2H_4 as molybdenum-treated electrodes. Iron-modified electrodes yielded only small quantities of C_2H_4. Small amounts of C_2H_6 and little or no H_2 or $CH_2{=}CHCH{=}CH_2$ were detected as by-products [14]. A comparison of the rate of conversion of acetylene to ethylene with that of chemical catalysis systems was made, see Table 42.

Table 42

Comparison of Rate of Conversion of Acetylene to Ethylene at Molybdenum-Modified $(SN)_x$ Electrodes with that of Chemical Catalysis Systems [14].

(cys = cysteine; ATP = adenosine-triphosphate)

system used	rate [a] (mol C_2H_2/mol Mo/min)
1 mmol thioglycerol, 2 mmol $Na_2S_2O_4$, 1 M NaOH, 1 mmol Na_2MoO_4	1.1×10^{-2}
$Na_2Mo_2O_4(cys)_2$ (0.42 mmol), $NaBH_4$ (1.0 mmol), t = 27°C, charge carriers ca. 0.017 mmol	2 to 9×10^{-2}
nitrogenase, ATP, $S_2O_4^{2-}$, pH 6 to 8	200
$(SN)_x$, pH 13, molybdate dipped, U = −1.46 V vs. SCE	1 to 10×10^4 [b]

[a] Based on coverage level on order of 10^{-11} mol/cm². – [b] References see [14].

The mechanism, which may be of general importance for the understanding of the nitrogenase mechanism, is not the same as that found for other Mo–S reduction systems [14].

Titration with $(SN)_x$ Electrodes

The potential response of $(SN)_x$ electrodes on variation of the pH value is neither Nernstian nor linear indicating that the $(SN)_x$ electrodes are not useful for pH measurements [24], however, $(SN)_x$ electrodes are more useful for aqueous weak acid-strong base (poorer for strong acid-strong base) titrations than glass electrodes. Comparative titration curves were presented for the titration of trifluoro acetic acid and glycine with sodium hydroxide. The titration curves of hydrochloric acid, perchloric acid, trifluoroacetic acid, propanoic acid, benzoic acid (in 80% ethanol), boric acid, glycine, resorcinol, phenol with sodium hydroxide were also presented together in a figure. Not all weak acids (for example resorcinol) exhibit improved titration curves with $(SN)_x$ electrodes [24]. The pAg-potential response of the $(SN)_x$ electrode (parallel and perpendicular) was examined [25]. Only electrodes pretreated (with $AgNO_3$–Na_2S) showed a nearly Nernstian behavior (at lower concentrations a non-Nernstian response was observed). Untreated and treated parallel $(SN)_x$ electrodes showed no advantages as endpoint indicators in potentiometric titrations over Orion commercial Ag–Ag_2S electrodes. The results of a potentiometric titration of Ag^+ with Cl^- and $K_4Fe(CN)_6$ as titrants were presented [25].

References:

[1] Chandler, G. K.; Pletcher, D. (Electrochemistry **10** [1985] 143/6).
[2] Nowak, R. J.; Mark, H. B., Jr.; MacDiarmid, A. G.; Weber, D. (J. Chem. Soc. Chem. Commun. **1977** 9/11).
[3] Nowak, R. J.; Kutner, W.; Mark, H. B., Jr.; MacDiarmid, A. G. (J. Electrochem. Soc. **125** [1978] 232/40).
[4] Mark, H. B., Jr.; Nowak, R. J.; Kutner, W.; Johnson, J. F.; MacDiarmid, A. G. (Bioelectrochem. Bioenerg. **5** [1978] 215/22).
[5] Nowak, R. J.; Kutner, W.; Mark, H. B., Jr.; MacDiarmid, A. G. (Ann. N.Y. Acad. Sci. **313** [1978] 767/70).
[6] Voulgaropoulos, A. N.; Nowak, R. J.; Kutner, W.; Mark, H. B., Jr. (J. Chem. Soc. Chem. Commun. **1978** 244/5).
[7] Nowak, R. J. (Diss. Univ. Cincinnati 1977 from Diss. Abstr. Intern. B **39** [1978] 1781/2 and [22]).
[8] Bernard, C.; Robert, G. (Bull. Soc. Chim. France **1978** 395/401).
[9] Voulgaropoulos, A. N. (Diss. Univ. Cincinnati 1980 from Diss. Abstr. Intern. B **41** [1980] 948).
[10] Czerwinski, A.; Voulgaropoulos, A. N.; Johnson, J. F.; Mark, H. B., Jr. (Anal. Letters **12** [1979] 1089/94).

[11] Nowak, R. J.; Kutner, W.; Rubinson, J. F.; Voulgaropoulos, A. N.; Mark, H. B., Jr.; MacDiarmid, A. G. (J. Electrochem. Soc. **128** [1981] 1927/31).
[12] Voulgaropoulos, A. N.; Mark, H. B., Jr. (Anal. Letters **13** [1980] 959/64).
[13] Rubinson, J. F. (Diss. Univ. Cincinnati 1981 from Diss. Abstr. Intern. B **42** [1981] 1876/7).
[14] Rubinson, J. F.; Behymer, T. D.; Mark, H. B., Jr. (J. Am. Chem. Soc. **104** [1982] 1224/9).
[15] Rubinson, J. F.; Behymer, T. D.; Mark, H. B., Jr. (J. Electrochem. Soc. **130** [1983] 121/4).
[16] Beaudoin, S.; Bernard, C.; Vallot, R.; Robert, G.; Yu, L. T. (Compt. Rend. C **286** [1978] 217/20).

[17] Nowak, R. J.; Joyal, C. L.; Weber, D. C. (J. Electroanal. Chem. Interfacial. Electrochem. **143** [1983] 413/7).
[18] Shenoy, K. P.; Mulligan, K. J.; Mark, H. B., Jr. (J. Electrochem. Soc. **130** [1983] 2391/3).
[19] Cheek, G.; Horine, P. A. (J. Electrochem. Soc. **132** [1985] 115/9).
[20] Bernard, C.; Tarby, C.; Robert, G. (Electrochim. Acta **25** [1980] 435/40).

[21] Tarby, C.; Bernard, C.; Robert, G. (Electrochim. Acta **26** [1981] 663/70).
[22] Mark, H. B., Jr.; Voulgaropoulos, A.; Meyer, C. A. (J. Chem. Soc. Chem. Commun. **1981** 1021/3).
[23] Fritz, H. P.; Bruchhaus, R. (Z. Naturforsch. **38b** [1983] 1375/82).
[24] Radic, N.; Mark, H. B., Jr. (Anal. Chim. Acta **144** [1982] 253/9).
[25] Radic, N.; Mulligan, K. J.; Mark, H. B., Jr. (Analyst **109** [1984] 963/4).

2.1.11.2 $(SN)_x$ as an Electrode Material in Primary Batteries

Several patents describe the use of $(SN)_x$ as an electrode material in batteries [1 to 8]. Primary batteries were fabricated. Current-voltage measurements show that $(SN)_x$ cannot be reversibly reduced [9]. $(SN)_x$ ("undoped", "doped" or as a conduction assistant to metal oxides, halides, or sulfides [8]) was used as a cathode material [3 to 8]. Anodes were based on alkali metals [3, 5 to 8] or their alloys [3, 5, 7] and on $(SN)_x$ "doped" with electron donors [4]. Alkali halides in propylene carbonate [7], $LiClO_4$ or NH_4ClO_4 [4, 5], AMF_4, $A_2M'F_6$ or $AM''F_6$ (A = alkali metal; M, M′, M″ = group III A, IV A, V A element, respectively) [5], and ammoniates of anode-metal salts [6] were employed as electrolytes. In a primary battery consisting of $(SN)_x$ ("undoped" or "doped") cathode and an Li anode and an LiBr in propylene carbonate electrolyte, an output voltage of 3.2 to 3.4 V and 2.4 to 2.7 V (two voltage plateaus) was measured [7]. Open-circuit voltages were 2.8 V in an $Li–LiClO_4 \cdot 4NH_3–(SN)_x$ (graphite) battery [6], 0.9 V in an $Ag–Br–(SN)_x$ battery [2] and 0.92 V in a light-chargeable silver-halogen-$(SN)_x$ battery [1]. A review of secondary batteries with polymer electrodes (including $(SN)_x$) has been published [10].

References:

[1] Asahi Chemical Industry Co., Ltd. (Japan. Kokai Tokkyo Koho 80-137671 [1979/80] from C.A. **94** [1981] No. 50297).
[2] Asahi Chemical Industry Co., Ltd. (Japan. Kokai Tokkyo Koho 83-197677 [1982/83] from C.A. **100** [1984] No. 128805).
[3] Asahi Chemical Industry Co., Ltd. (Japan. Kokai Tokkyo Koho 83-93164 [1981/83] from C.A. **99** [1983] No. 79008).
[4] Asahi Chemical Industry Co., Ltd. (Japan. Kokai Tokkyo Koho 83-93177 [1981/83] from C.A. **99** [1983] No. 79004).
[5] Asahi Chemical Industry Co., Ltd. (Japan. Kokai Tokkyo Koho 83-204478 [1982/83] from C.A. **100** [1984] No. 147461).
[6] Badoz, J.; Bardin, M.; Bernard, C.; Herlem, M.; Robert, G.; Thiebault, A. (Eur. Appl. 97078 [1982/83] from C.A. **100** [1984] No. 124146).
[7] Allied Corp. (Japan. Kokai Tokkyo Koho 84-49155 [1982/84] from C.A. **101** [1984] No. 113972).
[8] Hitachi Maxell, Ltd. (Japan. Kokai Tokkyo Koho 83-48359 [1981/83] from C.A. **98** [1983] No. 224187).
[9] Fritz, H. P.; Bruchhaus, R. (Z. Naturforsch. **38b** [1983] 1375/82).
[10] Kaneto, K. (Porina Daijesuto **35** [1983] 11/9 from C.A. **99** [1983] No. 215633).

2.1.12 Chemical Behavior

2.1.12.1 Thermal Behavior. Photosensitivity

The thermal behavior of "normal $(SN)_x$" [1, 2] and "red $(SN)_x$" [2] was studied by differential thermal and thermogravimetric analysis. Data for $(SN)_x$ are shown in **Fig. 45**.

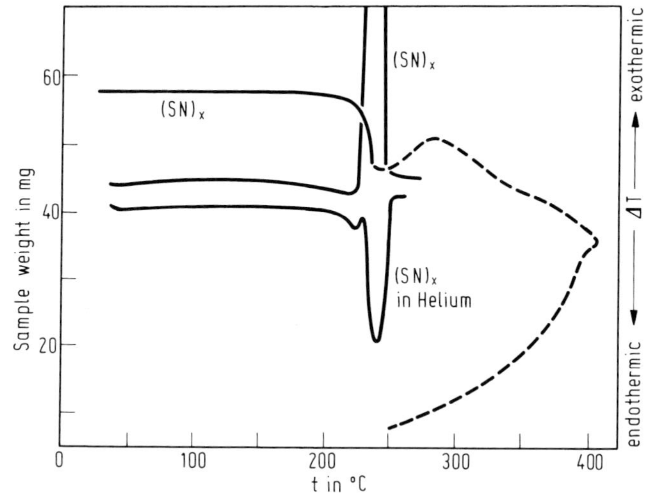

Fig. 45. Differential thermal analysis and thermogravimetric analysis of $(SN)_x$ in He and in air [1].

$(SN)_x$ is thermally stable between liquid nitrogen temperature and room temperature [1]. It is stable in vacuum and dry air up to about 120°C and in He atmosphere to about 125°C [3], where vaporization begins [4]. Other authors observed that $(SN)_x$ in vacuum begins to dissociate above 100°C [1]. Slow decomposition of the compound to S_8, N_2, and unidentified substances was observed when $(SN)_x$ was held at a temperature of 140 to 150°C in an evacuated vessel [5]. In an inert-gas atmosphere (He), endothermic decomposition starts at ca. 140°C for "normal $(SN)_x$" and at ca. 175°C for "red $(SN)_x$" [2]. The major endothermic peak was observed at 248°C for both "normal" and "red $(SN)_x$" [2] and at ~240°C for "normal $(SN)_x$" [1]. At 223°C another endothermic peak of minor intensity was detected [1]. Heating $(SN)_x$ in air led to a rapid decomposition at about 208°C [5] or explosions [1]. These observations are in accord with a very strong exothermic peak at ~240°C shown in Fig. 45 [1].

The thermal properties of chemically synthesized $(SN)_x$ powder were also studied using thermogravimetric analysis (TGA) under N_2, and differential scanning calorimetry (DSC) [6]. Times of total decomposition (TGA) for probes heated at 112, 118, 130, and 140°C are 3.5, 3.0, 1.0, and 0.5 h, respectively (weight loss started at ca. 90°C) [6]. The major thermal process is the irreversible exothermic conversion of $(SN)_x$ to S_4N_4 ($\Delta H = -(20 \pm 1)$ kJ/mol) according to DSC data [6].

Freshly prepared and outgassed $(SN)_x$ powder is photosensitive with rapid formation of S_4N_4, however when left in the evacuated vessel for several days in the darkness, the powder lost its photosensitivity, apparently owing to an aging process [6].

References:

[1] Street, G. B.; Greene, R. L. (IBM J. Res. Develop. **21** [1977] 99/110).
[2] Love, P.; Myer, G.; Kao, H. I.; Labes, M. M.; Junker, W. R.; Elbaum, C. (Ann. N.Y. Acad. Sci. **313** [1978] 744/58, 751/2).
[3] Pintschovius, L. (Colloid Polym. Sci. **256** [1978] 883/92).
[4] Street, G. B.; Arnal, H.; Gill, W. D.; Grant, P. M.; Greene, R. L. (Mater. Res. Bull. **10** [1975] 877/82).
[5] Mikulski, C. M.; Russo, P. J.; Saran, M. S.; MacDiarmid, A. G.; Garito, A. F.; Heeger, A. J. (J. Am. Chem. Soc. **97** [1975] 6358/63).
[6] Banister, A. J.; Hauptman, Z. V.; Passmore, J.; Wong, C.-M.; White, P. S. (J. Chem. Soc. Dalton Trans. **1986** 2371/9).

2.1.12.2 Stability to Air, Oxygen, and Humidity

$(SN)_x$ crystals, films, and powders are stable indefinitely in inert atmospheres or vacuum [1]. When $(SN)_x$ crystals are exposed to air at room temperature for several months a whitish gray powder covered the material [2]. An earlier work reported that $(SN)_x$ converts to SO_3^{2-}, $S_2O_3^{2-}$, and $S_3O_6^{2-}$. The same products are found when $(SN)_x$ is hydrolyzed by alkali hydroxides [3]. After exposure of $(SN)_x$ to atmospheric oxygen at 360 K, a decrease of the electrical conductivity occurred [4]. $(SN)_x$ crystals are not attacked during 7 d at room temperature by oxygen (dry or moist) at 1 atm pressure, by water vapor at a pressure of 4.6 Torr or by oxygen saturated with water vapor [2, 5]. $(SN)_x$ was also not attacked by degassed distilled water when completely immersed in it for 24 h at room temperature, however a very small amount of a white solid was observed after evaporation of the water [2,6]. The elemental analysis of the compound showed little difference before and after exposure [5]. $(SN)_x$ which contains unpolymerized S_2N_2 or other impurities is much more reactive to air [5]. $(SN)_x$ exposed for two weeks to the atmosphere showed no change in the X-ray diffraction pattern [6]. Surface oxygen contamination was reported in several papers [1, 2, 6 to 9], possibly owing to exposure to air or to reactions on glass surfaces [7]. The stability of $(SN)_x$ crystals against water is sufficient to permit use of $(SN)_x$ electrodes for cyclic voltammetric studies in aqueous solution [10]. High-quality crystals of "red $(SN)_x$" are stable in air at room temperature for a period of ca. five months. Not quite pure "brown $(SN)_x$" crystals decompose after two or fewer days in air at room temperature [11].

$(SN)_x$ powders prepared by chemical synthesis are less stable towards air or moisture than crystals [12, 13]. The powdered polymer is attacked by water or anhydrous formic acid [13]. Thin $(SN)_x$ films decompose readily [14]. Other workers declared that $(SN)_x$ films are stable for hundreds of days in dry air [15]. $(SN)_x$ films which had been exposed to air for two weeks showed a decreased conductivity and an increased thermopower [16]. The stability of $(SN)_x$ films to air was studied using ESCA (XPS). Wide-scan photoelectron spectra showed that pure $(SN)_x$ films are very slowly attacked by moist air. Films from powdered $(SN)_x$ samples are surface contaminated with (combined) oxygen after exposure to air [13]. Contamination by oxygen was also detected using X-ray photoelectron spectroscopy [8]. The degradation of epitaxial polymerized $(SN)_x$ (on various salt plates) by humidity was studied using Fourier-transform IR spectroscopy (FT-IR) [17]. Degradation of $(SN)_x$ is caused by water incorporated into S_2N_2 crystals before polymerization, and otherwise by atmospheric water. Infrared bands above the highest fundamental frequency of purer $(SN)_x$ ($\sim$1001 cm^{-1}, see Table 28, p. 129) were assigned at 1400 cm^{-1} to the NH group and at 1220 and 1090 cm^{-1} to two S=O groups. Spectroscopically pure $(SN)_x$ samples are expected in a vacuum of 10^{-11} Torr without the presence of a glass surface [17].

A study of stability of $(SN)_x$ vs. pressure up to about 40 kbar detected S_8 and nitrogen below 30 kbar and fibrous sulfur above that. A p/T diagram was presented [18].

References:

[1] Street, G. B.; Greene, R. L. (IBM J. Res. Develop. **21** [1977] 99/110).
[2] MacDiarmid, A. G.; Mikulski, C. M.; Saran, M. S.; Russo, P. J.; Cohen, M. J.; Bright, A. A.; Garito, A. F.; Heeger, A. J. (Advan. Chem. Ser. No. 150 [1976] 63/72).
[3] Becke-Goehring, M. (Ergebnisse und Probleme der Chemie der Schwefel-Stickstoffverbindungen, Akad.-Verlag, Berlin 1957, p. 21).
[4] Kaneto, K.; Tanimura, K.; Yoshino, K.; Inuishi, Y. (Solid State Commun. **22** [1977] 383/5).
[5] Mikulski, C. M.; MacDiarmid, A. G.; Garito, A. F.; Heeger, A. J. (Inorg. Chem. **15** [1976] 2943/5).
[6] Mikulski, C. M.; Russo, P. J.; Saran, M. S.; MacDiarmid, A. G.; Garito, A. F.; Heeger, A. J. (J. Am. Chem. Soc. **97** [1975] 6358/63).
[7] Smith, R. D.; Wyatt, J. R.; DeCorpo, J. J.; Saalfeld, F. E.; Moran, M. J.; MacDiarmid, A. G. (Chem. Phys. Letters **41** [1976] 362/4).
[8] Mengel, P.; Grant, P. M.; Rudge, W. E.; Schechtman, B. H.; Rice, D. W. (Phys. Rev. Letters **35** [1975] 1803/6).
[9] Walatka, V. V., Jr.; Labes, M. M.; Perlstein, J. H. (Phys. Rev. Letters **31** [1973] 1139/42).
[10] Novak, R. J.; Mark, H. B., Jr.; MacDiarmid, A. G.; Weber, D. (J. Chem. Soc. Chem. Commun. **1977** 9/11).

[11] Labes, M. M.; Love, P.; Nichols, L. F. (Chem. Rev. **79** [1979] 1/15, 6).
[12] Kennett, F. A.; MacLean, G. K.; Passmore, J.; Rao, M. N. (J. Chem. Soc. Dalton Trans. **1982** 851/7).
[13] Banister, A. J.; Hauptman, Z. V.; Passmore, J.; Wong, C.-M.; White, P. Y. (J. Chem. Soc. Dalton Trans **1986** 2371/9).
[14] Gill, W. D.; Clarke, T. C.; Street, G. B. (Appl. Phys. Commun. **2** [1982/83] 211/82, 212/4).
[15] Thomas, A. E.; Woods, J.; Hauptman, Z. V. (J. Phys. D **16** [1983] 1123/36, 1136).
[16] Beyer, W.; Gill, W. D.; Street, G. B. (Solid State Commun. **27** [1978] 343/8, 347).
[17] Ishida, H.; Rickert, S. E.; Hopfinger, A. J.; Lando, J. B.; Baer, E.; Koenig, J. L. (J. Appl. Phys. **51** [1980] 5188/93).
[18] Vezzoli, G. C.; Abel, J. E. (Mater. Res. Bull. **13** [1978] 151/8).

2.1.12.3 Reactions with Elements and Compounds

$(SN)_x$ forms intercalation compounds when exposed to Br_2 (see p. 181), I_2 (see p. 199), ICl (see p. 203), IBr (see p. 205), or immersed in a solution of Na^+ naphthalene$^{\bullet-}$ (see p. 206).

The reaction of $(SN)_x$ and BCl_3 at 20°C over a period of several days yields a material with a probable composition of $(S_2N_2BCl_3)_3$ on the basis of IR analysis. In contrast to the similar compound $(S_2N_2BCl_3)_x$, formed in the reaction of S_2N_2 with BCl_3, this material is not homogenous [1]. The reaction of $(SN)_x$ with C_4H_9Li in C_6H_{14} led to a product of the assumed composition $[SN(C_4H_9Li)_{0.2}]_x$. The Debye-Scherrer diagram is similar to that of an analogous electrochemically prepared compound, see p. 172 [2].

References:

[1] Patton, R. L.; Jolly, W. L. (Inorg. Chem. **8** [1969] 1392/5).
[2] Tarby, C.; Bernard, C.; Robert, G. (Electrochim. Acta **26** [1981] 663/70).

2.2 Modified Poly(sulfur nitride)

After the discovery of the extraordinary physical properties of $(SN)_x$, several modifications were prepared by reactions with halogen, interhalogens, and Na^+ naphthalene$^{\bar{\bullet}}$. The most widely studied compound is $(SNBr_{0.4})_x$. Marked changes in optical properties, an increase of the normal conductivity by an order of magnitude, and improved superconductivity characteristics were observed in comparison to pure $(SN)_x$. The incorporation of intercalates such as bromine into the $(SN)_x$ lattice increased the disorder, making a complete single crystal X-ray structure determination impossible. The properties of $(SNBr_{0.4})_x$ and other halogen-modified $(SN)_x$ materials are discussed in several review papers in connection with $(SN)_x$ [1 to 17].

Informations about the "hydride impurity" are given on p. 125.

References:

[1] Akhtar, M.; Chiang, C. K.; Cohen, M. J.; Heeger, A. J.; Kleppinger, J.; MacDiarmid, A. G.; Milliken, J.; Moran, M. J.; Peebles, D. L. (Organomet. Polym. Symp., New Orleans 1977 [1978], pp. 301/12).

[2] Akhtar, M.; Chiang, C. K.; Cohen, M. J.; Kleppinger, J.; Heeger, A. J.; Louis, E. J.; MacDiarmid, A. G.; Milliken, J.; Moran, M. J.; Peebles, D. L.; Shirakawa, H. (Ann. N.Y. Acad. Sci. **313** [1978] 726/36; AD-A041976 [1977] 1/24; C.A. **88** [1978] No. 44331; Solid State Commun. **23** [1977] 607/12).

[3] Street, G. B.; Etemad, S.; Geiss, R. H.; Gill, W. D.; Greene, R. L.; Kuyper, J.; Smith, R. D. (Ann. N.Y. Acad. Sci. **313** [1978] 737/44).

[4] Bernard, C.; Robert, G. (Bull. Soc. Chim. France **1978** 395/401, 398/9).

[5] Gill, W. D.; Bludau, W.; Geiss, R. H.; Grant, P. M.; Greene, R. L.; Street, G. B. (Phys. Rev. Letters **38** [1977] 1305/8).

[6] Gill, W. D.; Grant, P. M.; Geiss, R. H.; Greene, R. L.; Kwak, J. F.; Street, G. B. (Lect. Notes Phys. **96** [1979] 385/9).

[7] Street, G. B.; Gill, W. D. (NATO Conf. Ser. VI **1** [1978/79] 301/26, 306/26).

[8] Baughman, R. H.; Greene, R. L.; MacDiarmid, A. G.; Schmidt, M.; Zoos, Z.; Street, G. B.; Vogel, F. L., Jr. (NATO Conf. Ser. VI **1** [1978/79] 505/10).

[9] Street, G. B.; Clarke, T. C. (Advan. Chem. Ser. No. 186 [1980] 177/94).

[10] Gill, W. D.; Clarke, T. C.; Street, G. B. (Appl. Phys. Commun. **2** [1982/83] 211/82, 224/36).

[11] Marks, T. J.; Kalina, D. W. (in: Miller, J. S., Extended Linear Chain Compounds I, Plenum, New York 1982, pp. 197/331, 301/7).

[12] Kaneto, K.; Yoshino, K.; Inuishi, Y. (in: Monceau, P., Electronic Properties of Inorganic Quasi-One-Dimensional Compounds, Pt. II, Reidel, Dordrecht 1985, pp. 69/110).

[13] Street, G. B.; Clarke, T. C.; Wolmershäuser, G. (Org. Coat. Plast. Chem. Preprints **38** [1978] 623/4).

[14] Labes, M.; Love, P.; Nichols, L. F. (Chem. Rev. **79** [1979] 1/15, 12/13).

[15] Street, G. B.; Clarke, T. C. (IBM J. Res. Develop. **25** [1981] 51/7).

[16] Roch, K. M. (Macromol. Chem. [London] **2** [1982] 162/73, 170/3).

[17] Kamimura, H. (Theoretical Aspects of Band Structures and Electronic Properties of Pseudo-One-Dimensional Solids, Reidel, Dordrecht 1985, pp. 123/62).

2.2.1 Halogenated Poly(sulfur nitride), $(SNHal_y)_x$; Hal = Br, I, ICl, IBr

2.2.1.1 Brominated Poly(sulfur nitride), $(SNBr_y)_x$; y ≈ 0.04, 0.07, 0.25, 0.33, 0.4, 0.5, 0.55, 0.56, 0.6, 1.5

2.2.1.1.1 Preparation

Brominated $(SN)_x$ was prepared by treating crystals or films of $(SN)_x$ and S_4N_4 with bromine vapor. Samples of different bromine content were obtained by pumping these materials under corresponding experimental conditions. (It was not possible to obtain pure $(SN)_x$ in this way.) The bromine content was determined by the weight change of $(SN)_x$ or by elemental analysis.

Upon bromination, the $(SN)_x$ crystal swelled ~50% in volume. The color changed from a golden luster to blue-black, depending on the bromine content [1, 2]. A long exposure to bromine vapor caused the $(SN)_x$ crystal to burst and become brittle [3].

Bromine intercalation compounds are obtained only by reaction of bromine vapor with $(SN)_x$. Liquid bromine reacts with $(SN)_x$ to form $S_4N_3Br_3$ [1]. The first formation of a brominated poly(sulfur nitride) by addition of bromine vapor to $(SN)_x$ was reported by a french group [4]. The composition of the material, however, was not noted. Independently of [4] other groups published a route to polythiazylbromides, also by treatment of $(SN)_x$ crystals with bromine vapor [2, 5 to 9]. Some of the experimental conditions were presented in detail. $(SN)_x$ crystals were exposed to bromine vapor (dried over $CaCl_2$ and freed of HBr by treatment with anhydrous copper sulfate) at room temperature for about 24 h [5, 6]. Blue-black fibrous crystals with a composition of $(SNBr_{0.5})_x$ were obtained. Using similar experimental conditions, materials with compositions between $(SN)_x$ and $(SNBr_{0.6})_x$ were prepared [6].

The bromination of S_4N_4 with bromine was studied for the first time at the end of the 19th century [10, 11], see also "Schwefel" B 3, 1963, p. 1548. More recent studies [1, 6, 7, 12 to 15] were compiled in "Sulfur-Nitrogen Compounds" Part 2, 1985, pp. 165/6. As with $(SN)_x$, intercalation compounds are formed from solid S_4N_4 only by the reaction of bromine vapor. Treatment with liquid bromine yields $S_4N_3Br_3$ [1].

Preparation of $(SNBr_{0.07})_x$ and $(SNBr_{0.04})_x$

If brominated S_4N_4, $(SNBr_{0.4})_x$, is heated at 80°C for 18 h under reduced pressure (10^{-5} Torr) with continuous pumping, conducting films and a copper-bronze-colored material $(SNBr_{0.07})_x$ are formed. Heating brominated $(SN)_x$ under similar conditions at 135° also leads to the formation of a colored, conducting substance $(SNBr_{0.04})_x$. From this result it was concluded that brominated $(SN)_x$ is structurally very similar to brominated S_4N_4 [12].

Preparation of $(SNBr_{0.25})_x$

A copper-colored material of the composition $(SNBr_{0.25})_x$ is formed when crystals of $(SNBr_{0.4})_x$, obtained by the direct bromination of $(SN)_x$, were heated in vacuum at 86°C for ca. 30 h [2, 7]. Heating $(SNBr_{0.4})_x$ powder, prepared by bromination of S_4N_4 at 80°C for 4 h with constant pumping, yields a dark copper-colored powder of the same composition as the material above. The X-ray powder diagram of both materials is almost identical [7, 13, 14].

Preparation of $(SNBr_{0.33})_x$

Pumping of crystals of $(SNBr_{0.4})_x$ (from $(SN)_x$ and Br_2 vapor) at room temperature for 45 h or more leads to a material of composition $(SNBr_{0.33})$. Little change in appearance was observed [7].

Preparation of $(SNBr_{0.4})_x$

$(SNBr_{0.4})_x$ is obtained by evacuating $(SNBr_{0.5})_x$ (from $(SN)_x$ and Br_2) at 10^{-5} Torr for 30 min at ambient temperature [5, 6]. A similar procedure was published by another group. Analytically pure $(SN)_x$ crystals (ca. 1 to 10 mm^3) were allowed to react in 60 Torr of bromine vapor at room temperature. The initial product $(SNBr_{0.55})_x$ was pumped for 5 to 15 min to give shiny black crystals having a blue-purple tinge. The value of y in $(SNBr_y)_x$ is 0.38 to 0.42 [2, 7]. Other groups reported slightly different experimental conditions for the formation of $(SNBr_{0.4})_x$ [8, 9]. $(SN)_x$ needles of high purity yield tarnished grayish blue $(SNBr_{0.4})_x$ on exposure to bromine vapor for 0.5 h in a desiccator at 173 Torr [8]. The intercalation of $(SN)_x$ with Br_2 to give $(SNBr_{0.4})_x$ was also performed in an evacuated glass flask during 5 min at room temperature at vapor pressure equilibrium. The brominated material was desorbed for 2 h in vacuum [9]. When S_4N_4 was treated with Br_2 vapor at room temperature, after pumping the initial intercalation product $(SNBr_{1.5})_x$ in vacuum, a composition of $(SNBr_{0.4})_x$ was achieved [12 to 14]. It was shown by several analytical methods (IR, Raman, mass spectroscopy, thermal analysis) that the compound is identical to the bromination product of $(SN)_x$, however, the bromination product of S_4N_4 is poorly crystalline [1]. The reaction of $S_3N_3Cl_3$ and $BrSi(CH_3)_3$ in CH_2Cl_2 at $-60°C$ yields $(SNBr_{0.4})_x$ as a black-brown powder (yield 89%). The mass spectrum is in accord with that of samples formed by bromination of $(SN)_x$ or S_4N_4 [16].

Preparation of $(SNBr_{0.5})_x$, $(SNBr_{0.55})_x$, and $(SNBr_{0.56})_x$

The title compounds were detected as the initial products in the reaction of $(SN)_x$ with bromine vapor. In vacuum they lose bromine and $(SNBr_{0.4})_x$ forms [1, 15].

Preparation of $(SNBr_{1.5})_x$

The title compound was detected as the initial product in the bromination of S_4N_4 at room temperature with approximately 200 Torr of bromine [13 to 15], see also "Sulfur-Nitrogen Compounds" Part 2, 1985, p. 165.

References:

[1] Street, G. B.; Gill, W. D. (NATO Conf. Ser. VI **1** [1978/79] 301/26).
[2] Akhtar, M.; Kleppinger, J.; MacDiarmid, A. G.; Milliken, J.; Moran, M. J.; Chiang, C. K.; Cohen, M. J.; Heeger, A. J.; Peebles, D. L. (J. Chem. Soc. Chem. Commun. **1977** 473/4).
[3] Kaneto, K.; Yoshino, K.; Inuishi, Y. (in: Monceau, P., Electronic Properties of Inorganic Quasi-One-Dimensional Compounds, Pt. II, Reidel, Dordrecht 1985, pp. 69/110, 74).
[4] Bernard, C.; Herold, A.; Lelaurain, M.; Robert, G. (Compt. Rend. C **283** [1976] 625/7).
[5] Street, G. B.; Gill, W. D.; Geiss, R. H.; Greene, R. L.; Mayerle, J. J. (J. Chem. Soc. Chem. Commun. **1977** 407).
[6] Street, G. B.; Etemad, S.; Geiss, R. H.; Gill, W. D.; Greene, R. L.; Kuyper, J.; Smith, R. D. (Ann. N.Y. Acad. Sci. **313** [1978] 737/44).
[7] Akhtar, M.; Chiang, C. K.; Cohen, M. J.; Kleppinger, J.; Heeger, A. J.; Louis, E. J.; MacDiarmid, A. G.; Milliken, J.; Moran, M. J.; Peebles, D. L.; Shirakawa, H. (Ann. N.Y. Acad. Sci. **313** [1978] 726/36; Solid State Commun. **23** [1977] 607/12; AD-A041976 [1977] 1/24; C.A. **88** [1978] No. 44331).
[8] Philipp, A.; Seeger, K. (Phys. Status Solidi B **89** [1978] 187/92).
[9] Petermann, J.; Wagner, N.; Young, R. (Colloid Polym. Sci. **259** [1981] 247/51).
[10] Clever, A.; Muthmann, W. (Ber. Deut. Chem. Ges. **29** [1896] 340/3).

[11] Goehring, M. (Ergebnisse und Probleme der Chemie der Schwefelstickstoffverbindungen, Akademie-Verlag, Berlin 1957, pp. 68/9).

[12] Street, G. B.; Bingham, R. L.; Crowley, J. I.; Kuyper, J. (J. Chem. Soc. Chem. Commun. **1977** 464/5).
[13] Akhtar, M.; Chiang, C. K.; Heeger, A. J.; MacDiarmid, A. G. (J. Chem. Soc. Chem. Commun. **1977** 846).
[14] Akhtar, M.; Chiang, C. K.; Heeger, A. J.; Milliken, J.; MacDiarmid, A. G. (Inorg. Chem. **17** [1978] 1539/42).
[15] Monteil, Y.; Vincent, H.; Berthet, M. P. (Ann. Chim. [Paris] [15] **5** [1980] 431/41, 435/6).
[16] Demant, U.; Dehnicke, K. (Z. Naturforsch. **41 b** [1986] 929/30).

2.2.1.1.2 Structure

Bromination of $(SN)_x$ crystals increases disorder, making a complete single-crystal X-ray structure analysis impossible [1 to 3]. The assumption of additional microtwinning in brominated $(SN)_x$, however, could not be confirmed by a neutron scattering study [6]. Structural information about $(SNBr_{0.4})_x$ was obtained from X-ray [1, 3, 4, 8], electron [1, 5, 8], neutron diffraction [6], and also from EXAFS (Extended X-Ray Absorption Fine Structure) measurements [9, 10]. The first X-ray powder diffraction study showed, in agreement with a neutron diffraction study [6], that the repeat unit of $(SN)_x$ in the chain direction did not change on bromination [4]. Other workers reported that the $(\bar{1}02)$ reflection of $(SN)_x$ was broadened, but unshifted, indicating that the a and c lattice constants remained unchanged [1]. Only (0 k 0) reflections were observed in X-ray precession photographs of individual crystals of $(SNBr_{0.4})_x$ [1]. It was reported in another paper that at low-bromine concentration, $((SNBr_{0.049})_x)$, the observed X-ray powder reflections were consistent with a unit cell with parameters like those of unbrominated $(SN)_x$ [3]. In heavily brominated specimens (y = 0.25 and above) the unit cell parameters change. The important changes are: a decrease in the interplanar spacing for the closest packed plane in unbrominated $(SN)_x$, the $10\bar{2}$ plane; a decrease in the a axis length; an increase in c-axis length. Table 43, p. 184, compiles X-ray diffraction results of $(SNBr_y)_x$ compounds. An exact determination of lattice parameters could not be made from neutron diffraction results. The unit cell dimensions perpendicular to the chains are considerably greater than in pristine $(SN)_x$ [6], however, the chain axis length b is unchanged, in agreement with [4, 6] (b in $(SNBr_{0.4})_x$ is 4.431 Å and in $(SN)_x$ 4.429 Å [3]). Patterns typical for $(SN)_x$ were observed in electron-diffraction studies of $(SNBr_{0.4})_x$, however, streaking occurred perpendicular to b* which is associated with disorder [1, 2, 5]. In agreement with X-ray results, it was concluded that the $(SN)_x$ lattice remained unchanged. A fiber diameter of 30 Å in $(SNBr_{0.4})_x$ was determined from the width of the diffraction spots and from dark field electron micrographs which is in good agreement with a value of ~20 Å obtained from a neutron diffraction study [6] (for comparison: fiber diameter in $(SN)_x$: ~70 Å [2]). In addition to the $(SN)_x$-like diffraction pattern, diffuse lines perpendicular to b* at (0, k/2, 0) for all k and very diffuse, streaked reflections intersecting the b* axis slightly beyond the (0, k/2, 0) reflections for k = 3, 5, 7 were observed [5]. Bragg spacings of the extra diffuse streaks of brominated $(SN)_x$ observed by electron and X-ray diffraction method are compiled in a table [8]. The (0 k/2 0) reflections were attributed to a one-dimensional commensurate superlattice, oriented along the chain axis with a repeat unit twice that of pristine $(SN)_x$. An interesting temperature behavior was reported. The 2 b lines disappear at temperatures less than 145 K but the very diffuse streaks remained. The behavior was reversible. On heating the specimen above 410 K the diffuse streaks disappeared and the superlattice lines remained. This behavior was irreversible. The superlattice lines were confirmed by diffuse X-ray scattering [2, 7]. In the neutron scattering study they could not be detected [6]. The same study showed that there is a larger mosaic spread after bromination than in unbrominated $(SN)_x$. The small angle X-ray scattering which was observed for unbrominated $(SN)_x$ essentially disappeared in $(SNBr_{0.4})_x$. Bromination largely eliminates density fluctuations [3].

Table 43
X-Ray Diffraction Results on $(SNBr_y)_x$ [3].

y	diffraction technique[a]	a in Å	c in Å	β in °	d_{100} in Å	d_{002} in Å	$d_{10\bar{2}}$ in Å
0.049	P	—	—	—	3.873[b]	3.657[b]	3.231
0.25[c]	S	3.90	7.81	109.5	3.673	3.680	3.186
0.25[c,d]	S	3.92	7.74	109.5	3.692	2.649	3.179
0.27[c]	P	—	—	—	← 3.633[e] →		3.160
0.39	P	—	—	—	← 3.633[e] →		3.155
0.40	S	3.86	7.72	110.3	3.622	3.620	3.167
0.40	G	—	—	—	← 3.670[e] →		3.153

[a] S indicates single crystal data, P indicates powder diffraction data, and G indicates Gandolfi camera data. – [b] Poorly resolved reflections, spacings inaccurate. – [c] Obtained by thermally annealing $(SNBr_{0.40})_x$ while pumping at 85°C. – [d] Second crystal from same bromination run as for the above data. – [e] These reflections are unresolved.

Diffraction [3 to 5], Raman (in combination with IR) (see pp. 188/90), and mass spectrometric [1] studies showed that bromine is incorporated as Br_2 and Br_3^- species both in brominated $(SN)_x$ and S_4N_4. Paramagnetic species such as Br_2^- were not detected using magnetic susceptibility measurements [11]. X-ray absorption measurements around the Br Kα edge, utilizing synchrotron radiation were made to clarify the nature of intercalated bromine. The oscillatory part of the X-ray absorption spectra and the Fourier transform at 300, 170, and 86 K [9], and at 4.2 K [10] are presented. Br_3^- (or a more extended chain with a nearest neighbor spacing of 2.5 Å) was detected as the dominant species, increasing in intensity with decreasing temperature. On the basis of the observed polarization of the Raman peaks [3], from an EXAFS study with X-ray polarization perpendicular to the b axis [9, 10] and from ε_2 data (imaginary part of the dielectric function) [12], a parallel orientation of the bromine species relative to the $(SN)_x$ chains was evaluated. The linear Br_3^- ions which are ordered linearly along the $(SN)_x$ chains are responsible for the 2b superlattice seen in the X-ray and electron diffraction studies. Adsorbed Br_2 species are the origin of the observed diffuse streaks [5]. Bromine may be incorporated on the surface of the fibers and in the fibers themselves (occupying $(SN)_x$ chain sites) [2, 6]. A neutron scattering study indicates that the amount of bromine species which enter the $(SN)_x$ fibers is probably greater than assumed in earlier papers (~10% in [3]). There is also evidence for areas in brominated $(SN)_x$ which are scarcely modified with regard to pristine material. Brominated $(SN)_x$ may consist of two phases, the outerparts of the fibers with regularly incorporated bromine and the bromine free fiber cores (thinner films are believed to be brominated throughout) [6]. Br_3^- and Br_2 species could not be identified by IR and mass spectrometry in a $(SNBr_{0.4})_x$ sample prepared from $S_3N_3Cl_3$ and $BrSi(CH_3)_3$ [13].

References:

[1] Gill, W. D.; Bludau, W.; Geiss, R. H.; Grant, P. M.; Greene, R. L.; Mayerle, J. J.; Street, G. B. (Phys. Rev. Letters **38** [1977] 1305/8).

[2] Street, G. B.; Gill, W. D. (NATO Conf. Ser. VI **1** [1978/79] 301/26, 310, 311, 319).

[3] Iqbal, Z.; Baughman, R. H.; Kleppinger, J.; MacDiarmid, A. G. (Ann. N.Y. Acad. Sci. **313** [1978] 775/87).
[4] Bernard, C.; Herold, A.; Lelaurian, M.; Robert, G. (Compt. Rend. C **283** [1976] 625/7).
[5] Geiss, R. H.; Thomas, J.; Street, G. B. (Syn. Metals **1** [1979/80] 257/65).
[6] Pintschovius, L.; Kobbelt, M. (Syn. Metals **6** [1983] 25/9).
[7] Comes, R.; Megtert, S.; Pouget, J. P.; Schegoley, I. F.; Temkin, H. (Bull. Am. Phys. Soc. [2] **23** [1978] 424).
[8] Kawaguchi, A.; Isoda, S.; Petermann, J.; Katayama, K. (Colloid Polym. Sci. **263** [1985] 631/7).
[9] Morawitz, H.; Gill, W. D.; Grant, P.; Street, G. B.; Sayers, D. (Lect. Notes Phys. **96** [1979] 390/407).
[10] Morawitz, H.; Bagus, P.; Clarke, T.; Gill, W.; Grant, P.; Street, G. B.; Sayers, D. (Syn. Metals **1** [1979/80] 267/78).

[11] Scott, J. C.; Kulick, J. D.; Street, G. B. (Solid State Commun. **28** [1978] 723/7).
[12] Mitani, T.; Koyama, K.; Mori, H.; Koda, T.; Nakada, I. (J. Phys. Soc. Japan **51** [1982] 3197/205).
[13] Demant, U.; Dehnicke, K. (Z. Naturforsch. **41b** [1986] 929/30).

2.2.1.1.3 Density

The density of $(SNBr_{0.4})_x$ is 2.65 g/cm^3 (flotation method) (cf. $(SN)_x$: 2.32 g/cm^3). The volume perpendicular to the chain axis is expanded by ca. 50% compared to pristine $(SN)_x$.

Reference:

Street, G. B.; Gill, W. D. (NATO Conf. Ser. VI **1** [1978/79] 301/26, 306).

2.2.1.1.4 X-Ray Photoelectron Spectroscopy (XPS)

Valence band XPS spectra in the region 0 to 30 eV at 20°C were recorded for brominated $(SN)_x$ films with composition of $(SNBr_{0.25})_x$ and $(SNBr_{0.6})_x$ prepared by bromination of $(SN)_x$ and S_4N_4 [1, 2]. N (1s) and S ($2p_{1/2,\,3/2}$) core level spectra of brominated S_4N_4 films [1 to 3] and brominated $(SN)_x$ [3] together with Br (3s) and Br ($3d_{3/2,\,5/2}$) core level spectra of brominated $(SN)_x$ films via S_4N_4 [1, 2] were presented. The valence band spectra of brominated $(SN)_x$, and brominated S_4N_4 films are identical, implying that both materials are identical in electronic structure [1, 2]. The data in the valence band region of brominated $(SN)_x$ films are shifted to higher binding energies in comparison to normal $(SN)_x$ [2]. The peaks at 4.1 and 26.8 eV were assigned to the bromine $4p_{1/2,\,3/2}$ and 4s level, respectively [1]. The Br 4s line, unlike that in NaBr or CsBr, is more intense than the corresponding Br $4p_{1/2,\,3/2}$ doublet [1]. This fact was attributed to interaction between the intercalated bromine species and the $(SN)_x$ chain. The upper bromine level binding energies of brominated $(SN)_x$ are 254.9(3s), 188.2($3p_{1/2}$), 181.3($3p_{3/2}$), 68.2($3d_{3/2,\,5/2}$), 25.9(4s), and 3.0($4p_{1/2,\,3/2}$) eV [1]. The N (1s) and S ($2p_{1/2,\,3/2}$) core levels in brominated S_4N_4 films shift to higher binding energies by ca. 1.5 eV [1, 2]. This result is in contrast to a study showing more complicated, broad N (1s) and S (2p) envelopes than in [1, 2] and with no shift of maxima lines in comparison to normal $(SN)_x$ [3]. The differences in the results are attributed to the use of different spectrometers and to samples with different film compositions [3]. These groups also disagree in the assignment of the bromine levels (3d). The detection of a Br^- species is ambigous [1 to 3]. Experimental (and calculated) core level intensity ratios are 2.07 ± 0.20 (1.12), AlKα (1.26 MgKα) for S 2p : N 1s and 3.54 ± 0.30 (1.98) AlKα for Br 3d : N 1s. The changes, in comparison to data of $(SN)_x$, are interpreted by a model for bromine incorporation into the polymer in which the bromine resides on the outside of the $(SN)_x$ fibrils [3].

References:

[1] Sharma, J.; Iqbal, Z. (Chem. Phys. Letters **56** [1978] 373/6).

[2] Iqbal, Z.; Sharma, J.; Baughman, R. H.; Akhtar, M.; MacDiarmid, A. G. (Lect. Notes Phys. **96** [1979] 432/6).

[3] Brant, P.; Weber, D. C.; Ewing, C. T.; Carter, F. L. (Syn. Metals **1** [1979/80] 161/73).

2.2.1.1.5 Optical Properties

Reflectance Spectra

The reflectance spectra of brominated $(SN)_x$ for light polarized parallel (E ∥ b) and perpendicular (E ⊥ b) to the chain axis were measured in the Drude range [1 to 8], in one case [7] up to hν = 25 eV, and compared with that of pristine $(SN)_x$ [8], see **Fig. 46**.

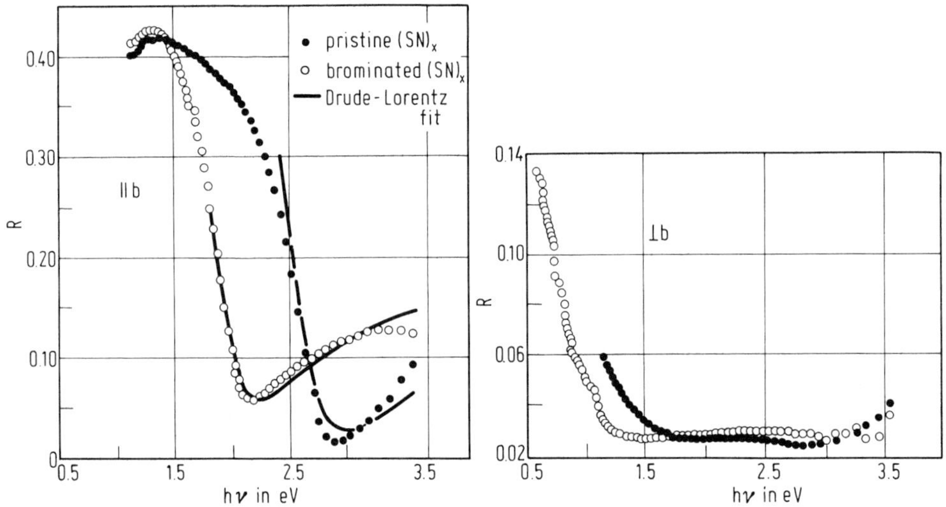

Fig. 46. Polarized reflectance of the same $(SN)_x$ crystal before and after bromination to $(SNBr_{0.4})_x$ [1].

The plasma edge is red shifted out of the visible region. Some workers observed no plasma edge for light polarized perpendicular to the chain axis [2 to 4]. They concluded that interchain coupling is weaker in $(SNBr_{0.4})_x$ than in $(SN)_x$. This assumption is in contrast to absorption spectra results [7], as well as conductivity and superconductivity behavior, see e.g. p. 193. The measured R(hν) curves were analyzed by a Drude-Lorentz fit [1 to 6]. Results are compiled in Table 44.

Table 44

Drude Parameters of Brominated $(SN)_x$, $(SNBr_y)_x$: plasma energy $\hbar\omega_p$, dielectric constant $\varepsilon_{core} \triangleq \varepsilon_\infty$, relaxation time of the free carrier τ, optical conductivity σ_{opt}, effective electron mass m*, carrier density n [6].

	y	ε_{core}	ω_p in 10^{15} s^{-1}	$\hbar\omega_p/\varepsilon_{core}^{1/2}$ in eV	τ in 10^{-15} s	$n/(m^*/m_e)$ in 10^{22} cm^{-3}	σ_{opt} in 10^3 $\Omega^{-1}\cdot cm^{-1}$	Ref.
	0.25	5.0	7.1	2.1	0.92	1.6	3	[5]
E∥b	0.4	6.48	7.6	2.0	1.2	1.8	5.5	[1]
	0.4	8.7	7.3	1.6	1.6	1.7	6.8	[2 to 4]
E⊥b	0.25	4.5	2.3	0.7	1.1	0.16	0.48	[5]

The values show that $\hbar\omega_p$ and σ_{opt} (for E∥b) of $(SN)_x$ changed only a little on bromination [1]. Therefore the spectral shift is due to an increase in the background dielectric constant presumably associated with included Br_2 molecules [1] and in a change of carrier density [5, 8]. One paper presented the ε_2 curve for brominated $(SN)_x$ derived from reflectivity data [7]. The spectrum is very similar to that of pristine $(SN)_x$ except for several additional features. The most prominent of them, appearing at ca. 3 (broad band) and 6 eV (sharp peak) predominantly for E∥b, are assigned to intramolecular transitions within Br_3^- ions aligned along the chain axis. The peak at 3 eV was also observed in polarized absorption spectra measured on oriented thin films of brominated $(SN)_x$ [7].

References:

[1] Gill, W. D.; Bludau, W.; Geiss, R. H.; Grant, P. M.; Greene, R. L.; Mayerle, J. J.; Street, G. B. (Phys. Rev. Letters **38** [1977] 1305/8).

[2] Akhtar, M.; Chiang, C. K.; Cohen, M. J.; Kleppinger, J.; Heeger, A. J.; Louis, E. J.; MacDiarmid, A. G.; Milliken, J.; Moran, M. J.; Peebles, D. L.; Shirakawa, H. (Solid State Commun. **23** [1977] 607/12).

[3] Akhtar, M.; Chiang, C. K.; Cohen, M. J.; Kleppinger, J.; Heeger, A. J.; Louis, E. J.; MacDiarmid, A. G.; Milliken, J.; Moran, M. J.; Peebles, D. L.; Shirakawa, H. (Ann. N.Y. Acad. Sci. **313** [1978] 726/36).

[4] Akhtar, M.; Chiang, C. K.; Cohen, M. J.; Kleppinger, J.; Heeger, A. J.; Louis, E. J.; MacDiarmid, A. G.; Milliken, J.; Moran, M. J.; Peebles, D. L.; Shirakawa, H. (AD-A041976 [1977] 1/24; C.A. **88** [1978] No. 44331).

[5] Yoshino, K.; Tanimura, K.; Yamamoto, Y.; Kaneto, K.; Inuishi, Y. (Technol. Rept. Osaka Univ. **28** [1978] 437/45).

[6] Kaneto, K.; Yoshino, K.; Inuishi, Y. (in: Monceau, P., Electrical Properties of Inorganic Quasi-One-Dimensional Compounds, Pt. II, Reidel, Dordrecht 1985, pp. 69/110, 103/6).

[7] Mitani, T.; Koyama, K.; Mori, H.; Koda, T.; Nakada, I. (J. Phys. Soc. Japan **51** [1982] 3197/205).

[8] Yoshino, K.; Kaneto, K.; Inuishi, Y. (Japan. J. Appl. Phys. II **22** [1983] L157/L158).

IR Spectrum

The transmission [1] and absorbance [2] IR (Fourier transform) spectra of brominated $(SN)_x$, $(SNBr_y)_x$, were recorded for films on KBr [1] and CsI [2] disks in the region 1000 to 400 cm^{-1} [1], 1100 to 400 cm^{-1} [2], and 500 to 150 cm^{-1} [2] at 90 and 300 K [2]. Table 45, p. 188, shows IR frequencies together with Raman frequencies (see p. 189) of brominated $(SN)_x$ and suggested assignments. The corresponding normal modes of $(SN)_x$ are illustrated in **Fig. 47**, p. 188.

Table 45

IR and Raman Frequencies of Brominated $(SN)_x$, $(SNBr_y)_x$ (y = 0,4?), [2].
(Frequencies in parentheses vary in relative intensity in spectra of different $(SN)_x$ samples; frequencies listed are those observed at 300 K.)

frequencies in cm^{-1} IR	Raman [5]	suggested assignments by class in symmetry group C_2 and approximate description
1018	~1010	ν_1A sym. str.
968 vw	—	$2\times\nu_3$ and $\nu_4+\nu_2$
830	~ 840	ν_5B antisym. str.
688	~ 670	ν_2A sym. str.
634	—	ν_6B in-plane deform.
550 sh	—	$2\times\nu_4$
468	~ 480	ν_3A in-plane deform.
(434)	~ 420	?
(386)	~385	ν_7B in-plane deform.
(362)	~350	ν_8B out-of-plane deform.
(320)	—	$2\times Br_3^-$ sym. str.
300	—	ν_4A out-of-plane deform.
—	222 vs	Br_2 str.
189	190	Br_3^- antisym. str.
158	154	Br_3^- sym. str.
140	130	Br_3^- sym. str.

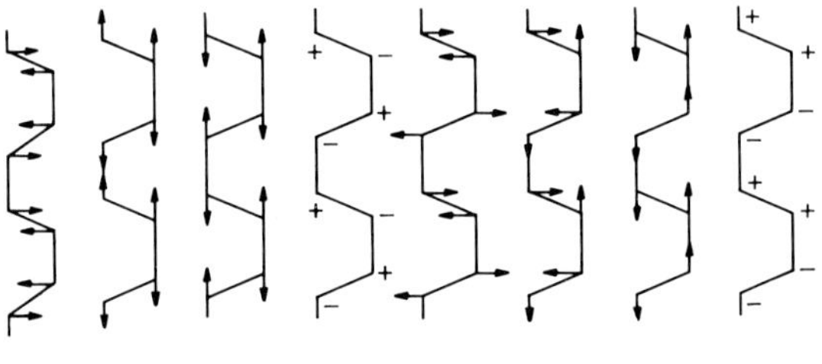

Fig. 47. Schematic of $(SN)_x$ normal modes [2].

An IR (Fourier transform) transmission spectrum of $(SNBr_{0.4})_x$ powder (from $S_3N_3Cl_3$ + $BrSi(CH_3)_3$) in Nujol was studied in the region 1500 to 400 cm^{-1}. (The region 400 to 200 cm^{-1} was recorded without Fourier transformation.) The data do not agree well with those above [3]. The absorptions (in cm^{-1}) at 1021(m), 849(vs), 704(vs), and 644(vs) are assigned to ν(S–N), the band at 475(vs) to $\delta(SN_2)$, the absorption at 1404(w) to 2×704, that at 1115 cm^{-1} to 644+475. The band at 305 cm^{-1} (vw) is assigned to ν(S–Br), assuming no loosely bound Br_2 molecules but S–Br bonds in the lattice [3]. The IR spectra of pristine $(SN)_x$ were compared to those of brominated S_4N_4 and $(SN)_x$ [1, 2]. The IR spectra of brominated $(SN)_x$ and pristine $(SN)_x$ are very similar, implying that the $(SN)_x$ chain remains nearly the same on bromination. The bands are broader and some are shifted upward by 20 to 60 cm^{-1} in comparison to $(SN)_x$. This is a measure of interaction between the $(SN)_x$ lattice and bromine [4]. The increase in frequency of SN stretching modes was explained by the oxidation of the $(SN)_x$ chain on bromination [4]. The IR spectrum of brominated S_4N_4 is also similar to that of brominated $(SN)_x$, indicating that polymerized sulfur-nitrogen chains exist. Much information about the structure was obtained from the interpretation of the IR spectral data of $(SNBr_y)_x$ below 400 cm^{-1} (see also p. 184). The band at 189 cm^{-1} and the bands at 158 and 140 cm^{-1} were assigned to one antisymmetric and two symmetric vibrations of the Br_3^- species, respectively. The number of bands observed in the IR and also in the Raman spectrum was interpreted with the assumption that Br_3^- (and Br_2) species are distributed among various environments in the disordered $(SN)_x$ lattice [2].

References:

[1] Street, G. B.; Etemad, S.; Geiss, R. H.; Gill, W. D.; Greene, R. L.; Kuyper, J.; Smith, R. D. (Ann. N.Y. Acad. Sci. **313** [1978] 737/44).
[2] Macklin, J. W.; Street, G. B.; Gill, W. D. (J. Chem. Phys. **70** [1979] 2425/30).
[3] Demant, U.; Dehnicke, K. (Z. Naturforsch. **41b** [1986] 929/30).
[4] Street, G. B.; Clarke, T. C. (Advan. Chem. Ser. No. 186 [1980] 177/94).
[5] Temkin, H.; Street, G. B. (Solid State Commun. **26** [1978] 455/60).

Raman Spectrum

A short review of the Raman spectroscopic studies on $(SNBr_y)_x$ is given in [8]. Raman spectra were recorded using a 514.5 nm laser line between 100 and 400 (350) cm^{-1} at 295 K: 1) from a freshly brominated $(SN)_x$ crystal of composition $(SNBr_{0.55})_x$ [1, 2]; 2) from the same crystal after pumping for 15 min [2], 1 h [2], or 2 h in a vacuum of 10^{-4} Torr [1]; 3) from a blue-black $(SNBr_{0.4})_x$ crystal which had been stored in a 10^{-2} to 10^{-3} Torr vacuum for 1 to 2 months [1]; 4) from copper-colored $(SNBr_{0.27})_x$ and $(SNBr_{0.25})_x$ crystals [1 to 3]. Also Raman spectra in the range 1100 to 300 cm^{-1} (using 488.0 nm excitation) [2] and two powder spectra of a sample prepared via S_4N_4 (using 488.0 and 632.0 nm excitation) [3] were measured. Polarized Raman spectra of brominated $(SN)_x$ crystals ($((SNBr_{0.4})_x)$) in the range 1100 to 50 cm^{-1} were recorded at 300 K (laser at 611.0 nm), at 80 K (laser at 457.9 nm [4, 5] and 611.6 nm [5]), and of $(SNBr_{0.4})_x$ and $(SNBr_{0.29})_x$ at 2 K (laser at 611.0 nm) [4, 6]. Other workers investigated $(SNBr_{0.4})_x$ crystals by polarized Raman spectra at temperatures of 15, 98, 146, 196, 245, and 268 K [1, 2] (laser at 514.5 nm) in the range ca. 400 to 50 cm^{-1}. The temperature dependence of the integrated intensities and of the line-width at half maximum for the 154 and 305 (304) cm^{-1} lines of an $(SNBr_{0.4})_x$ and of an $(SNBr_{0.27})_x$ crystal was studied [1 to 3]. The Raman spectrum of $(SNBr_{0.4})_x$ consists of weak bands at high frequencies quite similar to the LO modes of pristine $(SN)_x$, and two new Raman bands which have been only observed in brominated $(SN)_x$ and S_4N_4 at ca. 150 cm^{-1} (152 cm^{-1} [3], 154 cm^{-1} [1, 2, 4]) and at ca. 230 cm^{-1} [1 to 3] (222 cm^{-1} [4]). The broad band at 310 cm^{-1} [4] (304 cm^{-1} [1, 2]) was assigned to a second harmonic of the 154 cm^{-1} line. Further overtones of the 154 cm^{-1} line were observed at 315, 473, 628, 792, and 942 ($\pm$2) cm^{-1} at 80 K [4, 5], and at 305, 473, 638, and 788 cm^{-1} [1, 2] for the 457.9 and 488.0 nm laser lines, respectively. It was noted that the observed features at ~150 and 230 cm^{-1} are strongly polarized and are only seen with both incident and scattered light polarized in the direction of the polymer chains (α_{yy}) [4]. The weak and broad bands at 480, 670, and 840 cm^{-1} (450, 660, 840 cm^{-1} [5]) are also observed only in the α_{yy} polarization [4]. The Raman spectrum varies with temperature, bromine content, and laser line. A study of the temperature dependence of the integrated intensity showed that with decreasing temperature, the 154 cm^{-1} line decreased, whereas the ~230 cm^{-1} line increased [4]. In addition, a frequency change was observed. The 154 cm^{-1} line shifted to 159 [2] (160 cm^{-1} [4] at 2 K) and the ~230 to 219 cm^{-1} at 15 K [2]. The Raman spectra of samples with different bromine concentration (40, 29 [4], and 27% [2]), show the same features, but with weaker intensities in the debrominated crystals [2, 4]. In $(SNBr_{0.29})_x$ only a relative weak band of the ~230 cm^{-1} line (in $(SNBr_{0.4})_x$) remains at 215 cm^{-1} (at 2 K) [4]. The Raman spectrum of brominated $(SN)_x$ is also strongly dependent on the exciting laser frequency [2, 4]. The ~230 cm^{-1} line at 2 K is weaker when going from a red to blue laser. With a laser line at 457.9 nm the band disappeared and the ~160 cm^{-1} line is the dominant feature [4]. The 154 and ~230 cm^{-1} lines behave differently with respect to pressure changes. The line shape, intensity, and peak position of the 154 line does not change, up to 10 kbar at 295 K, in contrast to the 230 cm^{-1} line which showed an anomalous 0.8 cm^{-1} per kbar decrease in frequency with increasing pressure [2]. The ~150 cm^{-1} line was assigned unambiguously to the symmetric stretching mode (ν_1) of the tribromide, Br_3^- [2, 4]. The assignment of the ~230 cm^{-1} line is a subject of controversy [2, 4]. One group attributed the band to the antisymmetric stretching mode (ν_3) of the Br_3^- ion [4] and another to the stretching mode of Br_2 molecule [2] which shifts from its 325 cm^{-1} frequency in the gas phase to lower frequency by interaction with $(SN)_x$. Clarification came from an IR study [7]. No IR absorption was detected close to ~230 cm^{-1} which confirmed the assignment of the ~230 Raman band to Br_2 molecules. This result is also in agreement with the decreased relative intensity in brominated materials with decreased bromine concentration and with the strong polarization of both bands at ~150 cm^{-1} and ~230 cm^{-1} which indicated that both are derived from totally symmetric vibrations [7]. The temperature-dependent concentration ratio, Br_3^-/coordinated Br_2, recognizable in the intensity ratio of the corresponding Raman bands (increasing with decreasing temperature) agrees with EXAFS studies [8]. In a more recent publication, however, the assignment of the 230 cm^{-1} band to loosely bound Br_2 molecules is refuted, because no Br_2 vibration is expected in this range [9].

References:

[1] Iqbal, Z.; Baughman, R. H.; Kleppinger, J.; MacDiarmid, A. G. (Ann. N.Y. Acad. Sci. **313** [1978] 775/87).
[2] Iqbal, Z.; Baughman, R. H.; Kleppinger, J.; MacDiarmid, A. G. (Solid State Commun. **25** [1978] 409/13).
[3] Iqbal, Z.; Sharma, J.; Baughman, R. H.; Akhtar, M.; MacDiarmid, A. G. (Lect. Notes Phys. **96** [1979] 432/6).
[4] Temkin, H.; Street, G. B. (Solid State Commun. **26** [1978] 455/60).
[5] Temkin, H.; Fitchen, D. B.; Street, G. B.; Gill, W. D. (Ann. N.Y. Acad. Sci. **313** [1978] 771/4).
[6] Temkin, H.; Street, G. B. (Bull. Am. Phys. Soc. [2] **23** [1978] 305).
[7] Macklin, J. W.; Street, G. B.; Gill, W. D. (J. Chem. Phys. **70** [1979] 2425/30).
[8] Marks, T. J.; Kalina, D. W. (in: Miller, J. S., Extended Linear Chain Compounds, Pt. I, Plenum, New York 1982, pp. 197/331, 301/7).
[9] Demant, U.; Dehnicke, K. (Z. Naturforsch. **41b** [1986] 929/30).

2.2.1.1.6 Magnetic and Electrical Properties

Magnetic Susceptibility

The static magnetic susceptibility of brominated $(SN)_x$, $(SNBr_y)_x$, was measured as a function of both temperature (up to 300 K) and bromine content (up to $y = 0.4$) [1 to 3]. The low-temperature data are dominated by impurity (oxygen) and defect contributions. Below 30 K a small Curie-like contribution was observed which was used to put an upper limit of 2×10^{-4} molar on the concentration of spin-½ free radicals of bromine such as Br_2^-. Br_2^- species were therefore excluded as a molecular form of the intercalated bromine. At higher temperatures the susceptibility increases linearly with temperature. The conduction electron contribution is estimated to increase, at $y = 0.4$, by a factor of three in comparison to the temperature-independent paramagnetic contribution in $(SN)_x$. The magnitudes of both the paramagnetic susceptibility enhancement and the temperature coefficient depend monotonically on bromine content for concentrations in the range 20 to 40 at%. The increase was attributed to a band-structure change of $(SN)_x$ on bromination. Other possible reasons were also discussed [1].

References:

[1] Scott, J. C.; Kulick, J. D. (Solid State Commun. **28** [1978] 723/7).
[2] Scott, J. C.; Kulick, J. D.; Street, G. B. (Bull. Am. Phys. Soc. [2] **23** [1978] 304).
[3] Scott, J. C.; Kulick, J. D.; Street, G. B. (Lect. Notes Phys. **96** [1979] 438).

Electrical Conductivity

The electrical conductivity of $(SNBr_{0.4})_x$ is about an order of magnitude higher than that of $(SN)_x$ [1 to 4]. Values of the room temperature conductivity along the chain axis b, $\sigma_{\parallel}$, are $2\times10^4\,\Omega^{-1}\cdot cm^{-1}$ [1, 2], $3.8\times10^4\,\Omega^{-1}\cdot cm^{-1}$ (average value) [3], and ca. $4\times10^4\,\Omega^{-1}\cdot cm^{-1}$ [4]. The room temperature conductivity perpendicular to the chain axis b, $\sigma_{\perp}$, is enhanced by a factor of 3 [13]. Values of 15 to $40\,\Omega^{-1}\cdot cm^{-1}$ [2] and ca. $8\,\Omega^{-1}\cdot cm^{-1}$ [3] were measured, however these $\sigma_{\perp}$ values are believed not to be intrinsic because of visible cracks parallel to the b axis in the crystals [2, 3]. Resistance ratio values $\varrho_{\parallel}(300\,K)/\varrho_{\parallel}(4.2\,K)$ of 8 ± 2 [13], ca. 10 [1], ca. 12 to 90 [3], and ca. 50 [7] were reported.

The time dependence of the electrical resistivity during exposure to Br_2 vapor was studied and the results are presented in figures [3, 5, 6]. The conductivity became nearly constant after 40 min (Br_2:65 Torr). The length of the crystal in the b axis direction did not change but the volume increased by about 50% [5]. The dependence of normalized resistivity ϱ/ϱ_0 (ϱ_0 = resistivity before bromination) at room temperature of $(SN)_x$ crystals and films on vapor pressure of Br_2 was investigated [5]. The resistivity $\varrho_{\parallel}$ of brominated $(SN)_x$ at 290 K just after bromination, and in the steady state was studied as a function of bromine content y. There are two steps of bromination. In the steady state the ϱ decrease with increasing y is exponential and the ϱ decrease just after bromination is retarded up to $y\approx0.25$ [6].

The temperature dependence of the resistivity parallel and perpendicular to the chain axis of brominated $(SN)_x$ was measured by several groups [2 to 9, 13]. **Fig. 48**, p. 192, from [2] shows this dependence for $\varrho\parallel b$ and $\varrho\perp b$ of crystalline $(SNBr_{0.4})_x$. Additional results such as the temperature dependence of ϱ/ϱ_0 (ϱ_0 = resisitivity before bromination) for various y values [6], the anisotropy of the conductivity versus temperature [7], and the temperature dependence of an $(SNBr_{0.25})_x$ crystal [3] are presented by other workers. The parallel resistivity $\varrho_{\parallel}$ of the sample in Fig. 48, p. 192, decreased monotonically with decreasing temperature to 3 K. Below 3 K $\varrho_{\parallel}$ becomes temperature independent. No resistance minimum was detected. The perpendicular electrical conductivity is also metallic down to 3 K, in contrast to $(SN)_x$, below which it is temperature independent. A similar temperature dependence was observed on an $(SNBr_{0.25})_x$ crystal with somewhat larger residual resistivity [2, 3]. Unlike for $(SN)_x$, a weaker than T^2 behavior of the resistivity was observed in $(SNBr_{0.4})_x$ [1 to 4]. For y values 0, 0.35, 0.39, and 0.4 values of n = 2, 1.8, 1.5, and 1.4 in the expression $\varrho-\varrho_{res}\sim T^n$ (ϱ_{res} = residual resistivity) were calculated [6], see also [5, 8]. The deviation from the T^2 behavior of $(SN)_x$ (y = 0) was explained by a decreased electron-electron Umklapp scattering caused by a change of the Fermi surface [4]. In contrast, other workers found no deviation from the T^2 dependence [7].

The electrical conductivity of Br_2 doped $(SN)_x$ films and its temperature dependence was also examined [5, 8, 10]. The activation energy decreased with increasing degree of doping [8]. The conductance increased and the activation energy decreased with increasing frequency of an applied electric field. The resistance of a thin $(SN)_x$ film as a function of Br_2 vapor pressure was presented [5]. The dependence of the current on the applied field at various temperatures was also studied and the results interpreted in terms of contact barriers between $(SN)_x$ chains and the effect of dopant thereon [10]. The conductivity change on Br_2 treatment of S_4N_4 was also investigated [5].

A short publication reports the pressure dependence of the conductivity of $(SNBr_{0.4})_x$ crystals at 300 K both parallel and perpendicular to the chain axis, for hydrostatic pressure up to 17 kbar. In contrast to $(SN)_x$ where the conductivity increases by ca. 40%/kbar, $\sigma_{\parallel}$ of $(SNBr_{0.4})_x$ increases only about 1.3%/kbar. An explanation of this result was given in terms of an increase of the electron-phonon scattering lifetime [11].

The resistivity was calculated along the chain axis in $(SNBr_y)_x$ using the coupled Boltzmann equation, considering both carrier-carrier and carrier-phonon scattering mechanisms [4, 12].

Electron (hole)-phonon scattering became important on bromination of $(SN)_x$, causing the change in the temperature dependence and magnitude of the resistivity. The estimated net charge of a bromine atom in $(SN)_x$ is 0.12 to 0.17e [12].

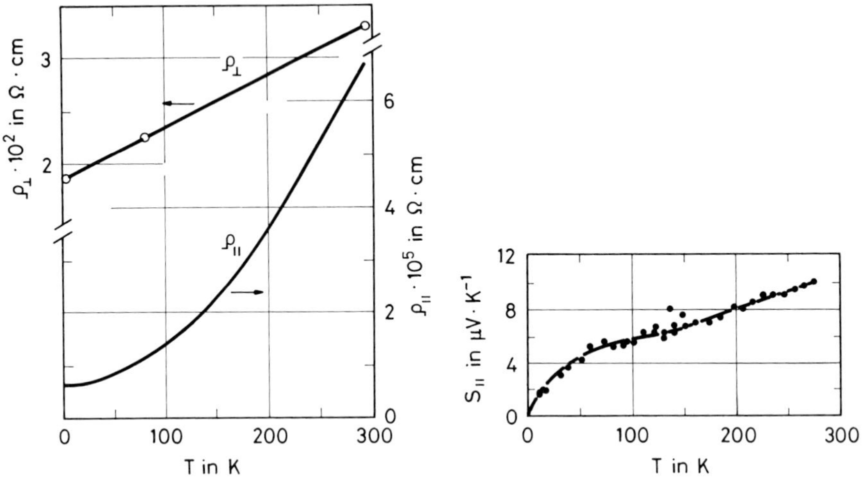

Fig. 48. Temperature dependence of the dc resistivities $\varrho_{\parallel}$ and $\varrho_{\perp}$ and the absolute thermoelectric power $S_{\parallel}$ for crystalline $(SNBr_{0.4})_x$ [2]. (Data for $\varrho_{\perp}$ were only obtained at points shown; a continuous curve for $\varrho_{\perp}$ is presented in [13].)

References:

[1] Street, G. B.; Gill, W. D.; Geiss, R. H.; Greene, R. L.; Mayerle, J. J. (J. Chem. Soc. Chem. Commun. **1977** 407).

[2] Gill, W. D.; Bludau, W.; Geiss, R. H.; Grant, P. M.; Greene, R. L.; Mayerle, J. J.; Street, G. B. (Phys. Rev. Letters **38** [1977] 1305/8).

[3] Akhtar, M.; Chiang, C. K.; Cohen, M. J.; Kleppinger, J.; Heeger, A. J.; Louis, E. J.; MacDiarmid, A. G.; Milliken, J.; Moran, M. J.; Peebles, D. L.; Shirakawa, H. (Ann. N.Y. Acad. Sci. **313** [1978] 726/36; AD-AO41976 [1977] 1/24; C.A. **88** [1978] No. 44331; Solid State Commun. **23** [1977] 607/12).

[4] Kaneto, K.; Yoshino, K.; Inuishi, Y. (in: Monceau, P., Electronic Properties of Inorganic Quasi-One-Dimensional Compounds, Pt. II, Reidel, Dordrecht 1985, pp. 69/111, 82/7).

[5] Yoshino, K.; Tanimura, K.; Yamamoto, Y.; Kaneto, K.; Inuishi, Y. (Technol. Rept. Osaka Univ. **28** [1978] 437/45).

[6] Kaneto, K.; Sasa, S.; Yoshino, K.; Inuishi, Y. (J. Phys. Soc. Japan **49** [1980] 1902/8; Technol. Rept. Osaka Univ. **30** [1980] 497/505).

[7] Philipp, A.; Seeger, K. (Phys. Status Solidi B **89** [1978] 187/92).

[8] Yoshino, K.; Yamamoto, Y.; Tanimura, K.; Kaneto, K.; Inuishi, Y. (Japan. J. Appl. Phys. **18** [1979] 841/2).

[9] Gill, W. D. (NRL-MR-3906 [1979]; AD-AO66201 [1979] Paper No. 3, pp. 1/13; C.A. **91** [1979] No. 220749).

[10] Yoshino, K.; Kaneto, K.; Inuishi, Y. (Technol. Rept. Osaka Univ. **31** [1981] 57/63).

[11] Gill, W. D.; Kwak, J. F.; Greene, R. L.; Seeger, K.; Street, G. B. (Bull. Am. Phys. Soc. [2] **23** [1978] 305, 382).

[12] Oshiyama, A. (J. Phys. C **14** [1981] 5109/25).

[13] Kwak, J. F.; Greene, R. L.; Fuller, W. W. (Phys. Rev. [3] B **20** [1979] 2658/64).

2.2.1.1.7 Electronic Structure. Conduction Mechanism

The origin of the increase of the electrical conductivity (at essentially unchanged plasma frequency) on bromination may be the charge transfer from SN chains to the bromine and the resulting change of the Fermi surfaces [1 to 7]. The Fermi surfaces were calculated in terms of the rigid band model [6, 7]. A charge transfer of ca. 0.1 electrons/SN unit by bromine from the conduction bands was assumed to lower the Fermi level (E_F) by about 1 eV [2 to 5]. Lowering of E_F leads to an expansion of the volume of the hole pockets of the Fermi surface and a shrinkage or even disappearance of the electron pockets. The consequent elimination of electron-hole scattering, which would be consistent with the Fermi surface geometry of $(SN)_x$ and $(SNBr_y)_x$, increases the dc scattering lifetime and therefore the conductivity. Now only phonon scattering channels limit the lifetime [1 to 7]. The resulting predominance of the electron (hole)-phonon scattering mechanism was confirmed by the temperature and pressure dependence of the conductivity [8, 9], see above.

References:

[1] Gill, W. D.; Bludau, W.; Geiss, R. H.; Grant, P. M.; Greene, R. L.; Mayerle, J. J.; Street, G. B. (Phys. Rev. Letters **38** [1977] 1305/8).

[2] Gill, W. D.; Grant, P. M.; Geiss, R. H.; Greene, R. L.; Kwak, J. F.; Street, G. B. (Lect. Notes Phys. **96** [1979] 385/9).

[3] Street, G. B.; Gill, W. D. (NATO Conf. Ser. VI **1** [1978/79] 301/26, 319/21).

[4] Street, G. B.; Clarke, T. C. (Advan. Chem. Ser. No. 186 [1980] 177/94).

[5] Gill, W. D.; Clarke, T. C.; Street, G. B. (Appl. Phys. Commun. **2** [1982/83] 211/82, 235/6).

[6] Oshiyama, A.; Kamimura, H. (J. Phys. C **14** [1981] 5091/107, 5103/5).

[7] Kamimura, H.; Oshiyama, A. (in: Kamimura, H., Theoretical Aspects of Band Structures and Electronic Properties of Pseudo-One-Dimensional Solids, Reidel, Dordrecht 1985, pp. 123/62).

[8] Gill, W. D.; Kwak, J. F.; Greene, R. L.; Seeger, K.; Street, G. B. (Bull. Am. Phys. Soc. [2] **23** [1978] 305).

[9] Grant, P. M.; Greene, R. L.; Gill, W. D.; Kwak, J. F. (Lect. Notes Phys. **96** [1979] 437).

2.2.1.1.8 Superconductivity

The superconducting properties of brominated $(SN)_x$ differs significantly from those of pristine $(SN)_x$ [1 to 5], an observation which had not been explained up to now. The fibers in $(SNBr_{0.4})_x$ are coupled to a greater extent than those in $(SN)_x$ (coherence length $\xi_\perp \gg$ fiber diameter, d), indicating that $(SNBr_{0.4})_x$ is more three-dimensional than $(SN)_x$ and causing it to behave more like a bulk superconductor than $(SN)_x$. The increased dimensionality is confirmed by the temperature behavior of perpendicular upper critical field H_{c2} ($\perp$b), which is quite different from that of $(SN)_x$, and additional arguments such as the other superconducting properties (see below) and also the behavior of the perpendicular conductivity [1].

The **resistivity transition temperature** is enhanced by about 20% and the transition region is greatly narrowed compared to that of $(SN)_x$, see **Fig. 49**, p. 194 [1 to 4]. As a possible explanation, it was suggested that the resistive transition is not fluctuation dominated. The pressure dependence of T_c is completely altered compared with that of $(SN)_x$ [2, 3]. The linear decrease of T_c with pressure to about 180 mK at 10 kbar observed for $(SNBr_{0.4})_x$ is typical for a 3D s–p superconductor.

The **magnetization** of $(SNBr_{0.4})_x$ was studied as a function of magnetic field and temperature and compared to that of $(SN)_x$ [6, 7]. The magnetization both in $(SNBr_y)_x$ and $(SN)_x$ is anisotropic, however the magnitude of the magnetization of $(SNBr_{0.4})_x$ for a given field is much larger than that of $(SN)_x$. Also a much narrower temperature range of the superconducting transition and a higher degree of flux trapping was reported for $(SNBr_{0.4})_x$. The full diamagnetic signal was observed (Meissner effect) for magnetic fields applied perpendicularly to the fiber axis, for H∥b only 57±4% of it. The magnetic critical temperature of $(SNBr_{0.4})_x$ is 308±10 mK [7]. Estimated values of the lower critical fields, $H_{c1\perp}$ and $H_{c1\parallel}$, derived from the relation $H_{c1} = H_p/(1-n_s)$ (H_p = field at which the magnetization reaches its peak value and n_s = demagnetization factor) are $H_{c1\perp} = 3$ to 4 Oe and $H_{c1\parallel} = 0.9$ to 1.3 Oe. Theoretical calculation yielded $H_{c1\perp} = 3.9$ Oe and $H_{c1\parallel} \approx 0.15$ Oe [7]. The perpendicular upper critical field $H_{c2\perp}$ exhibits in contrast to $(SN)_x$ a highly linear temperature dependence typical of a 3D superconductor [1 to 3]. The extrapolated zero-temperature value (for y = 0.4) is ~ 400 G [1]. Also the parallel upper critical field $H_{c2\parallel}$ was studied as a function of reduced temperature T/T_c. The magnitude (H(0) = 8 kG) of this field and its temperature dependence are not very different from those of $(SN)_x$. The measured field data exceed the paramagnetic limiting field $H_p = 18.4\ T_c \approx 5.9$ kOe for all $T<T_c$. Normalized critical field data of $(SNBr_{0.4})_x$ vs. small angles with respect to fiber axis were also presented. They are in agreement with theoretical 3D Ginzburg-Landau (GL) predictions [1].

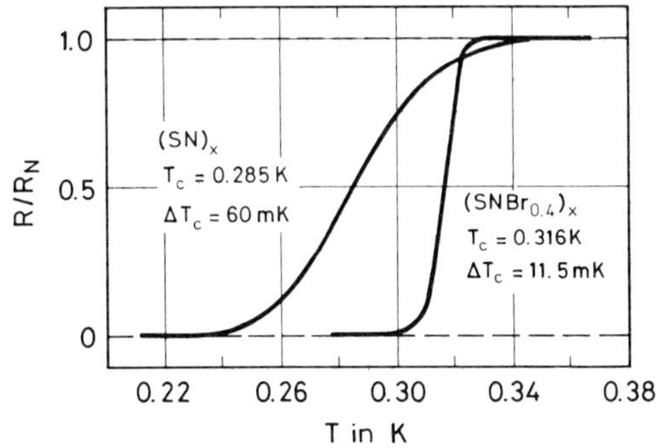

Fig. 49. The four-probe resistance of a typical crystal of $(SN)_x$ before and after bromination. The resistance is normalized to the value at 1 K (R_N) [1].

The **superconducting coherence length** (ξ) of $(SNBr_{0.4})_x$ was determined using the relation $\xi_i(0) = 10^{-6}(\varrho_i\gamma T_c)^{-1/2}$ (γ = electronic heat capacity constant, ϱ_i = normal resistivity, T_c = critical temperature). $\xi_\parallel(0) \approx 2500$ Å and $\xi_\perp(0) \approx 65$ Å. This is in good agreement with the values $\xi_\parallel(0) \approx 2000$ Å and $\xi_\perp(0) \approx 100$ Å, obtained by the GL theory for 3 D superconductors. This is not the case for $(SN)_x$ [1]. The **penetration depth** λ was determined experimentally from the magnetization and more theoretically via the thermodynamical critical field H_c. The experimental effective penetration depth $\lambda_e(H_\perp)$ lies in the range 0.001 to 0.01 mm and $\lambda_e(H_\parallel)$ is 0.05 mm (H = 0.025 Oe, T = 59 mK). The reasonable value of 3.5×10^{-4} mm was calculated for $\lambda_\perp$ from H_c but the value of $\lambda_\parallel$, 1.4×10^{-3} mm, is much too low. Possible reasons for this discrepancy are discussed [7].

References:

[1] Kwak, J. F.; Greene, R. L.; Fuller, W. W. (Phys. Rev. [3] B **20** [1979] 2658/64).
[2] Greene, R. L.; Kwak, J. F.; Fuller, W. W. (J. Phys. Colloq. [Paris] **39** [1978] C6-1401/C6-1403).
[3] Kwak, J. F.; Greene, R. L.; Street, G. B. (Bull. Am. Phys. Soc. [2] **23** [1978] 384).
[4] Gill, W. D. (NRL-MR 3906 [1979]; AD-AO66201 [1979] Paper No. 3, pp. 1/13; C. A. **91** [1979] No. 220749).

[5] Kaneto, K.; Yoshino, K.; Inuishi, Y. (in: Monceau, P., Electronic Properties of Inorganic Quasi-One-Dimensional Compounds, Pt. 2, Reidel, Dordrecht 1985, pp. 69/109, 87/9, 92/4).
[6] Dee, R. H.; Carolan, J. F.; Turrell, B. G.; Greene, R. L.; Street, G. B. (J. Phys. Colloq. [Paris] **39** [1978] C6-444/C6-445).
[7] Dee, R. H.; Carolan, J. F.; Turrell, B. G.; Greene, R. L. (Phys. Rev. [3] B **22** [1980] 174/9).

2.2.1.1.9 Thermoelectric Power

The temperature dependence of thermoelectric power, S, of $(SNBr_{0.4})_x$ was studied, see Fig. 48, p. 192. The behavior is typical for a metal. In comparison with pure $(SN)_x$ the value of S changes from negative to positive values. A possible phonon drag peak appears near 60 K.

Reference:

Gill, W. D.; Bludau, W.; Geiss, R. H.; Grant, P. M.; Greene, R. L.; Mayerle, J. J.; Street, G. B. (Phys. Rev. Letters **38** [1977] 1305/8).

2.2.1.1.10 Magnetoresistance

Transverse negative magnetoresistance, MR, was observed at low magnetic fields and at low temperatures in brominated $(SN)_x$ samples [1 to 4]. On bromination, the negative magnetoresitance becomes enhanced at 4.2 and appears also at 77 K, at which temperature negative magnetoresistance was detected in undoped $(SN)_x$. **Fig. 50** shows the magnetoresistance of

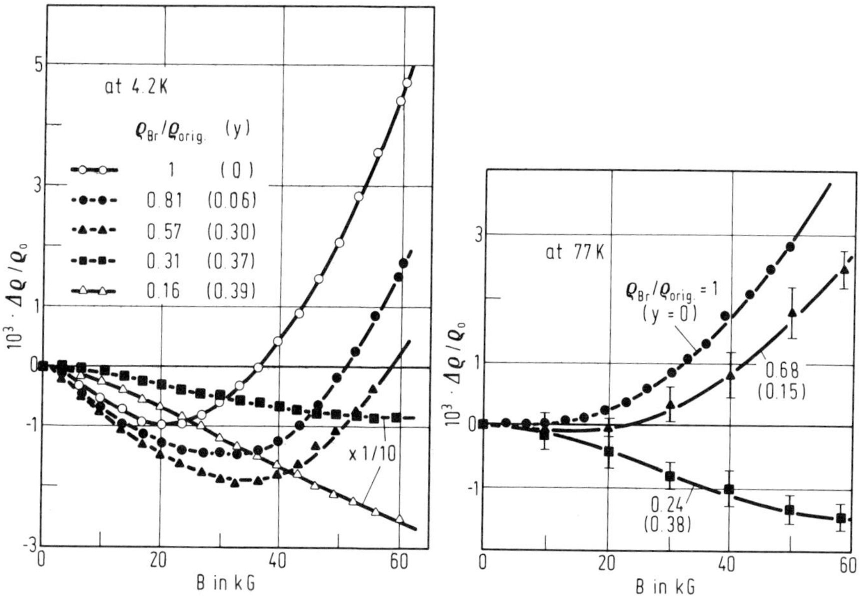

Fig. 50. Magnetoresistance of $(SNBr_y)_x$ against the applied magnetic field, B ⊥ J ∥ b, for samples just after bromination at 4.2 and 77 K [3].

$(SNBr_y)_x$ vs. magnetic field for different y values at 4.2 and 77 K just after bromination. From a comparison with results of steady state (after evacuation of the samples) it was concluded that the anomalous magnetoresistance is not due to irreversible decomposition of $(SN)_x$ chains. A negative comparable component of the longitudinal magnetoresistance was also observed at 4.2 K for the J∥b∥B configuration [3]. One of the possible origins for the negative magnetoresistance in $(SNBr_y)_x$ may be spin-dependent scattering at the surface of the fibers or at the bulk crystal with localized spins due to adsorbed bromine [3]. From the quadratic dependence of the positive magnetoresistance on magnetic field (presented in a figure) the carrier mobility in $(SNBr_y)_x$ was estimated to have a slightly larger value than that of pure $(SN)_x$ [2].

References:

[1] Philipp, A.; Seeger, K. (Phys. Status Solidi B **89** [1978] 187/92).
[2] Yoshino, K.; Tanimura, K.; Yamamoto, Y.; Kaneto, K.; Inuishi, Y. (Technol. Rept. Osaka Univ. **28** [1978] 437/45).
[3] Kaneto, K.; Sasa, S.; Yoshino, K.; Inuishi, Y. (J. Phys. Soc. Japan **49** [1980] 1902/8; Technol. Rept. Osaka Univ. **30** [1980] 497/505).
[4] Kaneto, K.; Yoshino, K.; Inuishi, Y. (in: Monceau, P., Electronic Properties of Inorganic Quasi-One-Dimensional Compounds, Pt. II, Reidel, Dordrecht 1985, pp. 103/6).

2.2.1.1.11 Schottky Barrier Solar Cell

$(SNBr_{0.4})_x$ was used to fabricate Schottky solar cells [1, 2]. No detailed results were published.

References:

[1] Cohen, M. J.; Harris, J. S., Jr. (Tech. Digest Intern. Electron Devices Meeting **1978** 247/9).
[2] Cohen, M. J.; Harris, J. S., Jr. (U.S. 4227943 [1979/80] 1/5; C.A. **94** [1981] No. 33758).

2.2.1.1.12 Mass Spectroscopic Study

Brominated $(SN)_x$ was investigated by different mass spectrometric techniques including electron impact [1, 2], phase angle [1, 2], field ionization [1], chemical ionization [2], and modulated molecular beam [2]. SNBr and Br_2 are the major bromine-containing molecular species volatilizing from $(SNBr_{0.43})_x$ identified by phase angle technique [1 to 6]. Table 46 shows electron impact (EI) mass spectral data of brominated $(SN)_x$ at different temperatures and, in addition, the EI spectra of brominated S_4N_4 at 90°C. Field ionization data are compiled in [1].

The temperature determines which bromine-containing ionic species are found [1 to 6]. The major bromine-containing species is Br_2^+ at lower temperatures and $SNBr^+$ at higher temperatures [5]. This result was interpreted on the basis of the structural model of brominated $(SN)_x$ consisting of intercalated Br_2 and Br_3^- molecules [6]. The mass spectrum of $(SNBr_{0.4})_x$ prepared by the reaction of $S_3N_3Cl_3$ with $BrSi(CH_3)_3$ is similar to that of brominated $(SN)_x$ and S_4N_4. In contrast to other workers, the origin of Br^+ and Br_2^+ species was explained by S–Br bond breaking [7]. The mass spectra of brominated $(SN)_x$ and S_4N_4 are nearly indistinguishable at higher temperatures [2]. The concentration of the hydride impurity detected in the mass spectrum of $(SN)_x$ is very small in both brominated $(SN)_x$ and S_4N_4 (< 0.1 [1]; 0.1 to 0.2 at%

hydrogen [2]; this is a factor of 10 less than in pristine $(SN)_x$). This result was interpreted with the assumption that bromine replaces hydrogen bound to $(SN)_x$ to form HBr [1 to 6]. One group concluded that bromination can be used to remove hydride from $(SN)_x$ [1].

Table 46
Electron Impact Mass Spectra (25 eV) of $(SNBr_{0.43})_x$ (brominated $(SN)_x$ and brominated S_4N_4).

ion[a)]	m/e	30°C [1][b)]	80°C [1][b)]	90°C [2][b)]	90°C [2][c)]	125°C [1][b)]
SN^+	46	56	100	100	100	100
S_2^+	64	3.2	3.7	2.0	1.5	1.0
S_2N^+	78	—	3.4	12.4	12.6	5.0
Br^+	79	32	21	3.3	1.8	8.4
HBr^+	80	14	4.0	1.9	1.5	1.8
$S_2N_2^+$	92	—	6.0	26	32	11
S_3N^+	110	—	0.6	1.0	0.9	1.2
SBr^+	111	4.1	8.7	1.0	0.5	3.9
$S_3N_2^+$	124	—	—	1.5	1.8	—
$SNBr^+$	125	9.7	33	4.4	2.1	13.2
$S_3N_3^+$	138	—	10	21	20	10
S_2Br^+	143	—	1.5	—	—	—
$S_4N_2^+$	156	—	—	0.5	0.5	0.4
Br_2^+	158	100	29	—	—	0.6
$S_4N_4^+$	184	—	3.8	1.8	0.6	0.6

a) Isotopic contributions were not considered in this table. – b) Brominated $(SN)_x$. – c) Brominated S_4N_4.

References:

[1] Allen, W. N.; DeCorpo, J. J.; Saalfeld, F. E.; Wyatt, J. R. (Chem. Phys. Letters **54** [1978] 524/6).
[2] Smith, R. D.; Street, G. B. (Inorg. Chem. **17** [1978] 938/41).
[3] Street, G. B.; Etemad, S.; Geiss, R. H.; Gill, W. D.; Greene, R. L.; Kuyper, J.; Smith, R. D. (Ann. N.Y. Acad. Sci. **313** [1978] 737/44, 738).
[4] Saalfeld, F. E.; DeCorpo, J. J.; Wyatt, J. R.; Mah, P. T.; Allen, W. N. (Ann. N. Y. Acad. Sci. **313** [1978] 759/66, 764).
[5] Saalfeld, F. E.; DeCorpo, J. J.; Wyatt, J. R. (in: Miller, J. S., Extended Linear Chain Compounds I, Plenum, New York 1982, pp. 37/72, 50/8).
[6] Street, G. B.; Gill, W. D. (NATO Conf. Ser. VI **1** [1978/79] 301/26, 309/11).
[7] Demant, U.; Dehnicke, K. (Z. Naturforsch. **41 b** [1986] 929/30).

2.2.1.1.13 Electrochemical Behavior

The electrochemical behavior of the bromine modified $(SN)_x$ electrode (prepared chemically, composition $(SNBr_{0.2})_x$, and electrochemically) was studied in solutions of LiBr and $LiClO_4$ in propylene carbonate by cyclic voltametry. In LiBr and $LiClO_4$ solutions it behaved like $(SN)_x$ with cathodic polarizations, but in $LiClO_4$ solution oxidation of bromide occurred with anodic polarization. The reaction was reversible. The quantity of electricity was higher in the oxidation reaction for the LiBr electrolyte because the Br^- ions of the solution were also oxidized. The quantity was identical for both electrolytes on reduction.

Reference:

Tarby, C.; Bernard, C.; Robert, G. (Electrochim. Acta **26** [1981] 663/70).

2.2.1.1.14 Chemical Behavior

$(SNBr_y)_x$ (y = 0.40 to 0.43 and 0.16 to 0.28) powders exploded when compressed to pellets [1].

The bromine content of $(SNBr_y)_x$ changes with Br_2 pressure and temperature, as described on pp. 181/2. Prolonged heating of brominated $(SN)_x$ at relatively low temperatures (<90°C) results in $(SN)_x$ crystals which are nearly free of bromine [1], but even on heating up to decomposition the bromine was not totally removed [2]. Thermogravimetric analysis data of $(SNBr_{0.4})_x$ which were obtained at a heating rate of 5°C/min and under an atmosphere of He, show rapid loss of bromine at temperatures of ca. 125°C, see **Fig. 51**. Differential thermal analysis data show a strong exothermal peak at ca. 165°C associated with the liberation of bromine and exothermic decomposition [3].

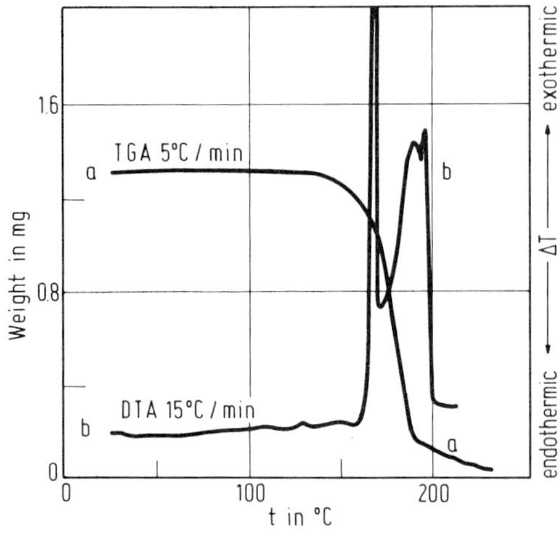

Fig. 51 Thermal analysis data for $(SNBr_{0.4})_x$ in He atmosphere (a = thermogravimetric analysis; b = differential thermal analysis) [3].

$(SNBr_{0.4})_x$ crystals react with air to a negligible extent during 1 h at room temperature. After 2 d no detectable change in weight was measured, but the crystals were visibly tarnished [4]. Mass spectrometric studies on brominated $(SN)_x$ and S_4N_4 exposed to air for 1 d, however, show that both materials react much more rapidly with air than pure $(SN)_x$ to yield several additional products including NH_3, SO_2, HNSO, S_x, S_4N_2, increased amounts of HBr, the hydride impurity and possibly also $S_2N_2H_2Br$, S_2NOBr, and S_3N_2O [2].

References:

[1] Akhtar, M.; Chiang, C. K.; Heeger, A. J.; Milliken, J.; MacDiarmid, A. G. (Inorg. Chem. **17** [1978] 1539/42).

[2] Smith, R. D.; Street, G. B. (Inorg. Chem. **17** [1978] 938/41).

[3] Street, G. B.; Etemad, S.; Geiss, R. H.; Gill, W. D.; Greene, R. L.; Kuyper, J.; Smith, R. D. (Ann. N. Y. Acad. Sci. **313** [1978] 737/44, 739).

[4] Akhtar, M.; Kleppinger, J.; MacDiarmid, A. G.; Milliken, J.; Moran, M. J.; Chiang, C. K.; Cohen, M. J.; Heeger, A. J.; Peebles, D. L. (J. Chem. Soc. Chem. Commun. **1977** 473/4).

2.2.1.2 Iodinated Poly(sulfur nitride), $(SNI_y)_x$; y = 0.02, 0.04, 0.14, 0.33. $(S_{3.00}N_{2.15}I_{0.53})_x$

Preparation

Iodinated poly(sulfur nitrides), $(SNI_y)_x$, cannot be synthesized as easily as brominated poly(sulfur nitrides), $(SNBr_y)_x$, by analogical pathways. Iodine does not react with either solid $(SN)_x$ or S_4N_4 at room temperature. At ca. 125°C both materials react with iodine to yield conducting, air-sensitive solids. No further information was reported [1]. A silver-colored material with the composition $(SNI_{0.02})_x$ was prepared by exposing $(SN)_x$ needles for 2 h to iodine vapor at a pressure of 20 Torr in an U-shaped tube at 350 K [2]. (At temperatures higher than 350 K sublimation of $(SN)_x$ was observed [2].) These results were confirmed by others who also reported that exposing $(SN)_x$ crystals for 12 h at 350 K (20 Torr) with subsequent pumping at 10^{-3} Torr for 24 h leads to a compound of composition $(SNI_{0.14})_x$ with an opaque silver color [3]. Treatment of $(SN)_x$ with iodine vapor in an evacuated glass flask at 90°C for 4 h followed by 2 h desorption in vacuum yielded a material with composition $(SNI_{0.04})_x$ [4]. A precipitate with the composition $(SNI_{0.33})_x$ forms in an inert solvent when S_4N_4 is exposed to I_2, the reaction is slow at room temperature, but it is appreciably accelerated when carried out in boiling solvents [7]. A black iodinated sample, $(S_3N_{2.15}I_{0.53})_x$, was prepared in a pyrolysis tube by treating S_4N_3Cl, vaporized at 130°C, with NaI supported on glass wool at 250°C [5, 6]. A detailed description and a figure of the pyrolysis apparatus was presented [5]. The iodinated material condensed during pumping ($<5\times10^{-3}$ Torr) over a period of 3 to 4 h in a cold zone (10°C) of the apparatus. Adequate safety precautions should be taken when heating or compressing the material [5, 6].

References:

[1] Street, G. B.; Bingham, R. L.; Crowley, J. I.; Kuyper, J. (J. Chem. Soc. Chem. Commun. **1977** 464/5).

[2] Philipp, A.; Seeger, K. (Phys. Status Solidi B **89** [1978] 187/92).

[3] Tubino, R.; Piseri, L. (Solid State Commun. **34** [1980] 173/8).

[4] Petermann, J.; Wagner, N.; Young, R. (Colloid Polym. Sci. **259** [1981] 247/51).

[5] Banister, A. J.; Smith, N. R. M. (J. Chem. Soc. Dalton Trans. **1980** 937/41).

[6] Banister, A. J.; Fielder, A. J.; Hauptman, Z. V.; Smith, N. R. M. (U.S. 4268491 [1981] 1/7; C.A. **95** [1981] No. 83144).

[7] Monteil, Y.; Vincent, H. (Z. Naturforsch. **31 b** [1976] 673/6).

Structure

The structure of iodinated $(SN)_x$ was studied by electron diffraction [1 to 4], X-ray diffraction [2, 3], and high-resolution electron microscopy [3, 4]. The crystal lattice of iodinated $(SN)_x$ is as heavily distorted as that of brominated $(SN)_x$ [4]. About 10 extra diffuse streaks perpendicular to the b* axis were observed between the layer lines of pristine $(SN)_x$ by electron diffraction [3]. An X-ray diffraction pattern shows only four extra diffuse streaks of Bragg spacings (0.452, 0.300, 0.216, and 0.154 nm). Only these streaks are believed to be intrinsic, the others are interpreted as due to double diffraction between the four streaks and the spotty diffraction of $(SN)_x$. This result led to the conclusion that iodine atoms are structured with a one-dimensional order parallel to the $(SN)_x$ chain. The repeat distance is about 0.9 nm [1 to 4].

The distance between neighboring iodine atoms is 0.3 nm which is in agreement with the assumption that I_3^- species are formed, but other iodine species are not excluded [4]. The content of iodine as a function of the distance from the end surface was measured with an X-ray microanalyzer [3]. The largest rate of diffusion of iodine atoms is that along the direction of the molecular chain. From high-resolution electron micrographs and their optical transforms, it was observed that iodine preferentially intercalates in the skin region of $(SN)_x$ specimens [2]. Heavy streaking was interpreted to mean that the crystal lattice of $(SN)_x$ was also distorted [4]. Iodine is not intercalated between $(\bar{1}02)$ planes (as Br in brominated $(SN)_x$) but is believed to be settled on the $(\bar{1}02)$ plane [4]. A structural model which fits the experimental results was presented. The main feature is two shifted chains of iodine, arranged together on the $(\bar{1}02)$ plane and holding an $(SN)_x$ chain between them. The structural model is illustrated [4]. X-ray powder diffraction data of iodinated poly(sulfur nitride) prepared from S_4N_3Cl and NaI before and after heating in vacuum were compiled. It was pointed out that the powder has a low degree of crystallinity [5, 6].

References:

[1] Petermann, J.; Wagner, N.; Young, R. (Colloid Polym. Sci. **259** [1981] 247/51).
[2] Kawaguchi, A.; Isoda, S.; Petermann, J.; Katayama, K. (Colloid Polym. Sci. **262** [1984] 429/34).
[3] Kawaguchi, A.; Isoda, S.; Petermann, J.; Katayama, K. (Colloid Polym. Sci. **263** [1985] 631/7).
[4] Isoda, S.; Kawaguchi, A.; Uemura, A.; Katayama, K. (Japan J. Appl. Phys. II **24** [1985] L341/L343; C.A. **103** [1985] No. 30620).
[5] Banister, A. J.; Smith, N. R. M. (J. Chem. Soc. Dalton Trans. **1980** 937/41).
[6] Banister, A. J.; Fielder, A. J.; Hauptman, Z. V.; Smith, N. R. M. (U.S. 4268491 [1981] 1/7; C.A. **95** [1981] No. 83144).

X-Ray Photoelectron Spectroscopy (XPS) of $S_{3.00}N_{2.15}I_{0.53}$

ESCA (XPS) data (in eV) of iodinated $(SN)_x$ are listed in the following table:

binding energy in eV	relative intensity in %	spin state	assignment
619.5	41	5/2	I 3d (anionic)
631.0	28	3/2	
621.2	14	5/2	I 3d (neutral)
632.8	10	3/2	
398.8	55		N 1s (neutral)
400.3	21.5		N 1s (cationic)
401.8	10		
164.0	73		S 2p (doublets)
166.1	21.5		

Binding energies of the I 3d molecular state suggest that iodine in iodinated $(SN)_x$ is probably present as an anion and as a neutral species.

Reference:

Banister, A. J.; Smith, N. R. M. (J. Chem. Soc. Dalton Trans. **1980** 937/41).

IR Spectrum of $S_{3.00}N_{2.15}I_{0.53}$

An IR spectrum of black powdered $S_{3.00}N_{2.15}I_{0.53}$ was recorded with the following absorption bands [1, 2]: ν (in cm^{-1}) = 3200 to 3100 w; 1390 m; 1190 vw, br; 1110 m, br; 1030 vw, br; 923 ms; 850 to 750 w, br; 705 ms, br; 610 m, br; 563 ms; 460 br; 348 ms. The absorptions at 3200 and 1390 cm^{-1} are due to ν(N–H). Some decomposition to S_4N_4 was also evident [1, 2].

References:

[1] Banister, A. J.; Smith, N. R. M. (J. Chem. Soc. Dalton Trans. **1980** 937/41).
[2] Banister, A. J.; Fielder, A. J.; Hauptman, Z. V.; Smith, N. R. M. (U.S. 4268491 [1981] 1/7; C.A. **95** [1981] No. 83144).

Raman Spectrum of $(SNI_{0.14})_x$

Resonance Raman spectra of iodinated $(SN)_x$ with composition $(SNI_{0.14})_x$ were recorded at 150 K using three different laser lines (λ = 488, 514.5, and 647.1 nm) [1, 2]. Two main features at 109 and 154 cm^{-1} were observed. The spectra are interpreted on the basis of a comparison with the Raman spectra of nonconducting iodine complexes such as the blue starch-iodine complex. It was concluded that iodine is intercalated as I_5^- ions and/or aggregates involving I_3^- linked to distorted I_2. The concentration of the two species depends on sample preparation and treatment [1, 2].

References:

[1] Tubino, R.; Piseri, L.; Carcano, G.; Pollini, I. (Solid State Commun. **34** [1980] 173/8).
[2] Carcano, G.; Piseri, L.; Pollini, I.; Tubino, R. (4th Convegno Ital. Sci. Macromol. Atti, Colleferro, Italy, 1979, pp. 164/6).

Electrical Conductivity

The temperature dependence of the electrical conductivity of iodinated $(SN)_x$ was studied [1 to 4]. I_2 vapor, in contrast to Br_2 vapor, does not change the electrical properties of $(SN)_x$ crystals whereas the conductivity of $(SN)_x$ films increases by several orders of magnitude on iodination. The activation energies decrease with increasing degree of doping, see **Fig. 52**, p. 202. A possible reason for the different behavior of iodine and bromine may be the larger size of iodine atoms, making it impossible for iodine atoms to diffuse into $(SN)_x$ single crystals [1 to 3]. Other workers reported that the dc conductivity of $(SN)_x$ crystals exposed 2 h to iodine vapor increased by a factor of ca. 2.5 [4]. The dc resistivity of a plug (3.1 mm diameter × 3.1 mm) $S_{3.00}N_{2.15}I_{0.53}$ between Pt electrodes is ca. 10 Ω at room temperature [5].

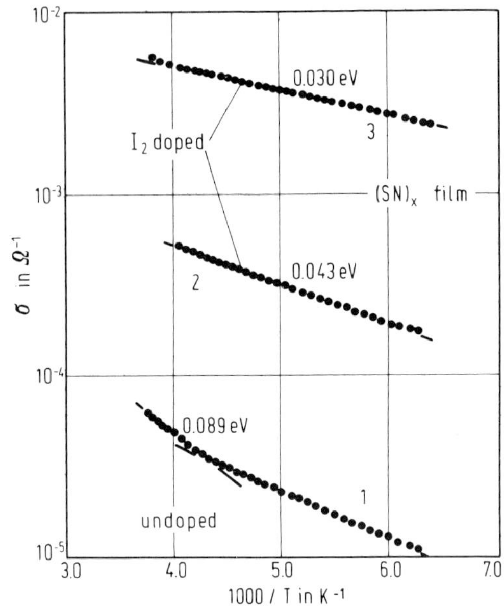

Fig. 52. Temperature dependence of conductivity of undoped and I_2-doped $(SN)_x$ thin films. Degree of doping (3)>(2) [2].

References:

[1] Yoshino, K.; Tanimura, K.; Yamamoto, Y.; Kaneto, K.; Inuishi, Y. (Technol. Rept. Osaka Univ. **28** [1978] 437/45).

[2] Yoshino, K.; Yamamoto, Y.; Tanimura, K.; Kaneto, K.; Inuishi, Y. (Japan. J. Appl. Phys. **18** [1979] 841/2).

[3] Yoshino, K.; Kaneto, K.; Inuishi, Y. (Technol. Rept. Osaka Univ. **31** [1981] 57/63).

[4] Philipp, A.; Seeger, K. (Phys. Status Solidi B **89** [1978] 187/92).

[5] Banister, A. J.; Smith, N. R. M. (J. Chem. Soc. Dalton Trans. **1980** 937/41).

Magnetoresistance

$(SNI_{0.02})_x$ shows a much larger negative part of the transverse magnetoresistance than pristine $(SN)_x$. Measurements were performed at 4.2 K for a current direction parallel to the b axis.

Reference:

Philipp, A.; Seeger, K. (Phys. Status Solidi B **89** [1978] 187/92).

Mass Spectrum of $S_{3.00}N_{2.15}I_{0.53}$

The mass spectral data shown in Table 47 are consistent with those of the decomposition products of the iodinated $(SN)_x$: sulfur, S_4N_4, and iodine. The absence of any sulfur-nitrogen-iodine peaks, such as SNI^+ and $S_2N_2I^+$, indicates that the polymer itself breaks down under the probe conditions.

Table 47
Mass Spectral Analysis of Black Iodinated Poly(sulfur nitride) at 70 eV and 280°C.

m/e	species	relative intensity in %	m/e	species	relative intensity in %
46	SN^+	100	127	I^+	22
48	SO^+	12	128	HI^+, S_4^+	10
64	S_2^+, SO_2^+	35	138	$S_3N_3^+$	57
78	S_2N^+	24	140	$S_3N_3H_2^+$	11
92	$S_2N_2^+$	83	156	$S_4N_2^+$	4
110	S_3N^+	7	184	$S_4N_4^+$	8
124	$S_3N_2^+$	5	254	I_2^+	62

Reference:

Banister, A. J.; Smith, N. R. M. (J. Chem. Soc. Dalton Trans. **1980** 937/41).

Chemical Behavior

The black iodinated poly(sulfur nitride), $(S_{3.00}N_{2.15}I_{0.53})_x$, is moisture-sensitive and decomposes above 40°C in vacuum (ca. 10^{-6} Torr) to S_8, S_4N_4, and I_2. When a sample was studied mass spectroscopically several days after preparation the hydrolysis products SO, SO_2, and HI were found. When left exposed to air for 3 to 4 d the material hydrolyzed fairly rapidly to give a greyish powder with release of iodine.

Reference:

Banister, A. J.; Smith, N. R. M. (J. Chem. Soc. Dalton Trans. **1980** 937/41).

2.2.1.3 Iodochlorinated Poly(sulfur nitride), $(SN(ICl)_y)_x$; y = 0.1, 0.125, 0.36, 0.4

Preparation

$(SN(ICl)_y)_x$ intercalation compounds were prepared by treatment of $(SN)_x$ or S_4N_4 with ICl vapor. If $(SN)_x$ had been exposed to ICl vapor for a short time only a surface reaction was observed. Longer exposure times led to decomposition [1]. A compound of composition $(SN(ICl)_{0.1})_x$ was formed by treating $(SN)_x$ crystals with ICl vapor at room temperature in a desiccator for 0.5 to 1 h. These crystals were similar in color to $(SNBr_{0.4})_x$ [2]. $(SN(ICl)_{0.125})_x$ was prepared by reacting $(SN)_x$ with ca. 28 Torr of ICl vapor during 20 h [3]. An intercalation material of unknown composition was obtained by exposing $(SN)_x$ during 5 min at room temperature in an evacuated glass flask followed by desorption for 2 h [4]. For the reaction of S_4N_4 with ICl vapor see "Sulfur-Nitrogen Compounds" Part 2, 1985, p. 172. Compositions of $(SN(ICl)_{0.4})_x$ [5] and $(SN(ICl)_{0.36})_x$ [6] were produced.

References:

[1] Wolmershäuser, G.; Street, G. B. (Inorg. Chem. **17** [1978] 2685/7).
[2] Philipp, A.; Seeger, K. (Phys. Status Solidi B **89** [1978] 187/92).

[3] Iqbal, Z.; Baughman, R. H.; Kleppinger, J.; MacDiarmid, A. G. (Solid State Commun. **25** [1978] 409/13).
[4] Petermann, J.; Wagner, N.; Young, R. (Colloid Polym. Sci. **259** [1981] 247/51).
[5] Akhtar, M.; Chiang, C. K.; Heeger, A. J.; Milliken, J.; MacDiarmid, A. G. (Inorg. Chem. **17** [1978] 1539/42).
[6] Street, G. B.; Gill, W. D. (NATO Conf. Ser. VI **1** [1978/79] 301/26).

Structure

Limited information about the structure of ICl-modified polymeric sulfur nitride was obtained by a transmission electron microscopy study [1, 2]. Electron diffraction patterns were taken at 115, 293, and 410 K [1]. Other workers also presented an electron-diffraction micrograph but with no new information [3]. ICl incorporates into the $(SN)_x$ lattice in an entirely different manner than bromine [1] but a biphasic system as in brominated $(SN)_x$ also exists [1, 2]. On the basis of their experimental results the workers proposed a structural model. Undissociated ICl (bond length ca. 2.8 Å) was assumed to be laid down on the exposed ($\bar{1}02$) faces of $(SN)_x$ whereas two forms of packing along the polymer chain may exist [1]. Evidence for an ICl_2^- species was not found [4].

References:

[1] Geiss, R. H.; Thomas, J.; Street, G. B. (Syn. Metals **1** [1979/80] 257/65).
[2] Geiss, R. H. (Proc. Ann. Meeting Electron Microsc. **38** [1980] 238/41).
[3] Petermann, J.; Wagner, N.; Young, R. (Colloid Polym. Sci. **259** [1981] 247/51).
[4] Iqbal, Z.; Baughman, R. H.; Kleppinger, J.; MacDiarmid, A.G. (Solid State Commun. **25** [1978] 409/13).

Raman Spectrum

$(SN(ICl)_{0.125})_x$ shows a Raman line at 100 cm^{-1} comparable to the 103 cm^{-1} line observed in 1.8 mol% ICl intercalated graphite.

Reference:

Iqbal, Z.; Baughman, R. H.; Kleppinger, J.; MacDiarmid, A.G. (Solid State Commun. **25** [1978] 409/13).

Electrical Conductivity

An ICl-treated S_4N_4 sample with the composition $(SN(ICl)_{0.43})_x$ has a conductivity of 11 $\Omega^{-1}\cdot cm^{-1}$ at 25°C [1]. A pellet conductivity of 50 $\Omega^{-1}\cdot cm^{-1}$ (at room temperature) was mentioned by another group [2]. The conductivity of $(SN)_x$ needles exposed to ICl vapor for 1/2 h (composition $(SN(ICl)_{0.1})_x$) is lowered by a factor of 0.7. This fact was explained with the assumption that ICl does not intercalate in $(SN)_x$ but yields a nonpolymeric substance after sufficient time [3]. This is in agreement with observations of others [4].

References:

[1] Akhtar, M.; Chiang, C. K.; Heeger, A. J.; Milliken, J.; MacDiarmid, A.G. (Inorg. Chem. **17** [1978] 1539/42).
[2] Street, G. B.; Bingham, R. L.; Crowley, J. I.; Kuyper, J. (J. Chem. Soc. Chem. Commun. **1977** 464/5).

[3] Philipp, A.; Seeger, K. (Phys. Status Solidi B **89** [1978] 187/92).
[4] Wolmershäuser, G.; Street, G. B. (Inorg. Chem. **17** [1978] 2685/7).

Magnetoresistance

The dependence of the transverse magnetoresistance on the magnetic field with current in b axis direction for $(SN)_x$ needles, before and after exposing to ICl vapor, was studied at 4.2 K. The magnetoresistance of ICl-modified $(SN)_x$ is negative at low fields.

Reference:

Philipp, A.; Seeger, K. (Phys. Status Solidi B **89** [1978] 187/92).

Mass Spectroscopic Study

The vapor phase species of ICl-treated S_4N_4 were examined using mass spectroscopic techniques. ICl, I_2, HCl, HI, S_4N_4, ClSN, and a very small amount of S_2Cl_2 was detected. Cl_2 and ISN were not observed.

Reference:

Smith, R. D.; Street, G. B. (Inorg. Chem. **17** [1978] 938/41).

2.2.1.4 Iodobrominated Poly(sulfur nitride), $(SN(IBr)_y)_x$; y = 0.06, 0.09, 0.21, 0.33, 0.34, 0.37, 0.40, 0.41

Preparation

Iodobrominated poly(sulfur nitrides) are prepared by reaction of $(SN)_x$ or S_4N_4 with IBr vapor or solution.

$(SN)_x$ was treated with IBr vapor in an U-tube at 10^{-4} to 10^{-5} Torr at 10°C. After 2 h an intercalation product with the composition $(SN(IBr)_{0.09})_x$ was isolated which was converted into blue $(SN(IBr)_{0.06})_x$ by desorption for 2 h in vacuum of 10^{-5} Torr. After a reaction time of ca. 100 h under the same conditions as above, a product with the composition $(SN(IBr)_{0.41})_x$ formed. A similar brown-black material with the formula $(SN(IBr)_{0.40})_x$ was obtained by reacting $(SN)_x$ and IBr vapor under an argon atmosphere at atmospheric pressure during 100 h. $(SN(IBr)_y)_x$ compounds with y = 0.21, 0.34, and 0.37 were obtained from $(SN(IBr)_{0.41})_x$ under various conditions. Iodine was more rapidly desorbed than bromine, which led to a formula, IBr_2, for the residual inserted molecule [1]. An intercalation reaction of $(SN)_x$ with IBr was also performed in an evacuated glass flask at 50°C during 2 h followed by 2 h desorption. No further details were presented [2]. Treatment of S_4N_4 with IBr vapor also yields conducting solids of unknown compositions [3, 4]; see "Sulfur-Nitrogen Compounds" Part 2, 1985, p. 172. The authors noted that IBr dissociated into I_2 and Br_2 at room temperature [4]. Other workers presented detailed experimental data [5]. The intercalation was performed in CCl_4 solution (ca. 6 mL IBr solution/ca. 250 mg S_4N_4). After a total reaction time of 40 h at room temperature the reaction was complete (monitored by X-ray powder diffraction pattern). A composition of $(SN(IBr)_{0.40})_x$ was determined for the product. Iodine and bromine are lost in equimolar amounts, apparently as IBr on heating this material at 80°C with pumping [5], in contrast to [1].

References:

[1] Bernard, C.; Touzain, P.; Robert, G. (Ann. Chim. [Paris] [15] **4** [1979] 591/8).
[2] Petermann, J.; Wagner, N.; Young, R. (Colloid Polym. Sci. **259** [1981] 247/51).
[3] Street, G. B.; Bingham, R. L.; Crowley, J. I.; Kuyper, J. (J. Chem. Soc. Chem. Commun. **1977** 464/5).
[4] Street, G. B.; Gill, W. D. (NATO Conf. Ser. VI **1** [1978/79] 301/26, 308).
[5] Akhtar, M.; Chiang, C. K.; Heeger, A. J.; Milliken, J.; MacDiarmid, A.G. (Inorg. Chem. **17** [1978] 1539/42).

Structure

The structure of two products of the intercalation of $(SN)_x$ with IBr vapor, $(SN(IBr)_y)_x$ with y = 0.06 and 0.33, was studied by X-ray diffraction (Debye-Scherrer method). $(SN(IBr)_{0.06})_x$ showed a disordered structure. Data corresponding to $(SNBr_{0.4})_x$ (monoclinic cell with a = 3.92, b = 4.44, and c = 7.85 Å; β = 111°) and $(SN)_x$ were observed. IBr molecules occupy the sites of the Br species. A monoclinic cell with parameters a = 8.50, b = 4.44, c = 7.95 Å, and β = 111.8° was reported for $(SN(IBr_{0.33}))_x$. A structural model consisting of alternating $(SN)_x$ and IBr planes with an interplanar distance of 6.8 Å was proposed.

Reference:

Bernard, C.; Touzain, P.; Robert, G. (Ann. Chim. [Paris] [15] **4** [1979] 591/8).

Electrical Conductivity

The conductivity of an $(SN(IBr)_{0.40})_x$ pellet, prepared by treatment of S_4N_4 with IBr solution, is 50 $\Omega^{-1}\cdot cm^{-1}$ at 25°C.

Reference:

Akhtar, M.; Chiang, C. K.; Heeger, A. J.; Milliken, J.; MacDiarmid, A.G. (Inorg. Chem. **17** [1978] 1539/42).

2.2.2 Na-Modified Poly(sulfur nitride)

Formation

For the intercalation of Na in $(SN)_x$ the $(SN)_x$ sample was immersed in a solution of Na^+ naphthalene$^{\overline{\bullet}}$ in THF (0.5 mol/L) at ca. 20 [1] and 30°C [2].

Electrical Conductivity

The time dependence of the electrical resistivity of $(SN)_x$ along the b axis after immersion in Na^+ ion solution was measured at ca. 20 [1] and 30°C [2]. The resistivity increases by several times but recovers to the values of pristine $(SN)_x$ on exposure to Br_2 vapor. The temperature dependence of the resistivity of Na^+-treated $(SN)_x$ is similar to that of pristine $(SN)_x$. The residual resistivity was only slightly increased. The transverse magnetoresistance of Na^+-treated $(SN)_x$ at 4.2 K was unchanged. These findings led to the conclusions that the $(SN)_x$ chains are undistorted. The increase in resistivity was explained by a larger electron-hole Umklapp scattering than in $(SN)_x$ [1, 2].

References:

[1] Kaneto, K.; Sasa, S.; Yoshino, K.; Inuishi, Y. (Mol. Cryst. Liq. Cryst. **86** [1982] 189/93).
[2] Kaneto, K.; Sasa, S.; Yoshino, K.; Inuishi, Y. (Solid State Commun. **40** [1981] 889/91).

2.3 Sulfur Nitride Compounds, Thiazyl Compounds

2.3.1 Sulfur Fluoride Nitride, Thiazyl Fluoride, FS≡N

2.3.1.1 Handling. Stability. Storage

FSN is a colorless, pungent gas at room temperature and at <0.4°C is a pale yellow liquid. The gas is extremely hygroscopic and storage time at pressures <760 Torr at room temperature is limited in a copper vessel. It decomposes completely during a few days or weeks when allowed to stand in a glass vessel at the same conditions. The reaction with glass at elevated temperature or elevated pressure gives S_4N_4, SOF_2, SO_2, SiF_4, and N_2 [1]. Liquid FSN decomposes as low as −80°C in glass vessels. If it is sealed in a Pyrex glass ampule the liquid becomes yellow-green at room temperature after a few minutes and deep red after 1 h. Crystals separate from the gas phase and the liquid phase. Decomposition is complete after about 14 d. The reaction products are N_2, OSF_2, SiF_4, crystalline $S_3N_3F_3$, and brown-red crystals of unknown composition. Liquid FSN stored in copper or teflon tubes trimerizes slowly at room temperature to give $S_3N_3F_3$. After 50 d only half of the amount was still monomeric [2, 3].

References:

[1] Glemser, O.; Richert, H. (Z. Anorg. Allgem. Chem. **307** [1961] 313/27).
[2] Glemser, O.; Meyer, H.; Haas, A. (Chem. Ber. **97** [1964] 1704/9).
[3] Glemser, O. (in: Jolly, W. L., Preparative Inorganic Reactions, Interscience, New York 1964, pp. 227/37).

2.3.1.2 Formation. Preparation

FSN can be synthesized by many methods, e.g., by reaction of sulfur with NF_3, SF_4 with NH_3, or by fluorination of S_4N_4. However, in these reactions numerous side products are formed, which are often difficult to separate. More suitable for the preparation of FSN is the reaction of SF_4 with $(C_6H_5)_3P{=}NH$ or $(C_6H_5)_3P{=}NSi(CH_3)_3$ [1], and of ClS≡N with CsF [2] as well as the cation-exchange reaction $SN^+AsF_6^- + Cs^+F^- \rightarrow FSN + Cs^+AsF_6^-$ [20]. For production on a preparative scale the reaction of $S_3N_3Cl_3$ with KF in sulfolane [19] and the decomposition of sulfur-nitrogen-fluorine compounds, e.g., $F_2S{=}NCOF$ [3] and $Hg(N{=}SF_2)_2$ [4] is preferred.

Formation from NF_3 and Sulfur or MoS_2

A mixture of 30% FSN, 60% SSF_2, and 10% SF_4, SF_6, and SOF_2 altogether formed by passing NF_3 into molten sulfur at 400°C with a 40 mm depth of molten sulfur and ca. 12 min contact time of NF_3. Using liver of sulfur at 250°C reaction temperature the ratio FSN to SSF_2 in the product mixture increased [5]. On photolyzing a mixture of COS, NF_3, and Ar in the ratio 1:2:20 formation of FSN occurs, probably by reaction of the excited $S(^1D)$ atom with NF_3 [6].

FSN formed in 5% yield by reaction of MoS_2 with NF_3 at 280°C. The other products were MoF_6 (50% yield) and SF_4, SF_6, and SOF_2 [7].

Formation from SF_4 and NH_3, $(C_6H_5)_3P{=}NH$, or $(C_6H_5)_3P{=}NSi(CH_3)_3$

FSN was obtained in 18% yield by reaction of excess SF_4 with NH_3, both diluted with N_2 in a continuous flow system at 20 [8] or 25°C [9]. The FSN was separated from the SF_4 on a low-temperature distillation column [8]. This formation of FSN was also reported by [10]. FSN and F_3SN were produced when NH_3 was passed into a suspension of sulfur and AgF_2 in CCl_4 [10].

FSN was prepared in ~78% yield by dropping 72 mmol of $(C_6H_5)_3P{=}NH$ in 200 mL toluene to 175 mmol of SF_4 in 200 mL toluene at −78°C within 1 h. After warming to −50°C the apparatus was evacuated to 0.01 Torr. FSN and SF_4 condensed in a trap cooled to −196°C when the reaction mixture warmed to −15°C within 1 h. When stoichiometric amounts were used, according to the equation $4(C_6H_5)_3P{=}NH + 3SF_4 \rightarrow 4(C_6H_5)_3PF_2 + 3FSN + NH_4F$ under similar reaction conditions, the yield of FSN was 40.5%.

FSN was synthesized by the same method in 71% yield from $(C_6H_5)_3P{=}NSi(CH_3)_3$ and SF_4. The reaction proceeds according to the equation $(C_6H_5)_3P{=}NSi(CH_3)_3 + SF_4 \rightarrow (C_6H_5)_3PF_2 + FSN + (CH_3)_3SiF$. FSN was separated from $(CH_3)_3SiF$, SF_4, and the solvent toluene by low-temperature condensation [1].

Preparation from $SN^+AsF_6^-$ and CsF

The cation-exchange reaction $SN^+AsF_6^- + Cs^+F^- \rightarrow FSN + Cs^+AsF_6^-$ was carried out for small-scale and a gram quantity scale preparation of FSN.

In the small-scale preparation a mixture of 0.52 mmol of $SN^+AsF_6^-$ and 1.56 mmol CsF was heated to 65°C at a rate of 14°C/min. The steady evolution of FSN between 40 and 50°C ceased after 3 min. FSN was produced in about 80% yield in ca. 95% purity. The impurity OSF_4 formed by hydrolysis of SF_4.

For the large-scale preparation in a Monel vessel, a mixture of 15.3 mmol $SN^+AsF_6^-$ and 46.1 mmol of CsF and a further charge of 19.35 mmol of CsF was heated to 110°C (1.5°C/min). The volatile products contained FSN (80±10% yield), OSF_2 (10±5% yield), SF_4 (10±5% yield), and N_2 [20].

Formation from S_4N_4

A mixture of FSN, F_3SN, and SF_4 was obtained in good yield by reaction of S_4N_4 with F_2 at −75°C [9].

FSN, SF_4, OSF_2, and N_2 formed when a reaction mixture of S_4N_4 and ClF in the mole ratio ~1:1 was held at −78°C for 4 h and at room temperature for 10 h. FSN possibly formed when S_4N_4 reacted with ClF in the mole ratio ~1:2 in CCl_3F or CCl_4 [2].

FSN was produced by decomposition of a white solid of composition $S_4N_4(NSF)_4$ at 50°C. This solid forms when S_4N_4 reacts with IF_5.

Only little FSN was obtained by reaction of S_4N_4 with SF_4 in a steel bomb under pressure at 160°C.

FSN formed in 15% yield by fluorinating S_4N_4 with excess of SeF_4 at −10°C in a glass apparatus.

FSN generated in 10% yield on melting a green solid of composition $S_4N_4(SbF_5)_4$ at 145°C. This adduct forms by reaction of S_4N_4 with SbF_5 at 25°C or at 100°C [9].

FSN was prepared in 35 to 47% yield by reaction of S_4N_4 with excess freshly prepared HgF_2 in CCl_4 at boiling temperature in a quartz vessel. A slow N_2 flow transports the volatile reaction products through the reflux condenser into a trap cooled to −80°C. The reaction stops after 5 to 6 h. FSN was separated from CCl_4, CCl_2F_2, SF_4, SOF_2, and SiF_4 by low-temperature condensation [10, 11].

FSN was also obtained in small yields by refluxing S_4N_4 and AgF_2 or CoF_3 in the mole ratio ~1:5 in CCl_4 for several hours. The volatile products were collected in a trap. FSN was separated from numerous side products by low-temperature refraction [12, 13]. The formation of FSN in very low yields by this reaction was observed in former experiments [14, 15].

Formation and Preparation from $S_3N_3Cl_3$ or ClS≡N

Small quantities of FSN were formed by reaction of $S_3N_3Cl_3$ with condensed ClF in the mole ratio ~1:1. The reaction mixture was warmed from −78°C to room temperature for 18 h, or it was held at 60°C for 35 h.

The formation of FSN in a gas-phase reaction of ClSN (18 Torr) with ClF (20 to 40 Torr) at ambient temperature was observed by IR spectroscopy [2].

In a convenient simplified method for the large-scale preparation of FSN, a suspension of 100 g (1.726 mol) of dried KF and 14.0 g (57.3 mmol) of $S_3N_3Cl_3$ in 80 mL sulfolane (tetrahydrothiophene-1,1-dioxide) was quickly heated to 80 to 85°C under stirring in a dynamic vacuum. The raw product was condensed in a trap at −125°C. Repeating the fractional condensation followed by purification from traces of Cl_2 with Hg at 0°C gave FSN in 69% yield and high purity [19].

The preparation of small quantities of FSN by reaction of ClSN with CsF at 110°C is described by [2]: Recrystallized $S_3N_3Cl_3$ (9.96 mmol) was placed in a dry Pyrex vessel connected to a Monel column packed with powdered anhydrous CsF (53.29 mmol). $S_3N_3Cl_3$ was slowly heated to 70°C and the volatiles passed through a column heated to 110°C to a collection vessel held at −196°C. Initially the main products were FSN and SF_4 but after 10 min the concentration of SF_4 was negligible. Besides FSN smaller amounts of SOF_2 and SO_2ClF were identified.

Formation from Sulfur-Nitrogen-Fluorine Compounds

FSN was produced in small amounts by refluxing $S_4N_4F_4$ in CCl_4 for 2 d [12]. It forms by decomposition of $S_4N_4F_4$ with excess BF_3 at room temperature according to $S_4N_4F_4 + 4BF_3 \rightarrow 4FSN + 4BF_3$ [16].

FSN and $SN^+AsF_6^-$ were obtained by pyrolysis of $S_3N_3F_2^+AsF_6^-$ at 85°C in the vacuum [17].

The formation of FSN and SF_4 was observed when $F_2S{=}NCl$ or $F_2S{=}NBr$ reacts with Hg below room temperature or at 0°C, respectively [4].

FSN and POF_3 were produced by thermal decomposition of $F_2S{=}NPOF_2$, even at room temperature [18].

FSN formed in 90% yield when $F_2S{=}NCOF$ reacts with excess CsF in a Monel cylinder at 25°C for 3 h. When this reaction was carried out in the presence of stainless steel balls, FSN was obtained in 93% yield after heating the reactor under vacuum to 120 to 130°C [3].

Pure FSN was prepared in quantitative yield by pyrolysis of $Hg(N{=}SF_2)_2$ at 110°C in the high vacuum for 2.5 h. The volatile products were fractioned and FSN collected in a trap cooled to −145°C [4].

FSN and IF_5 formed by reaction of $Hg(N{=}SF_2)_2$ with I_2 in the mole ratio 1:2 at 50°C in a steel bomb [4].

References:

[1] Appel, R.; Laßmann, E. (Chem. Ber. **104** [1971] 2246/9).
[2] Banister, A. J.; Hey, R. G.; Passmore, J.; Sudheendra Rao, M. N. (J. Fluorine Chem. **21** [1982] 429/36).
[3] Ruff, J. K. (Inorg. Chem. **5** [1966] 1787/91).
[4] Glemser, O.; Mews, R.; Roesky, H. W. (Chem. Ber. **102** [1969] 1523/8).
[5] Glemser, O.; Biermann, U.; Knaak, J.; Haas, A. (Chem. Ber. **98** [1965] 446/50).
[6] McCrath, W. D.; Morrow, T. (Nature **212** [1966] 746).
[7] Glemser, O.; Wegener, J.; Mews, R. (Chem. Ber. **100** [1967] 2474/83).
[8] Cohen, B.; Hooper, T. R.; Peacock, R. D. (J. Inorg. Nucl. Chem. **28** [1966] 919/20).
[9] Cohen, B.; Hooper, T. R.; Hugill, D.; Peacock, R. D. (Nature **207** [1965] 748/9).
[10] Glemser, O.; Meyer, H.; Haas, A. (Chem. Ber. **97** [1964] 1704/9).

[11] Glemser, O. (in: Jolly, W. L., Preparative Inorganic Reactions, Interscience, New York 1964, pp. 227/37).
[12] Glemser, O.; Richert, H. (Z. Anorg. Allgem. Chem. **307** [1961] 313/27).
[13] Glemser, O.; Richert, H.; Rogowski, F. (Naturwissenschaften **47** [1960] 94/5).
[14] Glemser, O.; Schröder, H.; Haeseler, H. (Z. Anorg. Allgem. Chem. **279** [1955] 28/37).
[15] Glemser, O.; Schröder, H.; Haeseler, H. (Naturwissenschaften **42** [1955] 44).
[16] Glemser, O.; Koch, W. (Anales Asoc. Quim. Arg. **59** [1971] 143/8).
[17] Mews, R.; Wagner, D. L.; Glemser, O. (Z. Anorg. Allgem. Chem. **412** [1975] 148/54).
[18] Glemser, O.; Roesky, H. W.; Heinze, P. R. (Angew. Chem. **79** [1967] 723/4; Angew. Chem. Intern. Ed. Engl. **6** [1967] 710/1).
[19] Haas, A.; Rieland, M. (Chimia [Switz.] **42** [1988] 67/8).
[20] Apblett, A.; Banister, A. J.; Biron, D.; Kendrick, A. G.; Passmore, J.; Schriver, M. J.; Stojanac, M. (Inorg. Chem. **25** [1986] 4451/2).

2.3.1.3 Molecular Properties. Spectra

Geometrical Structure

Analysis of microwave [1, 2] and IR data [3] indicates a bent structure of FSN for the **electronic ground state ($\tilde{X}^1A'$)**. The geometrical parameters are R(SN) = 1.448 ± 0.002 Å, r(SF) = 1.643 ± 0.002 Å, and α(FSN) = 116°55′ ± 2′ [2]. The geometrical structure has been confirmed by quantum chemical calculations; see Table 48. Although the stoichiometry of FSN was clear from the beginning, its structural formula has been debated. The first structure proposed was bent SNF, which was derived from IR [3] and electron-diffraction data [5, 6].

On the basis of a rotational analysis of the band system of the electronic transition $\tilde{A}^1A'' \leftarrow \tilde{X}^1A'$, the geometry of the **excited state $\tilde{A}^1A''$** in the (000) level was derived as R(SN) = $1.55_6 \pm 0.03$ Å, r(SF) = $1.62_5 \pm 0.03$ Å, and α(FSN) = 101°35′ ± 40′. The principal changes on excitation were an increase in SN bond length of 0.11 Å and a decrease in valence angle of 15.3°. This change in angle is contrasted with those for the lowest excited states of other molecules with 18 valence electrons [7].

Electronic Structure

On the basis of experimental ionization energies [32] and ab initio LCAO–MO SCF calculations [32, 33], the following electron configuration is proposed: $(1a')^2 \ldots (5a')^2\ (1a'')^2\ (6a')^2 \ldots (10a')^2\ (2a'')^2\ (11a')^2\ (3a'')^2\ (12a')^2\ (13a')^2$. The ab initio calculations indicate a valence structure approximately F–S≡N. The SN bond is formed mainly from 8a′, 11a′, 12a′, and 3a″. The SF bond is mainly localized in the 9a′ orbital, while 3a″ and 12a′ are SF antibonding [32].

Quantum Chemical Calculations

The electronic properties and geometric parameters have been the subject of several quantum chemical treatments. Table 48 gives a survey of publications. The following abbreviations and symbols are used:

MO	molecular orbital	MP	Møller-Plesset
SCF	self-consistent field	E_t	total energy
LCAO	linear combination of atomic orbitals	E_i	orbital energy or ionization energy
EH	extended Hückel	pop	population analysis
CNDO	complete neglect of differential overlap	q	charge distribution
CEPA	coupled electron pair approximation	μ	dipole moment
CPF	coupled pair functional		

Table 48
Quantum Chemical Calculations for FSN.

method	calculated molecular properties	Ref.
ab initio LCAO–MO SCF+MP	geometries, E_t, quadratic force constants, vibrational spectrum, μ, isomerization	[20]
ab initio LCAO–MO SCF+CPF	geometries, pop, q	[48]
ab initio LCAO–MO SCF+CEPA	geometries, pop, q, force constants	[45]
ab initio LCAO–MO SCF	E_t, E_i, pop, μ	[33]
ab initio LCAO–MO SCF	geometries, E_t, potential surface, pop, μ, isomerization	[40]
ab initio LCAO–MO SCF	geometries, E_t, pop	[41]
ab initio LCAO–MO SCF	geometries, E_t, isomerization	[42]
ab initio LCAO–MO SCF	E_i	[32]
ab initio LCAO–MO SCF	μ	[43]
CNDO/2	geometries, E_i, pop, μ	[44]
CNDO/2, EH	E_i	[35]
EH	E_i	[34]

Valence Level Ionization Energies

Adiabatic and vertical ionization energies, determined by HeI, HeII, HeIII, and NeI photoelectron spectroscopy, are given in Table 49, p. 212. The assignment to the molecular orbitals was made from the analysis of the vibrational band fine structure. For comparison, orbital energies obtained with ab initio LCAO–MO SCF calculations are listed [32]. Ionization energies determined by HeI photoelectron spectroscopy [34, 35, 36] are in accordance with those given in Table 49, p. 212. Interpretation differs in part, both with regard to vibrational assignments [35, 36], and to assignments of electronic states made by correlation with eigenvalues calculated by EH and CNDO/2 methods and Koopmans' theorem [34, 35]. Comparisons of the PE spectrum of FSN with those of SSO, O_3, and SO_2 are made by [33].

Table 49
Adiabatic and Vertical Ionization Energies, E_i, and Orbital Energies, ε_i, of FSN [32].

ionic state	MO	E_i (ad) in eV	E_i (vert) in eV	vibrational spacing in cm^{-1}	vibrational assignment	$-\varepsilon_i$ in eV
$\tilde{X}\,^2A'$	13a′	11.54(1)	11.89(3)	300(50)	ν_3	12.13
$\tilde{A}\,^2A'$	12a′	13.382(4)	13.543(4)	820(40)	ν_2	14.86
				460(50)	ν_3	
$\tilde{B}\,^2A''$	3a″	13.775(5)	13.906(2)	1060(40)	ν_1	14.23
				695(40)	ν_2	
				365(40)	ν_3	
$\tilde{C}\,^2A'$	11a′	14.93(1)	15.639(5)	895(30)	?	17.87
$\tilde{D}\,^2A''$	2a″	—	16.56(3)	—	—	18.62
$\tilde{E}\,^2A'$	10a′	—	17.24(8) (sh)	—	—	18.86
$\tilde{F}\,^2A'$	9a′	—	21.1(1)	—	—	24.15

The ionization energy $E_i = 11.36 \pm 0.03$ eV was obtained from a mass spectroscopic study [19].

Core Level Ionization Energies

The ionization energies of core electrons (core binding energies) $E_i\,(S2p_{3/2}) = 174.14(4)$ eV, $E_i\,(N1s) = 406.88(2)$ eV, and $E_i\,(F1s) = 692.74(2)$ eV of gaseous FSN were determined from X-ray photoelectron spectra [37].

Dipole Moment μ

$|\mu| = 1.902 \pm 0.012$ D was obtained from Stark effect measurements of the $5_{1,4} \rightarrow 4_{2,3}$ rotational transition in the microwave spectrum. The components along the principal inertial axes were $\mu_a = 0.242 \pm 0.012$ D and $\mu_b = 1.886 \pm 0.011$ D [2].

Polarizability

The molecular polarizabilities $\alpha = 35.98 \times 10^{-25}$ cm^3 [11] and 38.52×10^{-25} cm^3 [12] were calculated on the basis of the δ-function model of chemical binding.

^{14}N Quadrupole Coupling Constant

The ^{14}N quadrupole coupling constants of $^{19}F^{32}S^{14}N$ $eq_{aa}Q = -1.720 \pm 0.041$ MHz, $eq_{bb}Q = -3.990 \pm 0.036$ MHz, and $eq_{cc}Q = 5.710 \pm 0.046$ MHz were derived from microwave data [4]. Earlier, the value of $eq_{aa}Q = -1.66 \pm 0.05$ MHz and the asymmetry parameter $\eta = 5.86 \pm 0.05$ were obtained from the $1_{0,1} \rightarrow 1_{1,0}$ and the $1_{0,1} \rightarrow 2_{1,2}$ rotational transitions in the microwave spectrum [1].

Nuclear Magnetic Resonance

^{14}N NMR spectrum (1.05 M in SO_2/CH_3NO_2 external) at 22°C: $\delta = 196$ ppm [52]; (15% in C_6D_6/CH_3NO_2 external): $\delta = 115.5$ ppm [49].

^{19}F NMR spectrum (solvent?/$CFCl_3$ internal): $\delta = 238.5$ ppm [49]; ($SF_4/CFCl_3$ external) at −80°C: $\delta = 234$ ppm [53]; (neat liquid/saturated aqueous KF external) at ambient temperature: $\delta = 358$ ppm [3], referred to $CFCl_3$ external $\delta = 233.9$ ppm [53].

Rotational Constants. Moments of Inertia

Analysis of the microwave spectrum [1, 2] yielded the rotational constants A, B, C and the centrifugal distortion constants τ_{aaaa}, τ_{bbbb}, τ_{aabb}, τ_{abab}, defined by Kivelson and Wilson [54], for the **ground state $\tilde{X}^1A'$** of $F^{32}SN$ (in MHz): A = 49719.56, B = 8712.33, C = 7393.13; τ_{aaaa} = $-8.8343_{34} \pm 0.0158$, $\tau_{bbbb} = -0.05644_0 \pm 0.00054$, $\tau_{aabb} = 0.44799_6 \pm 0.00346$, $\tau_{abab} = -0.08956_5$ ± 0.00149 [2]. Almost identical values, based on microwave data from [1], are given in [8]. A fit of observed transition frequencies of the ν_2 band and the 73 ground-state rotational transitions already available [1, 2, 8] to Watson's A-reduced Hamiltonian [51], produced the values for the ground-state parameters [50]: A = 49719.632 ± 0.014, B = 8712.4490 ± 0.0025, C = 7393.0332 ± 0.0022 MHz; $\Delta_J = 9.3803 \pm 0.0082$ kHz, $\Delta_{JK} = -0.12408 \pm 0.00013$ MHz, $\Delta_K = 2.34209 \pm 0.00046$ MHz; $\delta_J = 2.29107 \pm 0.00044$ kHz, $\delta_K = 0.037250 \pm 0.000039$ MHz; $\Phi_J = -0.0088 \pm 0.0025$ Hz, $\Phi_{JK} = -0.43 \pm 0.10$ Hz, $\Phi_{KJ} = -0.0361 \pm 0.0033$ kHz, $\Phi_K = 0.483$ ± 0.029 kHz. A review on the calculation and interpretation of centrifugal distortion constants using the published microwave spectra of FSN gives Watson's [55, 56] determinable parameters as A″ = 49719.607 ± 0.017, B″ = 8712.259 ± 0.003, C″ = 7392.998 ± 0.003; $\tau_1 = 0.3772$ ± 0.0027, $\tau_2 = 0.0254 \pm 0.0004$, $\tau_3 = 1.276 \pm 0.009$, $\tau_{aaaa} = -8.897 \pm 0.004$, $\tau_{bbbb} = -0.05626$ ± 0.00015, $\tau_{cccc} = -0.01943 \pm 0.00011$ MHz [9]; see also [4].

For $F^{34}SN$ the rotational constants were obtained from a rigid-rotor analysis after correction for centrifugal distortion effects, and the centrifugal distortion constants were calculated from a force field based on microwave and IR data (see p. 215): A = 48298.98, B = 8701.26, C = 7352.78; $\tau_{aaaa} = -8.46707$, $\tau_{bbbb} = -0.05713$, $\tau_{aabb} = 0.42993$, $\tau_{abab} = -0.08346$ MHz [2].

The moments of inertia for $F^{32}SN$ and $F^{34}SN$ were obtained from the rotational constants using the conversion factor $h/8\pi = 505531$ MHz·u·Å²: $I_a = 10.1680$, $I_b = 58.0242$, and I_c = 68.3789 u·Å² for $F^{32}SN$; $I_a = 10.4672$, $I_b = 58.0982$, $I_c = 68.7538$ u·Å² for $F^{34}SN$ [1].

Rotational constants for the **vibrationally excited states** $v_2 = 1$ and $v_1 = 1$ were obtained from an analysis of the rotational transitions of the ν_2 and ν_1 bands measured by diode laser spectrometry [50]. A fit of 265 transition frequencies with $J \leq 49$ and $K \leq 19$ and the 73 ground-state microwave frequencies already available [1, 2, 8] to Watson's A-reduced Hamiltonian [51], gives for $v_2 = 1$: A = 49673.86 ± 0.29, B = 8656.295 ± 0.093, C = 7349.211 ± 0.074 MHz; $\Delta_J = 9.327 \pm 0.028$ kHz, $\Delta_{JK} = -0.12628 \pm 0.00026$ MHz, $\Delta_K = 2.3794 \pm 0.0011$ MHz, $\delta_J = 2.276$ ± 0.019 kHz, $\delta_K = 0.0357 \pm 0.0024$ MHz. For $v_1 = 1$, only the quantities (B + C)/2 = 8037.40 ± 0.38 and (A − (B + C))/2 = 41393.08 ± 0.75 MHz could be determined from a-type transitions with $K \geq 5$ of the ν_1 band [50].

The rotational constants of the (001) level of the $\tilde{X}^1A'$ state were derived from the analysis of the (001)–(001) band of the electronic transition $\tilde{A}^1A'' \leftarrow \tilde{X}^1A'$: A = 50755 ± 30 MHz (1.693 ± 0.001 cm⁻¹), B = 8514 ± 150 (0.284 ± 0.005), C = 7375 ± 150 (0.246 ± 0.005); $\tau_{aaaa} = -9.68$ ± 0.39 $(-(323 \pm 13) \times 10^{-6})$, $\tau_{bbbb} = -0.048 \pm 0.006$ $(-(1.6 \pm 0.2) \times 10^{-6})$, $\tau_{aabb} = 0.483 \pm 0.027$ $((16.1 \pm 0.9) \times 10^{-6})$, $\tau_{abab} = 0 \pm 0.18$ $((0 \pm 6) \times 10^{-6})$. $I_a = 9.957 \pm 0.007$, $I_b = 59.4 \pm 1.0$, $I_c = 68.4$ ± 1.4 u·Å² [7].

From the rotational analysis of the band system with origin at 25487 cm⁻¹ of the electronic transition $\tilde{A}^1A'' \leftarrow \tilde{X}^1A'$, the rotational constants for the (000) level of the **$\tilde{A}^1A''$ state** of FSN were determined [7] (for (001) and (002), see the original paper): A = 31645.8 ± 3.9 MHz (1.05559 ± 0.00013 cm⁻¹), B = 10051.02 ± 0.69 (0.335266 ± 0.000023), C = 7552.193 ± 0.210 (0.251914 ± 0.000007); $\tau_{aaaa} = -2.95 \pm 0.07$ $((-98.5 \pm 2.5) \times 10^{-6})$, $\tau_{bbbb} = -0.102 \pm 0.006$ $((-3.4 \pm 0.2) \times 10^{-6})$, $\tau_{aabb} = 0.68 \pm 0.06$ $((22.8 \pm 2.1) \times 10^{-6})$, $\tau_{abab} = -0.30 \pm 0.04$ $((-10.0 \pm 1.5)$ $\times 10^{-6})$. $I_a = 15.970 \pm 0.002$, $I_b = 50.281 \pm 0.003$, $I_c = 66.925 \pm 0.002$ u·Å².

Fundamental Vibrations

The gas-phase IR spectrum gave the following frequencies of the three fundamentals of FSN (point group symmetry C_s) [3].

Table 50
Frequencies of the Fundamental Vibrations of FSN [3, 14].

symmetry class	No.	approximate type of mode	activity	frequency in cm^{-1}
A′	ν_1	SN stretch.	IR, Ra	1372
A′	ν_2	SF stretch.	IR, Ra	640
A′	ν_3	bend.	IR, Ra	366

The fundamental frequencies $\nu_1=1375.636$ and $\nu_2=644.996$ cm^{-1} of gaseous FSN were determined from measurements with Doppler-limited resolution, using a tunable diode laser spectrometer [50].

From vibrational analyses of electronic spectra, fundamental frequencies of the excited states $\tilde{a}^3A''$ and $\tilde{A}^1A''$ were obtained. $\tilde{a}^3A''$ state: $\nu_1=864.0$ [15], 862 cm^{-1} [16], $\nu_2=673$ cm^{-1} [17], $\nu_3=303.1$ [15], 297 cm^{-1} [16]. $\tilde{A}^1A''$ state: $\nu_1=974.6$ [15], 984 cm^{-1} [18], $\nu_2=707$ cm^{-1} [17], $\nu_3=317.1$ [15], 315 cm^{-1} [18].

Vibrational Amplitudes

The mean amplitudes of vibration given in Table 51 were calculated from various sets of force constants. For an earlier calculation, see [10].

Table 51
Mean Amplitudes of Vibration u (in Å) of FSN at 298.16 K.

u(S–N)	u(S–F)	u(N···F)	Ref.
0.0355	0.0490	0.0780	[21]
0.0355	0.0489	0.0708	[22]
0.0367	0.0496	0.0637	[25]

Coriolis Coupling Constants

Two sets were derived from different force fields: $\zeta_{12}=0.10557$, $\zeta_{13}=0.76147$, $\zeta_{23}=0.63754$ [25] and $|\zeta_{12}|=0.3595$, $|\zeta_{13}|=0.7598$, $|\zeta_{23}|=0.5417$ [21], both obeying the ζ sum rule $\zeta_{12}^2+\zeta_{13}^2+\zeta_{23}^2=1$.

Force Constants

The constants (f) of the general valence force field (GVFF) and the symmetry force constants (F_{ij}) given in Table 52 were calculated by using molecular parameters [2], vibrational data [3], and/or centrifugal distortion constants [1, 2]. The constants correspond to SN bond stretching ($f_R=F_{11}$), SF bond stretching ($f_r=F_{22}$), FSN bond bending ($f_\alpha=F_{33}\cdot R\cdot r$), interaction between bond stretching SN and SF ($f_{Rr}=F_{12}$), and interactions between bond stretching and bond bending ($f_{R\alpha}=F_{13}\cdot R$, $f_{r\alpha}=F_{23}\cdot r$). Values of force constants, in addition to those compiled in Table 52, are given by [1, 8, 22, 24]. A general relation $f_R=145\cdot R^{-7.00}$ with the bond length

R = 1.446 Å [1] yields f_R = 10.71 mdyn/Å [19]. Quadratic force constants were calculated by an ab initio Hartree-Fock method [20].

Table 52
Symmetry Force Constants F_{ij} (all in mdyn/Å) of FSN.

F_{11}	F_{22}	F_{33}	F_{12}	F_{13}	F_{23}	Ref.
10.94(6)	2.88(4)	0.41	0.86(10)	0.0	0.0	[23][a)]
10.703	2.872	0.411	0.0	0.0	0.014	[2][b)]
10.851	2.932	0.401	0.358	0.079	0.072	[21][c)]
11.3822	2.9006	0.412	1.21	0.925	—	[25][d)]
10.89	2.98	0.40	0.33	0.17	0.12	[12][e)]

[a)] f_α = 0.98(1) mdyn·Å. F_{13} and F_{23} constrained to zero. – [b)] $F_{23} = f_{r\alpha}/(R \cdot r)^{1/2}$. F_{12} and F_{13} constrained to zero. – [c)] Approximate method due to [57]. – [d)] Approximate method due to [58]. f_α = 0.9802 mdyn·Å, $f_{R\alpha}$ = 1.34 mdyn in original paper. – [e)] Approximate method due to [59].

Cartesian force constants were used to calculate "bond asymmetry parameters" [26].

Bond Dissociation Enthalpies

The S≡N bond dissociation enthalpy D°_{298}(FS–N) = 125.7 ± 3.3 kcal/mol [27] is calculated from the appearance potential a.p. (SF^+) = 15.45 ± 0.1 eV for the process FSN + e^- → SF^+ + N + $2e^-$ [19] and the ionization potential E_i (SF) = 10 eV. That ionization potential was based upon a (corrected) theoretical value (10.0 ± 0.4 eV) from [28] and an experimental value (10.09 ± 0.10 eV) from [29] (see also "Schwefel" Erg.-Bd. 2, 1978, p. 202). The obsolete values D°_{298}(FS–N) = 71 ± 5 kcal/mol [31] and 81 kcal/mol [19] were based on erroneous FSN formation enthalpies.

The S–F bond dissociation enthalpy D°_{298}(F–SN) = 91 ± 6 kcal/mol [27] was obtained by combining the enthalpies of formation $\Delta H^\circ_{f,298}$(FSN, g) = –5.0 ± 0.9 kcal/mol [27], $\Delta H^\circ_{f,298}$(SN, g) = 67.1 ± 6.0 kcal/mol [30], and $\Delta H^\circ_{f,298}$(F, g) = 18.9 ± 0.1 kcal/mol.

Microwave Spectrum

Rotational transitions of $F^{32}SN$ were measured in the region 8 to 40 GHz by using a Stark effect modulated microwave spectrometer. The assignment was based upon measurements of the ^{14}N quadrupole splitting, the observed Stark effect, and comparisons with values calculated from a rigid asymmetric rotor model [1, 2]. Rotational transitions for values of J from 0 to 18 [1] and 0 to 39 [2] were observed and assigned. This yielded the rotational constants and thus the structural parameters [1, 2], and information for force field determination [2]. Hyperfine structure arising from the quadrupole effect was observed and yielded the ^{14}N quadrupole coupling constant (see p. 212) [1]. Stark effect measurements were performed and gave the molecular dipole moment (see p. 212) [2]. A survey on microwave spectra is given by [4].

IR Absorption

The gas-phase IR spectrum of FSN given in Table 53, p. 216, was studied between 263 and 4000 cm^{-1} using 10 cm gas cells at 10, 50, and ~700 Torr and ambient temperature [3].

Table 53
IR Spectrum of Gaseous FSN [3].

branch	ν in cm^{-1}	intensity	assignment
P	356	s	
minimum	366		ν_3
R	380	s	
P	627	vs	
(Q)	640	vs	ν_2
R	651	vs	
	1000	sh	$\nu_2+\nu_3$
P	1270	m	
Q	1283	m	$2\nu_2$
R	1295	m	
P	1361	s	
Q	1372	s	ν_1
R	1387	s	
	1720	vvvw	$\nu_1+\nu_3$
	1920	vvw	$3\nu_2$
	2020	vvw	$\nu_1+\nu_2$
	2730	vw	$2\nu_1$

The ν_2 band at 644.996 cm^{-1} and the ν_1 band at 1375.636 cm^{-1} of gaseous FSN were measured with Doppler-limited resolution using a diode laser spectrometer with a path length of 20 m and gas pressures in the region 0.5 to 1 Torr. The ν_2 a-type Q branch shows 265 vibration-rotation transitions with J values between 3 and 49, and K values between 0 and 19. For the a-type Q branch of ν_1, 135 transitions with J values between 3 and 24 and K values between 3 and 12 were assigned. The ν_1 band was found to be perturbed as the result of a Coriolis-type interaction between the states (1, 0, 0) and (0, 1, 2) [50]. Absorption cross sections were measured for 15 CO_2 laser frequencies from 986.568 to 1035.474 cm^{-1} [47].

The IR spectrum of solid FSN at low temperatures (liquid N_2) shows frequencies of the fundamental vibrations at 1380 (ν_1), 588 (ν_2), and 372 cm^{-1} (ν_3) and a band at 1305 cm^{-1} [46].

IR Emission

Optical pumping of FSN vapor at 220 to 250 K with the output from a tunable transversely excited atmosphere (TEA) CO_2 laser with frequencies of 986 to 1032 cm^{-1} stimulates emission in the region 618 to 658 cm^{-1}. The CO_2 laser output pumps the $\nu_2+\nu_3$ combination band at 1006 cm^{-1}, and stimulated emission is obtained on the transition $\nu_2+\nu_3 \rightarrow \nu_3$ [47].

Raman Effect

The Raman spectrum of solid FSN at −196°C shows bands at 1385 (s), 650 (w), and 380 (m) cm^{-1} [49].

Electronic Absorption Spectra

Three electronic transitions of FSN are known in the gas state. The lowest energy transition at 536.0 nm (18662 cm^{-1}) has been shown by vibrational [15, 16, 17] and partial rotational [38] analysis and by Zeeman effect experiments [38] to be an $\tilde{a}\,^3A'' \leftarrow \tilde{X}\,^1A'$ transition. The vibrational analysis gave the fundamental frequencies of the $\tilde{a}\,^3A''$ state (see p. 214). Long hot-band progressions with assignments $1_0^m 3_1^n$ were identified [17].

The singlet transition $\tilde{A}\,^1A'' \leftarrow \tilde{X}\,^1A'$ occurs in the near UV with an origin at 392.3 nm (25487 cm^{-1}) [7, 15, 18, 38]. The vibrational structure of this system consists of long progressions in upper state vibrational frequencies (see p. 214), together with hot bands involving up to three quanta of a lower state frequency of 370 cm^{-1} (ν_3) [17, 18]. Because of the near equality of ν_1' and $3\nu_3'$ most of the bands in the system are heavily overlapped [7].

A further electronic transition [18] with one short progression in $\nu_1'=1090$ cm^{-1} begins at 236.2 nm (42341 cm^{-1}). It has also been studied under high resolution [7]. The diffuseness in the bands has not allowed the transition symmetry to be assigned.

Fluorescence Emission Spectrum of the $\tilde{A}\,^1A''$ State

The fluorescence spectrum excited in the 3_0^3 band at 378.05 nm of vibrationally equilibrated FSN vapor (~1 Torr FSN in ~120 Torr Ar) was observed between 380 and 550 nm. Most of the twenty-nine major band heads ranging from 397.8 to 519.3 nm could be assigned to ground-state vibrational levels $n\nu_1 + m\nu_3$ ($n=0$ to 3, $m \geq 1$; assignments in a table differ from those in a figure) [39].

Fluorescence lifetimes of $\tilde{A}^1A''$ levels with ≤ 4100 cm^{-1} excess vibrational energy lay around 10 μs. The fluorescence lifetime for excitation of levels with $\gtrsim 4500$ cm^{-1} excess vibrational energy became controlled by a unimolecular radiationless process, which is likely photodissociation. The dependence of this radiationless rate on energy and vibrational mode was investigated [39].

References:

[1] Kirchhoff, W. H.; Wilson, E. B., Jr. (J. Am. Chem. Soc. **85** [1963] 1726/9).
[2] Cook, R. L.; Kirchhoff, W. H. (J. Chem. Phys. **47** [1967] 4521/7).
[3] Richert, H.; Glemser, O. (Z. Anorg. Allgem. Chem. **307** [1961] 328/44).
[4] Lovas, F. J. (J. Phys. Chem. Ref. Data **7** [1978] 1445/750, 1550/4).
[5] Glemser, O.; Richert, H.; Rogowski, T. (Naturwissenschaften **47** [1960] 94/5).
[6] Rogowski, F. (Z. Physik. Chem. [Frankfurt] **27** [1961] 277/96).
[7] Barrow, T.; Dixon, R. N. (Mol. Phys. **26** [1973] 109/17).
[8] Mirri, A.; Guarnieri, A. (Spectrochim. Acta A **23** [1967] 2159/65).
[9] Kirchhoff, W. H. (J. Mol. Spectrosc. **41** [1972] 333/80).
[10] Nagarajan, G. (Indian J. Pure Appl. Phys. **4** [1966] 244/50).

[11] Lippincott, E. R.; Nagarajan, G.; Stutman, J. M. (J. Phys. Chem. **70** [1966] 78/84).
[12] Chopra, J. R.; Verma, U. P.; Pandey, A. N. (Indian J. Pure Appl. Phys. **20** [1982] 500/2).
[13] Glemser, O.; Richert, H. (Z. Anorg. Allgem. Chem. **307** [1961] 313/27).
[14] Shimanouchi, T. (J. Phys. Chem. Ref. Data **6** [1977] 993/1102, 1009).
[15] Craig, D. P.; Fischer, G. (Chem. Phys. Letters **4** [1969/70] 227/8).
[16] Barrow, T.; Dixon, R. N. (Chem. Phys. Letters **4** [1969/70] 547/9).
[17] Fischer, G. (J. Mol. Spectrosc. **51** [1974] 208/15).
[18] Barrow, T.; Dixon, R. N.; Glemser, O.; Mews, R. (Trans. Faraday Soc. **65** [1969] 2295/9).
[19] Glemser, O.; Müller, A.; Böhler, D.; Krebs, B. (Z. Anorg. Allgem. Chem. **357** [1968] 184/9).
[20] Schaad, L. J.; Hess, B. A., Jr.; Čársky, P.; Zahradník, R. (Inorg. Chem. **23** [1984] 2428/30).

[21] Namasivayam, R.; Mayilavelan, S. (Z. Naturforsch. **34a** [1979] 716/20).
[22] Peacock, C. J.; Heidborn, U.; Müller, A. (J. Mol. Spectrosc. **30** [1969] 338/44).
[23] Müller, A.; Mohan, N.; Cyvin, S. J.; Weinstock, N.; Glemser, O. (J. Mol. Spectrosc. **59** [1976] 161/70).
[24] Sawodny, W.; Fadini, A.; Ballein, K. (Spectrochim. Acta **21** [1965] 995/1006).
[25] Gopinath, C. R.; Rao, K. S. R. (Indian J. Pure Appl. Phys. **19** [1981] 89/91).
[26] Ebenezer, J. S. P.; Srinivasamoorthy, R.; Savariraj, G. A. (Indian J. Pure Appl. Phys. **19** [1981] 249/53).

[27] Larson, J. W.; Johnson, G. K.; O'Hare, P. A. G.; Hubbard, W. N.; Glemser, O. (J. Chem. Thermodyn. **5** [1973] 689/97).
[28] O'Hare, P. A. G.; Wahl, A. C. (J. Chem. Phys. **53** [1970] 2834/46).
[29] Hildenbrand, D. L. (J. Phys. Chem. **77** [1973] 897/902).
[30] O'Hare, P. A. G. (J. Chem. Phys. **52** [1970] 2992/6).

[31] O'Hare, P. A. G.; Hubbard, W. N.; Glemser, O.; Wegener, J. (J. Chem. Thermodyn. **2** [1970] 71/6).
[32] DeKock, R. L.; Lloyd, D. R.; Breeze, A.; Collins, G. A. D.; Cruickshank, D. W. J.; Lempka, H. J. (Chem. Phys. Letters **14** [1972] 525/7).
[33] Rosmus, P.; Dacre, P.; Solouki, B.; Bock, H. (Theor. Chim. Acta **35** [1974] 129/39).
[34] Cowan, D. O.; Gleiter, R.; Glemser, O.; Heilbronner, E. (Helv. Chim. Acta **55** [1972] 2418/21).
[35] Cowan, D. O.; Gleiter, R.; Glemser, O.; Heilbronner, E.; Schäublin, J. (Helv. Chim. Acta **54** [1971] 1559/64).
[36] Dixon, R. N.; Duxbury, G.; Fleming, G. R.; Hugo, J. M. V. (Chem. Phys. Letters **14** [1972] 60/3).
[37] Beach, D. B.; Jolly, W. L.; Mews, R.; Waterfeld, A. (Inorg. Chem. **23** [1984] 4080/4).
[38] Fischer, G. (Chem. Phys. Letters **11** [1971] 356/8).
[39] McDonald, J. R. (Chem. Phys. **13** [1976] 339/47).
[40] Seeger, R.; Seeger, U.; Bartetzko, R.; Gleiter, R. (Inorg. Chem. **21** [1982] 3473/7).

[41] So, Suk Ping; Richards, W. G. (J. Chem. Soc. Faraday Trans. II **74** [1978] 1743/5).
[42] Collins, M. P. S.; Duke, B. J. (J. Chem. Soc. Dalton Trans. **1978** 277/9).
[43] Burton, P. G.; Carlsen, N. R.; Magnusson, E. A. (Mol. Phys. **32** [1976] 1687/94).
[44] Gonbeau, D.; Pfister-Guillouzo, G. (J. Chim. Phys. Physicochim. Biol. **73** [1976] 787/94).
[45] Zirz, C.; Ahlrichs, R. (Inorg. Chem. **23** [1984] 26/31).
[46] Miller, K. V.; Emken, W. C.; Duncan, L. C. (J. Fluorine Chem. **26** [1984] 125/32).
[47] Fischer, T. A.; Tiee, J. J.; Wittig, C. (Appl. Phys. Letters **37** [1980] 592/4).
[48] Ehrhardt, C.; Ahlrichs, R. (Chem. Phys. **108** [1986] 417/28).
[49] Haas, A.; Rieland, M. (Chimia [Switz.] **42** [1988] 67/8).
[50] Magg, U.; Lindenmayer, J.; Jones, H. (J. Mol. Spectrosc. **126** [1987] 270/81).

[51] Watson, J. K. G. (in: Durig, J. R., Vibrational Spectra and Structure, Vol. 6, Dekker, New York 1977).
[52] Passmore, J.; Schriver, M. J. (Inorg. Chem. **27** [1988] 2749/51).
[53] Appel, R.; Laßmann, E. (Chem. Ber. **104** [1971] 2246/9).
[54] Kivelson, D.; Wilson, E. B., Jr. (J. Chem. Phys. **20** [1952] 1575/9).
[55] Watson, J. K. G. (J. Chem. Phys. **46** [1967] 1935/49).
[56] Watson, J. K. G. (J. Chem. Phys. **48** [1968] 4517/24).
[57] Redington, R. L.; Aljibury, A. L. K. (J. Mol. Spectrosc. **37** [1971] 494/516).
[58] Herranz, J.; Castaño, F. (Spectrochim. Acta **22** [1966] 1965/7).
[59] Sharma, D. K.; Verma, U. P.; Pandey, A. N. (Z. Physik. Chem. [Leipzig] **260** [1979] 209/13).

2.3.1.4 Thermal Properties

Phase Transitions

Experimental values of the vapor pressure p (given in selection) [1]:

T in K	197.77	198.88	204.45	205.41	214.78	221.86	226.07	230.55	238.79	239.14
p in Torr	7.2	8.2	13.4	14.2	30.2	50.1	65.9	88.4	143.6	146.4

The boiling point at 760 Torr $+0.4\pm2$°C is calculated from vapor pressure values [1 to 4]. The melting point is -89°C [1, 2, 3] or -84°C [4].

The enthalpy of evaporation at the boiling point is $\Delta H_v = 6.052 \pm 0.030$ kcal/mol which yields Trouton's constant $\Delta H_v/T_v = 22.1\ cal \cdot mol^{-1} \cdot K^{-1}$ [1].

References:

[1] Glemser, O.; Richert, H. (Z. Anorg. Allgem. Chem. **307** [1961] 313/27).
[2] Glemser, O.; Meyer, H.; Haas, A. (Chem. Ber. **97** [1964] 1704/9).
[3] Glemser, O.; Richert, H.; Rogowski, F. (Naturwissenschaften **47** [1960] 94/5).
[4] Larson, J. W.; Johnson, G. K.; O'Hare, P. A. G.; Hubbard, W. N.; Glemser, O. (J. Chem. Thermodyn. **5** [1973] 689/97).

2.3.1.5 Enthalpy of Formation. Heat Capacity. Thermodynamic Functions

The enthalpy of formation of gaseous FSN from the elements in the standard states $\Delta H^\circ_{f,298} = -5.0 \pm 0.9$ kcal/mol was determined from calorimetric measurements of the heat of combustion of FSN in F_2: FSN (g) + $5/2 F_2$ (g) $\rightarrow$ $1/2 N_2$ (g) + SF_6 (g) [1, 2].

The values $\Delta H^\circ_{f,298} = -9.3 \pm 1.7$ kcal/mol and -9.5 ± 3.5 kcal/mol, obtained from combination of ionization energy $E_i(FSN^+) = 11.89 \pm 0.03$ eV [3] and mass spectrometric results [4], are in reasonable agreement with the calorimetric enthalpy of formation [1].

The value $\Delta H^\circ_{f,298} = +41 \pm 2$ kcal/mol [5], previously obtained from combination of $\Delta H^\circ_{f,298}(F_3SN, gas) = -85.2 \pm 0.5$ kcal/mol [5] and the appearance potential a.p. $(FSN^+) = 11.36 \pm 0.03$ eV [4], and the value $\Delta H^\circ_{f,298} = +30$ kcal/mol, calculated from mass spectroscopic results [4], are obsolete.

The thermodynamic functions for the ideal gaseous state at 1 atm pressure were calculated on the basis of a rigid rotator, harmonic oscillator model [6] with the fundamental vibrations $\nu_1 = 1372$, $\nu_2 = 640$, $\nu_3 = 366\ cm^{-1}$ [7] and the structural parameters R(SN) = 1.446, r(SF) = 1.646 Å, α(FSN) = 116°52′ [8].

Table 54
Selected Values of the Thermodynamic Functions C°_p, S°, $-(G^\circ - H^\circ_0)/T$, and $(H^\circ - H^\circ_0)/T$ of FSN (ideal gas) [6].

T in K	C°_p	S° in $cal \cdot mol^{-1} \cdot K^{-1}$	$-(G^\circ - H^\circ_0)/T$	$(H^\circ - H^\circ_0)/T$
100	8.254	51.933	43.930	8.003
150	8.891	55.394	47.206	8.188
200	9.537	58.041	49.595	8.446
273.16	10.326	61.133	52.287	8.846

Table 54 (continued)

T in K	C_p°	S°	$-(G^\circ-H_0^\circ)/T$	$(H^\circ-H_0^\circ)/T$
		in $cal \cdot mol^{-1} \cdot K^{-1}$		
298.16	10.552	62.044	53.065	8.979
400	11.320	65.258	55.776	9.482
600	12.311	70.049	59.776	10.273
800	12.868	73.680	62.819	10.861
1000	13.192	76.589	65.290	11.299
1500	13.562	82.039	70.035	12.004
2000	13.708	85.970	73.556	12.414
3000	13.816	90.857	77.999	12.858
4000	13.857	95.554	82.440	13.114
5000	13.875	98.590	85.326	13.264
6000	13.884	101.127	87.763	13.364

References:

[1] Larson, J. W.; Johnson, G. K.; O'Hare, P. A. G.; Hubbard, W. N.; Glemser, O. (J. Chem. Thermodyn. **5** [1973] 689/97).
[2] Hubbard, W. N.; O'Hare, P. A. G.; Johnson, G. K.; Ader, M.; Settle, J. L.; Larson, J. W.; Tevebaugh, A. D.; Vogel, R. C. (ANL-7876 [1972] 1/33 from N.S.A. **26** [1972] No. 33190).
[3] DeKock, R. L.; Lloyd, D. R.; Breeze, A.; Collins, G. A. D.; Cruickshank, D. W. J.; Lempka, H. J. (Chem. Phys. Letters **14** [1972] 525/7).
[4] Glemser, O.; Müller, A.; Böhler, D.; Krebs, B. (Z. Anorg. Allgem. Chem. **357** [1968] 184/9).
[5] O'Hare, P. A. G.; Hubbard, W. N.; Glemser, O.; Wegener, J. (J. Chem. Thermodyn. **2** [1970] 71/6).
[6] Nagarajan, G. (Indian J. Pure Appl. Phys. **4** [1966] 244/50).
[7] Richert, H.; Glemser, O. (Z. Anorg. Allgem. Chem. **307** [1961] 328/44).
[8] Kirchhoff, W. H.; Wilson, E. B., Jr. (J. Am. Chem. Soc. **85** [1963] 1726/9).

2.3.1.6 Chemical Reactions

2.3.1.6.1 Isomerization

The isomerization of FSN to give SNF was investigated with use of ab initio Hartree-Fock methods. The calculated potential surface shows that the isomers are separated by a barrier of about 29 [1] or 60 kcal/mol [2] relative to FSN. Intramolecular isomerization is energetically favored compared to fragmentation [1]. FSN is more stable than SNF by 21 [1] or 26 kcal/mol [2].

2.3.1.6.2 Trimerization

FSN trimerizes in the liquid phase and in the gas phase at room temperature in glass ampules within 14 d to give crystalline $S_3N_3F_3$. In copper or teflon tubes the reaction is even slower. About half of the substance is still monomeric after storing for 50 d at room temperature [3].

2.3.1.6.3 Thermolysis

Decomposition of gaseous FSN at about 110°C occurs according to the equation $4\,FSN \rightarrow 2\,N_2 + SF_4 + 3/8\,S_8$ [10].

2.3.1.6.4 Ionization and Fragmentation by Electron Impact

Ionization and fragmentation by 70 eV electron impact was observed in the mass spectrum [4, 5]. The following relative abundances were measured at m/e = 65 FSN^+ (100%), 46 SN^+ (64%), 23 SN^{2+} (3.2%), 51 SF^+ (5.6%), 32 S^+ (9.6%) [5]. The appearance potentials are a.p. $(FSN^+) = 11.36 \pm 0.03$ eV, a.p. $(SN^+) = 11.80 \pm 0.05$ eV, and a.p. $(SF^+) = 15.45 \pm 0.1$ eV [4].

A lower limit of the barrier of a possible decay channel to the fragments SN^+ and F^- is calculated from the total energies of the fragments and the lowest minimum of FSN with use of Hartree-Fock methods. Depending on the basis sets, the barrier is 224.1 kcal/mol (9.7 eV) and 217.3 kcal/mol (9.4 eV) [1].

2.3.1.6.5 Reactions with Nonmetals and Nonmetal Compounds

F atoms. When a solid solution of FSN in SF_6 is γ-irradiated at 110 K the radicals, $F_2SN^{\bullet}$ and probably $FSNF^{\bullet}$, form by attack of F atoms generated from SF_6 on FSN. The radicals were detected by ESR spectroscopy [6].

F_2. FSN reacts with F_2 at a pressure of 10 atm in a combustion bomb to give SF_6 and N_2 in stoichiometric amounts [7]. Fluorination in the presence of CsF at −78°C yields SF_5NF_2, SF_6, and NF_3 in appreciable quantities [8].

Cl_2. Gaseous FSN (10 Torr, 0.033 mol) does not react with Cl_2 added in three aliquots giving total pressures of Cl_2 of 20, 100, and 300 Torr over a period of 10 h [9]. When FSN (112 Torr) and Cl_2 (225 Torr) are stored in a glass vessel for 10 d, ClSN forms which trimerizes to give crystalline $S_3N_3Cl_3$. In the gas phase only SiF_4, SOF_2, and $SOCl_2$ could be detected [10, 11]. A mixture of FSN (5 Torr, 0.0165 mmol), Cl_2 (100 Torr, 0.34 mmol), and SO_2 (50 Torr, 0.0105 mmol) reacts at ambient temperature for 65 h to give ClSN, SO_2Cl_2, and SO_2ClF [9]. Condensed FSN (1.03 mmol) and Cl_2 (1.13 mmol) in a Monel cylinder containing 5 g of dried CsF react on warming to ambient temperature within 3 h to give $F_2S{=}NCl$ in 96% yield [8].

Br_2. A mixture of FSN (3.81 mmol) and Br_2 (2.48 mmol), condensed into a Monel bomb containing 5 g of CsF, reacts on warming to ambient temperature for 3 h to produce an unstable material tentatively formulated as $F_2S{=}NBr$ on the basis of its IR spectrum [8].

ClF. The gas-phase reaction of FSN (10 Torr, 0.033 mmol) with ClF (30 Torr, 0.099 mmol) gives SF_4 [9].

H_2O. Hydrolysis of FSN gas in a microwave absorption cell at −78°C gives *cis*-O=SNH [12]. The hydrolysis of FSN gas at room temperature yields primarily gaseous O=S=NH and finally NH_4^+, F^-, and SO_3^{2-} are formed [10, 13]. Liquid FSN reacts with concentrated aqueous NH_3 at −50°C to give $S(OH)_2$. Reaction of FSN with an aqueous $(C_6H_5)_4P^+Cl^-$ solution gives a precipitate of $((C_6H_5)_4P^+)_2S_6O_6^{2-}$ and $((C_6H_5)_4P^+)_2S_3O_6^{2-}$ and the ions NH_4^+, SO_3^{2-}, F^-, and Cl^- in the aqueous solution. Hydrolysis in aqueous Na_2SO_3 at pH 5.5 yields H_2NOH [14]. A mechanism for the hydrolysis of FSN is derived from these experimental results [14]. This mechanism includes the hydrolysis of O=S=NH [15].

HNF_2. Reaction of a solution of 10.8 mmol FSN in 5 mL $CFCl_3$ or CH_2Cl_2 with 9.4 mmol HNF_2 at −78°C for 5 d gives N_2F_4 in 62% yield, S_4N_4, and HF. The occurrence of the reactive intermediate $F_2NS{\equiv}N$ was postulated. The reaction takes place in analogous way in the presence of KF [37].

$(CF_3)_2NO^•$. FSN does not react with the radical $(CF_3)_2NO^•$ at room temperature within 3 weeks [16].

$(CF_3S)_2NH$. A solution of FSN (1.54 mmol) and $(CF_3S)_2NH$ (0.92 mmol) in $CFCl_3$ reacts on heating to 80°C over a period of 3 h with a 10% conversion to give the final product $CF_3SN{=}S{=}NSCF_3$ and HF [37].

$(CF_3)_2S{=}NH$. When FSN is treated with equimolar amounts of $(CF_3)_2S{=}NH$ in CH_2Cl_2–CD_2Cl_2 (8:2) at ambient temperature, the compounds CF_3SCF_3, S_4N_4, and N_2 form. It was shown by ^{19}F and ^{14}N NMR spectroscopy that the unstable intermediate $(CF_3)S{=}NS{\equiv}N$ is produced [37].

BF_3, PF_5, AsF_5, SbF_5. FSN reacts with an excess of BF_3 and PF_5 (mole ratio ~1:2) at −23°C to give the ionic solids $SN^+BF_4^-$ and $SN^+PF_6^-$ [17]. The formation of colorless crystals from the reaction of FSN and BF_3 in the gas phase was previously observed by [10]. Reaction with AsF_5 at −23°C in the liquid phase [17] and at room temperature in the gas phase in a quartz flask [18, 19] in the mole ratio ~1:1 produces $SN^+AsF_6^-$. Simultaneous introduction of FSN and SbF_5, each with 1 Torr pressure into a quartz flask at room temperature, leads to $SN^+SbF_6^-$ [18, 19]. FSN can be reacted with AsF_5 and SbF_5 in liquid SO_2 in a preparative scale to give $SN^+AsF_6^-$ and $SN^+SbF_6^-$, respectively [20].

$F_3CC{\equiv}CX$ (X = F, Br, CF_3). Reaction of FSN with $F_3CC{\equiv}CX$ (X = F, Br, CF_3) in the presence of CsF yields the 3,4-disubstituted 1,2,5-thiadiazoles I [21, 22]. On the reaction with $F_3CC{\equiv}CCF_3$ the dithiadiazole II is also formed [22].

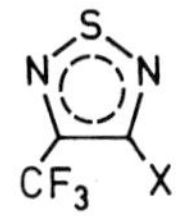

I

X = F, Br, CF_3

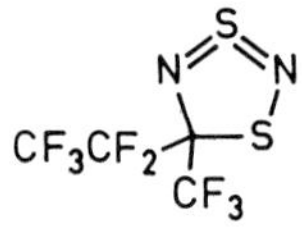

II

$CF_3CF{=}CF_2$. Irradiation of a mixture of 214 mmol of FSN and 133 mmol of CF_3–$CF{=}CF_2$ in a quartz bomb with a high pressure Hg lamp for 12 h at room temperature gives $F_2S{=}N$–S–$\overline{N\text{–}CF_2\text{–}C}FCF_3$ in 28% yield and i-$C_3F_7N{=}S{=}N$–S–$\overline{N\text{–}CF_2\text{–}C}FCF_3$ in 12% yield [23]. The reaction of FSN with $CF_3CF{=}CF_2$ in the mole ratio 1:1 in the presence of equimolar amounts of CsF in a Monel cylinder at 130°C for 24 h gives i-$C_3F_7SSC_3F_7$-i (48 wt%), i-$C_3F_7SC_3F_7$-i (4 wt%), $(CF_3)_2C{=}NSC_3F_7$-i (22 wt%), and i-C_3F_7S–$N{=}S{=}N$–C_3F_7-i (17 wt%). When the reaction is carried out with 0.3 mol of FSN and 0.43 mol of $CF_3CF{=}CF_2$ in the presence of catalytic amounts of CsF (0.09 mol) at 130°C for 16 h the following products are obtained: i-$C_3F_7SSC_3F_7$-i (43 wt%), i-$C_3F_7SC_3F_7$-i (5 wt%), $(CF_3)_2C{=}NSC_3F_7$-i (13 wt%), i-C_3F_7S–$N{=}S{=}N$–C_3F_7-i (21 wt%), and i-$C_3F_7SN{=}S{=}NSC_3F_7$-i (2 wt%). On repetition of the experiment with 0.39 mol of FSN and 0.39 mol of $CF_3CF{=}CF_2$ in the presence of ~0.09 mol CsF formerly used the products are: i-$C_3F_7SSC_3F_7$-i (29 wt%), i-$C_3F_7SC_3F_7$-i (21 wt%), i-$C_3F_7SN{=}SF_2$ (22 wt%), i-$C_3F_7SN{=}S{=}NC_3F_7$-i (9 wt%), and traces of i-$C_3F_7SN{=}S{=}NSC_3F_7$-i [24].

$(CF_3)_2C{=}CF_2$. Reacting 42 mmol of FSN with 46 mmol of $(CF_3)_2C{=}CF_2$ in the presence of 11 g CsF at 130°C for 24 h in a Monel cylinder gives t-$C_4F_9N{=}S{=}N$–SC_4F_9-t in 75% yield [25].

$CF_2{=}CFCF{=}CF_2$. FSN reacts with $CF_2{=}CFCF{=}CF_2$ in an exothermic [4 + 2]addition to give the cyclic compound III in almost quantitative yield [22].

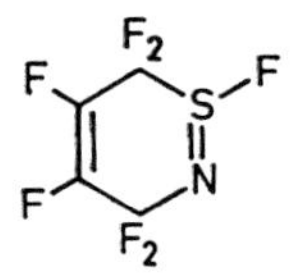

III

CH_2=C(CH_3)C(CH_3)=CH_2. FSN reacts with alkylbutadienes, e.g. CH_2=C(CH_3)C(CH_3)=CH_2 explosively [22].

$(CH_3)_3SiN(CH_3)_2$. The reaction of FSN with $(CH_3)_3SiN(CH_3)_2$ in n-C_5H_{12} affords $(CH_3)_2NSN(CH_3)_2$, S_4N_4, S_8, and N_2. The unstable intermediate product $(CH_3)_2$NS≡N was detected by mass spectrometry [26].

$(CH_3)_3$SiN=S=N–C_4H_9-t. The title compound reacts with FSN, mole ratio 1:1, in n-C_5H_{12} at room temperature for 4 h to give t-C_4H_9–N=S=NSN=S=N–C_4H_9-t in 91% yield [26].

$(CH_3)_3$SiN=S=NSi$(CH_3)_3$. Allowing to warm to room temperature a mixture of condensed FSN and $(CH_3)_3$SiN=S=NSi$(CH_3)_3$ in n-C_5H_{12} (mole ratio 1:1) cooled to −196°C gives the explosive solid $(S_3N_4)_n$ of unknown structure in 67% yield, S_4N_4, and volatile products, e.g. $(CH_3)_3SiF$. Spontaneous violent explosions occurred when the mole ratio 2:1 was chosen in this reaction [26].

$((CH_3)_2N)_3S^+(CH_3)_3SiF_2^-$. FSN reacts with $((CH_3)_2N)_3S^+(CH_3)_3SiF_2^-$ in the mole ratio 2:1 in CH_3CN at −35°C for 10 min to give $((CH_3)_2N)_3S^+F_2SN^-$ in almost quantitative yield and $(CH_3)_3SiF$. The salt is crystallized by addition of ether at −196°C [35].

$B(OTeF_5)_3$. Reaction of 4.0 mmol of FSN with 1.3 mmol of $B(OTeF_5)_3$ at −35°C for 1 h produces F_5TeO–S≡N·BF_3 in small amounts, O=S=$NTeF_5$ in 29% yield, and F_5TeN=S=$NTeF_5$ in 15% yield [27].

$F_5TeOM(CH_3)_3$ (M = Si, Sn). When FSN is reacted with $F_5TeOSi(CH_3)_3$ or $F_5TeOSn(CH_3)_3$, O=S=$NTeF_5$, polymers, and $(CH_3)_3SiF$ and $(CH_3)_3SnF$, respectively, are formed [27].

$[S=N=S]^+AsF_6^-$ reacts with FSN in the mole ratio 1:1 in SO_2 at −80°C in a [2+3]cycloaddition to give $S_3N_2F^+AsF_6^-$ in quantitative yield [34].

2.3.1.6.6 Reactions with Metal Compounds

$LiOC_4H_9$-t. Passing 10.9 mmol of FSN in an N_2 stream into a solution of 30.5 mmol of $LiOC_4H_9$-t in 40 mL n-C_5H_{12} at −60°C with subsequent stirring of the mixture at room temperature for 12 h yields the crystalline compound $\{Li_8[t\text{-}C_4H_9OS(N)F]_4F_2\}[Li_4(OC_4H_9\text{-}t)_5]_2 \cdot 4\,n\text{-}C_5H_{12}$.

If a deficiency of FSN is used, the intermediate compound Li^+ t-C_4H_9O–S(F)N^- is partially maintained. Li^+ t-C_4H_9O–S(F)N^- eliminates LiF to give t-C_4H_9O–S≡N which reacts to give unknown products [28].

LiN=C$(CF_3)_2$. A CsF-catalyzed reaction of FSN with equimolar amounts of LiN=C$(CF_3)_2$ in CH_2Cl_2 at 70 to 80°C over a period of 18 h provides $(CF_3)_2$C=NSN=S=NCF$(CF_3)_2$ in 10% yield and $(CF_3)_2$C=NS_xN=C$(CF_3)_2$ with x=1,2 [37].

LiN(Si$(CH_3)_3$)C_4H_9-t. FSN condensed into a solution of LiN(Si$(CH_3)_3$)C_4H_9-t in n-C_5H_{12} and cooled to −196°C (mole ratio 2:1) reacts on warming to room temperature to give t-C_4H_9N=S=NSN=S=NC_4H_9-t in 79% yield along with amounts of t-C_4H_9N=S=NC_4H_9-t, t-C_4H_9N=S=NSN=S=NSi$(CH_3)_3$, and S_4N_4. A reaction mechanism is proposed [26].

$LiN(Si(CH_3)_3)_2$. When condensed FSN and a solution of $LiN(Si(CH_3)_3)_2$ in n-C_5H_{12} cooled to −196°C is allowed to warm to room temperature over a period of 6 h and is stirred at room temperature for 2 h, the volatile compounds $(CH_3)_3SiF$ and $(CH_3)_3SiN{=}S{=}NSi(CH_3)_3$ and crystalline $((CH_3)_3Si)_2NSN{=}S{=}NSi(CH_3)_3$ form. If the same reaction mixture is allowed to warm to room temperature without stirring and is allowed to stand for 12 h at this temperature, colorless crystals of the bicyclic compound $S_4N_5^+F^-$ are separated (yield <5%). Spontaneous explosions were repeatedly observed on carrying out this reaction by both ways. Attention is necessary [26].

CsF. Condensing 4.23 mmol of FSN into a Monel bomb containing 10 g of CsF and stainless steel balls then warming the mixture to ambient temperature with shaking for 3 h gives only a trace amount of volatile material on opening the bomb at 25°C. However, a recovery of 4.19 mmol of FSN was made by pumping on the reactor at 125°C under vacuum [8].

$Hg(ON(CF_3)_2)_2$. Reaction of FSN (2.62 mmol) with excess $Hg(ON(CF_3)_2)_2$ (1.69 mmol) at room temperature gives $(CF_3)_2NOS{\equiv}N$ in 1% yield and very small amounts of a pale yellow oil which was tentatively identified from its mass spectrum as $FSN(ON(CF_3)_2)_2$. Preparing $Hg(ON(CF_3)_2)_2$ in situ by distilling excess of $(CF_3)_2NO^\bullet$ into a mixture of Hg (15 mmol) and FSN (2.14 mmol) permits the strongly exothermic reaction of FSN with $Hg(ON(CF_3)_2)_2$ at −78°C to give $(CF_3)_2NOS{\equiv}N$ in >80% yield [16].

$Hg(N(CF_3)_2)_2$. FSN does not react with $Hg(N(CF_3)_2)_2$ in $CFCl_3$ at 20°C over a period of 3 h [37].

$Hg(N(SCF_3)_2)_2$. FSN reacts with $Hg(N(SCF_3)_2)_2$ in the mole ratio 2:1 in $CFCl_3$ at 70°C for 18 h to give $CF_3SN{=}S{=}NSCF_3$ in 68% yield. The intermediate $(CF_3S)_2NS{\equiv}N$ was supposed to form [37].

$[Ni(SO_2)_x]^{2+}(AsF_6^-)_2$ and $[Co(SO_2)_x]^{2+}(AsF_6^-)_2$. Condensing a 10 to 20% excess of FSN into $[Ni(SO_2)_x]^{2+}(AsF_6^-)_2$ or $[Co(SO_2)_x]^{2+}(AsF_6^-)_2$ in SO_2 at −196°C and stirring the mixture at −20°C for 2 to 3 h yields $[Ni(NSF)_6]^{2+}(AsF_6^-)_2$ [29, 30] and $[Co(NSF)_6]^{2+}(AsF_6^-)_2$ [29, 31].

$[Re(CO)_5(SO_2)]^+AsF_6^-$. When equimolar amounts of FSN are condensed into $[Re(CO)_5(SO_2)]^+$-AsF_6^- in SO_2 at −196°C followed by stirring at −30°C for 1 h $[Re(CO)_5(NSF)]^+AsF_6^-$ forms [32].

AgF_2. FSN reacts with excess AgF_2 in a static system at 25 [8] or 100°C [10] and in a flow system at 20 [33] or 100°C [10] to give F_3SN in an essentially quantitative yield. Side products are SF_4, SOF_2, SF_6, and SO_2F_2 [10].

References:

[1] Seeger, R.; Seeger, U.; Bartetzko, R.; Gleiter, R. (Inorg. Chem. **21** [1982] 3473/7).
[2] Schaad, L. J.; Hess, B. A., Jr.; Čársky, P.; Zahradník, R. (Inorg. Chem. **23** [1984] 2428/30).
[3] Glemser, O.; Meyer, H.; Haas, A. (Chem. Ber. **97** [1964] 1704/9).
[4] Glemser, O.; Müller, A.; Böhler, D.; Krebs, B. (Z. Anorg. Allgem. Chem. **357** [1968] 184/9).
[5] Appel, R.; Laßmann, E. (Chem. Ber. **104** [1971] 2246/9).
[6] Boate, A. R.; Preston, K. F. (Inorg. Chem. **17** [1978] 1669/70).
[7] Larson, J. W.; Johnson, G. K.; O'Hare, P. A. G.; Hubbard, W. N.; Glemser, O. (J. Chem. Thermodyn. **5** [1973] 689/97).
[8] Ruff, J. K. (Inorg. Chem. **5** [1966] 1787/91).
[9] Banister, A. J.; Hey, R. G.; Passmore, J.; Sudheendra Rao, M. N. (J. Fluorine Chem. **21** [1982] 429/36).
[10] Glemser, O.; Richert, H. (Z. Anorg. Allgem. Chem. **307** [1961] 313/27).

[11] Glemser, O.; Perl, H. (Naturwissenschaften **48** [1961] 620).
[12] Kirchhoff, W. H.; Wilson, E. B., Jr. (J. Am. Chem. Soc. **85** [1963] 1726/9).

[13] Glemser, O.; Richert, H.; Rogowski, F. (Naturwissenschaften **47** [1960] 94/5).
[14] Roesky, H. W.; Glemser, O.; Hoff, A. (Chem. Ber. **101** [1968] 1215/22).
[15] Goehring, M.; Messner, J. (Z. Anorg. Allgem. Chem. **268** [1952] 47/56).
[16] Emeléus, H. J.; Poulet, R. J. (J. Fluorine Chem. **1** [1971/72] 13/21).
[17] Miller, K. V.; Emken, W. C.; Duncan, L. C. (J. Fluorine Chem. **26** [1984] 125/32).
[18] Glemser, O.; Koch, W. (Angew. Chem. **83** [1971] 145; Angew. Chem. Intern. Ed. Engl. **10** [1971] 127).
[19] Glemser, O.; Koch, W. (Anales Asoc. Quim. Arg. **59** [1971] 143/8; C.A. **76** [1972] No. 30285).
[20] Mews, R. (Angew. Chem. **88** [1976] 757/8; Angew. Chem. Intern. Ed. Engl. **15** [1976] 691/2).

[21] Geisel, M.; Mews, R. (Chem. Ber. **115** [1982] 2135/40).
[22] Bludssus, W.; Mews, R. (J. Chem. Soc. Chem. Commun. **1979** 35/6).
[23] Bludssus, W.; Mews, R. (Chem. Ber. **114** [1981] 1539/41).
[24] Bludssus, W.; Mews, R.; Glemser, O.; Alange, G. G. (Israel J. Chem. **17** [1978] 137/40).
[25] Waterfeld, A.; Bludssus, W.; Mews, R.; Glemser, O. (Z. Anorg. Allgem. Chem. **464** [1980] 268/72).
[26] Isenberg, W.; Mews, R.; Sheldrick, G. M. (Z. Anorg. Allgem. Chem. **525** [1985] 54/66).
[27] Hoppenheit, R.; Mews, R. (Chem. Ber. **118** [1985] 4276/80).
[28] Isenberg, W.; Mews, R.; Sheldrick, G. M. (Angew. Chem. **96** [1984] 816/7).
[29] Buß, B.; Jones, P. G.; Mews, R.; Noltemeyer, M.; Sheldrick, G. M. (Angew. Chem. **91** [1979] 253; Angew. Chem. Intern. Ed. Engl. **18** [1979] 231).
[30] Hartmann, G.; Jones, P. G.; Mews, R.; Sheldrick, G. M. (Z. Naturforsch. **35b** [1980] 1191/2).

[31] Buss, B.; Jones, P. G.; Mews, R.; Noltemeyer, M.; Sheldrick, G. M. (Acta Cryst. B **36** [1980] 141/2).
[32] Mews, R.; Liu, C. S. (Angew. Chem. **95** [1983] 156; Angew. Chem. Intern. Ed. Engl. **22** [1983] 158).
[33] Glemser, O.; Schröder, H. (Z. Anorg. Allgem. Chem. **284** [1956] 97/100).
[34] Padma, D. K.; Mews, R. (Z. Naturforsch. **42b** [1987] 699/702).
[35] Heilemann, W.; Mews, R. (Chem. Ber. **121** [1988] 461/3).
[36] Apblett, A.; Banister, A. J.; Biron, D.; Kendrick, A. G.; Passmore, J.; Schriver, M.; Stojanac, M. (Inorg. Chem. **25** [1986] 4451/2).
[37] Haas, A.; Rieland, M. (Z. Anorg. Allgem. Chem. **571** [1989] 82/90).

2.3.2 Sulfur Chloride Nitride, Thiazyl Chloride, ClS≡N

2.3.2.1 Formation. Preparation

Formation from SCl_2

ClS≡N forms in quantitative yield by reaction of gaseous SCl_2 with atomic nitrogen generated by microwave discharge in N_2 according to $SCl_2 + N \rightarrow ClSN + 1/2\,Cl_2$. In three runs with atomic nitrogen flow rates around 1.0 μmol/s, ClSN was formed at the rates 0.037, 0.045, and 0.23 μmol/s, corresponding to SCl_2 flow rates of 0.075, 0.075, and 1.28 μmol/s, respectively [1, 2].

Formation from S_2Cl_2

ClSN, Cl_2, SCl_2, S_8, and traces of $(SN)_x$ and S_4N_4 formed when a stream of S_2Cl_2 vapor was passed into a stream of atomic nitrogen generated by microwave discharge of N_2 [1 to 3]. The rate of formation of ClSN increases linearly with increasing S_2Cl_2 flow rates at low S_2Cl_2 flow rates, and becomes constant at high S_2Cl_2 flow rates. The rate of formation of ClSN at high S_2Cl_2 flow rates increases linearly with increasing N atom flow rates. It also increases with increasing temperature as seen from the following table.

Table 55

Rate of Formation of ClSN, F(NSCl), for the $S_2Cl_2 + N$ Reaction at a Flow Rate of N f(N) = 0.9 μmol/s and Various Flow Rates of S_2Cl_2, $f(S_2Cl_2)$, at Elevated Temperatures [1].

t in °C	$f(S_2Cl_2)$ in μmol/s	F(ClSN) in μmol/s	t in °C	$f(S_2Cl_2)$ in μmol/s	F(ClSN) in μmol/s
30	0.19	0.037	320	0.24	0.088
100	0.37	0.073	320	1.29	0.135
120	0.33	0.077			

The following mechanism for the formation of ClSN is consistent with the kinetic data: $N + S_2Cl_2 \rightleftharpoons N \cdot S_2Cl_2$, $N + N \cdot S_2Cl_2 \rightarrow 2\,ClSN$ [1, 2].

ClSN formation by an electric discharge through a mixture of N_2 and S_2Cl_2 vapor was also observed [4].

Passing gaseous NH_3 into S_2Cl_2 yields ClSN according to the equation $4\,S_2Cl_2 + 2\,NH_3 \rightarrow 2\,ClSN + 3/4\,S_8 + 6\,HCl$ [5]. When a suspension of NH_4Cl in excess S_2Cl_2 was heated under reflux for an extended period, ClSN and HCl evolve and a solution of sulfur in S_2Cl_2 remains. ClSN is probably formed by the reaction $NH_4Cl + 2\,S_2Cl_2 \rightarrow ClSN + 4\,HCl + 3/8\,S_8$ [3].

Formation from $SOCl_2$

ClSN was detected as a by-product in the reaction of $SOCl_2$ with NH_3 (mole ratio 1:3) in the gas phase. In the initial reaction $SOCl_2 + 2\,NH_3 \rightarrow O{=}S{=}NH + 2\,NH_4Cl$, OSNH is formed, which most likely reacts according to the equation $OSNH + SOCl_2 \rightarrow ClSN + SO_2 + HCl$ [6].

Formation from FSN

ClSN forms initially when FSN (112 Torr) is reacted with Cl_2 (225 Torr) at room temperature for 10 d. ClSN trimerizes to give crystalline $S_3N_3Cl_3$ [7, 8]. It was produced when a mixture of FSN (5 Torr), Cl_2 (100 Torr), and SO_2 (50 Torr) was stored at ambient temperature for 65 h [9].

Preparation from $S_3N_2Cl^+Cl^-$

When $S_3N_2Cl^+Cl^-$ is heated in vacuum to 80 to 100°C ClSN in 84% yield, SCl_2 and $(S_3N_2Cl)_2$ form according to the equation $3\,S_3N_2Cl^+Cl^- \rightarrow (S_3N_2Cl)_2 + 2\,ClSN + SCl_2$ [3, 10].

ClSN and SCl_2 formed by the reaction of Cl_2 gas with $S_3N_2Cl^+Cl^-$ at room temperature in the absence of a solvent [2, 3].

Formation from $(S_3N_2Cl)_2$

ClSN, SCl_2, and $S_4N_3^+Cl^-$ were obtained when $(S_3N_2Cl)_2$ was heated in vacuum to 120 to 140°C [10] or 130 to 150°C [11].

Preparation and Formation from $S_3N_3Cl_3$

ClSN evolves by heating $S_3N_3Cl_3$ at 70°C [4], 90°C [12], or 110°C [7] in the high vacuum. Thermodynamic data of [13] are given in "Sulfur-Nitrogen Compounds" Part 2, 1985, p. 94.

ClSN in solution is formed when $S_3N_3Cl_3$ is dissolved in an inert solvent as CCl_4 at room temperature after 3 d, or at reflux temperature after 0.5 h [12]. ClSN is in equilibrium with $S_3N_3Cl_3$ in solutions of $S_3N_3Cl_3$ in $CHCl_3$ at 25°C [20], in SO_2Cl_2–$CHCl_3$ at 0 to 60°C [21, 22], in CCl_4 at 22 to 70°C, and in SO_2 at 22°C [23]. This was concluded from ^{14}N NMR spectra [20 to 23].

Preparation from $S_4N_3^+Cl^-$

ClSN was obtained by thermal decomposition of $S_4N_3^+Cl^-$ at about 120°C at low pressures according to the equation $2S_4N_3^+Cl^- \rightarrow 2ClSN + S_4N_4 + 1/4S_8$ [14].

Formation from $S_4N_4Cl_2$ and S_4N_4

ClSN forms by decomposition of $S_4N_4Cl_2$ in CCl_4 solution at 29°C in a first-order reaction with the rate constant $k = 0.031\ min^{-1}$ according to the equation $2S_4N_4Cl_2 \rightarrow S_4N_4 + S_3N_3Cl_3 +$ ClSN. It also forms when $S_4N_4Cl_2$ is reacted with Cl_2 in CCl_4 at 31°C according to the equation $S_4N_4Cl_2 + Cl_2 \rightarrow S_3N_3Cl_3 + ClSN$. The rate constant of the first-order reaction is $k = 0.024\ min^{-1}$ [15, 16]. Both of these reactions of $S_4N_4Cl_2$ occur when S_4N_4 is chlorinated in CS_2 at about 30°C [17] or in CCl_4 at ambient temperatures [18] to give $S_4N_4Cl_2$ as an intermediate. ClSN is observed in the gas phase when the product of the reaction of S_4N_4 with ICl is heated at 60 to 90°C under vacuum [19].

References:

[1] Smith, J. J.; Jolly, W. L. (Inorg. Chem. **4** [1965] 1006/10).
[2] Smith, J. J. (UCRL-11763 [1965] 1/74; N.S.A. **19** [1965] No. 10706).
[3] Maguire, K. D.; Smith, J. J.; Jolly, W. L. (Chem. Ind. [London] **1963** 1589).
[4] Beppu, T.; Hirota, E.; Morino, Y. (J. Mol. Spectrosc. **36** [1970] 386/97, 387).
[5] Becke-Goehring, M.; Schläfer, D. (Z. Naturforsch. **21b** [1966] 492/3).
[6] DeKock, R. L.; Haddad, M. S. (Inorg. Chem. **16** [1977] 216/7).
[7] Glemser, O.; Perl, H. (Naturwissenschaften **48** [1961] 620).
[8] Glemser, O.; Richert, H. (Z. Anorg. Allgem. Chem. **307** [1961] 313/27, 316).
[9] Banister, A. J.; Hey, R. G.; Passmore, J.; Sudheendra Rao, M. N. (J. Fluorine Chem. **21** [1982] 429/36).
[10] Jolly, W. L.; Maguire, K. D. (Inorg. Syn. **9** [1967] 102/11, 109/11).

[11] Jolly, W. L.; Maguire, K. D.; Rabinovich, D. (Inorg. Chem. **2** [1963] 1304/5).
[12] Berthet, M. P.; Vincent, H.; Monteil, Y. (Z. Naturforsch. **35b** [1980] 329/31).
[13] Patton, R. L.; Jolly, W. L. (Inorg. Chem. **9** [1970] 1079/83).
[14] Peake, S. C.; Downs, A. J. (J. Chem. Soc. Dalton Trans. **1974** 859/64).
[15] Zbořilová, L.; Gebauer, P. (Z. Anorg. Allgem. Chem. **448** [1979] 5/10).
[16] Zbořilová, L.; Gebauer, P. (Z. Chem. [Leipzig] **19** [1979] 32/3).
[17] Nelson, J.; Heal, H. G. (Inorg. Nucl. Chem. Letters **6** [1970] 429/33).
[18] Vincent, H.; Monteil, Y. (Syn. React. Inorg. Metal-Org. Chem. **8** [1978] 51/63).
[19] Smith, R. D.; Street, G. B. (Inorg. Chem. **17** [1978] 938/41).
[20] Schriver, M. J.; Passmore, J. (personal communication cited in [21]).

[21] Apblett, A.; Chivers, T. (J. Chem. Soc. Chem. Commun. **1987** 1889/90).
[22] Apblett, A.; Chivers, T. (J. Chem. Soc. Chem. Commun. **1988** 508).
[23] Passmore, J.; Schriver, M. J. (Inorg. Chem. **27** [1988] 2749/51).

2.3.2.2 Molecular Properties. Spectra

Geometrical Structure

From microwave [1], electron diffraction [2, 3], and IR [2, 4, 5, 6] studies, a bent molecular structure for the electronic ground state of ClSN (C_s symmetry) was established. The geometrical parameters (r_s structure) calculated from the rotational constants of $^{35}Cl^{32}S^{14}N$, $^{37}Cl^{32}S^{14}N$, $^{35}Cl^{34}S^{14}N$, and $^{35}Cl^{32}S^{15}N$, all in the ground vibrational states, are $r_s(SN) = 1.450 \pm 0.001$ Å, $r_s(SCl) = 2.161 \pm 0.001$ Å, and $\alpha_s(ClSN) = 117°42' \pm 3'$ [1]. They agree with the values obtained from electron-diffraction studies: $r_a(SN) = 1.448 \pm 0.003$ Å, $r_a(SCl) = 2.159 \pm 0.003$ Å, $\alpha_a(ClSN) = 117.5° \pm 1.0°$ [2, 3]. Molecular parameters (r_0 structure) have also been calculated from the rotational constants of the normal species combined with those of isotopic species (in parentheses): 1.458 Å, 2.157 Å, 118°8′ ($^{37}Cl^{32}S^{14}N$); 1.450 Å, 2.163 Å, 117°48′ ($^{35}Cl^{34}S^{14}N$); 1.448 Å, 2.164 Å, 117°45′ ($^{35}Cl^{32}S^{15}N$) [1].

Quantum Chemical Calculations

Molecular properties have been calculated by various quantum chemical methods (abbreviations, see p. 211). Ab initio LCAO–MO SCF: geometry, force fields [17]; eqQ (^{14}N, ^{35}Cl) [11]; ab initio SCF + CPF: geometry, pop [28]; ab initio CI: μ, eqQ (^{14}N, ^{35}Cl) [26]; CNDO/2: geometry, ε_i, pop, μ [23]; ε_i [22]; EH: ε_i [21].

Ionization Energies

Adiabatic (ad) and vertical (vert) ionization energies determined by HeI [21, 22], HeII, and NeI [22] photoelectron spectroscopy are given in Table 56. The assignment was carried out by comparison of the spectra with those of related molecules and by use of Koopmans' theorem in connection with CNDO/2 [22] and EH calculations [21].

Table 56

Adiabatic and Vertical Ionization Energies, E_i, and Orbital Energies, ε_i, of ClSN.

	MO							Ref.
	16a′	15a′	4a″	14a′	13a′	3a″	12a′	
E_i (ad) in eV	10.61(1)	11.38(2)	11.38(2)	13.73(2)	14.28(2)	—	16.5(1)	[22]
E_i (vert) in eV	10.95(1)	11.82(2)	11.82(2)	13.83(2)	14.49(1)	15.0(1)	16.96(4)	[22]
vibrational spacing in cm^{-1}	—	—	—	250(30)	970(40)	—	—	[22]
vibrational assignment	—	—	—	ν_3	ν_1	—	—	[22]
E_i (vert) in eV	10.96	11.80	11.80	13.77	14.46	—	—	[21]
$-\varepsilon_i$ in eV (EH)	11.69	12.89	13.04	14.33	15.25	—	—	[21]

Dipole Moment μ

$|\mu| = 1.836 \pm 0.036$ D for $ClS^{15}N$ [7] and 1.87 ± 0.02 D for $ClS^{14}N$ [9] was obtained from Stark effect measurements of the $1_{1,1} \leftarrow 0_{0,0}$ [7, 9], $2_{1,1} \leftarrow 2_{0,2}$, and $3_{1,2} \leftarrow 3_{0,3}$ [9] rotational transitions in the microwave spectra. The components along the principal axes are $|\mu_a| = 0.5645 \pm 0.0085$ D [7] and 0.57 ± 0.03 D [9] and $|\mu_b| = 1.747 \pm 0.035$ D [7] and 1.77 ± 0.02 D [9]. For a discussion of signs see [7].

Polarizability

The molecular polarizability $\alpha = 58.82 \times 10^{-25}$ cm^3 [10] and 62.04×10^{-25} cm^3 [5] was calculated on the basis of a semiempirical δ-function model of chemical binding.

Quadrupole Coupling Constants eqQ in MHz

The ^{14}N and ^{35}Cl nuclear quadrupole coupling constants of $^{35}Cl^{32}S^{14}N$ have been determined from an analysis of the hyperfine structure of six transitions in the microwave spectrum as $eq_{aa}Q(^{14}N) = -1.42(6)$, $eq_{bb}Q(^{14}N) = -4.16(6)$; $eq_{aa}Q(^{35}Cl) = -39.00(3)$, $eq_{bb}Q(^{35}Cl) = 23.84(4)$. Principal values of the eqQ(^{35}Cl) tensor: $eq_{yy}Q = 30(3)$, $eq_{zz}Q = -45(6)$ (angle between z and a: 17(4)°) [11]. The ^{35}Cl nuclear quadrupole coupling constants of $^{35}Cl^{32}S^{14}N$, $eq_{aa}Q = -39.1$, $eq_{bb}Q = 23.8$ ($eq_{cc}Q = 15.3$), had been determined from the hyperfine structure of rotational transitions, neglecting the splittings due to the nitrogen nucleus. From the hyperfine structure of four rotational transitions of $^{35}Cl^{32}S^{15}N$, the following ^{35}Cl nuclear quadrupole coupling constants and the asymmetry parameter η were calculated: $eq_{aa}Q = -38.51$, $eq_{bb}Q = 23.51$ ($eq_{cc}Q = 15.00$); $eq_{yy}Q = 28.36$, $eq_{zz}Q = -43.36$ ($eq_{xx}Q = 15.00$); $\eta = 0.3081$ [1].

Nuclear Magnetic Resonance

^{14}N NMR spectrum (1.05 M in SO_2/CH_3NO_2 external) at 22°C: $\delta = 323$ ppm [32].

ClSN, which is in equilibrium with $S_3N_3Cl_3$ in solutions of $S_3N_3Cl_3$, shows the following ^{14}N chemical shifts referred to external CH_3NO_2: in $CHCl_3$ at 25°C, $\delta = 338$ ppm [29]; in SO_2Cl_2–$CHCl_3$ at 0 to 60°C, $\delta = 338$ ppm [30, 31]; 0.25 M and 0.26 M $S_3N_3Cl_3$ solutions in CCl_4 at 22 to 70°C, $\delta = 348$ to 352 ppm; 0.25 M and 0.28 M $S_3N_3Cl_3$ solutions in SO_2 at 22°C, $\delta = 323$ ppm [32].

Rotational Constants. Moments of Inertia. Inertia Defects

Twenty-eight rotational transitions of $^{35}Cl^{32}S^{14}N$ in the ground state have been analyzed to determine the rotational constants $A = 41723.34 \pm 0.16$, $B = 4114.154 \pm 0.026$, and $C = 3738.605 \pm 0.021$ MHz and the four centrifugal distortion constants $\tau_{aaaa} = -6.99 \pm 0.30$, $\tau_{bbbb} = -0.0110 \pm 0.0013$, $\tau_{aabb} = 0.176 \pm 0.014$, and $\tau_{abab} = -0.0334 \pm 0.0068$ MHz [7]. The following constants ("Watson's determinable parameters") were later given (all in MHz) by [19]: $A'' = 41723.36 \pm 0.14$, $B'' = 4114.1091 \pm 0.0238$, $C'' = 3738.5576 \pm 0.0225$, $\tau_1 = 0.1986 \pm 0.0314$, $\tau_2 = (0.9526 \pm 0.2940) \times 10^{-2}$, $\tau_3 = 0.9 \pm 0.2$, $\tau_{aaaa} = -7.071 \pm 0.284$, $\tau_{bbbb} = -0.01074 \pm 0.00110$, $\tau_{cccc} = (-5.358 \pm 1.010) \times 10^{-3}$.

An earlier analysis of the rotational spectrum of ClSN neglected centrifugal distortion effects. The b-type Q branch series gave the difference A − C and the asymmetry parameter ϰ, whereas A + C was determined by the b-type R branch lines. In addition to the ground vibrational state spectra, several sets of vibrational satellites were observed, from which rotational constants were calculated for vibrationally excited states. Table 57, p. 230, presents the rotational constants of isotopic species of ClSN in the ground vibrational state and Table 58, p. 230, shows the rotational constants of $^{35}Cl^{32}S^{14}N$ in the singly excited vibrational states of ν_2 and ν_3 [1].

Table 57

Rotational Constants A_0, B_0, C_0, Asymmetry Parameters $\varkappa$, Moments of Inertia I_a, I_b, I_c, and Inertia Defects Δ_0 of $^{35}Cl^{32}S^{14}N$, $^{37}Cl^{32}S^{14}N$, $^{35}Cl^{34}S^{14}N$, and $^{35}Cl^{32}S^{15}N$ in the Ground Vibrational State [1].

	$^{35}Cl^{32}S^{14}N$	$^{37}Cl^{32}S^{14}N$	$^{35}Cl^{34}S^{14}N$	$^{35}Cl^{32}S^{15}N$
A_0 in MHz	41721.61	41623.10	40583.07	40355.96
B_0 in MHz	4114.26	3997.51	4079.15	4019.65
C_0 in MHz	3738.61	3641.10	3700.26	3649.36
$\varkappa$	−0.980220	−0.981233	−0.979455	−0.979824
I_a in u·Å^2	12.11676_6	12.14544_3	12.45669_7	12.52679_9
I_b in u·Å^2	122.872_9	126.461_5	123.930_5	125.764_9
I_c in u·Å^2	135.219_0	138.840_2	136.620_4	138.525_9
Δ_0 in u·Å^2	0.229_3	0.233_3	0.232_9	0.234_2

Table 58

Rotational Constants A, B, C, Asymmetry Parameter $\varkappa$, Moments of Inertia I_a, I_b, I_c, and Inertia Defects Δ of $^{35}Cl^{32}S^{14}N$ in Excited Vibrational States [1].

	$^{35}Cl^{32}S^{14}N$	
	ν_2	ν_3
A in MHz	41835.87	42436.12
B in MHz	4097.62	4104.81
C in MHz	3722.47	3726.52
$\varkappa$	−0.980314	−0.980455
I_a in u·Å^2	12.08367_4	11.91275_3
I_b in u·Å^2	123.371_9	123.155_8
I_c in u·Å^2	135.805_3	135.657_7
Δ in u·Å^2	0.349_7	0.589_2

Sets of centrifugal distortion constants for $^{35}Cl^{32}S^{14}N$ in the vibrational ground state were also calculated from the force constants [4, 8].

Inertia defects for $^{35}Cl^{32}S^{14}N$ in the ground state and vibrationally excited states deduced from various sets of force constants:

Table 59

Inertia Defects Δ_0 of $^{35}Cl^{32}S^{14}N$ in the Ground State and Inertia Defect Increments $\Delta(\nu_i)-\Delta_0$ of $^{35}Cl^{32}S^{14}N$ in the Excited Vibrational States of ν_1, ν_2, and ν_3 Based on Force Constants.

Δ_0 in u·Å^2	$\Delta(\nu_1)-\Delta_0$ in u·Å^2	$\Delta(\nu_2)-\Delta_0$ in u·Å^2	$\Delta(\nu_3)-\Delta_0$ in u·Å^2	Ref.
0.231	−0.0066	0.1173	0.3502	[4]
0.240	−0.0059	0.0750	0.4110	[4]
0.2652	—	—	—	[8]
—	−0.0058	0.0903	0.4031	[1]

Fundamental Vibrations

The frequencies of the three fundamentals of ClSN (point-group symmetry C_s) were obtained from the gas-phase IR [2, 4] and Raman [12] spectra. The averages of the values corresponding to the R and P branches were taken as the IR fundamental frequencies in [4]. The Raman spectrum was observed at 90°C [12].

Table 60
Frequencies of the Fundamental Vibrations of ClSN [2, 4, 12].

symmetry class	No.	approximate type of mode	activity	frequency in cm^{-1} IR [4]	IR [2]	Ra [12]
A′	ν_1	SN stretch.	IR, Ra	1324.5	1324.0	1321
A′	ν_2	SCl stretch.	IR, Ra	414.8	417.8	414
A′	ν_3	bend.	IR, Ra	271.65	271	269

Frequencies of the fundamental vibrations for isotopic species of ClSN in an Ar matrix at 15 K were observed in the IR spectra [6], see Table 61.

Table 61
Frequencies of the Fundamental Vibrations of Isotopic Species of ClSN (Ar Matrix IR Spectrum) [6].

isotopic species	ν_1 in cm^{-1}	ν_2 in cm^{-1}	ν_3 in cm^{-1}
$^{35}Cl^{32}S^{14}N$	1327.3	403.75	267.4
$^{37}Cl^{32}S^{14}N$	—	400.5	265.0
$^{35}Cl^{32}S^{15}N$	1296.6	402.55	262.6
$^{37}Cl^{32}S^{15}N$	—	399.3	260.3
$^{35}Cl^{34}S^{14}N$	1315.5	—	—
$^{35}Cl^{34}S^{15}N$	1284.3	—	—

Comparisons of the observed fundamental frequencies with calculated values obtained by various force fields were made [4, 6, 13].

An isotopic sum rule was applied to the fundamental vibrations of ClSN [27].

Vibrational Amplitudes

Mean amplitudes of vibration of ClSN were measured by electron diffraction [2, 3] and were calculated using various sets of force constants [2, 3, 4, 8, 14]. (Structural data obtained by microwave studies [1] and fundamental frequencies from [4] were used in [4, 8, 14], the author's own electron-diffraction structural data and fundamental frequencies were used in [2, 3].) The values are compiled in Table 62, p. 232.

Further values of mean amplitudes of vibration using an assumed geometry from [5] and fundamental frequencies from [5] were obtained by [5, 15, 16].

Mean amplitudes of vibration of the isotopic species of ClSN at 0, 298.15, and 500 K given in Table 63, p. 233, were calculated [13] on the basis of structural data from [1] and fundamental frequencies (in an Ar matrix) from [6].

Table 62
Mean Amplitudes of Vibration u of ClSN.

T in K	u(S–N) in Å	u(S–Cl) in Å	u(N···Cl) in Å	Ref.
363 to 373	0.0378(29)	0.0526(30)	0.0766(121)	[2, 3]
298	0.0362	0.0593	0.0883	[4]
298	0.0362	0.0560	0.0900	[8]
298(?)	0.0354	0.0506	0.0749	[2, 3]
298	0.0374	0.0568	0.0734	[14]
373	0.0364	0.0638	0.0962	[4]

Coriolis Coupling Constants

Three sets were calculated from different force fields: $|\zeta_{12}|=0.4489$, $|\zeta_{13}|=0.8068$, $|\zeta_{23}|=0.3841$ [8]; $\zeta_{12}=0.6283$, $\zeta_{13}=0.6815$, $\zeta_{23}=0.3752$ [1]; $\zeta_{12}=-0.10512$, $\zeta_{13}=0.57634$, $\zeta_{23}=0.80934$ [14]. The ζ sum rule $\zeta_{12}^2+\zeta_{13}^2+\zeta_{23}^2=1$ is obeyed.

Force Constants

For the definition of the constants (f) of the general valence force field and the symmetry force constants (F_{ij}), see p. 214. The sets of valence force constants of the quadratic force field in Table 64 were calculated using the structural data R(SN) = 1.450 Å, r(SCl) = 2.161 Å, α(ClSN) = 117.7° [1].

Further constants of the general valence force field were calculated using an assumed geometry [5, 15, 16] and on the basis of frequencies of matrix-isolated ClSN [6, 13]. Three constants only are given by [2].

The Urey-Bradley force field was calculated by employing the fundamental frequencies measured for matrix-isolated ClSN by [6]: stretching constants K(S–N) = 9.964, K(S–Cl) = 1.411, bending constant H(ClSN) = 0.093, and repulsion constants F = 0.365, F′ = −0.1·F = −0.037 mdyn/Å [13].

Comparison of an ab initio quadratic force field [17] with an empirical one [4] shows close accordance. Cubic and diagonal quartic force constants are also calculated [17].

Cartesian force constants were used to calculate bond asymmetry parameters [18].

Two sets of compliance constants have been computed [20].

Microwave Spectrum

Rotational transitions of $^{35}Cl^{32}S^{14}N$, $^{37}Cl^{32}S^{14}N$, and $^{35}Cl^{34}S^{14}N$ (all in natural abundance) and $^{35}Cl^{32}S^{15}N$ (enriched) were observed. Assignment for the b-type Q branch ($J_{1,J-1} \leftarrow J_{0,J}$) and R branch series yielded the rotational constants and thus the structural parameters. In addition to the ground state spectra, several sets of vibrational satellites were observed and assigned to the ν_2, ν_3, 2 ν_2, and 2 ν_3 states. The rotational constants of $^{35}Cl^{32}S^{14}N$ in the excited states ν_2 and ν_3 could be derived [1].

Table 63

Mean Amplitudes of Vibration u of the Isotopic Species of ClSN [13].

	u(S–N) in Å			u(S–Cl) in Å			u(N···Cl) in Å		
	0 K	298.15 K	500 K	0 K	298.15 K	500 K	0 K	298.15 K	500 K
$^{35}Cl^{32}S^{14}N$	0.0361	0.0362	0.0369	0.0494	0.0570	0.0682	0.0678	0.0860	0.1056
$^{37}Cl^{32}S^{14}N$	0.0361	0.0362	0.0369	0.0490	0.0566	0.0677	0.0676	0.0860	0.1057
$^{35}Cl^{32}S^{15}N$	0.0357	0.0357	0.0365	0.0495	0.0571	0.0683	0.0671	0.0856	0.1052
$^{37}Cl^{32}S^{15}N$	0.0357	0.0357	0.0365	0.0490	0.0567	0.0679	0.0668	0.0855	0.1052
$^{35}Cl^{34}S^{14}N$	0.0359	0.0360	0.0368	0.0491	0.0569	0.0682	0.0678	0.0862	0.1059
$^{35}Cl^{34}S^{15}N$	0.0355	0.0356	0.0364	0.0494	0.0574	0.0689	0.0671	0.0858	0.1054

Table 64

Constants (f) of the General Valence Force Field of ClSN.

$f_R(F_{11})$ in mdyn/Å	$f_r(F_{22})$ in mdyn/Å	$f_\alpha(R\cdot r\cdot F_{33})$ in mdyn·Å	$f_{Rr}(F_{12})$ in mdyn/Å	$f_{R\alpha}(R\cdot F_{13})$ in mdyn	$f_{r\alpha}(r\cdot F_{23})$ in mdyn	Ref.
10.031	1.465	0.767	0.0	0.0	−0.067	[1] [a)]
10.04(7)	1.49(15)	0.83(2)	0.04(23)	0.0	−0.03(1)	[4] [b)]
10.41(10)	1.65(13)	0.81(3)	1.19(40)	0.40(8)	0.03(5)	[4] [c)]
10.032(3)	1.444(13)	0.749(36)	0.0	0.0	−0.086(22)	[7] [d)]
10.090	1.761	0.715	0.188	0.096	0.153	[8] [e)]
10.5333	1.5102	0.8998	1.24	1.35	—	[14] [f)]

[a)] From ν_1, ν_2 [5] and two inertia defect increments, $\Delta(\nu_1)$ and $\Delta(\nu_2)$. f_{Rr} and $f_{R\alpha}$ constrained to zero. – [b)] From ν_i [4], isotopic shifts [6], centrifugal distortion constants, and inertia defects. $f_{R\alpha}$ constrained to zero. – [c)] As in [b)], but without inertia defects. The element C_{23} of the compliance matrix was constrained to zero. – [d)] From ν_1, ν_2 [5], two inertia defect increments, and centrifugal distortion constants [7]. f_{Rr} and $f_{R\alpha}$ constrained to zero. – [e)] From ν_i [4] only, making use of an approximate method due to [33]. Original paper gives F_{33}, F_{13}, and F_{23}. – [f)] As in [e)], with a method from [34].

The rotational transitions for ClSN in the ground state for values of J from 1 to 20 have been analyzed to determine four centrifugal distortion constants [7]. For a correction ($12_{2,10} \leftarrow 13_{1,13}$), see [19]. The analysis of the hyperfine structure of rotational transitions for $^{35}Cl^{32}S^{14}N$ yielded the ^{14}N [11] and ^{35}Cl quadrupole coupling constants [1, 11]. For $^{35}Cl^{32}S^{15}N$, the ^{35}Cl quadrupole coupling constants and the asymmetry parameter were calculated [1]. Stark effect measurements of the $1_{1,1} \leftarrow 0_{0,0}$ [7, 9], $2_{1,1} \leftarrow 2_{0,2}$, and $3_{1,2} \leftarrow 3_{0,3}$ [9] rotational transitions gave the molecular dipole moment.

IR Spectrum

The gas-phase IR spectrum of ClSN was studied between 4000 and 200 cm^{-1} using a 5 cm gas cell at 60 to 80°C. The fundamental vibrations ν_i (in cm^{-1}) exhibited well-separated PQR structures. ν_1: 1333.0 (R), 1323.8 (Q), 1316.0 (P); ν_2: 423.9 (R), 417.7 (Q?), 414.4 (Q?), 405.6 (P); ν_3: 281.5 (R) (all $\pm$1 cm^{-1}), 275.2 $\pm$ 3 (Q), 261.7 $\pm$ 1 (P) [4]. The three fundamental vibrations were also observed in an earlier gas-phase IR spectrum from 4000 to 200 cm^{-1} [2]. In the region 4000 to 300 cm^{-1}, the fundamentals ν_1 = 1325 cm^{-1} and ν_2 = 414 cm^{-1}, the combination band $\nu_2 + \nu_3$ = 687 cm^{-1}, and the overtone $2\nu_3$ = 545 cm^{-1} had been observed [5].

The fundamental frequencies observed for isotopic species of ClSN in an Ar matrix at 15 K in the region 4000 to 200 cm^{-1} [6] are given in Table 61, p. 231.

Aggregates of ClSN in the Ar matrix (at 15 K) are responsible for extra bands in the IR spectrum at 1342, 1331.5 (m, sh), 391, 387, and 384 (s, br) cm^{-1}. The aggregates may assume a variety of forms, e.g., N$\equiv$S–Cl···N$\equiv$S–Cl, in which electron transfer occurs to the SCl but not to the SN unit [6].

The IR spectra of solid films of ClSN on various substrates cooled with liquid N_2 were studied from 4000 to 95 cm^{-1}. Two forms of the solid ClSN films were observed: one form with main peaks at 1397.0, 205, and 140 cm^{-1}; and another form with main peaks at 1346, 350, and 256 cm^{-1} [2].

The IR spectra of solutions of $S_3N_3Cl_3$ in CS_2 [24] and CCl_4 [32] show an absorption of ClSN (ν_1) at 1322 [24] or 1323 cm^{-1} [32].

Raman Spectrum

In the gas-phase Raman spectrum of ClSN at 90°C, excited by an Ar ion laser (488 nm), the fundamental vibrations, see p. 231, were observed [12].

UV-Visible Spectrum

The vapor spectrum at 23°C consisted of a strong peak at λ_{max} = 214 nm. The extinction coefficient was calculated (at 5°C) as ε = 11300 $L \cdot mol^{-1} \cdot cm^{-1}$. At 60°C at least 59 distinct absorptions with regular distances appeared in the range 360 to 490 nm [25].

References:

[1] Beppu, T.; Hirota, E.; Morino, Y. (J. Mol. Spectrosc. **36** [1970] 386/97).

[2] Emken, W. C. (Diss. Oregon State Univ. 1970 from Diss. Abstr. Intern. B **31** [1971] 3952).

[3] Emken, W. C.; Hedberg, K. (J. Chem. Phys. **58** [1973] 2195/6).

[4] Müller, A.; Mohan, N.; Cyvin, S. J.; Weinstock, N.; Glemser, O. (J. Mol. Spectrosc. **59** [1976] 161/70).

[5] Müller, A.; Nagarajan, G.; Glemser, O.; Cyvin, S. J.; Wegener, J. (Spectrochim. Acta A **23** [1967] 2683/9).

[6] Peake, S. C.; Downs, A. J. (J. Chem. Soc. Dalton Trans. **1974** 859/64).

[7] Mizumoto, S.; Izumi, J.; Beppu, T.; Hirota, E. (Bull. Chem. Soc. Japan **45** [1972] 786/9).
[8] Namasivayam, R.; Mayilavelan, S. (Z. Naturforsch. **34a** [1979] 716/20).
[9] Guarnieri, A. (Z. Naturforsch. **26a** [1971] 1246).
[10] Chopra, J. R.; Verma, U. P.; Pandey, A. N. (Indian J. Pure Appl. Phys. **20** [1982] 500/2).

[11] Cervellati, R.; Lister, D. G.; Degli Esposti, A.; Ortolani, F. (Nuovo Cimento Soc. Ital. Fis. D **3D** [1984] 475/88).
[12] Berthet, M. P.; Vincent, H.; Monteil, Y. (Z. Naturforsch. **35b** [1980] 329/31).
[13] Sanyal, N. K.; Goel, R. K.; Pandey, A. N. (Spectrosc. Letters **9** [1976] 681/8).
[14] Gopinath, C. R.; Rao, K. S. R. (Indian J. Pure Appl. Phys. **19** [1981] 89/91).
[15] Namasivayam, R.; Nair, K. V. (Z. Physik. Chem. [Leipzig] **259** [1978] 1117/20).
[16] Peacock, C. J.; Heidborn, U.; Müller, A. (J. Mol. Spectrosc. **30** [1969] 338/44).
[17] Cervellati, R.; Lister, D. G.; Palmieri, P. (J. Mol. Struct. **108** [1984] 321/3 [THEOCHEM 17]).
[18] Ebenezer, J. S. P.; Srinivasamoorthy, R.; Savariraj, G. A. (Indian J. Pure Appl. Phys. **19** [1981] 249/53).
[19] Lovas, F. J. (J. Phys. Chem. Ref. Data **7** [1978] 1445/750, 1525/7).
[20] Gupta, S. L.; Verma, U. P.; Pandey, A. N. (Acta Ciencia Indica **3** [1977] 340/2).

[21] Cowan, D. O.; Gleiter, R.; Glemser, O.; Heilbronner, E. (Helv. Chim. Acta **55** [1972] 2418/21).
[22] DeKock, R. L.; Shehfeh, M. A.; Lloyd, D. R.; Roberts, P. J. (J. Chem. Soc. Faraday Trans. II **72** [1976] 807/13).
[23] Gonbeau, D.; Pfister-Guillouzo, G. (J. Chim. Phys. Physicochim. Biol. **73** [1976] 787/94).
[24] Nelson, J.; Heal, H. G. (Inorg. Nucl. Chem. Letters **6** [1970] 429/33).
[25] Lipp, S. A.; Jolly, W. L. (Inorg. Chem. **10** [1971] 33/7).
[26] Bendazzoli, G. L.; Palmieri, P.; Fano, G.; Ortolani, F. (Nuovo Cimento Soc. Ital. Fis. D **3D** [1984] 489/97).
[27] Müller, A.; Mohan, N.; Königer, F. (J. Mol. Struct. **30** [1976] 297/310).
[28] Ehrhardt, C.; Ahlrichs, R. (Chem. Phys. **108** [1986] 417/28).
[29] Schriver, M. J.; Passmore, J. (private communication cited in [30]).
[30] Apblett, A.; Chivers, T. (J. Chem. Soc. Chem. Commun. **1987** 1889/90).

[31] Apblett, A.; Chivers, T. (J. Chem. Soc. Chem. Commun. **1988** 508).
[32] Passmore, J.; Schriver, M. J. (Inorg. Chem. **27** [1988] 2749/51).
[33] Redington, R. L.; Aljibury, A. L. K. (J. Mol. Spectrosc. **37** [1971] 494/516).

2.3.2.3 Heat Capacity. Thermodynamic Functions

The thermodynamic functions for the ideal gaseous state at 1 atm pressure were calculated on the basis of a rigid rotator, harmonic oscillator model using the fundamental frequencies $\nu_1=1325$, $\nu_2=414$, $\nu_3=273\ cm^{-1}$ and the assumed structural parameters r(SN) = 1.45 Å, r(SCl) = 2.00 Å, and α(ClSN) = 116°.

Table 65
Selected Values of the Thermodynamic Functions C_p°, S°, $-(G^\circ-H_0^\circ)/T$, and $(H^\circ-H_0^\circ)/T$ of ClSN (ideal gas).

T in K	C_p°	S°	$-(G^\circ-H_0^\circ)/T$ in $cal \cdot mol^{-1} \cdot K^{-1}$	$(H^\circ-H_0^\circ)/T$
200	10.420	59.341	50.433	8.908
273.16	11.069	62.683	53.284	9.399
298.16	11.243	63.658	54.112	9.546
400	11.833	67.058	56.998	10.060
600	12.607	72.023	61.228	10.795
800	13.055	75.692	64.390	11.302
1000	13.316	78.641	66.960	11.681
1500	13.623	84.115	71.827	12.288
2000	13.743	88.005	75.369	12.636

Reference:

Müller, A.; Nagarajan, G.; Glemser, O.; Cyvin, S. F.; Wegener, J. (Spectrochim. Acta A **23** [1967] 2683/9).

2.3.2.4 Chemical Reactions

ClSN vapor trimerizes to give $S_3N_3Cl_3$. The pressure of ClSN vapor in equilibrium with solid $S_3N_3Cl_3$ was measured in a static system at 31 to 60°C and can be represented by the equation $\log (p/Torr) = 12.321 - 3360/T$ [1]. The trimerization of ClSN in CS_2 solution at about 30°C within days was observed in IR spectroscopic studies [2]. Thermolysis of ClSN vapor at 120°C gives S_2Cl_2 and N_2 [3]. Microwave discharge of ClSN–He mixtures at low pressures produces SN radicals [12].

The reaction of ClSN with H_2O vapor gives NH_4Cl and SO_2. No evidence was found for the formation of OSNH with the aid of IR spectroscopy [4]. ClSN reacts with excess ClF in the gas phase to give FSN, F_2SNCl, SF_4, and NCl_3 [5]. With S_2Cl_2, ClSN reacts in the gas phase [6, 7] and in the liquid phase at room temperature or lower [4, 7, 8, 9] according to the equation $2ClSN + S_2Cl_2 \rightarrow SCl_2 + S_3N_2Cl^+Cl^-$. When $S_4N_4Cl_2$ or a mixture of $S_3N_3Cl_3$ and S_4N_4 is heated at 65°C, the intermediate ClSN occurs, which reacts with S_4N_4 to give $S_5N_5^+Cl^-$. This reaction is also observed in HCOOH solution [10, 11].

The reaction of ClSN vapor with Cu yields S_4N_4 and CuCl [3]. When ClSN vapor is passed through a column packed with powdered anhydrous CsF at 110°C, FSN and smaller amounts of SF_4, SOF_2, and SOClF are formed [5].

The chemical reactions of the trimeric $S_3N_3Cl_3$ are described in "Sulfur-Nitrogen Compounds" Part 2, 1985, pp. 94/106. ClSN is in equilibrium with $S_3N_3Cl_3$ in solutions of $S_3N_3Cl_3$ in inert solvents, see p. 227.

References:

[1] Patton, R. L.; Jolly, W. L. (Inorg. Chem. **9** [1970] 1079/83).
[2] Nelson, J.; Heal, H. G. (Inorg. Nucl. Chem. Letters **6** [1970] 429/33).
[3] Berthet, M. P.; Vincent, H.; Monteil, Y. (Z. Naturforsch. **35b** [1980] 329/31).

[4] Smith, J. J. (UCRL-11763 [1965] 1/74; N.S.A. **19** [1965] No. 10706).
[5] Banister, A. J.; Hey, R. G.; Passmore, J.; Sudheendra Rao, M. N. (J. Fluorine Chem. **21** [1982] 429/36).
[6] Becke-Goehring, M.; Schläfer, D. (Z. Naturforsch. **21b** [1966] 492/3).
[7] Jolly, W. L.; Maguire, K. D.; Rabinovich, D. (Inorg. Chem. **2** [1963] 1304/5).
[8] Maguire, K. D.; Smith, J. J.; Jolly, W. L. (Chem. Ind. [London] **1963** 1589).
[9] Smith, J. J.; Jolly, W. L. (Inorg. Chem. **4** [1965] 1006/10).
[10] Zbořilová, L.; Gebauer, P. (Z. Chem. [Leipzig] **19** [1979] 32/3).

[11] Zbořilová, L.; Gebauer, P. (Z. Anorg. Allgem. Chem. **483** [1981] 44/50).
[12] Dyke, J. M.; Morris, A.; Trickle, I. R. (J. Chem. Soc. Faraday Trans. II **73** [1977] 147/51).

2.3.3 Sulfur Bromide Nitride, Thiazyl Bromide, BrS≡N

2.3.3.1 Formation. Preparation

BrS≡N was prepared by carefully controlled pyrolysis of $S_4N_3^+Br^-$ at about 90°C [1, 2]. The volatile pyrolysis product was passed through a packed Pyrex furnace maintained at 120°C in order to promote the decomposition of oligomeric species. BrSN forms according to the equation $2S_4N_3^+Br^- \rightarrow S_4N_4 + 2BrSN + 2S$ [1]. Gaseous BrSN was produced by passing ClSN vapor in the vacuum over KBr heated to 900 to 950 K. ClSN vapor was generated from $S_3N_3Cl_3$ by gentle heating to 80 to 85°C in the vacuum [6]. BrSN forms probably as an intermediate in the reaction of S_4N_4 with excess Br_2, which leads to $S_4N_3^+Br_3^-$ [3]. BrSN vapor evolves by heating polymeric $(SNBr_{0.43})_x$ or $(SNBr_{0.25})_x$ at 30, 80, and 125°C under vacuum [4], and by heating $(SNBr_{0.4})_x$ obtained by bromination of $(SN)_x$ and S_4N_4 at 60 to 90°C under vacuum [5]. BrSN was detected in these experiments by 25 eV electron-impact [4, 5] and phase-angle mass spectrometry [5].

References:

[1] Peake, S. C.; Downs, A. J. (J. Chem. Soc. Dalton Trans. **1974** 859/64).
[2] Vincent, H.; Monteil, Y.; Berthet, M. P. (J. Inorg. Nucl. Chem. **42** [1980] 5/7).
[3] Wolmershäuser, G.; Street, G. B. (Inorg. Chem. **17** [1978] 2685/7).
[4] Allen, W. N.; DeCorpo, J. J.; Saalfeld, F. E.; Wyatt, J. R. (Chem. Phys. Letters **54** [1978] 524/6).
[5] Smith, R. D.; Street, G. B. (Inorg. Chem. **17** [1978] 938/41).
[6] Allaf, A. W.; Matti, G. Y.; Suffolk, R. J.; Watts, J. D. (J. Electron Spectrosc. Relat. Phenom. **48** [1989] 411/9).

2.3.3.2 Molecular Properties. Spectra

Geometrical Structure

The geometrical parameters were not determined experimentally. The bond distances R(SN) = 1.45 Å and r(SBr) = 2.24 Å were assumed by analogy with ClSN and S_2Br_2, respectively. The bond angle α(BrSN) = 118° ± 3° was obtained from a fit of a general valence force field to the observed fundamental frequencies [1].

Inertia Defect. Centrifugal Distortion Constants

The inertia defect $\Delta_0 = 0.2886$ u·Å² and the centrifugal distortion constants $\tau_{aaaa} = -3.6412$, $\tau_{bbbb} = -0.0045$, $\tau_{cccc} = -0.0028$, $\tau_{aabb} = +0.0492$, and $\tau_{abab} = -0.0221$ MHz were calculated from a general valence force field [5].

Ionization Energies

Vertical ionization energies E_i (vert) = 10.45 (7a′), 10.85 (2a″), 11.09 (6a′), 12.87 (1a″), 13.45 eV (5a′) were determined from HeI photoelectron spectroscopy. The assignment was made by comparison of the spectra with those of FSN and ClSN and orbital energies calculated by the ab initio LCAO-MO SCF method [7].

Polarizability

The molecular polarizability $\alpha = 70.02 \times 10^{-25}$ cm³ was calculated on the basis of the δ-function model of chemical binding [3].

Fundamental Vibrations

The fundamental vibration ν_1 (SN stretch.) at 1311 cm⁻¹ was observed in the gas-phase Fourier transform IR spectrum [7]. The frequencies of the fundamental vibrations of isotopic species of BrSN (point group symmetry C_s) were obtained from the IR spectrum in an Ar matrix at 15 K [1].

Table 66

Frequencies of the Fundamental Vibrations of Isotopic Species of BrSN (Ar Matrix IR Spectrum), all of Symmetry Class A′: ν_1 (SN stretch.), ν_2 (SBr stretch.), ν_3 (BrSN bend.) [1].

isotopic species	ν_1 in cm⁻¹	ν_2 in cm⁻¹	ν_3 in cm⁻¹
$Br^{32}S^{14}N$	1312.9	346.1	226.2
$Br^{34}S^{14}N$	1301.0	—	—
$Br^{32}S^{15}N$	1282.7	344.2	222.0
$Br^{34}S^{15}N$	1270.5	—	—

An isotopic sum rule was applied to the fundamental vibration ν_1 of BrSN [2].

Vibrational Amplitudes

Mean amplitudes of vibration u of the isotopic species of BrSN at 0, 298.15, and 500 K given in Table 67 were calculated [4] on the basis of structural data and fundamental frequencies from [1].

Table 67

Mean Amplitudes of Vibration u of the Isotopic Species of BrSN [4].

	u(S-N) in Å			u(S-Br) in Å			u(N···Br) in Å		
	0 K	298 K	500 K	0 K	298 K	500 K	0 K	298 K	500 K
$Br^{32}S^{14}N$	0.0363	0.0364	0.0371	0.0454	0.0548	0.0666	0.0670	0.0913	0.1136
$Br^{34}S^{14}N$	0.0361	0.0362	0.0370	0.0449	0.0545	0.0664	0.0670	0.0914	0.1138
$Br^{32}S^{15}N$	0.0359	0.0359	0.0368	0.0454	0.0550	0.0669	0.0661	0.0907	0.1131
$Br^{34}S^{15}N$	0.0357	0.0358	0.0366	0.0450	0.0547	0.0668	0.0660	0.0907	0.1131

Later, for $Br^{32}S^{14}N$ at 298 K, the following mean amplitudes of vibration were calculated: u(S–N) = 0.0363, u(S–Br) = 0.0546, u(N···Br) = 0.0973 Å [5].

Coriolis Coupling Constants

The following set was calculated from a general valence force field: $|\zeta_{12}| = 0.4882$, $|\zeta_{13}| = 0.8148$, $|\zeta_{23}| = 0.3124$. The ζ sum rule $\zeta_{12}^2 + \zeta_{13}^2 + \zeta_{23}^2 = 1$ is obeyed [5].

Force Constants

Sets of symmetry force constants (see p. 214) calculated from structural data and fundamental frequencies given by [1] are compiled in Table 68.

Table 68
Symmetry Force Constants F_{ij} (all in mdyn/Å) of BrSN.

$F_{11} = f_R$	$F_{22} = f_r$	$F_{33} = f_\alpha/R \cdot r$	$F_{12} = f_{Rr}$	$F_{13} = f_{R\alpha}/R$	$F_{23} = f_{r\alpha}/r$	Ref.
9.875	1.294	0.220	0.010	0.076	−0.032	[1]*)
9.96	1.75	0.17	0.22	0.13	0.10	[3]
9.955	1.759	0.161	0.287	0.085	0.084	[4]
9.924	1.760	0.166	0.221	0.062	−0.082	[5]

*) α(BrSN) = 122°

The Urey-Bradley force field was calculated on the basis of data given by [1]: stretching constants K(S–N) = 9.733, K(S–Br) = 1.441, bending constant H(BrSN) = 0.037, and repulsion constants F = 0.386, F′ = −0.1 F = −0.039 mdyn/Å [4].

A set of compliance constants has been computed [6].

References:

[1] Peake, S. C.; Downs, A. J. (J. Chem. Soc. Dalton Trans. **1974** 859/64).
[2] Müller, A.; Mohan, N.; Königer, F. (J. Mol. Struct. **30** [1976] 297/310).
[3] Chopra, J. R.; Verma, U. P.; Pandey, A. N. (Indian J. Pure Appl. Phys. **20** [1982] 500/2).
[4] Sanyal, N. K.; Goel, R. K.; Pandey, A. N. (Spectrosc. Letters **9** [1976] 681/8).
[5] Namasivayam, R.; Mayilavelan, S. (Z. Naturforsch. **34a** [1979] 716/20).
[6] Gupta, S. L.; Verma, U. P.; Pandey, A. N. (Acta Ciencia Indica **3** [1977] 340/2).
[7] Allaf, A. W.; Matti, G. Y.; Suffolk, R. J.; Watts, J. D. (J. Electron Spectrosc. Relat. Phenom. **48** [1989] 411/9).

2.3.4 Sulfur Iodide Nitride, Thiazyl Iodide, IS≡N

ISN is not known. Attempts were made to prepare ISN by pyrolysis of $S_4N_3^+I^-$ at 0 to 30°C. No spectroscopic evidence was obtained for even transitory formation of ISN. The identificable products of the reaction were S_4N_4 and I_2.

Reference:

Peake, S. C.; Downs, A. J. (J. Chem. Soc. Dalton Trans. **1974** 859/64).

2.3.5 *cis*-Sulfur Hydroxide Nitride, *cis*-Thiazyl S-Hydroxide, *cis*-HOS≡N

cis-HOSN forms by UV photolysis of *cis*-HN=S=O in an Ar matrix (M/A = 400 [1, 2] or >1000 [5]) using light of a medium- (λ>200 nm) [1, 2] or high-pressure Hg lamp (λ = 250 nm) [5]. The reaction is very fast with light of λ>200 nm but considerably slower when only light of λ>300 nm is used. *cis*-HOSN and its D and ^{15}N species were identified by their IR absorptions (see below) [1, 2].

Based upon comparisons with similar molecules, the structure of *cis*-HOSN was approximated by the parameters r(HO) = 0.96 Å, r(OS) = 1.62 Å, r(SN) = 1.46 Å, α(HOS) = 100°, and α(OSN) = 115° [1]. This assumption was confirmed by ab initio SCF MO calculations [5, 6].

HOSN is predicted from ab initio SCF MO calculations to possess no stable trans conformer [5]. The energy difference between the stable cis and the trans ground state was calculated to be ~20 kJ/mol [5] and 18.4 kJ/mol [6].

The IR spectrum of normal *cis*-HOSN, of *cis*-DOSN, and of *cis*-HOS^{15}N is given in Table 69.

Table 69
IR Spectrum of *cis*-HOSN, *cis*-DOSN, and *cis*-HOS^{15}N in an Ar Matrix [1].

cis-HOSN ν in cm^{-1}	*cis*-DOSN ν in cm^{-1}	*cis*-HOS^{15}N ν in cm^{-1}	intensity	assignment
3519.8	2597.1	3519.3	m	OH or OD stretch.
1320.8	1319.0	1294.3	w	SN stretch.
991.8	—	991.1	w	HOS bend.
673.9	671.3	673.3	vs	SO stretch.
417.5	325.2	416.5	m	torsion
374	—	368	vw	OSN bend.

The following force field for *cis*-HOSN was calculated on the basis of the estimated structural parameters and measured fundamental frequencies: f(HO) = 6.91(0), f(OS) = 2.97(2), f(SN) = 9.90(1) mdyn/Å, f(HOS) = 0.500(1), f(OSN) = 1.03(1), f(torsion) = 0.0782(6) mdyn·Å/rad^2, f(HOS-OS) = 0.259(21) mdyn/rad [1]. The potential energy distribution for *cis*-HOSN is given in the original paper [1].

The principal moments of inertia for *cis*-HOSN, 52.197, 11.95, and 6.141 u·Å^2, were calculated from the approximate structure. Values of the thermodynamic functions for the ideal gaseous state at 1 atm pressure compiled in Table 70 were calculated from the complete set of IR frequencies [1].

Table 70
Values of the Thermodynamic Functions C_p°, S°, $-(G^\circ-H_0^\circ)/T$, and $(H^\circ-H_0^\circ)/T$ of HOSN (ideal gas) [1].

T in K	C_p°	S°	$-(G^\circ-H_0^\circ)/T$ in cal·mol^{-1}·K^{-1}	$(H^\circ-H_0^\circ)/T$
100.00	8.40	57.22	49.19	8.02
200.00	10.52	63.69	54.95	8.73
273.15	11.89	67.18	57.77	9.40
298.10	12.31	68.24	58.60	9.63

Table 70 (continued)

T in K	C_p°	S°	$-(G^\circ-H_0^\circ)/T$ in $cal\cdot mol^{-1}\cdot K^{-1}$	$(H^\circ-H_0^\circ)/T$
400.00	13.71	72.06	61.56	10.50
500.00	14.71	75.24	63.99	11.25
600.00	15.45	77.99	66.09	11.89
700.00	16.01	80.41	67.97	12.44
800.00	16.46	82.58	69.66	12.91
900.00	16.83	84.54	71.21	13.33
1000.00	17.14	86.33	72.63	13.70
1100.00	17.41	87.98	73.96	14.02
1200.00	17.65	89.51	75.19	14.31
1300.00	17.86	90.93	76.35	14.58
1400.00	18.04	92.26	77.43	14.82
1500.00	18.21	93.51	78.47	15.04

Molecular properties were calculated by quantum chemical methods. A survey of publications is given below.

method	calculated molecular properties	Ref.
ab initio SCF MO	E_t, geometry, population analysis, atomic charges	[3]
ab initio SCF MO+CI	relative energy to *cis*-HNSO, geometry, vibrational frequencies, cis-trans isomerization, excitation energies, oscillator strengths, μ	[5]
ab initio SCF MO	relative energy to *cis*-HNSO, geometry, population analysis	[6]
semiempirical MNDO MO	geometry, ΔH_f°, atomic charges	[4]

Cis$\rightleftharpoons$trans isomerization is not possible, because no energy minimum was found for the trans configuration (see above) [5]. *cis*-HOS≡N generated by photolysis of *cis*-HN=S=O in an Ar matrix using a medium- ($\lambda>200$ nm) [1] or high-pressure Hg lamp ($\lambda=250$ nm) [5] reacts to give *cis*-HSNO and *trans*-HSNO [1, 5]. This follows from the intensity of absorptions of *cis*-HOSN, *cis*-HNSO, and *trans*-HSNO as a function of photolysis time for $\lambda>200$ nm and $\lambda>300$ nm [1].

References:

[1] Tchir, P. O.; Spratley, R. D. (Can. J. Chem. **53** [1975] 2318/30).
[2] Tchir, P. O. (Diss. Univ. British Columbia, Vancouver, B. C., 1972 from Diss. Abstr. Intern. B **33** [1973] 3587).
[3] Turner, A. G. (Inorg. Chim. Acta **84** [1984] 85/7).
[4] Oleksik, J. J.; Turner, A. G. (Inorg. Chim. Acta **59** [1982] 165/9).
[5] Nonella, M.; Huber, J. R.; Ha, Tae-Kyu (J. Phys. Chem. **91** [1987] 5203/9).
[6] Ehrhardt, C.; Ahlrichs, R. (Chem. Phys. **108** [1986] 417/28, 419).

2.3.6 Sulfur Bis(trifluoromethyl)nitroxide Nitride, Thiazyl Bis(trifluoromethyl)nitroxide, $(CF_3)_2NOS{\equiv}N$

$(CF_3)_2NOS{\equiv}N$ formed in about 1% yield by reaction of FSN with $Hg(ON(CF_3)_2)_2$ in the mole ratio ~3:2 at room temperature [1, 2]. Yields of $(CF_3)_2NOS{\equiv}N$ in excess of 80% were obtained by preparing $Hg(ON(CF_3)_2)_2$ in situ. In a typical experiment $FS{\equiv}N$ (0.139 g, 2.14 mmol) was condensed on Hg (3 g, 15 mmol). Excess of the radical $(CF_3)_2NO^{\bullet}$ was distilled in. The strongly exothermic reaction at room temperature was moderated by cooling to −78°C. Traces of $(CF_3)_2NOS{\equiv}N$ probably formed when $S_4N_4H_4$ was reacted with $(CF_3)_2NO^{\bullet}$ at room temperature.

The compound is volatile in the vacuum. The molecular ion at m/e = 214 was observed in the mass spectrum.

^{19}F NMR spectrum (solvent ?/$CFCl_3$): $\delta = -69.4$ ppm (s) [1].

On standing in vacuum for several days at room temperature, crystals of the trimer $S_3N_3(ON(CF_3)_2)_3$ were deposited [1, 2]. When 0.149 mmol of $(CF_3)_2NOS{\equiv}N$ is treated with 22.2 mmol of H_2O $(CF_3)_2NOH$ (0.134 mmol) and white crystals, the polymer $(-NHS(O)-)_x$ probably forms. $(CF_3)_2NOS{\equiv}N$ (0.024 mmol) reacts with gaseous HCl (2.00 mmol) to give $(CF_3)_2NOH$ (0.023 mmol) and $S_3N_3Cl_3$ (0.086 mmol) [1].

A mixture of excess $(CF_3)_2NOS{\equiv}N$, $[M(SO_2)_2]^{2+}(AsF_6^-)_2$ with M = Fe, Co, Ni and SO_2 condensed at −196°C reacts on warming to −40°C to give $[M(N{\equiv}SON(CF_3)_2)_6]^{2+}(AsF_6^-)_2$ in quantitative yield. The corresponding reaction of $[Cu(SO_2)_2]^{2+}(AsF_6^-)_2$ with excess $(CF_3)_2NOS{\equiv}N$ in SO_2 at −40°C yields $[Cu(N{\equiv}SON(CF_3)_2)_4]^{2+}(AsF_6^-)_2$ [3]. When a mixture of excess $(CF_3)_2NOS{\equiv}N$, $[M(CO)_5(SO_2)]^+AsF_6^-$ with M = Mn, Re and SO_2 condensed at −196°C is reacted on warming to −40°C, the stable complexes $[M(CO)_5(N{\equiv}SON(CF_3)_2)]^+AsF_6^-$ form in quantitative yield [3, 4].

References:

[1] Emeléus, H. J.; Poulet, R. J. (J. Fluorine Chem. **1** [1971/72] 13/21).
[2] Emeléus, H. J.; Forder, R. A.; Poulet, R. J.; Sheldrick, G. M. (J. Chem. Soc. Chem. Commun. **1970** 1483).
[3] Hartmann, G.; Mews, R.; Sheldrick, G. M. (J. Organometal. Chem. **252** [1983] 195/200).
[4] Mews, R. (unpublished results in: Glemser, O.; Mews, R., Angew. Chem. **92** [1980] 904/21; Angew. Chem. Intern. Ed. Engl. **19** [1980] 883/99).

2.3.7 Thiazyl Pentafluorooxotellurate, $F_5TeOS{\equiv}N$

$F_5TeOS{\equiv}N$ was prepared in 87% yield by reaction of $SN^+SbF_6^-$ with $CsOTeF_5$ in equimolar amounts in liquid SO_2 at −35°C for 0.75 h. It was isolated in a trap cooled to −90°C by fractional condensation in a dynamic vacuum. Only small amounts of F_5TeOSN were obtained when $FS{\equiv}N$ (4.0 mmol) was reacted with $B(OTeF_5)_3$ (1.3 mmol) at −35°C for 1 h. Fractional condensation gave the main products $O{=}S{=}NTeF_5$ in 29% yield, $F_5TeN{=}S{=}NTeF_5$ in 15% yield, and the adduct $BF_3 \cdot TeF_5OS{\equiv}N$ in low yield.

The compound is extremely instable. The colorless crystals can be sublimed at −50°C in a dynamic vacuum.

^{19}F NMR spectrum (SO_2/$CFCl_3$): δ (in ppm) = −37.8 (F(A)), −32.9 ($F(B_4)$); $^2J(A, B_4) = 171.9$ Hz. Gas-phase IR spectrum of an $F_5TeOS{\equiv}N{-}BF_3$ mixture: $\nu(SN) = 1312.5$ cm^{-1}.

On warming in condensed phase to temperatures above −50°C, isomerization occurs to give $O{=}S{=}NTeF_5$ and polymeric products. In the gas phase this isomerization could not be observed. A mixture of F_5TeOSN and excess of AsF_5 condensed at −196°C reacts at −60°C for 24 h to give $SN^+F_5AsOTeF_5^-$.

Reference:

Hoppenheit, R.; Mews, R. (Chem. Ber. **118** [1985] 4276/80).

2.3.8 S-Trifluoromethylsulfur Nitride, Trifluoromethylthiazyl, $CF_3S{\equiv}N$

Monomeric CF_3SN could not be prepared and isolated until now. From the reaction of CF_3SCl with excess $(CH_3)_3SiN_3$ in CCl_3F at −50 to −30°C, only the oligomeric $(CF_3SN)_x$ was obtained (x = 1, 2, 3; at t = 20°C: x = 4). The reaction of $CF_3S(F){=}NCOF$ with HgF_2 at 25 and 55°C gives polymeric $(CF_3SN)_x$ with $x \geqq 7$.

Reference:

Bielefeldt, D.; Haas, A. (Chem. Ber. **116** [1983] 1257/60).

2.4 Metal Complexes of Sulfur Nitride (Thiazyl) Compounds

2.4.1 Thiazyl Halide-, Thiazyl Bis(trifluoromethyl)nitroxide-, and Thiazyl Amide Complexes

2.4.1.1 Thiazyl Fluoride Complexes

2.4.1.1.1 $[M(N{\equiv}SF)_6]^{2+}(AsF_6^-)_2$, M = Co, Ni

The title compounds were prepared in almost quantitative yield by reaction of $[M(SO_2)_x]^{2+}$-$(AsF_6^-)_2$, M = Co, Ni, with 10 to 20% excess of $FS{\equiv}N$ in SO_2 below −20°C for 2 to 3 h [1].

The crystal structures determined by X-ray diffraction are given in Table 71.

Table 71
Crystallographic Data for $[Co(N{\equiv}SF)_6]^{2+}(AsF_6^-)_2$ [1, 2] and $[Ni(N{\equiv}SF)_6]^{2+}(AsF_6^-)_2$ [3].

	$[Co(N{\equiv}SF)_6]^{2+}(AsF_6^-)_2$	$[Ni(N{\equiv}SF)_6]^{2+}(AsF_6^-)_2$
system	monoclinic	monoclinic
space group	$P2_1/n-C_{2h}^5$ (No. 14)	$P2_1/n-C_{2h}^5$ (No. 14)
a in Å	13.105(9)	13.016(6)
b in Å	9.201(7)	9.062(4)
c in Å	9.366(6)	9.048(4)
β in °	90.43(4)	90.55(2)
V in Å³	1129	1067
D_x in g/cm³	2.432	2.476
Z	2	2

The molecular structure of the $[Co(N{\equiv}SF)_6]^{2+}$ cation is shown in **Fig. 53**, p. 244. The Co atom lies at the origin and the cation is thus centrosymmetric; the bond angles at Co approximate closely to octahedral values [2]. The most notable feature of the structure is the shortening of the N–S and N–F bonds from 1.448 Å and 1.643 Å, respectively in the free ligand [4] to average values of 1.403(11) Å and 1.571(11) Å in the complex [2]. The structure of $[Co(N{\equiv}SF)_6]^{2+}(AsF_6^-)_2$ is isomorphous with that of $[Ni(N{\equiv}SF)_6]^{2+}(AsF_6^-)_2$ [3]. Bond lengths and bond angles of the cations $[M(N{\equiv}SF)_6]^{2+}$, M = Co, Ni, are compiled in Table 72. For atom coordinates and effective isotropic temperature factors see the papers [2, 3].

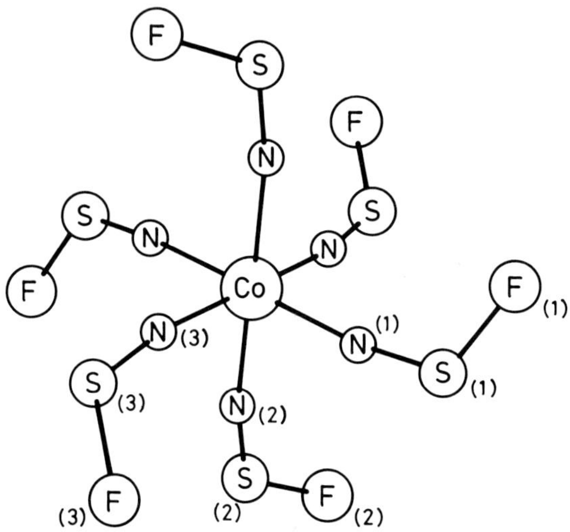

Fig. 53. Molecular structure of the cation in $[Co(N{\equiv}SF)_6]^{2+}(AsF_6^-)_2$ [2].

IR (solvent ?): ν (in cm^{-1}) = 1429 ν(SN), 655 ν(SF) for $[Co(NSF)_6]^{2+}$; 1438 ν(SN), 655 ν(SF) for $[Ni(NSF)_6]^{2+}$ [1].

The salts are stable in dry glass vessels at room temperature, but ligand exchange takes place in SO_2 solution according to $[M(N{\equiv}SF)_6]^{2+} + x\,SO_2 \rightleftharpoons [M(N{\equiv}SF)_{6-x}(SO_2)_x]^{2+} + x\,N{\equiv}SF$, with M = Co, Ni [1].

Table 72

Bond Lengths and Bond Angles in the Cation $[M(N{\equiv}SF)_6]^{2+}$ with M = Co [2], Ni [3].

	$[Co(N{\equiv}SF)_6]^{2+}(AsF_6^-)_2$	$[Ni(N{\equiv}SF)_6]^{2+}(AsF_6^-)_2$
bond length in Å		
M–N(1)	2.088(11)	2.056(5)
M–N(2)	2.098(12)	2.056(5)
M–N(3)	2.084(10)	2.050(5)
S(1)–N(1)	1.400(11)	1.402(5)
S(2)–N(2)	1.400(11)	1.410(5)
S(3)–N(3)	1.408(11)	1.401(5)
S(1)–F(1)	1.587(10)	1.582(4)
S(2)–F(2)	1.575(10)	1.594(5)
S(3)–F(2)	1.552(11)	1.568(5)

Table 72 (continued)

	$[Co(N{\equiv}SF)_6]^{2+}(AsF_6^-)_2$	$[Ni(N{\equiv}SF)_6]^{2+}(AsF_6^-)_2$
bond angle in °		
M–N(1)–S(1)	172.3(9)	171.3(3)
M–N(2)–S(2)	169.1(9)	166.4(3)
M–N(3)–S(3)	170.0(9)	168.2(3)
N(1)–M–N(2)	88.7(5)	88.6(2)
N(2)–M–N(3)	89.4(5)	88.5(2)
N(1)–M–N(3)	90.8(5)	91.7(2)
N(1)–S(1)–F(1)	115.8(7)	115.5(3)
N(2)–S(2)–F(2)	114.6(8)	115.7(3)
N(3)–S(3)–F(3)	115.5(7)	116.9(3)

References:

[1] Buß, B.; Jones, P. G.; Mews, R.; Noltemeyer, M.; Sheldrick, G. M. (Angew. Chem. **91** [1979] 253; Angew. Chem. Intern. Ed. Engl. **18** [1979] 231).

[2] Buß, B.; Jones, P. G.; Mews, R.; Noltemeyer, M.; Sheldrick, G. M. (Acta Cryst. B **36** [1980] 141/2).

[3] Hartmann, G.; Jones, P. G.; Mews, R.; Sheldrick, G. M. (Z. Naturforsch. **35b** [1980] 1191/2).

[4] Cook, R. L.; Kirchhoff, W. H. (J. Chem. Phys. **47** [1967] 4521/7).

2.4.1.1.2 $[Re(CO)_5(N{\equiv}SF)]^+AsF_6^-$

The title compound formed by reaction of $[Re(CO)_5SO_2]^+AsF_6^-$ with equimolar amounts of FS≡N in SO_2 at −30°C for 1 h.

The orange-red crystals decompose at 115°C. ^{19}F NMR spectrum (solvent ?/$CFCl_3$): δ (in ppm) = +219.7 (SF), −100.0 (AsF_6^-); $^1J(As,F)$ = 930 Hz. IR spectrum (solvent ?): ν (in cm^{-1}) = 2179 m, 2146 vs, br, ν(CO); 1390 vw, ν(SN); 647 m, ν(SF); 585 s, 450 w, 341 w, δ(Re–C–O), ν(Re–C); 702 vs, 678 vw, 400 s, ν(As–F). Raman spectrum (SO_2): 1396 cm^{-1} s, ν(SN) [1].

The reaction of $[Re(CO)_5(N{\equiv}SF)]^+AsF_6^-$ with equimolar amounts of H_2O in SO_2 at −60°C gives probably $[Re(CO)_5(HNSO)]^+AsF_6^-$ and HF. The reaction is difficult to control because HF will attack $[Re(CO)_5(HNSO)]^+AsF_6^-$ [2] further. The title compound reacts with $SiCl_4$ in the mole ratio 4:1 in SO_2 at −15°C for 3 h to give $[Re(CO)_5(N{\equiv}SCl)]^+AsF_6^-$ in 95% yield and SiF_4. The analogous reaction with $SiBr_4$ but at −40°C gives $[Re(CO)_5(N{\equiv}SBr)]^+AsF_6^-$ in 91% yield and SiF_4 [3]. Treatment with a slight excess of AsF_5 in SO_2 at 30°C for 1 h gives $[Re(CO)_5(NS)]^{2+}(AsF_6^-)_2$ in quantitative yield [1]. The reaction with equimolar amounts of $(CH_3)_3SiN(CH_3)_2$ in SO_2 at −30°C for 1 h gives $[Re(CO)_5(N{\equiv}SN(CH_3)_2)]^+AsF_6^-$ in quantitative yield and $(CH_3)_3SiF$. When the title compound is treated with equimolar amounts of $((CH_3)_3Si)_2NCH_3$ in SO_2 at −30°C the primary product $[Re(CO)_5(N{\equiv}SN(CH_3)Si(CH_3)_3)]^+AsF_6^-$ could not be isolated in pure form. The Si–N bond in the complex reacts even above −15°C with AsF_6^-. But the reaction in the mole ratio 2:1 in SO_2 at −30°C for 12 h produces $[(CO)_5ReN{\equiv}SN(CH_3)S{\equiv}NRe(CO)_5]^{2+}(AsF_6^-)_2$ in 80% yield and $(CH_3)_3SiF$ [4, 5].

When the complex is reacted with SnI_4 in SO_2 at low temperatures $[Re(CO)_5(NS)]^{2+}(AsF_6^-)_2$ in 40% yield, SnF_4, I_2, and compounds containing $Re(CO)_5$ units could be isolated [3]. When

$[Re(CO)_5(N{\equiv}SF)]^+AsF_6^-$ is reacted with equimolar amounts of $(CH_3)_3SnOH$ in SO_2 at −30°C for 2 h the reaction forms $[Re(CO)_5(HNSO)]^+AsF_6^-$ in quantitative yield and $(CH_3)_3SnF$ [2].

References:

[1] Mews, R.; Liu, C. S. (Angew. Chem. **95** [1983] 156; Angew. Chem. Intern. Ed. Engl. **22** [1983] 158).
[2] Hartmann, G.; Hoppenheit, R.; Mews, R. (Inorg. Chim. Acta **76** [1983] L201/L202).
[3] Hartmann, G.; Mews, R. (Z. Naturforsch. **40b** [1985] 343/6).
[4] Hartmann, G.; Mews, R.; Sheldrick, G. M. (Angew. Chem. **95** [1983] 753; Angew. Chem. Intern. Ed. Engl. **22** [1983] 723).
[5] Hartmann, G.; Mews, R.; Sheldrick, G. M. (Angew. Chem. Suppl. **1983** 945/56).

2.4.1.2 Thiazyl Chloride Complexes

In this section metal complexes with ClS≡N are considered which are bonded in a relatively loose manner via the nitrogen atom to the metal atom by a donor-acceptor interaction $\overset{\ominus}{M}-\overset{\oplus}{N}{\equiv}\underline{S}-Cl$. Transition metal complexes with chlorothionitrene ligands of the type $M{=}\overset{\oplus}{N}{=}\overset{\ominus}{\underline{\bar{S}}}-Cl$ with M = Mo, W, Re, and Os are dealt with in the next section, pp. 256/71.

2.4.1.2.1 $(VCl_3(N{\equiv}SCl)_2)_2$

$(VCl_3(N{\equiv}SCl)_2)_2$ was obtained in almost quantitative yield by dropping a solution of VCl_4 in CCl_4 into a suspension of $S_3N_3Cl_3$ in CCl_4 (mole ratio 1.5:1) and stirring the mixture for 5 h at room temperature.

The black microcrystalline solid is sensitive to humidity.

From the IR spectrum given in Table 73, a dimeric ClS≡N, chloro-bridged adduct with VCl_3 is deduced. The ClS≡N molecules are in cis position. The proposed molecular structure is given in **Fig. 54**.

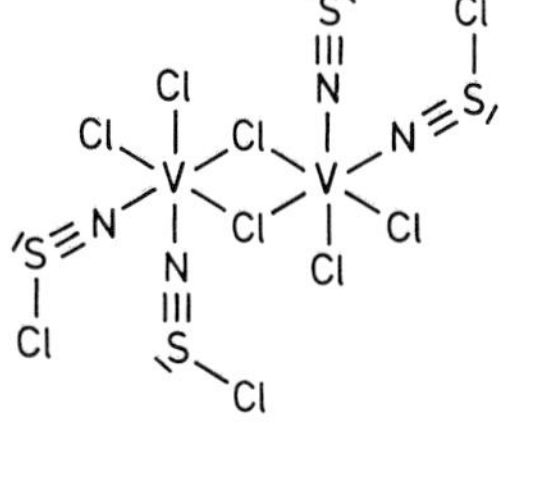

Fig. 54. Proposed molecular structure of $(VCl_3(N{\equiv}SCl)_2)_2$.

Table 73
IR Spectrum (Nujol) of $(VCl_3(N{\equiv}SCl)_2)_2$.

ν in cm^{-1}	intensity	assignment	ν in cm^{-1}	intensity	assignment
1357	vs	ν(S≡N)	390	s	$\nu_s(VCl_2)$
1290	vs	ν(S≡N)	360	m to w	$\nu(VCl_2V)$
500	m	ν(S–Cl)	338	vs	$\nu(VCl_2V)$
440	s	ν(V–N)	304	m	δ(NSCl)
428	m	ν(V–N)	282	s	δ(NSCl)
410	s	$\nu_{as}(VCl_2)$			

Thermolysis of a CCl_4 suspension of $(VCl_3(N{\equiv}SCl)_2)_2$ in a closed tube at 105°C yields $((ClS)_2N^+)_3V_2Cl_9^{3-}$ as a main product. Treatment with $POCl_3$ gives $(VCl_3(OPCl_3)_2)_2$. The reaction with $(C_6H_5)_4As^+Cl^-$ (mole ratio 1:2) affords $(C_6H_5)_4As^+VCl_5^-$ and S_4N_4.

Reference:

Beber, G.; Hanich, J.; Dehnicke, K. (Z. Naturforsch. **40b** [1985] 9/12).

2.4.1.2.2 $NbCl_5(N{\equiv}SCl)$

$NbCl_5(N{\equiv}SCl)$ was prepared by dropping a solution of $S_3N_3Cl_3$ in CCl_4 into a suspension of sublimed $(NbCl_5)_2$ (mole ratio 1:1.5).

The red crystals melt at 75°C, evolving Cl_2 and becoming green-colored.

The crystal structure was determined by X-ray diffraction. The crystals are triclinic, space group $P\bar{1}-C_i^1$ (No. 2), with a = 636.7(3), b = 1215.5(5), c = 1266.8(5) pm, α = 83.32(3)°, β = 78.52(3)°, γ = 82.26(3)°; Z = 4; R = 0.019 for 1879 reflexes. There are two crystallographically independent molecules which differ somewhat in their geometry. Bond lengths and bond angles are compiled in Table 74, p. 248. The structure of molecule I is shown in **Fig. 55**. The atom group Nb–N–S–Cl lies almost exactly in a plane which is almost eclipsed to an Nb–Cl group

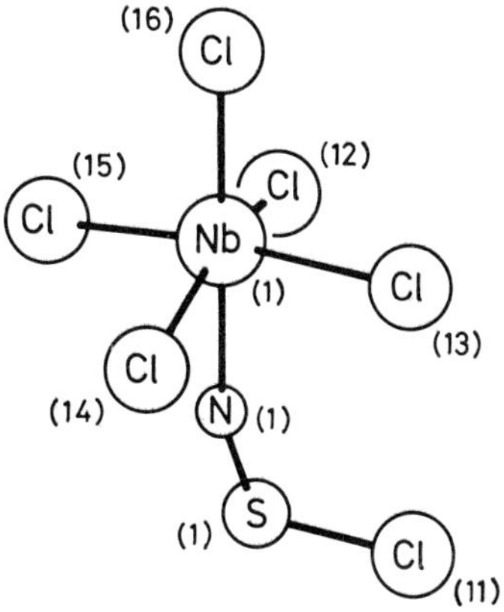

Fig. 55. Molecular structure of $NbCl_5(N{\equiv}SCl)$ (molecule I) [1].

in the molecule I but which forms a dihedral angle of 14.6° in the molecule II. The structure of the ClS≡N ligand corresponds closely to that of the monomeric ClS≡N. The most striking difference is a shortening of the S–Cl bond. For atom coordinates and isotropic temperature factors see the paper [1].

Table 74

Bond Lengths and Bond Angles of the Two Crystallographically Independent Molecules of $NbCl_5(N{\equiv}SCl)$ [1]. (Standard deviations: Nb–Cl 0.1, Nb–N 0.4, S–N 0.3, S–Cl 0.2 pm; Cl–Nb–Cl <0.1°, Nb–N–S 0.2°, N–S–Cl 0.2°.)

bond length	in pm	bond angle	in °
	molecule I		
Nb(1)–Cl(12)	229.0	Cl(12)–Nb(1)–Cl(16)	100.9
Nb(1)–Cl(13)	232.5	Cl(13)–Nb(1)–Cl(16)	93.2
Nb(1)–Cl(14)	233.9	Cl(14)–Nb(1)–Cl(16)	98.6
Nb(1)–Cl(15)	232.7	Cl(15)–Nb(1)–Cl(16)	94.6
Nb(1)–Cl(16)	225.3	Cl(12)–Nb(1)–Cl(14)	160.4
Nb(1)–N(1)	226.3	Cl(13)–Nb(1)–Cl(15)	171.9
N(1)–S(1)	144.1	N(1)–Nb(1)–Cl(12)	81.5
S(1)–Cl(11)	206.7	N(1)–Nb(1)–Cl(13)	86.0
		N(1)–Nb(1)–Cl(14)	78.9
		N(1)–Nb(1)–Cl(15)	86.1
		N(1)–Nb(1)–Cl(16)	177.5
		Nb(1)–N(1)–S(1)	157.7
		N(1)–S(1)–Cl(11)	116.4
dihedral angle Nb(1)–N(1)–S(1)–Cl(11)/Nb(1)–Cl(13)–Cl(15)–Cl(16) 3.9°			
	molecule II		
Nb(2)–Cl(22)	231.7	Cl(22)–Nb(2)–Cl(26)	98.9
Nb(2)–Cl(23)	228.9	Cl(23)–Nb(2)–Cl(26)	92.2
Nb(2)–Cl(24)	228.9	Cl(24)–Nb(2)–Cl(26)	97.8
Nb(2)–Cl(25)	238.9	Cl(25)–Nb(2)–Cl(26)	95.1
Nb(2)–Cl(26)	225.8	Cl(22)–Nb(2)–Cl(24)	162.7
Nb(2)–N(2)	226.4	Cl(23)–Nb(2)–Cl(25)	167.7
N(2)–S(2)	143.4	N(2)–Nb(2)–Cl(22)	82.3
S(2)–Cl(21)	207.8	N(2)–Nb(2)–Cl(23)	85.6
		N(2)–Nb(2)–Cl(24)	81.0
		N(2)–Nb(2)–Cl(25)	82.1
		N(2)–Nb(2)–Cl(26)	177.2
		Nb(2)–N(2)–S(2)	163.6
		N(2)–S(2)–Cl(21)	116.4
dihedral angle Nb(2)–N(2)–S(2)–Cl(21)/Nb(2)–Cl(23)–Cl(25)–Cl(26) 14.6°			

The vibrational spectrum of crystalline $NbCl_5(N{\equiv}SCl)$ is given in Table 75.

Table 75
Vibrational Spectrum of Crystalline $NbCl_5(N{\equiv}SCl)$ (in mineral oil) [1].

ν in cm^{-1}	IR	Raman	assignment
1332	m	—	ν(S≡N)
1310	m	—	
(850	w	—	decomposition product)
505	m, br	—	ν(S–Cl)
465	vw	—	
415	—	vs	ν(Nb–Cl)
405	sh	—	
388	sh, br	m to w	
359	—	s	
345	vs	—	
300	sh	—	
270	vw	—	δ(NSCl)
260	—	vw	
187	—	m	δ(NSCl)
165	—	m	
128	—	m	
115	—	m	
100	—	m	

When the red crystals are stored at room temperature for a longer time the color changes to yellow-green. The adduct is sensitive to humidity [1]. Thermolysis at 160°C under normal pressure for 5 h gives $(NbCl_5)_2 \cdot S_2N_2$ in >60% yield [2]. The reaction with $(C_6H_5)_4P^+Cl^-$ [1] or $(C_6H_5)_4As^+Cl^-$ [2] in CH_2Cl_2 produces $(C_6H_5)_4P^+NbCl_6^-$ [1] and $(C_6H_5)_4As^+NbCl_6^-$ [2], respectively.

References:

[1] Hanich, J.; Klingelhöfer, P.; Müller, U.; Dehnicke, K. (Z. Anorg. Allgem. Chem. **506** [1983] 68/74).
[2] Hanich, J.; Dehnicke, K. (Z. Naturforsch. **39b** [1984] 1467/71).

2.4.1.2.3 $TaCl_5(N{\equiv}SCl)$

$TaCl_5(N{\equiv}SCl)$ is obtained in 60% yield by addition of a solution of $S_3N_3Cl_3$ in CH_2Cl_2 to a suspension of $TaCl_5$ in CCl_4 (mole ratio 1:3) and by refluxing the resulting solution for 2 h.

The adduct forms moisture-sensitive red crystals. From the IR spectrum given in Table 76, p. 250, it is concluded that $TaCl_5(N{\equiv}SCl)$ has a structure analogous to $NbCl_5(N{\equiv}SCl)$.

Table 76
IR Spectrum (Nujol) of $TaCl_5(N{\equiv}SCl)$.

ν in cm^{-1}	intensity	assignment
1340	sh	ν(S≡N)
1320	m	
493	w	ν(S–Cl)
450	sh	ν(Ta–N)
432	m	
355	sh	ν(Ta–Cl trans)
323	vs	$\nu(TaCl_4)$
300	sh	
280	sh	
228	w to m	δ(NSCl)

Thermolysis at 160°C under normal pressure for 5 h gives $(TaCl_5)_2 \cdot S_2N_2$ in >60% yield and Cl_2. The reaction with $(C_6H_5)_4As^+Cl^-$ in CH_2Cl_2 produces $(C_6H_5)_4As^+TaCl_6^-$.

Reference:

Hanich, J.; Dehnicke, K. (Z. Naturforsch. **39b** [1984] 1467/71).

2.4.1.2.4 $(C_6H_5)_4P^+[RuCl_4(NO)(N{\equiv}SCl)]^-$

$(C_6H_5)_4P^+[RuCl_4(NO)(N{\equiv}SCl)]^-$ was prepared by treating $((C_6H_5)_4P^+)_2[RuCl_4(NO)]_2^{2-}$ with equimolar amounts of $S_3N_3Cl_3$ in CH_2Cl_2 at room temperature and allowing the solution to stand for 7 d. A red-brown crystalline precipitate was obtained in 90% yield.

The crystal structure was determined by X-ray diffraction at 22°C. The complex crystallizes in the monoclinic space group $P2_1/n$ (standard setting $P2_1/c$)–C_{2h}^5 (No. 14), with a = 982.6(2), b = 1700.0(2), c = 1772.3(2) pm; β = 104.79(1)°; V = 2862×10^{-30} m^3; Z = 4, D_x = 1.61 g/cm^3. R = 0.046 for 2548 observed reflections. The molecular structure of the $[RuCl_4(NO)(N{\equiv}SCl)]^-$ ion is shown in **Fig. 56**. Bond lengths and bond angles are compiled in Table 77. Atomic coordinates and parameters of the thermic vibration are given in the paper. In the $[RuCl_4(NO)(N{\equiv}SCl)]^-$ ion the NO and N≡SCl ligands occupy cis positions. Bond lengths indicate

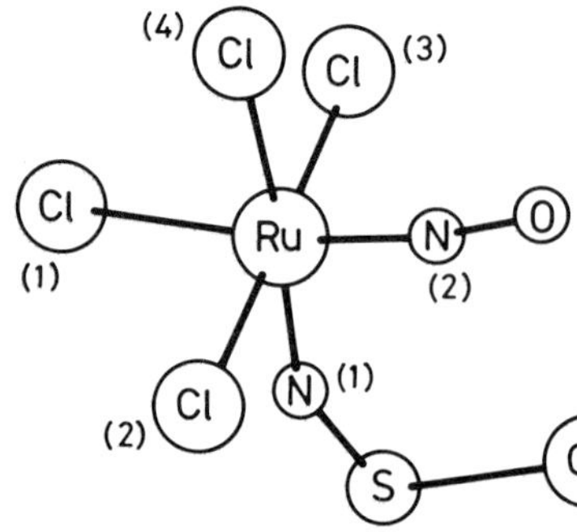

Fig. 56. Molecular structure of the anion in $(C_6H_5)_4P^+[RuCl_4(NO)(N{\equiv}SCl)]^-$.

double bonds in the linear Ru=N=O group, whereas the N≡SCl molecule is attached by a donor-acceptor interaction $\overset{\ominus}{Ru}$–$\overset{\oplus}{N}$≡SCl, although with a rather short Ru–N bond of 198 pm. The N≡S and S–Cl bond lengths of 144 and 213 pm, respectively, show only small differences from those of the free molecule (see p. 228) and the ligand in $NbCl_5$(N≡SCl) (see p. 248). (In the corresponding Os complex $(C_6H_5)_4P^+[OsCl_4(NO)(NSCl)]^-$ however, NSCl acts as a chlorothionitrene ligand of the type Os=$\overset{\oplus}{N}$–$\overset{\ominus}{S}$Cl; see p. 271.)

Table 77

Bond Lengths and Bond Angles of the Anion in $(C_6H_5)_4P^+[RuCl_4(NO)(N{\equiv}SCl)]^-$.

bond length	in pm	bond angle	in °	bond angle	in °
Ru–Cl(1)	234.9(2)	Cl(1)–Ru–Cl(2)	88.6(1)	N(1)–Ru–Cl(3)	89.3(2)
Ru–Cl(2)	237.1(2)	Cl(1)–Ru–Cl(3)	88.2(1)	N(1)–Ru–Cl(4)	176.5(2)
Ru–Cl(3)	237.8(2)	Cl(1)–Ru–Cl(4)	88.9(1)	N(2)–Ru–Cl(1)	173.3(2)
Ru–Cl(4)	237.1(2)	Cl(2)–Ru–Cl(3)	176.4(1)	N(2)–Ru–Cl(2)	95.9(2)
Ru–N(1)	198.0(5)	Cl(2)–Ru–Cl(4)	89.9(1)	N(2)–Ru–Cl(3)	87.4(2)
Ru–N(2)	171.0(8)	Cl(3)–Ru–Cl(4)	91.7(1)	N(2)–Ru–Cl(4)	86.2(2)
N(1)–S	144.2(6)	N(1)–Ru–N(2)	97.2(3)	Ru–N(1)–S	149.2(5)
S–Cl	212.6(4)	N(1)–Ru–Cl(1)	87.8(2)	N(1)–S–Cl	113.0(3)
N(2)–O	116.5(8)	N(1)–Ru–Cl(2)	88.8(2)	Ru–N(2)–O	170.1(7)

The IR spectrum (Nujol) confirms the structure obtained by X-ray diffraction analysis: ν (in cm^{-1}) = 1903 s, ν(NO); 1342 s, ν(NS); 1292 s, ν(NS); 395 m, ν(RuN); 322 br, ν(RuCl); 290 m, δ(NSCl); 277 m, δ(RuNS).

Vacuum pyrolysis at 200 to 220°C/10^{-4} Torr over a period of about 8 h gives $((C_6H_5)_4P^+)_2[RuCl_4(NS)]_2^{2-}$ and NOCl in quantitative yield.

Reference:

Willing, W.; Müller, U.; Demant, U.; Dehnicke, K. (Z. Naturforsch. **41 b** [1986] 560/6).

2.4.1.2.5 $OsCl_4(N{\equiv}SCl)_2$

$OsCl_4(N{\equiv}SCl)_2$ formed in 92% yield by reaction of $OsCl_5$ with $S_3N_3Cl_3$ (mole ratio 1.5:1) in $POCl_3$ at room temperature for 24 h. An analytically pure product was obtained in quantitative yield when the reaction was carried out in CCl_4 suspension over a period of 14 d. Addition of some $POCl_3$ as a solubilizer accelerated the reaction.

The fine black powder melts at 130°C with decomposition. It is extraordinarily sensitive to humidity. The solubility in CCl_4 and CH_2Cl_2 is extremely low.

IR spectrum (Nujol): ν (in cm^{-1}) = 1335 vs, br, ν(S≡N); 508/492 s, ν(S–Cl); 327 vs, structured, ν(Os–Cl).

The IR spectrum suggests that both N≡SCl ligands are in cis position with local C_{2v} symmetry at the Os center.

Reaction with equimolar amounts of $(C_6H_5)_4As^+Cl^-$ in CH_2Cl_2 at room temperature for some hours gives $(C_6H_5)_4As^+[OsCl_4((NS)_2Cl)]^-$ in almost quantitative yield along with Cl_2.

Reference:

Weber, R.; Müller, U.; Dehnicke, K. (Z. Anorg. Allgem. Chem. **504** [1983] 13/22).

2.4.1.2.6 $[Re(CO)_5(N{\equiv}SCl)]^+AsF_6^-$

$[Re(CO)_5(N{\equiv}SCl)]^+AsF_6^-$ was prepared in 95% yield by reaction of $[Re(CO)_5(N{\equiv}SF)]^+AsF_6^-$ with $SiCl_4$ in the mole ratio 4:1 in SO_2 at −15°C for 3 h. It also formed in nearly quantitative yield by stirring equimolar amounts of $Re(CO)_5Cl$ and $SN^+AsF_6^-$ in SO_2 at −15°C for 5 h or by reacting equimolar amounts of $[Re(CO)_5(NS)]^{2+}(AsF_6^-)_2$ and CsCl in SO_2 at −15°C for 0.5 h.

The green-black crystals decompose at 124°C.

The IR spectrum (Nujol) shows twelve absorptions in the range 2170 to 330 cm^{-1}; 1308 cm^{-1} m, ν(S≡N).

Reference:

Hartmann, G.; Mews, R. (Z. Naturforsch. **40b** [1985] 343/6).

2.4.1.3 Thiazyl Bromide Complex

$[Re(CO)_5(N{\equiv}SBr)]^+AsF_6^-$ was produced in 91% yield by stirring a mixture of $[Re(CO)_5$-$(N{\equiv}SF)]^+AsF_6^-$ and $SiBr_4$ in the mole ratio 4:1 in SO_2 at −40°C. The complex was also obtained by reaction of equimolar amounts of $Re(CO)_5Br$ and $SN^+AsF_6^-$ in SO_2 at −40°C for 3 h in 48% yield and by reaction of equimolar amounts of $[Re(CO)_5(NS)]^{2+}(AsF_6^-)_2$ and CsBr in SO_2 at −15°C for 0.5 h in 65% yield. The complex has to be quickly isolated at −40°C.

The light green solid decomposes at 130°C. Slow decomposition occurs even at room temperature.

The IR spectrum (Nujol) exhibits thirteen absorptions in the range 2178 to 395 cm^{-1}; 1299 cm^{-1} m, ν(S≡N).

Reference:

Hartmann, G.; Mews, R. (Z. Naturforsch. **40b** [1985] 343/6).

2.4.1.4 Thiazyl Iodide Complexes

Thiazyl iodide complexes are unknown. The attempt to prepare $[Re(CO)_5(N{\equiv}SI)]^+AsF_6^-$ by the reaction of $[Re(CO)_5(N{\equiv}SF)]^+AsF_6^-$ with SnI_4 in the mole ratio 4:1 in SO_2 at low temperature was unsuccessful.

Reference:

Hartmann, G.; Mews, R. (Z. Naturforsch. **40b** [1985] 343/6).

2.4.1.5 Thiazyl Bis(trifluoromethyl)nitroxide Complexes

2.4.1.5.1 $[M(N{\equiv}SON(CF_3)_2)_6]^{2+}(AsF_6^-)_2$, M = Fe, Co, Ni

The compounds formed in quantitative yield by reaction of excess $(CF_3)_2NOS{\equiv}N$ with $[M(SO_2)_2]^{2+}(AsF_6^-)_2$, M = Fe, Co, Ni in SO_2 at −40°C.

$[Fe(N{\equiv}SON(CF_3)_2)_6]^{2+}(AsF_6^-)_2$: rust-red solid, decomposition at 212°C; IR spectrum (Nujol): 15 absorptions in the range 1340 to 400 cm^{-1}.

$[Co(N{\equiv}SON(CF_3)_2)_6]^{2+}(AsF_6^-)$: ocherous solid, decomposition at 120°C; IR spectrum (Nujol): 14 absorptions in the range 1345 to 400 cm^{-1}.

$[Ni(N{\equiv}SON(CF_3)_2)_6]^{2+}(AsF_6^-)_2$: light green solid, decomposition at 110°C; IR spectrum (Nujol): 15 absorptions in the range 1340 to 395 cm^{-1}. The crystals are monoclinic, space group $P2_1/c-C_{2h}^5$ (No. 14), with a = 1447.7(1.6), b = 1670.6(1.4), c = 1118.9(2.1) pm, and β = 103.3(2)°; V = 2633 × 10^6 pm^3, Z = 2, D_x = 2.17 g/cm^3; R = 0.12 for 1820 independent reflexes. The molecular structure of the cation is given in **Fig. 57**. Selected bond lengths and bond angles are compiled in Table 78, p. 254. The Ni central atom is octahedrally surrounded. The thiazyl ligands are coordinated over the N atoms.

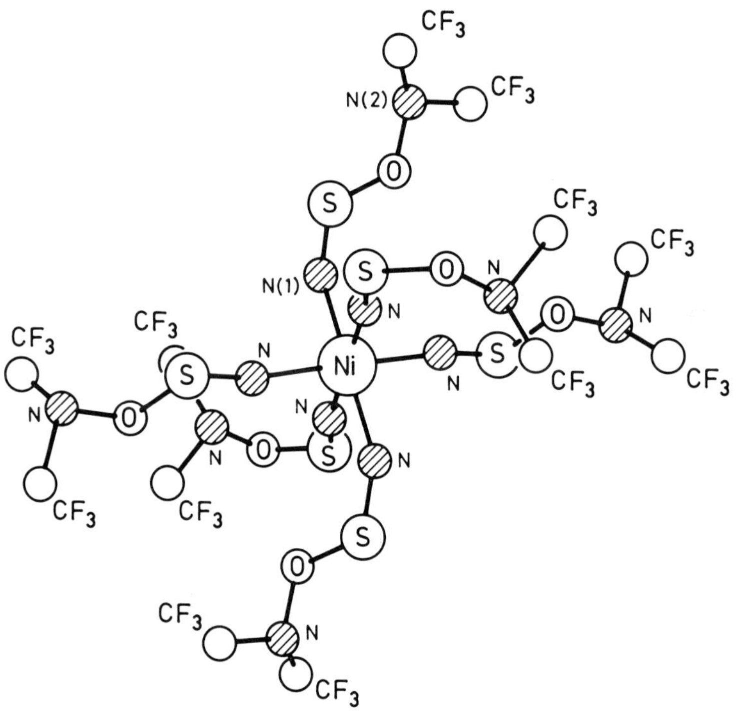

Fig. 57. Molecular structure of the cation in $[Ni(N{\equiv}SON(CF_3)_2)_6]^{2+}(AsF_6^-)_2$.

Table 78
Selected Average Bond Lengths and Bond Angles of the Cation in $[Ni(N{\equiv}SON(CF_3)_2)_6]^{2+}(AsF_6^-)_2$.

bond length	in pm	bond angle	in °
Ni–N(1)	204.7(2.2)	Ni–N(1)–S	161.7(8.5)
N(1)–S	144.1(1.8)	N(1)–S–O	112.9(4.4)
S–O	170.1(2.8)	N(2)–C–F	111.1(5.7)
O–N(2)	144.6(5.0)	S–O–N(2)	109.1(3.3)
N(2)–C	140.4(4.9)	O–N(2)–C	108.0(2.6)
C–F	128.0(3.4)		

Reference:

Hartmann, G.; Mews, R.; Sheldrick, G. M. (J. Organometal. Chem. **252** [1983] 195/200).

2.4.1.5.2 $[Cu(N{\equiv}SON(CF_3)_2)_4]^{2+}(AsF_6^-)_2$

$[Cu(N{\equiv}SON(CF_3)_2)_4]^{2+}(AsF_6^-)_2$ formed in quantitative yield by treating $[Cu(SO_2)_2]^{2+}(AsF_6^-)_2$ with large excess of $(CF_3)_2NOS{\equiv}N$ in SO_2 at −40°C. The green solid decomposes at 130°C. IR spectrum (Nujol): 12 absorptions in the range 1328 to 400 cm^{-1}.

Reference:

Hartmann, G.; Mews, R.; Sheldrick, G. M. (J. Organometal. Chem. **252** [1983] 195/200).

2.4.1.5.3 $[M(CO)_5(N{\equiv}SON(CF_3)_2)]^+AsF_6^-$, M = Mn, Re

The compounds were obtained in quantitative yield by reaction of excess $(CF_3)_2NOS{\equiv}N$ with $[M(CO)_5SO_2]^+AsF_6^-$, M = Mn, Re in SO_2 at −40°C.

$[Mn(CO)_5(N{\equiv}SON(CF_3)_2)]^+AsF_6^-$: light brown solid, decomposition at 165°C; IR spectrum (Nujol): 16 absorptions in the range 2165 to 400 cm^{-1}.

$[Re(CO)_5(N{\equiv}SON(CF_3)_2)]^+AsF_6^-$: light red solid, decomposition at 135°C; IR spectrum (Nujol): 16 absorptions in the range 2155 to 400 cm^{-1}.

Reference:

Hartmann, G.; Mews, R.; Sheldrick, G. M. (J. Organometal. Chem. **252** [1983] 195/200).

2.4.1.6 Thiazyl Amide Complexes

2.4.1.6.1 $[Re(CO)_5(N{\equiv}SN(CH_3)_2)]^+AsF_6^-$

$[Re(CO)_5(N{\equiv}SN(CH_3)_2)]^+AsF_6^-$ was prepared in quantitative yield by reaction of equimolar amounts of $[Re(CO)_5(N{\equiv}SF)]^+AsF_6^-$ and $(CH_3)_3SiN(CH_3)_2$ in SO_2 at −30°C.

The crystal structure was determined by X-ray diffraction. The crystals are orthorhombic, space group Pbca−D_{2h}^{15} (No. 61), with a = 1136.8(4), b = 2497.6(6), c = 1139.8(4) pm; Z = 8; D_x = 2.486 g/cm³. R = 0.044 for 2173 independent reflexes. The molecular structure of the cation is shown in **Fig. 58.** Selected bond lengths and bond angles are compiled in Table 79. From the bond lengths r(N(1)–S) = 144.7 pm and r(S–N(2)) = 158.1 pm it follows that the new ligand functions as a thiazyl dimethylamide. The bond length r(Re–N(1)) = 216.1 pm corresponds to a single bond [1, 2].

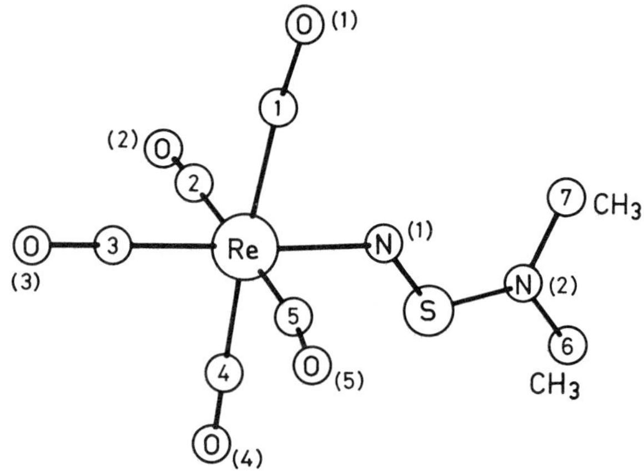

Fig. 58. Molecular structure of the cation in $[Re(CO)_5(N{\equiv}SN(CH_3)_2)]^+AsF_6^-$.

The light red solid decomposes at 73°C. ^{1}H NMR spectrum (50% in $(CD_3)_2CO$/TMS): δ = 3.49 ppm (s, $N(CH_3)_2$). The IR spectrum (Nujol) shows 17 absorptions in the range 2165 to 395 cm^{-1} [2].

Table 79
Selected Bond Lengths and Bond Angles of the Cation in $[Re(CO)_5(N{\equiv}SN(CH_3)_2)]^+AsF_6^-$ [2].

bond length	in pm	bond angle	in °	bond angle	in °
Re–N(1)	216.1(8)	N(1)–Re–C(1)	83.7(4)	C(2)–Re–C(3)	88.6(5)
Re–C(1)	200.1(10)	C(1)–Re–C(2)	92.5(4)	C(1)–Re–C(4)	174.7(4)
Re–C(3)	201.0(11)	C(1)–Re–C(3)	88.8(4)	C(3)–Re–C(4)	90.9(4)
Re–C(5)	202.2(10)	N(1)–Re–C(4)	91.0(4)	C(1)–Re–C(5)	90.7(4)
N(1)–S	144.7(8)	C(2)–Re–C(4)	92.8(5)	C(3)–Re–C(5)	175.5(4)
N(2)–C(7)	144.3(16)	N(1)–Re–C(5)	90.4(4)	Re–N(1)–S	130.0(5)
C(1)–O(1)	113.3(13)	C(2)–Re–C(5)	87.0(4)	S–N(2)–C(6)	117.2(8)
C(3)–O(3)	112.2(13)	C(4)–Re–C(5)	90.0(4)	C(6)–N(2)–C(7)	119.3(10)
C(5)–O(5)	111.3(12)	N(1)–S–N(2)	113.2(5)	Re–C(2)–O(2)	176.9(10)

Table 79 (continued)

bond length	in pm	bond angle	in °	bond angle	in °
Re–C(2)	191.9(11)	S–N(2)–C(7)	123.4(8)	Re–C(4)–O(4)	178.3(9)
Re–C(4)	199.1(10)	Re–C(1)–O(1)	173.5(9)		
N(2)–S	158.1(9)	Re–C(3)–O(3)	176.8(11)		
N(2)–C(6)	145.5(14)	Re–C(5)–O(5)	172.8(10)		
C(2)–O(2)	114.7(15)	N(1)–Re–C(2)	175.4(4)		
C(4)–O(4)	166.4(13)	N(1)–Re–C(3)	94.0(4)		

2.4.1.6.2 $[(CO)_5ReN{\equiv}SN(CH_3)S{\equiv}NRe(CO)_5]^{2+}(AsF_6^-)_2$

The title compound was produced in 80% yield when $[Re(CO)_5(N{\equiv}SF)]^+AsF_6^-$ is reacted with $((CH_3)_3Si)_2NCH_3$ in the mole ratio 2:1 in SO_2 at −30°C for 12 h.

The following reaction sequence is proposed: $[Re(CO)_5(N{\equiv}SF)]^+AsF_6^-$ (I) + $((CH_3)_3Si)_2NCH_3$

$$\rightarrow [(CO)_5ReN{\equiv}SN(CH_3)Si(CH_3)_3]^+AsF_6^-\ (II) + (CH_3)_3SiF,$$

$$II + I \rightarrow [(CO)_5ReN{\equiv}SN(CH_3)S{\equiv}NRe(CO)_5]^{2+}(AsF_6^-)_2 + (CH_3)_3SiF$$

The primary product II could not be isolated in pure form [1, 2].

The deep red solid decomposes at 83°C. ^1H NMR spectrum (50% in $(CD_3)_3CO$/TMS): $\delta = 2.65$ ppm (s, $N(CH_3)$). The IR spectrum (Nujol) exhibits nine absorptions in the range 2160 to 340 cm^{-1} [1, 2].

References:

[1] Hartmann, G.; Mews, R.; Sheldrick, G. M. (Angew. Chem. **95** [1983] 753; Angew. Chem. Intern. Ed. Engl. **22** [1983] 723).

[2] Hartmann, G.; Mews, R.; Sheldrick, G. M. (Angew. Chem. Suppl. **1983** 945/56).

2.4.2 Halothionitrene Complexes

2.4.2.1 Chlorothionitrene Complexes

In chlorothionitrene complexes of Mo, W, Re, and Os, NSCl is bonded via its N atom to the metal to form an approximately linear M–N–S group; the M–N and N–S distances correspond to double bonds according to the formula $M{=}\overset{\oplus}{N}{=}\overset{\ominus}{S}Cl$. The bond angle at the S atom is about 104°. The chlorothionitrene ligand can formally be derived from the $|N{\equiv}\underline{S}Cl$ molecule by the absorption of two electrons to give $[<N{=}\underline{\overline{S}}Cl]^{2-}$. An excellent review about "Chlorothionitrene Complexes of Transition Metals" is given by Dehnicke, K.; Müller, U. (Comments of Inorganic Chem. **4** [1985] 213/28). Some complexes which posses the chlorothionitrene ligand in addition to the thionitrosyl ligand are dealt with in the Chapter 1.4 "Thionitrosyl Complexes": Re complexes on pp. 66/8, Ru complex on p. 76, and Os complex on pp. 84/5. Complexes which contain the chlorothionitrene ligand in addition to metal-bonded sulfur-nitrogen compounds, e.g., S_2N_2, are not described in this volume.

2.4.2.1.1 $(MoCl_4(NSCl))_2$

The complex was prepared in 90% yield by adding a solution of $S_3N_3Cl_3$ in CH_2Cl_2 dropwise to a suspension of $MoCl_5$ in CH_2Cl_2 (mole ratio 1:3) with stirring at 35°C. The reaction was finished after 20 h.

The black microcrystalline solid melts at 120°C. It is very sensitive to humidity. IR (Nujol): ν (in cm^{-1}) = 999 m to w, ν_{as}(MoNS); 503 vs, ν(SCl); 495 sh, ν(SCl); 480 sh, ν(SCl); 418 sh, 412 m, ν_s(MoNS); 385 vs, 358 s, 330 vs, 262 m to w, 255 sh, all ν(MoCl); 210 s, 205 s, all ν($MoCl_2Mo$).

Two bands in the IR at 210 and 205 cm^{-1}, which can be assigned to the valence vibrations of an $MoCl_2Mo$ bridge, suggest that the complex can be formulated as dimer.

EIMS (70 eV): fragment ion (rel. int. in %) = $MoNCl_3^+$ (100), $MoNCl_2^+$ (63), $MoNCl^+$ (11), MoN^+ (6), $NSCl^+$ (16).

$(MoCl_4(NSCl))_2$ is slightly soluble in CH_2Cl_2 to give a black solution. Thermolysis at 130°C in the dynamic vacuum gives $Mo(N)Cl_3$ and SCl_2 in high yield [1]. Fluorination of a few grams of finely powdered $(MoCl_4(NSCl))_2$ in a quartz flask with F_2–N_2 gas mixtures of increasing volume ratio of initially 1:2, then 1:1 for ca. 10 h, and 2:1 for 6 h affords $MoF_4(NCl)$ in almost quantitative yield along with gaseous products [2]. Dropping a solution of $S_3N_3Cl_3$ in CH_2Cl_2 into a suspension of $MoCl_4(NSCl)$ in CH_2Cl_2 (mole ratio 1:2.5) cooled to −5°C followed by stirring the reaction mixture for 16 h at room temperature gives μ-$(S_2N_2)(MoCl_4(NSCl))_2$ in 20% yield, $(ClS)_2N^+[MoCl_5(NSCl)]^-$, and $(ClS)_2N^+[Mo(N)Cl_4]^-$ [3]. Dissolution in $POCl_3$ yields the solvate $MoCl_4(NSCl)(OPCl_3)$ and some $(Mo(N)Cl_3 \cdot POCl_3)_4$. The latter forms by cleavage of SCl_2 [1]. A suspension of $MoCl_4(NSCl)$ in CH_2Cl_2 reacts with a solution of $P(C_6H_5)_3$ in CH_2Cl_2 (mole ratio 1:4) at ambient temperature for 10 h to give $MoCl_4(NP(C_6H_5)_3)(P(C_6H_5)_3)$ in 63% yield, $(C_6H_5)_3PS$, and $(C_6H_5)_3PCl_2$ [4].

Reactions with chlorides of cations such as $(C_6H_5)_4As^+$ and $(C_6H_5)_3(CH_3)P^+$, which are soluble in CH_2Cl_2, proceed according to the equation $MoCl_4(NSCl) + Cl^- \rightarrow [Mo(N)Cl_4]^- + SCl_2$ [1]. When $(MoCl_4(NSCl))_2$ is treated with $(CH_3)_3SiBr$ in the mole ratio 1:11 in CH_2Br_2 at room temperature for about 12 h, $Mo(N)Br_3$ forms in 93% yield according to the equation $(MoCl_4(NSCl))_2 + 10(CH_3)_3SiBr \rightarrow 2Mo(N)Br_3 + S_2Br_2 + Br_2 + 10(CH_3)_3SiCl$ [5].

References:

[1] Kynast, U.; Dehnicke, K. (Z. Anorg. Allgem. Chem. **502** [1983] 29/34).
[2] Fenske, D.; Völp, K.; Dehnicke, K. (Z. Naturforsch. **42b** [1987] 1398/402).
[3] Kynast, U.; Klingelhöfer, P.; Müller, U.; Dehnicke, K. (Z. Anorg. Allgem. Chem. **515** [1984] 61/8).
[4] Schmidt, I.; Kynast, U.; Hanich, J.; Dehnicke, K. (Z. Naturforsch. **39b** [1984] 1248/51).
[5] Hösler, K.; Dehnicke, K. (Z. Anorg. Allgem. Chem. **554** [1987] 108/12).

2.4.2.1.2 $MoCl_4(NSCl)(OPCl_3)$

The complex formed by dropping $S_3N_3Cl_3$ into $MoCl_5$ (mole ratio 1:3) in $POCl_3$ under stirring and cooling the reaction mixture with an ice bath. The reaction, finnished after 2 h, is quantitative. The compound formed also when $MoCl_4(NSCl)$ reacted with $POCl_3$. The black crystalline solvate is very sensitive to humidity.

IR (Nujol): ν (in cm^{-1}) = 1255 vs, 1215 vs, all ν(PO) Fermi resonance; 1011 m, ν_{as}(MoNS); 998 s, ν_{as}(MoNS); 963 w; 825 m; 810 sh; 618 vs, $\nu_{as}(PCl_3)$; 510 sh, $\nu_s(PCl_3)$; 495 s, ν(SCl); 420 w, ν_s(MoNS); 382 sh, δ(POCl); 370 sh, 355 w, 330 vs, 320 s, 300 sh, all ν(MoCl) and/or ν(PCl_3); 240 m to s, ν(Mo–O).

A solution of 6.56 mmol of $MoCl_4(NSCl)(OPCl_3)$ in CH_2Cl_2 reacts with 5.98 mmol of $CH_3(C_6H_5)_3P^+Cl^-$ in CH_2Cl_2 to give $CH_3(C_6H_5)_3P^+[MoCl_5(NSCl)]^-$ in about 25% yield. The complex reacts with $(C_6H_5)_4As^+Cl^-$ in a similar way to give $(C_6H_5)_4As^+[MoCl_5(NSCl)]^-$. If the reaction is carried out with an excess of the chloride the product reacts to give $[Mo(N)Cl_4]^-$ and SCl_2.

Reference:

Kynast, U.; Dehnicke, K. (Z. Anorg. Allgem. Chem. **502** [1983] 29/34).

2.4.2.1.3 Salts of $[MoCl_5(NSCl)]^-$

$(ClS)_2N^+$ $[MoCl_5(NSCl)]^-$

The complex was prepared in 9% yield by dropping a solution of $S_3N_3Cl_3$ in CH_2Cl_2 to a suspension of $MoCl_4$ in CH_2Cl_2 (mole ratio 2:3) at 0°C followed by stirring the mixture at room temperature for 30 h. The filtrate was cooled to −10°C because of the appreciable solubility of the complex in CH_2Cl_2. After some days laminated crystals appeared. In an analogous way the compound was obtained in 13% yield by reaction of $MoCl_5$ with $S_3N_3Cl_3$ (mole ratio 3:4) in CH_2Cl_2 [1]. It was observed as a by-product in the reactions of $MoCl_5$ with $S_3N_3Cl_3$ (mole ratio 3:2) in CH_2Cl_2 and of $MoCl_4(NSCl)$ with $S_3N_3Cl_3$ (mole ratio 2.5:1) in CH_2Cl_2 [2]. The compound formed in very small yields when $MoCl_5$ was reacted with $S_3N_2Cl^+Cl^-$ and Cl_2 [1].

The laminated black crystals are monoclinic, space group $P2_1/n\text{–}C^5_{2h}$ (No. 14), with a=716.3(2), b=1627.4(6), c=1178.9(4) pm, β=100.90(3)°; Z=4; R=0.067 for 746 reflexes. The molecular structures of the $(ClS)_2N^+$- and $[MoCl_5(NSCl)]^-$ ion are given in **Fig. 59**. In the $[MoCl_5(NSCl)]^-$ ion the Mo atom is octahedrally coordinated by five Cl atoms and the N atom of the chlorothionitrene ligand. The MoN and SN bond lengths of 175 and 159 pm, respectively, correspond to double bonds. The MoNS group has an angle of 171°. The bond angle at the S atom is 102°. Bond lengths, bond- and dihedral angles are given in Table 80. The crystal structure is shown in **Fig. 60**. Each ion is surrounded by six counterions. Atomic coordinates and parameters of the thermal vibration are given in the paper [1]. The IR spectrum is compiled in Table 81, p. 260.

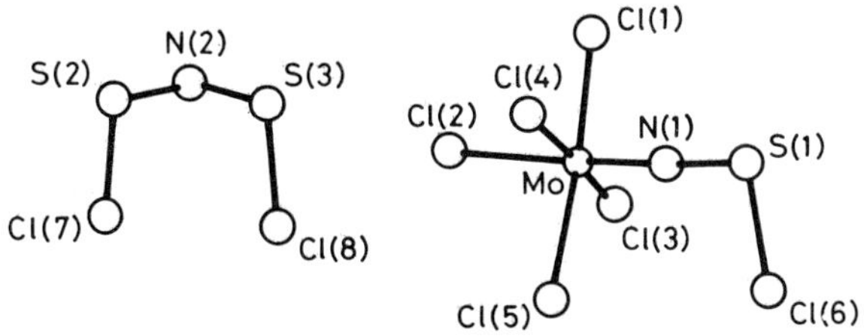

Fig. 59. Molecular structures of the $(ClS)_2N^+$ ion (left) and the $[MoCl_5(NSCl)]^-$ ion (right) [1].

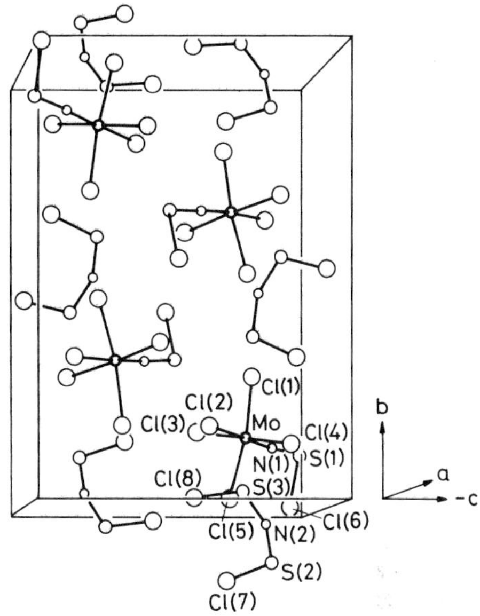

Fig. 60. Crystal structure of $(ClS)_2N^+$-$[MoCl_5(NSCl)]^-$. Perspectivic view of the unit cell [1].

Table 80

Bond Lengths, Bond Angles, and Dihedral Angles in $(ClS)_2N^+[MoCl_5(NSCl)]^-$ [1].

bond length	in pm	bond angle	in °
	cation		
N(2)–S(2)	155(1)	Cl(7)–S(2)–N(2)	109.8(5)
N(2)–S(3)	152(1)	S(2)–N(2)–S(3)	150.9(9)
S(2)–Cl(7)	198.5(7)	N(2)–S(3)–Cl(8)	110.2(5)
S(3)–Cl(8)	199.1(6)		
	dihedral angle N(2)–S(2)–Cl(7)/N(2)–S(2)–Cl(8)		2.40°
	anion		
Mo–N(1)	175(1)	Cl(1)–Mo–Cl(2)	86.6(2)
Mo–Cl(1)	232.4(4)	Cl(1)–Mo–Cl(3)	88.3(1)
Mo–Cl(2)	246.8(4)	Cl(1)–Mo–Cl(4)	92.8(1)
Mo–Cl(3)	236.6(5)	Cl(1)–Mo–Cl(5)	170.4(2)
Mo–Cl(4)	234.8(5)	Cl(2)–Mo–Cl(3)	89.1(1)
Mo–Cl(5)	235.8(4)	Cl(2)–Mo–Cl(4)	85.9(2)
N(1)–S(1)	159(1)	Cl(2)–Mo–Cl(5)	85.3(2)
S(1)–Cl(6)	202.5(6)	Cl(3)–Mo–Cl(4)	174.9(2)
		Cl(3)–Mo–Cl(5)	86.3(1)
		Cl(4)–Mo–Cl(5)	91.9(1)
		Cl(1)–Mo–N(1)	95.3(4)
		Cl(2)–Mo–N(1)	174.5(4)
		Cl(3)–Mo–N(1)	96.1(4)
		Cl(4)–Mo–N(1)	88.8(4)

Table 80 (continued)

bond length	in pm	bond angle	in °
		Cl(5)–Mo–N(1)	93.2(4)
		Mo–N(1)–S(1)	171.2(9)
		N(1)–S(1)–Cl(6)	101.9(5)
	dihedral angle Mo–N(1)–S(1)–Cl(6)/Mo–N(1)–Cl(5)		9.30°

Table 81
IR Spectrum (Nujol) of $(ClS)_2N^+[MoCl_5(NSCl)]^-$ [1].

ν in cm^{-1}	intensity	assignment*)
1110	m to s	$\nu_{as}(NS_2)$ (C)
980	sh	2 × 509
945	m	ν_{as}(MoNS) (A)
656	vw	δ(SNS) (C)
638	m	
509	s	ν_{as}(SCl) (C)
492	m to s	
480	m to s	ν(SCl) (A)
450	vw	ν_s(SCl) (C)
438	vw	ν_s(MoNS) (A)
388	w	δ_{as}(NSCl) (C)
330	sh	
316	vs	$\nu(MoCl_4)$ (A)
300	sh	
242	m	$\nu(Mo–Cl_{trans})$ (A)

*) C = cation; A = anion

$CH_3(C_6H_5)_3P^+[MoCl_5(NSCl)]^-$

The complex was prepared in about 25% yield by reaction of $MoCl_4(NSCl)(OPCl_3)$ with $CH_3(C_6H_5)_3P^+Cl^-$ (mole ratio 1:1) in CH_2Cl_2 at room temperature. Crystals were obtained on addition of CCl_4 and subsequent cooling to −18°C.

The dark violet compact crystals are stable for a reasonably long time at room temperature.

IR (Nujol): ν (in cm^{-1}) = 962 s, ν_{as}(MoNS); 506 w, ν(SCl); 430 w, 400 w, all ν_s(MoNS); 348 vs, 341 vs, 325 sh, 320 sh, all ν(MoCl); 258 s, $\nu(MoCl_{trans})$.

The anion $[MoCl_5(NSCl)]^-$ in CH_2Cl_2 solution decomposes rapidly in the presence of Cl^- ions and decomposes slowly in the absence of Cl^- ions to give $Mo(N)Cl_4^-$ and SCl_2 [3].

$(C_6H_5)_4As^+[MoCl_5(NSCl)]^-$

The complex was produced from $MoCl_4(NSCl)(OPCl_3)$ and $(C_6H_5)_4As^+Cl^-$ in a way analogous to that of the phosphonium salt [3]. The crystal structure was not established. From the sharp diffraction pattern, a rhombic unit cell in the space group $Pmn2_1-C_{2v}^7$ (No. 31) or possibly $Pmnm-D_{2h}^{12}$ (No. 58) is derived: a = 4253(4), b = 907.7(4), c = 1491.2(5) pm; Z = 8 [1].

References:

[1] Müller, U.; Klingelhöfer, P.; Kynast, U.; Dehnicke, K. (Z. Anorg. Allgem. Chem. **520** [1985] 18/24).

[2] Kynast, U.; Klingelhöfer, P.; Müller, U.; Dehnicke, K. (Z. Anorg. Allgem. Chem. **515** [1984] 61/8).

[3] Kynast, U.; Dehnicke, K. (Z. Anorg. Allgem. Chem. **502** [1983] 29/34).

2.4.2.1.4 $(WCl_4(NSCl))_2$

The complex was prepared in 77% yield by reaction of WCl_6 with $S_3N_3Cl_3$ in the mole ratio 3:1 in CH_2Cl_2 under stirring at room temperature for 2 d. When the reaction was carried out in CCl_4 the yield decreased to about 30%.

The red-brown powder sublimes at 145°C in high vacuum ($p < 10^{-4}$ Torr). The complex forms black crystals by crystallization from CH_2Cl_2 solution at −18°C or by sublimation. The crystal structure was determined by X-ray diffraction at 18°C. The crystals are monoclinic, space group $P2_1/n-C_{2h}^5$ (No. 14), with a = 1523(1), b = 904(1), c = 583(1) pm, β = 91.35°; Z = 2; R = 0.055 for 1059 reflexes. In the centrosymmetric molecules the W atoms are linked via two chloro bridges, see **Fig. 61**, p. 262. Short and long W–Cl distances (244 and 265 pm) alternate in the W_2Cl_2 ring. The NSCl groups are found in the trans positions to the longer W–Cl bonds. The WNS bond angle (175°) and short distances correspond to a formulation $W{=}\overset{\oplus}{N}{=}\overset{\ominus}{\underline{\bar{S}}}{-}Cl$. Bond lengths, bond angles, and a dihedral angle are compiled in Table 82.

Table 82

Bond Lengths, Bond Angles, and a Dihedral Angle in $(WCl_4(NSCl))_2$ [1].

bond length	in pm	bond angle	in °	bond angle	in °
Cl(1)–S	201.3(8)	Cl(1)–S–N	104.0(7)	Cl(2)–W–Cl(2′)	79.4(2)
S–N	158(2)	S–N–W	175(1)	Cl(2)–W–Cl(3)	86.8(2)
W–N	176(2)	N–W–Cl(2)	93.2(6)	Cl(2)–W–Cl(5)	87.5(2)
W–Cl(2)	244.1(4)	N–W–Cl(2′)	172.6(7)	Cl(2′)–W–Cl(3)	82.7(2)
W–Cl(2′)	265.0(4)	N–W–Cl(3)	96.4(6)	Cl(2′)–W–Cl(4)	88.0(2)
W–Cl(3)	228.1(5)	N–W–Cl(4)	99.4(6)	Cl(2′)–W–Cl(5)	83.6(2)
W–Cl(4)	228.8(6)	N–W–Cl(5)	96.8(6)	Cl(3)–W–Cl(4)	91.3(2)
W–Cl(5)	227.5(5)	Cl(2)–W–Cl(4)	167.4(2)	Cl(4)–W–Cl(5)	91.4(2)
W···W	391.8(1)	Cl(3)–W–Cl(5)	165.8(2)	W–Cl(2)–W′	100.6(2)

dihedral angle Cl(1)–S–N/N–W–Cl(3) 5.1°

Fig. 62, p. 262, shows the packing of the $(WCl_4(NSCl))_2$ molecules in the crystal. Atomic coordinates and parameters of the thermal vibration are compiled in the paper [1]. The IR spectrum, which is given in Table 83, p. 262, confirms the dimeric structure of the complex.

Table 83
IR Spectrum (Nujol) of $(WCl_4(NSCl))_2$ [1].

ν in cm^{-1}	intensity	assignment
1058	m	$\nu_{as}(WNS)$
508	s	ν(SCl)
498	w	ν(SCl)
398	w	—
378	sh	ν(WCl terminal)
366	vs	
338	s	$\nu(WCl_2W$ bridge)
328	sh	
271	m	δ(NSCl)
248	vw	—

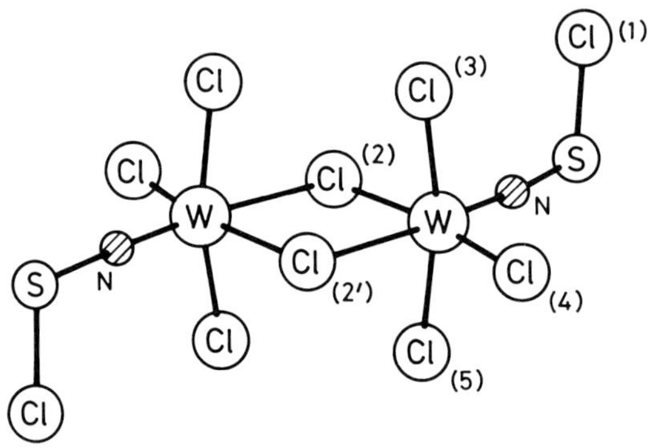

Fig. 61. Molecular structure of $(WCl_4(NSCl))_2$ [1].

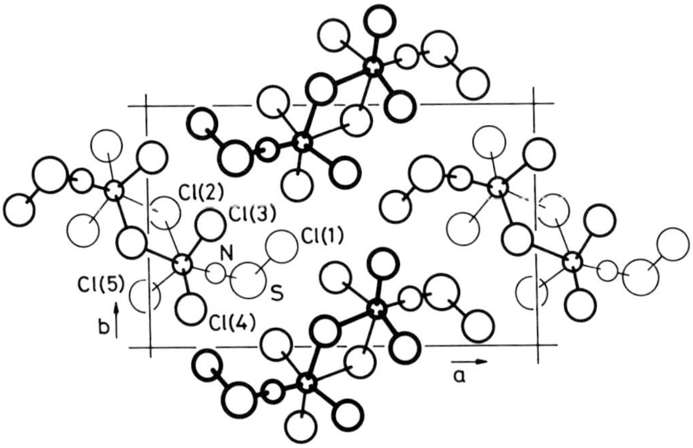

Fig. 62. Projection of the crystal structure of $(WCl_4(NSCl))_2$ down the c axis [1].

The mass spectrum (70 eV) under ion impact conditions shows the molecular ion $WCl_4(NSCl)^+$ and the fragments $WCl_3(NSCl)^+$ and $WCl_2(NSCl)^+$ with relatively high intensities.

The complex is very sensitive to humidity. The reaction of $(WCl_4(NSCl))_2$ with $(C_6H_5)_4As^+Cl^-$ in the mole ratio 1:2 in CH_2Cl_2 gives $(C_6H_5)_4As^+[WCl_5(NSCl)]^-$ in 94% yield [1]. The complex reacts with an excess of $(C_6H_5)_4As^+Cl^-$ (mole ratio 1:4) in CH_2Cl_2 at room temperature for 12 h to give $((C_6H_5)_4As^+)_2[Cl_5W{=}N{=}S{=}N{=}WCl_5]^{2-}$ in 57% yield. A reaction mechanism was proposed [3]. Treatment with excess SCl_2 in boiling CCl_4 gives $(ClS)_2N^+[WCl_5(NSCl)]^-$ [2]. When $(WCl_4(NSCl)_2)_2$ is reacted with an excess of $(CH_3)_3SiBr$ (mole ratio 1:16) in CH_2Br_2 at 0°C to room temperature for 12 h, complete bromination occurs according to the equation: $(WCl_4(NSCl))_2 + 10(CH_3)_3SiBr \rightarrow (WBr_4(NSBr))_2 + 10(CH_3)_3SiCl$. Preferred bromination at the S atom is observed if the complex is treated with $(CH_3)_3SiBr$ in the mole ratio 1:2 in CH_2Br_2 at room temperature for 12 h to give a product of the approximate composition $(WCl_4(NSBr))_2$ in 94% yield and $(CH_3)_3SiCl$. Partial Cl exchange at the W atom by Br was observed in all experiments with stoichiometric amounts of $(CH_3)_3SiBr$ [4].

References:

[1] Kynast, U.; Müller, U.; Dehnicke, K. (Z. Anorg. Allgem. Chem. **508** [1984] 26/32).

[2] Kynast, U.; Klingelhöfer, P.; Müller, U.; Dehnicke, K. (Z. Anorg. Allgem. Chem. **515** [1984] 61/8).

[3] Kynast, U.; Willing, W.; Müller, U.; Dehnicke, K. (Z. Anorg. Allgem. Chem. **529** [1985] 129/36).

[4] Hösler, K.; Dehnicke, K. (Z. Anorg. Allgem. Chem. **554** [1987] 108/12).

2.4.2.1.5 Salts of $[WCl_5(NSCl)]^-$

$(ClS)_2N^+[WCl_5(NSCl)]^-$

The compound was prepared in 89% yield by reaction of WCl_6 with $S_3N_2Cl^+Cl^-$ in CCl_4 (mole ratio 1:1) saturated with Cl_2. Cl_2 was passed into the reaction mixture for 3.5 h and stirring was continued at room temperature for 12 h. The salt was also obtained by treating $(WCl_4(NSCl))_2$ with excess SCl_2 in boiling CCl_4.

The IR spectrum (Nujol) shows 13 absorptions in the range 1120 to 254 cm^{-1}.

The brown-violet crystalline powder is very sensitive to humidity. It is very slightly soluble in CCl_4. Thermolysis in boiling CCl_4 for 6 h yields μ-$(S_2N_2)(WCl_4(NSCl))_2$ in 93% yield according to the equation $2(ClS)_2N^+[WCl_5(NSCl)]^- \rightarrow \mu\text{-}(S_2N_2)(WCl_4(NSCl))_2 + 2SCl_2 + Cl_2$ [1].

$(C_6H_5)_4P^+[WCl_5(NSCl)]^-$

The salt was obtained by reaction of μ-$(S_2N_2)(WCl_4(NSCl))_2$ with $(C_6H_5)_4P^+Cl^-$ in CH_2Cl_2 according to the equation $\mu\text{-}(S_2N_2)(WCl_4(NSCl))_2 + 2(C_6H_5)_4P^+Cl^- \rightarrow 2(C_6H_5)_4P^+[WCl_5(NSCl)]^- + S_2N_2$ [1].

$(C_6H_5)_4As^+[WCl_5(NSCl)]^-$

The salt was prepared in 94% yield by addition of a solution of $(C_6H_5)_4As^+Cl^-$ in CH_2Cl_2 to a suspension of $(WCl_4(NSCl))_2$ in CH_2Cl_2 (mole ratio 2:1). A dark green crystalline powder precipitated from the CH_2Cl_2 solution by addition of CCl_4.

IR (Nujol): ν (in cm^{-1}) = 958 w, $\nu_{as}(WNS)$; 493 m, $\nu(SCl)$; 320 vs, $\nu(WCl)$; 268 m to s, $\delta(NSCl)$ [2].

References:

[1] Kynast, U.; Klingelhöfer, P.; Müller, U.; Dehnicke, K. (Z. Anorg. Allgem. Chem. **515** [1984] 61/8).

[2] Kynast, U.; Müller, U.; Dehnicke, K. (Z. Anorg. Allgem. Chem. **508** [1984] 26/32).

2.4.2.1.6 Neutral Re–NSCl Complexes

$(ReCl_3(NSCl)_2)_2$

The complex was prepared in 78% yield by stirring a suspension of equimolar amounts of $ReCl_5$, $S_3N_2Cl^+Cl^-$, and some drops of $POCl_3$ in CH_2Cl_2 at room temperature for 5 d. The isolated powder was washed with boiling CH_2Cl_2 [1]. $(ReCl_3(NSCl)_2)_2$ of low purity was obtained by reaction of $(C_6H_5)_4As^+[ReCl_4(NSCl)_2]^-$ with $GaCl_3$ in CH_2Cl_2 [2, 3].

The brown powder is very sensitive to humidity. It is very slightly soluble in CH_2Cl_2 [1, 2, 3].

From the IR spectrum given in Table 84 it is concluded that the complex is a dimer with a centrosymmetric structure. The molecular structure is given schematically in **Fig. 63** [1].

A suspension of $(ReCl_3(NSCl)_2)_2$ in CH_2Cl_2 reacts with excess CH_3CN to give $ReCl_3(NSCl)_2(CH_3CN)$ [4]. Reaction with $(C_6H_5)_4As^+[GaCl_4]^-$ in CH_2Cl_2 yields $(C_6H_5)_4As^+$-$[ReCl_4(NSCl)_2]^- \cdot CH_2Cl_2$ [2].

Table 84
IR Spectrum (Nujol) of $(ReCl_3(NSCl)_2)_2$ [1].

ν in cm^{-1}	intensity	assignment
1115	m	$\nu_{as}(ReNS)$
525	m	$\nu_s(ReNS)$
475	vs	$\nu_{as}(SCl)$
452	m	$\nu_s(SCl)$
440	sh	$\nu_s(^{34}SCl)$
403	vw	—
375	w	$\delta(ReNS)$
348	s	$\nu_{as}(ReCl_{2\,terminal})$
325	s	$\nu_s(ReCl_{2\,terminal})$
295	m	$\nu(ReCl_2Re)$
263	m	$\nu(ReCl_2Re)$
245	vw	$\delta(NSCl)$

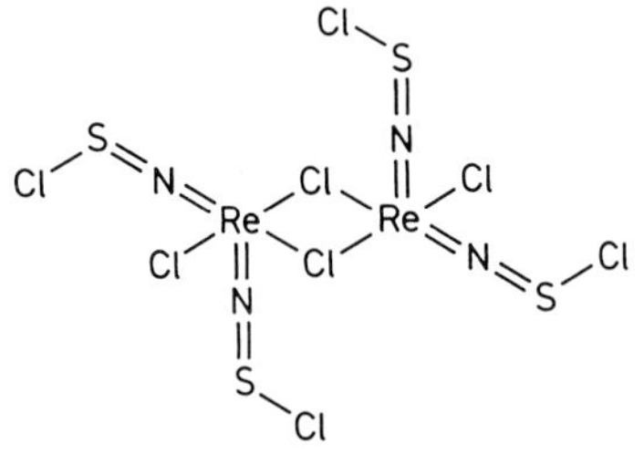

Fig. 63. Molecular structure of $(ReCl_3(NSCl)_2)_2$ drawn schematically [1].

$ReCl_3(NSCl)_2(POCl_3)$

The complex was obtained in 78% yield by addition of equimolar amounts of $S_3N_2Cl^+Cl^-$ to an ice-cooled suspension of $ReCl_5$ in $POCl_3$ followed by stirring the reaction mixture and warming slowly to room temperature [1]. It formed also in almost quantitative yield when a solution of $S_3N_3Cl_3$ in $POCl_3$ was dropped into a solution of $ReCl_5$ in $POCl_3$ at 0°C (mole ratio 1:1.5). The reaction mixture was stirred on warming to room temperature for 12 h [2, 3].

The olive-green crystalline powder or black crystals show the following absorptions in the IR spectrum (Nujol): ν (in cm^{-1}) = 1258 s, ν(PO); 1210 vs, ν(PO); 1110 vw; 948 m, ν(ReN); 618 vs, $\nu_{as}(PCl_3)$; 582 vw; 550 w, $\nu_s(PCl_3)$; 478 s, ν(SCl); 448 s, ν(SCl); 372 m, δ(OPCl); 340 sh, 328 s, all ν(ReCl); 296 m; 268 m [2].

When $P(C_6H_5)_3$ in CH_2Cl_2 is dropped to an equimolar amount of $ReCl_3(NSCl)_2(POCl_3)$ suspended in CH_2Cl_2 under stirring at 0°C $Cl(C_6H_5)_3P^+[ReCl_4(N_2S_2)]^-$ is produced in almost quantitative yield [5]. The reaction with equimolar amounts of $(C_6H_5)_4As^+Cl^-$ in CH_2Cl_2 at room temperature for about 1 h yields $(C_6H_5)_4As^+[ReCl_4(NSCl)_2]^- \cdot CH_2Cl_2$ [2]. Reduction with equimolar amounts of $Sb(C_6H_5)_3$ in CH_2Cl_2 at room temperature gives $(C_6H_5)_4P^+[ReCl_4(N_2S_2)]^-$ [6].

$ReCl_4(NSCl)(POCl_3)$

The complex was prepared in quantitative yield by dropping a solution of $S_3N_3Cl_3$ in $POCl_3$ into a solution of $ReCl_5$ in $POCl_3$ (mole ratio 1:3) at 0°C. The reaction mixture was stirred on warming to room temperature for 12 h. The reaction proceeds according to the equation $ReCl_5 \cdot OPCl_3 + 1/3\ S_3N_3Cl_3 \rightarrow ReCl_4(NSCl)(POCl_3) + 1/2\ Cl_2$.

The dark brown powder is very sensitive to humidity. It is highly soluble in $POCl_3$, CH_2Cl_2, and CCl_4.

IR (Nujol): ν (in cm^{-1}) = 1270 vs, ν(PO); 1225 s, ν(PO); 1122 w; 1059 w; 1022 w; 930 w, ν(ReN); 660 vw; 649 w; 615 vs, $\nu_{as}(PCl_3)$; 585 sh; 560 vw, $\nu_s(PCl_3)$; 503 s, ν(SCl); 490 sh; 444 w; 429 m; 355 sh, δ(OPCl); 340 sh, 328 vs, 298 sh, all ν(ReCl); 230 vw [2].

$ReCl_4(NSCl)(POCl_3)$ reacts with a molar excess of $P(C_6H_5)_3$ in CH_2Cl_2 at room temperature for 12 h to give $Re(N)Cl_2(P(C_6H_5)_3)_2$ in 63% yield according to the equation $2ReCl_4(NSCl)(POCl_3) + 9P(C_6H_5)_3 \rightarrow 2Re(N)Cl_2(P(C_6H_5)_3)_2 + 3Cl(C_6H_5)_3P^+Cl^- + 2(C_6H_5)_3PS + 2POCl_3$ [1]. $ReCl_4(NSCl)(POCl_3)$ in slight excess reacts with $(C_6H_5)_4As^+Cl^-$ in CH_2Cl_2 at room temperature for 1 h to give $(C_6H_5)_4As^+[Re(N)Cl_4]^-$ in high yield, $POCl_3$, and SCl_2 [2]. The reaction with equimolar amounts of $(CH_3)_3SiN{=}S{=}NSi(CH_3)_3$ in CH_2Cl_2 at reflux temperature for 2 h gives a brown precipitate which has the composition $ReCl_3N_3S_2$ after extraction with hot CH_2ClCH_2Cl [7].

$ReCl_3(NSCl)_2(CH_3CN)$

The complex was prepared in 83% yield by dropping 5 mL CH_3CN into a suspension of 0.9 mmol of $(ReCl_3(NSCl)_2)_2$ in 30 mL CH_2Cl_2 under stirring and cooling to 0°C. Black crystals formed on storing the solution at 5°C for several days.

The black crystals are moisture-sensitive. The crystal structure was determined by X-ray diffraction at 25°C. The crystals are triclinic, space group $P\bar{1}-C_i^1$ (No. 2), with a = 681.8(2), b = 899.9(3), c = 1104.0(7) pm, α = 110.50(4)°, β = 97.90(5)°, γ = 101.34(3)°; V = 606×10^{-30} m^3; Z = 2. D_x = 2.72 g/cm^3. R = 0.035 for 2483 observed independent reflexions.

The molecular structure is shown in **Fig. 64**, p. 266. Bond lengths and bond angles are compiled in Table 85, p. 266. In the monomeric molecule the Re atom has a distorted octahedral coordination with meridional arrangement of the Cl atoms. The chlorothionitrene ligands have nearly linear ReNS groups with bond lengths that correspond to double bonds (r(ReN) = 178 and 179 pm); this agrees with the NReN bond angle of 100°. Atomic coordinates and parameters of the thermic vibration are given in the paper [4].

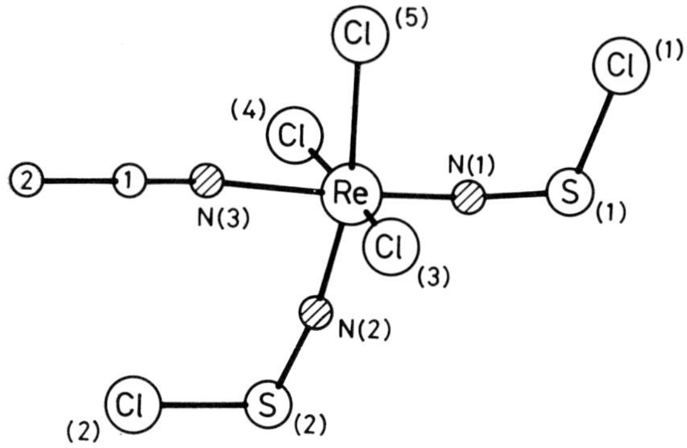

Fig. 64. Molecular structure of $ReCl_3(NSCl)_2(CH_3CN)$ [4].

Table 85
Bond Lengths and Bond Angles in $ReCl_3(NSCl)_2(CH_3CN)$ [4].

bond length	in pm	bond angle	in °	bond angle	in °
Re–Cl(3)	238.7(2)	Cl(3)–Re–Cl(4)	169.1(1)	N(1)–Re–N(3)	175.5(3)
Re–Cl(4)	237.1(2)	Cl(3)–Re–Cl(5)	86.2(1)	N(2)–Re–N(3)	84.1(3)
Re–Cl(5)	234.7(2)	Cl(4)–Re–Cl(5)	87.3(1)	N(1)–Re–N(2)	100.2(3)
Re–N(3)	217.4(6)	Cl(3)–Re–N(3)	84.7(2)	Re–N(3)–C(1)	176.4(7)
Re–N(1)	178.2(6)	Cl(3)–Re–N(1)	93.6(2)	N(3)–C(1)–C(2)	178(1)
Re–N(2)	179.4(7)	Cl(3)–Re–N(2)	94.6(2)	Re–N(1)–S(1)	173.5(5)
N(3)–C(1)	113.0(10)	Cl(4)–Re–N(3)	85.7(2)	Re–N(2)–S(2)	171.1(4)
C(1)–C(2)	145.0(12)	Cl(4)–Re–N(1)	95.6(2)	N(1)–S(1)–Cl(1)	106.2(3)
S(1)–N(1)	153.2(7)	Cl(4)–Re–N(2)	89.5(2)	N(2)–S(2)–Cl(2)	107.8(2)
S(1)–Cl(1)	204.1(5)	Cl(5)–Re–N(3)	81.4(2)		
S(2)–N(2)	154.0(8)	Cl(5)–Re–N(1)	94.4(2)		
S(2)–Cl(2)	205.6(4)	Cl(5)–Re–N(2)	165.3(2)		

The IR spectrum is compiled in Table 86.

Table 86
IR Spectrum (Nujol) of $ReCl_3(NSCl)_2(CH_3CN)$ [4].

ν in cm^{-1}	intensity	assignment	ν in cm^{-1}	intensity	assignment	ν in cm^{-1}	intensity	assignment
2300	m to w	ν(C≡N)	945	s	ν(CC)	349	sh	ν(ReCl)
1360	w	δ(CH_3)	480	sh	ν(SCl)	335	sh	
1130	m to w	ν(NS)	462	s		321	s	
1023	m	ϱ(CH_3)	423	w	δ(CCN)	310	sh	
968	s	ν(NS)	405	vw		285	sh	

References:

[1] Hauck, H. G.; Klingelhöfer, P.; Müller, U.; Dehnicke, K. (Z. Anorg. Allgem. Chem. **510** [1984] 180/8).
[2] Müller, U.; Kafitz, W.; Dehnicke, K. (Z. Anorg. Allgem. Chem. **501** [1983] 69/78).
[3] Müller, U.; Kafitz, W.; Dehnicke, K. (Naturwissenschaften **69** [1982] 503/4).
[4] Hauck, H. G.; Patt-Siebel, U.; Müller, U.; Dehnicke, K. (Z. Anorg. Allgem. Chem. **546** [1987] 177/82).
[5] Hiller, W.; Mohyla, J.; Strähle, J.; Hauck, H. G.; Dehnicke, K. (Z. Anorg. Allgem. Chem. **514** [1984] 72/8).
[6] Conradi, E.; Hauck, H. G.; Müller, U.; Dehnicke, K. (Z. Anorg. Allgem. Chem. **539** [1986] 39/49).
[7] Hauck, H. G.; Willing, W.; Müller, U.; Dehnicke, K. (Z. Naturforsch. **41 b** [1986] 825/30).

2.4.2.1.7 Salts of $[ReCl_4(NSCl)_2]^-$

$(C_6H_5)_4P^+[ReCl_4(NSCl)_2]^- \cdot CH_2Cl_2$

The formation is not reported. The complex salt reacts with $C_6H_5C{\equiv}CC_6H_5$ in CH_2Cl_2 at room temperature for 2 d to give $(C_6H_5)_4P^+[ReCl_4(N_2S_2)]^-$ in 86.5% yield and $C_6H_5ClC{=}CClC_6H_5$ [1].

$(C_6H_5)_4As^+[ReCl_4(NSCl)_2]^- \cdot CH_2Cl_2$

The compound was obtained by reaction of $ReCl_3(NSCl)(POCl_3)$ with equimolar amounts of $(C_6H_5)_4As^+Cl^-$ in CH_2Cl_2 at room temperature for 1 h. CCl_4 was added to the filtrated solution until crystallization began [2, 3]. The complex salt formed also by treatment of $(ReCl_3(NSCl)_2)_2$ with $(C_6H_5)_4As^+Cl^-$ in CH_2Cl_2 [3].

The small bronze-colored leafs are moisture-sensitive [2, 3].

The crystal structure was determined by X-ray diffraction at 24°C. The lustrous brown needles are triclinic, space group $P\bar{1}$–C_i^1 (No. 2), with a = 1119(1), b = 1144(1), c = 1473(1) pm, $\alpha = 77.59°$, $\beta = 70.79°$, $\gamma = 71.18° \pm 0.5°$; Z = 2. R = 0.031 for 2785 reflections. The molecular structure of the $[ReCl_4(NSCl)_2]^-$ ion is shown in **Fig. 65**. Bond lengths, bond angles, and a dihedral angle are given in Table 87, p. 268. The two NSCl ligands have a cis arrangement with nearly linear Re=N=S groups, with interatomic distances corresponding to double bonds. The Re–Cl bonds are somewhat longer than usual and show no trans effect. This is possibly due to Cl···H–C bridges; see **Fig. 66**, p. 268, showing a section of the crystal structure. Atomic coordinates and parameters of the thermic vibration are given in the paper [2].

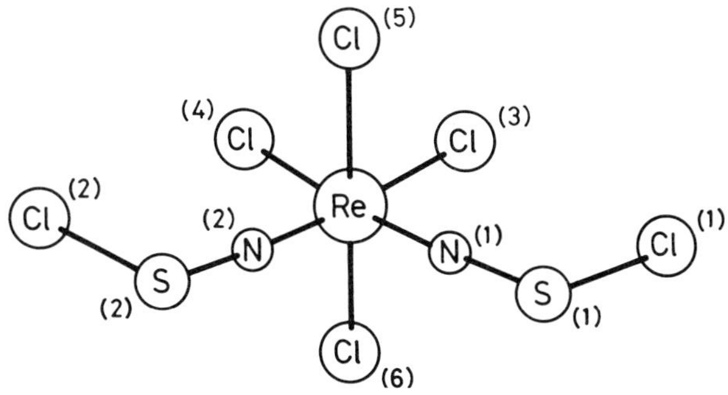

Fig. 65. Molecular structure of the anion in $(C_6H_5)_4As^+[ReCl_4(NSCl)_2]^- \cdot CH_2Cl_2$ [2].

Table 87

Bond Lengths, Bond Angles, and a Dihedral Angle in $(C_6H_5)_4As^+[ReCl_4(NSCl)_2]^- \cdot CH_2Cl_2$ [2].

bond length	in pm	bond angle	in °
Re–N(1)	179(1)	N(1)–Re–Cl(3)	87.9(3)
N(1)–S(1)	153(1)	N(1)–Re–Cl(4)	172.2(3)
S(1)–Cl(1)	209.0(5)	N(1)–Re–Cl(5)	91.7(4)
Re–N(2)	176.6(10)	N(1)–Re–Cl(6)	93.5(4)
N(2)–S(2)	155.5(10)	N(1)–Re–N(2)	97.4(5)
S(2)–Cl(2)	206.5(6)	N(1)–S(1)–Cl(1)	108.2(4)
Re–Cl(3)	240.6(3)	N(2)–S(2)–Cl(2)	107.9(4)
Re–Cl(4)	237.5(3)	Re–N(1)–S(1)	175.3(6)
Re–Cl(5)	241.5(4)	Re–N(2)–S(2)	175.9(6)
Re–Cl(6)	238.4(4)	Cl(3)–Re–Cl(4)	84.3(1)
As–C	189.1 to 191.1	Cl(3)–Re–Cl(5)	87.4(1)
		Cl(3)–Re–Cl(6)	87.7(1)
		Cl(4)–Re–Cl(5)	87.4(1)
		Cl(4)–Re–Cl(6)	86.7(1)
		Cl(5)–Re–Cl(6)	172.7(1)
		C–As–C	108.7 to 110.8

dihedral angle Re–N(1)–N(2)/Re–Cl(3)–Cl(4) 4.30°

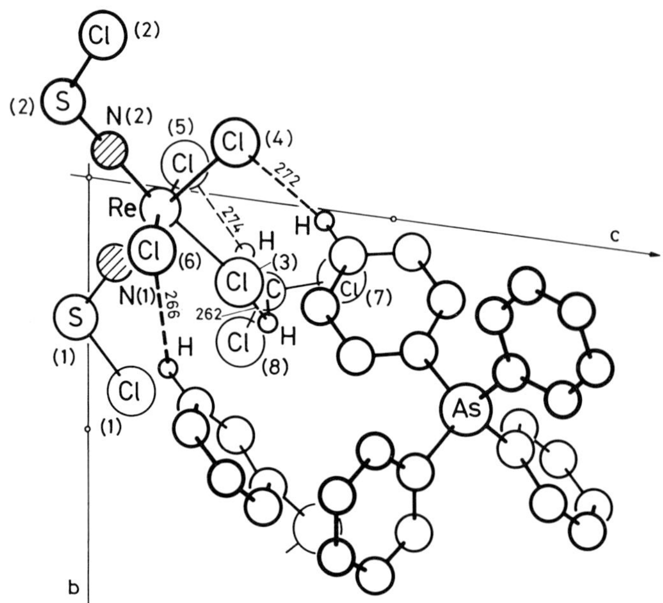

Fig. 66. Section of the crystal structure of $(C_6H_5)_4As^+[ReCl_4(NSCl)_2]^- \cdot CH_2Cl_2$ with view down the a axis. Only H atoms with strikingly short contact distances to Cl atoms are considered (dashed lines, distances in pm) [2].

The IR spectrum (Nujol) of the $[ReCl_4(NSCl)_2]^-$ ion confirms the molecular structure established by X-ray diffraction: ν (in cm^{-1}) = 940 s, ν(ReN); 920 w, ν(ReN); 431 m to s, ν(SCl); 338 w, 322 s, 302 s, all ν(ReCl); 280 m [2].

Dropping a solution of $S(NSi(CH_3)_3)_2$ in CH_2Cl_2 into a solution of $(C_6H_5)_4As^+[ReCl_4(NSCl)_2]^-$ in CH_2Cl_2 under stirring at room temperature gives $(C_6H_5)_4As^+[ReCl_4(N_2S_2)]^-$ in 50.7% yield and $(CH_3)_3SiCl$. The complex salt crystallizes after addition of some CCl_4 and cooling to 5°C [1]. The reaction with $GaCl_3$ in CH_2Cl_2 gives $(ReCl_3(NSCl)_2)_2$ of low purity and $(C_6H_5)_4As^+GaCl_4^-$ [2, 3].

$S_4N_3^+[ReCl_4(NSCl)_2]^-$

The title compound was formed as a by-product in the reaction of $Re_2(CO)_{10}$ with excess $S_3N_3Cl_3$ in CH_2Cl_2. The main product $Re(CO)_5Cl$ crystallized from the reaction solution after standing for a week at room temperature. The blue-violet needles of the title compound were obtained along with some yellow crystals of $S_4N_3^+Cl^-$ on allowing the filtrate to stand for another week at −5°C.

The compound is characterized by an X-ray crystal structure analysis at 20°C. It crystallizes in the noncentrosymmetric orthorhombic space group $P2_12_12_1-D_2^4$ (No. 19) with a = 979.9(1), b = 1204.6(2), c = 1361.5(3) pm; Z = 4. R = 0.076 for 2376 reflections. The molecular structure of the $[ReCl_4(NSCl)_2]^-$ ion is shown in **Fig. 67**. Bond lengths and bond angles are given in Table 88, p. 270. Atomic coordinates and parameters of the thermic vibration are given in the paper [4]. The complex compound consists of the well-known cyclic planar $S_4N_3^+$ cations (see "Sulfur-Nitrogen Compounds" Part 2, 1985, pp. 108/26) and $[ReCl_4(NSCl)_2]^-$ anions, in which the Re atom is coordinated octahedral by four Cl atoms and two cis-positioned NSCl ligands. The mean Re–N and N–S bond lengths of 177 and 158 pm, respectively, correspond to double bonds. The bond lengths and angles are much like those in the structure of $(C_6H_5)_4As^+[ReCl_4(NSCl)_2]^-\cdot CH_2Cl_2$ (see above); however, the Cl atoms of the NSCl ligands are turned to each other.

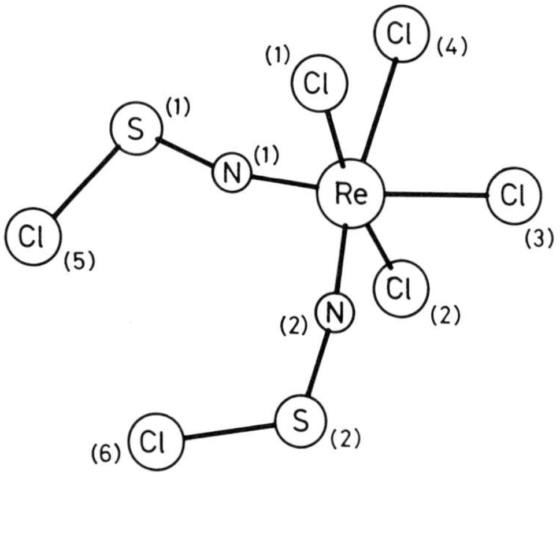

Fig. 67. Molecular structure of the anion in $S_4N_3^+[ReCl_4(NSCl)_2]^-$ [4].

The IR spectrum shows the characteristic absorptions of the $S_4N_3^+$ ion which overlap the absorptions of the Re–N and N–S stretching vibrations. The S–Cl stretching vibration is observed at 450 cm^{-1}. Re–Cl stretching vibrations occur at 328, 305, and 282 cm^{-1} [4].

Table 88
Bond Lengths and Bond Angles of the Anion in $S_4N_3^+[ReCl_4(NSCl)_2]^-$ [4].

bond length	in pm	bond angle	in °	bond angle	in °
Re–Cl(1)	239.7(4)	Cl(1)–Re–Cl(2)	172.1(2)	N(2)–Re–Cl(1)	95.9(6)
Re–Cl(2)	239.7(4)	Cl(1)–Re–Cl(3)	96.5(2)	N(2)–Re–Cl(2)	89.4(6)
Re–Cl(3)	239.4(5)	Cl(1)–Re–Cl(4)	87.6(2)	N(2)–Re–Cl(3)	89.3(6)
Re–Cl(4)	245.1(5)	Cl(2)–Re–Cl(3)	87.7(2)	N(2)–Re–Cl(4)	170.6(6)
Re–N(1)	178(2)	Cl(2)–Re–Cl(4)	95.3(2)	N(1)–Re–N(2)	99.5(9)
Re–N(2)	175(2)	Cl(3)–Re–Cl(4)	82.2(2)	Re–N(1)–S(1)	166(1)
N(1)–S(1)	156(2)	N(1)–Re–Cl(1)	92.1(5)	Re–N(2)–S(2)	170(1)
N(2)–S(2)	159(2)	N(1)–Re–Cl(2)	92.9(6)	N(1)–S(1)–Cl(5)	104.7(7)
S(1)–Cl(5)	202.9(9)	N(1)–Re–Cl(3)	171.1(6)	N(2)–S(2)–Cl(6)	108.5(7)
S(2)–Cl(6)	205.0(9)	N(1)–Re–Cl(4)	89.0(5)		

References:

[1] Conradi, E.; Hauck, H. G.; Müller, U.; Dehnicke, K. (Z. Anorg. Allgem. Chem. **539** [1986] 39/49).
[2] Müller, U.; Kafitz, W.; Dehnicke, K. (Z. Anorg. Allgem. Chem. **501** [1983] 69/78).
[3] Müller, U.; Kafitz, W.; Dehnicke, K. (Naturwissenschaften **69** [1982] 503/4).
[4] Fenske, D.; Berg, A.; Weller, F.; Dehnicke, K. (Z. Anorg. Allgem. Chem. **527** [1985] 105/10).

2.4.2.1.8 $(C_6H_5)_4P^+[OsCl_4(NO)(NSCl)]^-$

The complex was obtained from a mixture of $S_4N_3^+[OsCl_4(NO)(NSCl)]^-$ and $S_4N_3^+Cl^-$ by treating it with a solution of $(C_6H_5)_4P^+Cl^-$ in CH_2Cl_2 for 10 h at room temperature. Precipitated $S_4N_3^+Cl^-$ was filtered off and some CCl_4 was added to the dark red filtrate. After some days, dark olive-green crystals of $((C_6H_5)_4P^+)_2[OsCl_5(NO)]^{2-} \cdot 2CH_2Cl_2$ formed which were filtered off. Repeated addition of some mL of CCl_4 to the filtrate and allowing the solution to stand for some days causes formation of dark red compact single crystals of $(C_6H_5)_4P^+[OsCl_4$-$(NO)(NSCl)]^-$ in about 70% yield. The starting mixture of $S_4N_3^+[OsCl_4(NO)(NSCl)]^-$ and $S_4N_3^+Cl^-$ was produced by reaction of $OsCl_3(NO)$ with excess molten $S_3N_3Cl_3$ at 80°C for 40 min.

The crystal structure was determined by X-ray diffraction at 21°C. The crystals are monoclinic, space group $P2_1/c$–C_{2h}^5 (No. 14), with a = 1715.8(3), b = 1054.2(2), c = 1588.4(5) pm; β = 96.25(2)°; V = 2843 × 10^{-30} m^3; Z = 4. D_x = 1.83 g/cm^3. R = 0.048 for 3001 observed reflections. The molecular structure of the $[OsCl_4(NO)(NSCl)]^-$ ion is shown in **Fig. 68**. Bond lengths and bond angles are compiled in Table 89. Atomic coordinates and parameters of the thermic vibration are given in the paper.

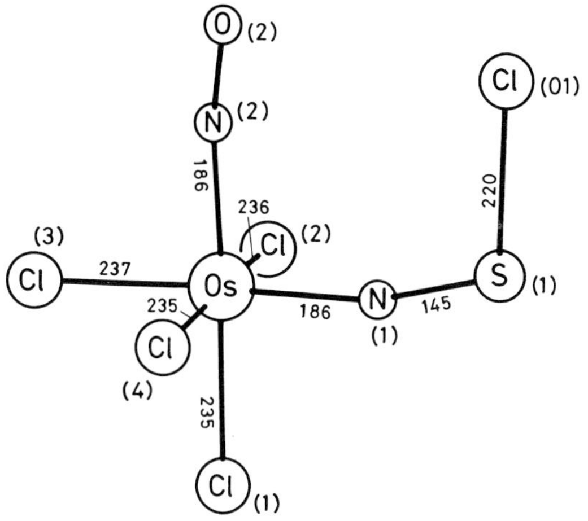

Fig. 68. Molecular structure of the anion in $(C_6H_5)_4P^+[OsCl_4(NO)(NSCl)]^-$ (bond lengths in pm).

Table 89

Bond Lengths and Bond Angles of the Anion in $(C_6H_5)_4P^+[OsCl_4(NO)(NSCl)]^-$.
(The NO and NSCl groups are statistically superimposed in the crystal.)

bond length	in pm	bond angle	in °	bond angle	in °
Os–Cl(1)	235.2(2)	Cl(1)–Os–Cl(2)	88.4(1)	N(2)–Os–Cl(2)	90.3(3)
Os–Cl(2)	236.3(3)	Cl(1)–Os–Cl(3)	88.0(1)	N(2)–Os–Cl(3)	87.5(3)
Os–Cl(3)	236.8(3)	Cl(1)–Os–Cl(4)	87.2(1)	N(2)–Os–Cl(4)	93.9(3)
Os–Cl(4)	235.2(3)	Cl(2)–Os–Cl(3)	89.2(1)	N(1)–Os–N(2)	95.3(4)
Os–N(1)	185.6(9)	Cl(2)–Os–Cl(4)	175.3(1)		
Os–N(2)	185.9(9)	Cl(3)–Os–Cl(4)	88.8(1)	Os–N(1)–S(1)[a]	165.4(7)
N(1)–S(1)[a]	145(1)	N(1)–Os–Cl(1)	89.2(3)	Os–N(1′)–S(1′)[b]	167.9(8)
N(1′)–S(1′)[b]	151(1)	N(1)–Os–Cl(2)	89.7(3)	Os–N(2′)–O(2′)[b]	172(2)
S(1)–Cl(01)[a]	220(1)	N(1)–Os–Cl(3)	177.0(3)	Os–N(2)–O(2)[a]	177(2)
S(1′)–Cl(01′)[b]	216(1)	N(1)–Os–Cl(4)	92.0(3)	N(1)–S(1)–Cl(01)[a]	105.3(5)
N(2′)–O(2′)[a]	112.5[c]	N(2)–Os–Cl(1)	175.3(3)	N(1′)–S(1′)–Cl(01′)[b]	104.9(6)
N(2)–O(2)[b]	112.5[c]				

[a] Occupation probability 62(1)%. – [b] Occupation probability 38(1)%. – [c] Distance not refined.

The NO and NSCl ligands have a cis arrangement at the Os atom which is distortedly octahedrally coordinated. The NS bond length of ~148 pm of the NSCl ligand is shorter than that in other chlorothionitrene complexes which have NS bond lengths of about 155 pm. The bonding of the NSCl ligand can be represented by the two formulas $\overset{\ominus}{Os}-\overset{\oplus}{N}\equiv SCl$ and $Os=\overset{\oplus}{N}=\overset{\ominus}{S}Cl$.

IR spectrum (Nujol): ν (in cm^{-1}) = 1857 s, ν(NO); 1310 s, ν(NS); 1220 s, ν(NS); 365 m, 358 m, 327 vs, 295 m, all ν(OsCl).

Reference:

Figge, R.; Patt-Siebel, U.; Conradi, E.; Müller, U.; Dehnicke, K. (Z. Anorg. Allgem. Chem. **558** [1988] 107/13).

2.4.2.2 Bromothionitrene Complexes

In analogy to the chlorothionitrene complexes, the bromothionitrene complexes of Re and W have an NSBr group which is coordinated to the metal atom via its N atom. The Re–N and N–S distances of 186 and 148 pm, respectively, in $((C_6H_5)_3P^+)_2[ReBr_4(NS)(NSBr)]_2^{2-} \cdot CH_2Br_2$, correspond to double bonds according to the formula $Re{=}\overset{\oplus}{N}{=}\overset{\ominus}{S}Br$. The complexes $((C_6H_5)_3P^+)_2[ReBr_4(NS)(NSBr)]_2^{2-} \cdot CH_2Br_2$ and $ReBr_2(NS)(NSBr)(NC_5H_5) \cdot CH_2Br_2$ are described in the Chapter 1.4 "Thionitrosyl Complexes" on pp. 68/9.

2.4.2.2.1 $(WCl_4(NSBr))_2$

A product with the approximate composition $WCl_4(NSBr)$ formed when $(CH_3)_3SiBr$ was added slowly to a stirred suspension of $(WCl_4(NSCl))_2$ in CH_2Br_2 in the mole ratio 2:1. After stirring 12 h at room temperature a black precipitate was obtained in 94% yield.

The diamagnetic crystalline powder is very sensitive to humidity and sparingly soluble in CH_2Br_2.

A dimeric structure analogous to those of $(WCl_4(NSCl))_2$ and $(WBr_4(NSBr))_2$ was proposed from comparison of the IR spectra. IR (Nujol): ν (in cm^{-1}) = 1060 m, ν_{as} (WNS); 440 m, ν(SBr); 398 sh, 383 sh, δ(WNS); 370 vs, 346 s, $\nu(WCl_{terminal})$ and δ(NSBr); 270 m, $\nu(WCl_{bridge})$ [1].

The title compound (14.37 mmol) reacts with $(C_6H_5)_4P^+Cl^-$ (15.15 mmol) in boiling CH_2Cl_2 over a period of 5 h according to the equation [2] $(WCl_4(NSBr))_2 + 2(C_6H_5)_4P^+Cl^- \xrightarrow{CH_2Cl_2} ((C_6H_5)_4P^+)_2[Cl_5WN{=}S{=}NWCl_5]^{2-} \cdot 2CH_2Cl_2 + {}^1\!/_2 S_2Br_2 + {}^1\!/_2 Br_2$.

2.4.2.2.2 $(WBr_4(NSBr))_2$

The complex was prepared in 97% yield by reaction of $(WCl_4(NSCl))_2$ with excess $(CH_3)_3SiBr$ (mole ratio 1:6) in CH_2Br_2 at 0°C to room temperature over a period of 12 h.

The black diamagnetic crystalline powder is very sensitive to humidity and is sparingly soluble in CH_2Br_2.

The dimeric structure (see **Fig. 69**) was proposed from comparison of the IR spectrum with that of $(WCl_4(NSCl))_2$ whose crystal structure is known, see pp. 261/2.

IR (Nujol): ν (in cm^{-1}) = 1050 m, ν_{as}(WNS); 442 s, $\nu(^{32}S^{79}Br)$; 439 m, $\nu(^{32}S^{81}Br)$; 432 w, $\nu(^{34}S^{81}Br)$; 395 m, 370 sh, δ(WNS); 360 w, 348 m, δ(NSBr); 255 m, 245 vs, $\nu(WBr_{terminal})$; 225 m, $\nu(WBr_{bridge})$ [1].

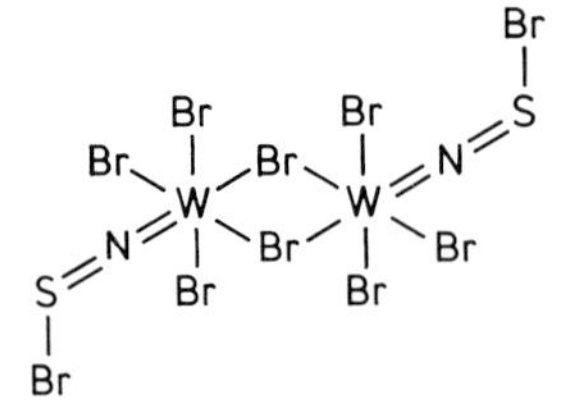

Fig. 69. Molecular structure of $(WBr_4(NSBr))_2$ drawn schematically [1].

References:

[1] Hösler, K.; Dehnicke, K. (Z. Anorg. Allgem. Chem. **554** [1978] 108/12).
[2] Willing, W.; Hösler, K.; Müller, U.; Dehnicke, K. (Acta Cryst. C **43** [1987] 218/21).

2.5 Halogeno-imidosulfur Ions (1+), $[XS{=}NR]^+$, X = F, Cl; R = H, CH_3, C_6F_5

2.5.1 Fluoro-imidosulfur Ion (1+), $[FS{=}NH]^+$, and Fluoro-(methylimido)sulfur Ion (1+), $[FS{=}NCH_3]^+$

The cations have not been obtained experimentally until now. Geometrical structure parameters (see Table 90) and electronic structures have been computed by SCF MO calculations for isomers with a cis, trans, and linear RNS arrangement. The molecules were considered as having C_s symmetry in the equilibrium structure. The MO's 13a', 3a" for $[FS{=}NH]^+$ and 16a', 4a" for $[FS{=}NCH_3]^+$ are occupied in the electronic ground states leading in both cases to an $^1A'$ ground state. In both cases the cis form was found to be the most stable one [1]. A consideration of the linear arrangement only is given in [2]. For atomic charges from Mulliken population analysis, see the paper [1].

Table 90

Equilibrium Geometries and Relative Stabilities of *cis-*, *trans-*, and *linear*-$[FS{=}NH]^+$ and -$[FS{=}NCH_3]^+$. (Distances in Å, angles in °, energy values ΔE in kJ/mol relative to the most stable structure with ΔE = 0.0 kJ/mol.)

	d(N–H(C))	d(N–S)	d(S–F)	∢(SNH(C))	∢(NSF)	ΔE	Ref.
cis-$[FS{=}NH]^+$	1.013	1.447	1.534	132.1	111.3	0.0	[1]
trans-$[FS{=}NH]^+$	1.014	1.465	1.520	120.8	104.5	17.3	[1]
linear-$[FS{=}NH]^+$	1.004	1.413	1.535	180.0	111.1	28.1	[1]
	1.02	1.38	1.51	—	117	—	[2]
cis-$[FS{=}NCH_3]^+$	1.464	1.454	1.550	139.8	110.8	0.0	[1]
trans-$[FS{=}NCH_3]^+$	1.468	1.445	1.540	144.8	106.2	29.7	[1]
linear-$[FS{=}NCH_3]^+$	1.457	1.419	1.553	180.0	111.1	23.7	[1]
	1.44	1.39	1.60	—	117	—	[2]

References:

[1] Ehrhardt, C.; Ahlrichs, R. (Chem. Phys. **108** [1986] 417/28).
[2] Zirz, C.; Ahlrichs, R. (Inorg. Chem. **23** [1984] 26/31).

2.5.2 Salts of the Halogeno-(pentafluorophenyl-imido)sulfur Ion (1+), $[XS{=}NC_6F_5]^+Y^-$, X = F, Cl; Y = SO_3F, Sb_nF_{5n+1}, $SbCl_6$

$[FS{=}NC_6F_5]^+Y^-$, Y = SO_3F, Sb_nF_{5n+1}, salts were formed by dissolving $C_6F_5N{=}SF_2$ in HSO_3F at −40°C or in an excess of freshly distilled SbF_5 at 10°C (40°C?) but were not isolated. Analogously, $[ClS{=}NC_6F_5]^+SbCl_6^-$ was obtained from the reaction of $C_6F_5N{=}SCl_2$ with $SbCl_5$, but was not isolated. Identity and molecular structure of the salts were deduced from conductivity and spectral data. For an $[FS{=}NC_6F_5]^+Sb_nF_{5n+1}^-$ solution in SbF_5, the specific conductance is higher by a factor of 10^3 in comparison with SbF_5 itself ($\varkappa_{SbF_5} = 1.2 \times 10^{-8}\ \Omega^{-1}\ cm^{-1}$).

^{19}F NMR spectra of the reaction mixtures (see the paper) show the absence of signals of the original compound as well as significant changes in the spin-spin coupling constants (higher values for the salt). Such changes in the ^{19}F NMR spectral data are in agreement with data concerning charge distribution, obtained from quantum-chemical calculations by the CNDO/2 method, −0.1780, 0.5664, and −0.0806 at N, S, and F, respectively. From the same calculation for the cation $[FS{=}NC_6F_5]^+$ the following values were obtained: −124.27 eV bonding energy, 17.75 eV first ionization potential, and −8.97 eV electron affinity.

Reference:

Zibarev, A. V.; Gol'ding, I. R.; Sladkov, A. M.; Furin, G. G.; Yakobson, G. G. (Izv. Akad. Nauk SSSR Ser. Khim. **1979** 2536/41; Bull. Acad. Sci. USSR Div. Chem. Sci. **28** [1979] 2352/7).

Physical Constants and Conversion Factors

Avogadro constant N_A (or L) $= 6.02214 \times 10^{23}\ mol^{-1}$	Planck constant $h = 6.62608 \times 10^{-34}\ J \cdot s$
Faraday constant $F = 9.64853 \times 10^{4}$ C/mol	elementary charge $e = 1.60218 \times 10^{-19}$ C
molar gas constant $R = 8.31451\ J \cdot mol^{-1} \cdot K^{-1}$	electron mass $m_e = 9.10939 \times 10^{-31}$ kg
molar volume (ideal gas) $V_m = 2.24141 \times 10^{1}$ L/mol (273.15 K, 101325 Pa)	proton mass $m_p = 1.67262 \times 10^{-27}$ kg

1 kg $= 2.205$ pounds

1 m $= 3.937 \times 10^{1}$ inches $= 3.281$ feet

1 $m^3 = 2.642 \times 10^{2}$ gallons (U.S.)

1 $m^3 = 2.200 \times 10^{2}$ gallons (Imperial)

Force	N	dyn	kp
1 N	1	10^{5}	1.019716×10^{-1}
1 dyn	10^{-5}	1	1.019716×10^{-6}
1 kp	9.80665	9.80665×10^{5}	1

Pressure	Pa	bar	kp/m²	at	atm	Torr	lb/in²
1 Pa = 1 N/m²	1	10^{-5}	1.019716×10^{-1}	1.019716×10^{-5}	9.86923×10^{-6}	7.50062×10^{-3}	1.450378×10^{-4}
1 bar = 10^{6} dyn/cm²	10^{5}	1	1.019716×10^{4}	1.019716	9.86923×10^{-1}	7.50062×10^{2}	1.450378×10^{1}
1 kp/m² = 1 mm H_2O	9.80665	9.80665×10^{-5}	1	10^{-4}	9.67841×10^{-5}	7.35559×10^{-2}	1.422335×10^{-3}
1 at (technical)	9.80665×10^{4}	9.80665×10^{-1}	10^{4}	1	9.67841×10^{-1}	7.35559×10^{2}	1.422335×10^{1}
1 atm = 760 Torr	1.01325×10^{5}	1.01325	1.033227×10^{4}	1.033227	1	7.60×10^{2}	1.469595×10^{1}
1 Torr = 1 mmHg	1.333224×10^{2}	1.333224×10^{-3}	1.359510×10^{1}	1.359510×10^{-3}	1.315789×10^{-3}	1	1.933678×10^{-2}
1 lb/in² = 1 psi	6.89476×10^{3}	6.89476×10^{-2}	7.03069×10^{2}	7.03069×10^{-2}	6.80460×10^{-2}	5.17149×10^{1}	1

Work, Energy, Heat	J	kW·h	kcal	Btu	eV
1 J = 1 W·s = 1 N·m = 10^7 erg	1	2.778×10^{-7}	2.39006×10^{-4}	9.4781×10^{-4}	6.242×10^{18}
1 kW·h	3.6×10^{6}	1	8.604×10^{2}	3.41214×10^{3}	2.247×10^{25}
1 kcal	4.1840×10^{3}	1.1622×10^{-3}	1	3.96566	2.6117×10^{22}
1 Btu (British thermal unit)	1.05506×10^{3}	2.93071×10^{-4}	2.5164×10^{-1}	1	6.5858×10^{21}
1 eV	1.602×10^{-7}	4.450×10^{-14}	3.8289×10^{-11}	1.51840×10^{-10}	1

1 cm^{-1} = 1.239842×10^{-4} eV
1 hartree = 27.2114 eV
1 Hz = 4.135669×10^{-15} eV
1 eV ≙ 23.0578 kcal/mol

Power	kW	hp	$kp\cdot m\cdot s^{-1}$	kcal/s
1 kW = 10^3 J	1	1.35962	1.01972×10^{2}	2.39006×10^{-1}
1 hp (horsepower, metric)	7.3550×10^{-1}	1	7.5×10^{1}	1.7579×10^{-1}
1 $kp\cdot m\cdot s^{-1}$	9.80665×10^{-3}	1.333×10^{-2}	1	2.34384×10^{-3}
1 kcal/s	4.1840	5.6886	4.26650×10^{2}	1

References:

International Union of Pure and Applied Chemistry, Manual of Symbols and Terminology for Physicochemical Quantities and Units, Pergamon, London 1979; Pure Appl. Chem. **51** [1979] 1/41.
The International System of Units (SI), National Bureau of Standards Spec. Publ. 330 [1972].
Landolt-Börnstein, 6th Ed., Vol. II, Pt. 1, 1971, pp. 1/14.
ISO Standards Handbook 2, Units of Measurement, 2nd Ed., Geneva 1982.
Cohen, E. R., Taylor, B. N., Codata Bulletin No. 63, Pergamon, Oxford 1986.